AF361946

Mathematics and Engineering

Mathematics and Engineering

Editors

Mingheng Li
Hui Sun

MDPI • Basel • Beijing • Wuhan • Barcelona • Belgrade • Manchester • Tokyo • Cluj • Tianjin

Editors
Mingheng Li
California State Polytechnic University
USA

Hui Sun
California State University
USA

Editorial Office
MDPI
St. Alban-Anlage 66
4052 Basel, Switzerland

This is a reprint of articles from the Special Issue published online in the open access journal *Mathematics* (ISSN 2227-7390) (available at: https://www.mdpi.com/journal/mathematics/special_issues/MaE).

For citation purposes, cite each article independently as indicated on the article page online and as indicated below:

LastName, A.A.; LastName, B.B.; LastName, C.C. Article Title. *Journal Name* **Year**, *Article Number*, Page Range.

ISBN 978-3-03943-256-1 (Hbk)
ISBN 978-3-03943-257-8 (PDF)

Contents

About the Editors

Mingheng Li (Professor) received his B.E. from Beijing University of Chemical Technology, M.E. from Tsinghua University, and Ph.D. from UCLA, all in Chemical Engineering. He worked as a senior research engineer at PPG Industries (Pittsburgh, PA) for two and a half years before joining the faculty at California State Polytechnic University, Pomona, in 2007. Dr. Li's research interests lie in the general areas of process systems engineering and applications to water, energy, environmental engineering, and materials processing.

Hui Sun (Assistant Professor) received his B.Sc. from the Chinese University of Hong Kong in Mathematics, and Ph.D. from UCLA in Applied Mathematics. He worked as a postdoctoral researcher in UCSD for three and a half years before joining the faculty at California Statement University, Long Beach, in 2017. Dr. Sun's research interests lie in scientific computation, mathematical modeling, and computational biophysics.

Preface to "Mathematics and Engineering"

Engineering problems arising in energy, environment, and agriculture, amongst others, with enormous scale and complexity are featured, and these have posed challenges and provided opportunities for the development of advanced mathematical tools to ensure sound decision making. For example, with the breakthrough of computational power over the last few decades, modeling and numerical linear algebra have been intensely utilized and developed to simulate various engineering processes. More recently, data sciences and machine learning have emerged in a diverse array of engineering fields.

This book consists of a compilation of works covering a wide variety of application domains, including seashore sulfuric acid erosion, industrial cyberphysical systems, vehicle target detection, agrohydrological systems, nanofluid flow, next-generation manufacturing, and smart grids. The purpose of this book is to assemble a collection of articles covering the current progress in mathematics applied in complex engineering problems, which includes but is not limited to modeling and simulation, computations, analysis, control, optimization, data science, and machine learning.

We would like to thank those who have contributed to this book. We would also like to thank those who performed reviews of the manuscripts—the feedback of these reviewers is invaluable. We would like to thank Dr. Jean Wu for her great support as the Managing Editor throughout the process of putting together this Special Issue of *Mathematics*. We would like to thank our colleagues at California State Polytechnic University, Pomona, and at California State University, Long Beach, for their continuous support. Finally, our deepest gratitude is extended to our families and friends for their constant encouragement and support. Without them, this work would not have been possible.

Mingheng Li, Hui Sun
Editors

Article

Mathematical Models for Stress–Strain Behavior of Nano Magnesia-Cement-Reinforced Seashore Soft Soil

Wei Wang [1], Yong Fu [2,*], Chen Zhang [1], Na Li [1] and Aizhao Zhou [3]

[1] School of Civil Engineering, Shaoxing University, Shaoxing, Zhejiang 312000, China; wellswang@usx.edu.cn (W.W.); golenchen@outlook.com (C.Z.); lina@usx.edu.cn (N.L.)

[2] Department of Ocean Science and Engineering, Southern University of Science and Technology, Shenzhen 518055, China

[3] Department of Civil and Architecture Engineering, Jiangsu University of Science and Technology, Zhenjiang 212003, Jiangsu, China; zhouaizhao@126.com

* Correspondence: fuyong@u.nus.edu

Received: 11 February 2020; Accepted: 21 March 2020; Published: 23 March 2020

Abstract: The stress–strain behavior of nano magnesia-cement-reinforced seashore soft soil (Nmcs) under different circumstances exhibits various characteristics, e.g., strain-hardening behavior, falling behavior, S-type falling behavior, and strong softening behavior. This study therefore proposes a REP (reinforced exponential and power function)-based mathematical model to simulate the various stress–strain behaviors of Nmcs under varying conditions. Firstly, the mathematical characteristics of different constitutive behaviors of Nmcs are explicitly discussed. Secondly, the conventional mathematical models and their applicability for modeling stress–strain behavior of cemented soil are examined. Based on the mathematical characteristics of different stress–strain curves and the features of different conventional models, a simple mathematical REP model for simulating the hardening behavior, modified falling behavior and strong softening behavior is proposed. Moreover, a CEL (coupled exponential and linear) model improved from the REP model is also put forth for simulating the S-type stress–strain behavior of Nmcs. Comparisons between conventional models and the proposed REP-based models are made which verify the feasibility of the proposed models. The proposed REP-based models may facilitate researchers in the assessment and estimation of stress–strain constitutive behaviors of Nmcs subjected to different scenarios.

Keywords: seashore soft soil; cement; sulfuric acid erosion; stress–strain behavior; mathematical model

---

1. Introduction

Soft soil is widely distributed in coastal areas with many defects such as large natural moisture content, excessive compression capacity, and poor bearing capacity [1–3]. In geotechnical engineering, deep mixing method is generally adopted to improve the strength of soil by adding cement to the soil [4–8]. The soft soil is reinforced to impart higher strength through a series of reactions between raw materials and curing agent [9–16]. In addition to the traditional cement curing agent, researchers are constantly looking for some novel materials such as nano materials to improve the bearing capacity of the soft soil layer [2,3,14,15,17–20].

Nano cemented soil refers to the cement-soil mixture which is improved by adding nano materials as admixtures into the mixture of water, cement and soil [20–22]. At present, the nano materials for enhancing cemented soil mainly include nano titanium oxide, nano montmorillonite, nano magnesia, nano silicon, nano alumina, etc. Previous experimental results show that adding nano admixtures can improve the performances of cemented soil such as its soil strength and anticorrosive properties.

Based on some recent studies [2,3], nanometer magnesia (Nm) can be added into cemented soil to improve its mechanical performance.

In the past few years, mathematical models have been adopted to study the stress–strain response of soils [23–31]. In term of the stress–strain behaviors of cemented soil, a large number of researches have been reported [2,3,5,6,13–18,32–37]. However, the reported mathematical models have some limitations particularly in the modeling of the constitutive behavior of nano magnesia-cement-reinforced seashore soft soil (Nmcs) which manifests different stress–strain behaviors under varying conditions [2–4,17,18,38,39].

This study aims to propose a REP (reinforced exponential and power function)-based mathematical model to simulate the various stress–strain behaviors of Nmcs under varying circumstances. The mathematical characteristics of different constitutive behaviors are firstly examined. Then, the conventional mathematical models for stress–strain behavior of cemented soil are discussed. Based on the mathematical characteristics of various stress–strain curves and the features of different conventional models, a new REP model for characterizing hardening behavior, modified falling behavior and strong softening behavior is proposed. Furthermore, a CEL (coupled exponential and linear) model improved from the REP model is also proposed which is able to simulate the S-type stress–strain behavior of Nmcs. Comparisons between conventional models and the proposed REP-based models are made which verifies the feasibility of the proposed models.

2. Mathematical Characteristics of Stress–Strain Constitutive Relations of Nmcs

2.1. Materials and Samples

The seashore soft soil discussed in this study was collected from coastal areas in Shaoxing, Zhejiang Province, China. Its specific gravity, liquid limit, and plastic limit were 2.6, 43.5%, and 30%, respectively. According to American Society of Testing Materials, ASTM (1994), it belongs to silty clay [2]. In direct shear test, the thickness of the shear plane is assumed to be zero which means the corresponding shear strain cannot be calculated; thus in this case shear displacement is applied to represent the strain behavior. In the following, a more general symbol δ is therefore used to represent shear deformation characteristics, i.e., shear displacement or shear strain.

The soil samples were prepared under a standard maintenance temperature of 20 °C with a relative humidity of 95%. The mechanical tests were performed after 28 days' standard curing time. For the tests considering sulfuric acid erosion, after the standard curing the samples were immersed in sulfuric acid solution for another 14 days.

2.2. Mathematical Characteristics of τ-δ Behavior

The shear stress–shear strain (τ-δ) behavior of Nmcs typically comprises four types [2,3,17,18]. They are strain-hardening behavior, falling behavior, S-type falling behavior and strain-softening behavior. According to Wang et al. [17] the strain-softening behavior can be well captured using a generalized mathematical model, so it will not be discussed in this study. In order to establish the constitutive model characterizing the shear stress-displacement curve, it is necessary to analyze the mathematical characteristics.

2.2.1. Strain-Hardening τ-δ Behavior

The typical shear stress-shear strain (τ-δ) curve for strain-hardening behavior of Nmcs is shown in Figure 1. As can be seen, the strain-hardening process can be divided into three stages: elastic stage (OA), plastic stage (AB), and failure stage (BC). In the elastic stage, the τ-δ curve shows a straight line and the shear stress increases linearly with the shear strain at a gradient of initial elastic modulus E_0. In the plastic stage, the tangent modulus E_i reduces gradually as strain accumulates, leading to a nonlinear τ-δ curve. In the failure stage, the τ-δ curve flattens out and the shear stress reaches the ultimate shear strength τ_p with a corresponding shear strain δ_p. In this stage, the shear stress is mainly contributed by the friction resistance at the failure surface of the soil sample. In sum, the strain-hardening curve includes four mathematical features: through the origin, monotone increasing, convex to τ-axis, and infinite convergence.

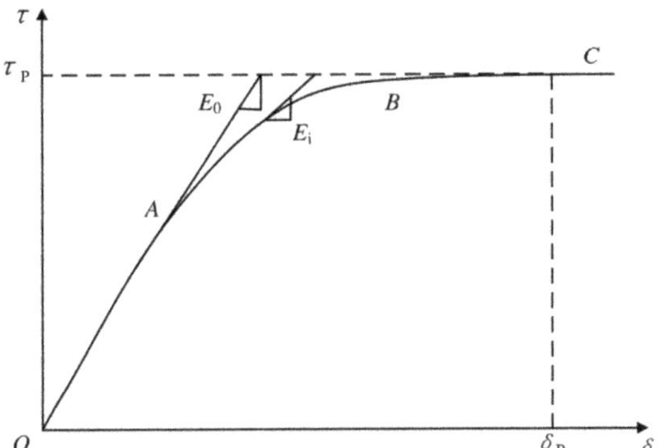

Figure 1. Strain-hardening τ-δ curve.

2.2.2. Falling τ-δ Behavior

In the direct shear test of seashore soft soil after adding Nm, the cohesive force was lost after failure and hence the shear stress decreased dramatically. The falling τ-δ curve after the elastic stage shows an evident softening behavior. As shown in Figure 2, the typical τ-δ curve can be divided into four stages. Stage 1 is the elastic stage (OA) which is similar to that of strain-hardening curve. In this stage, the shear stress increases with the gradient of initial elastic modulus until reaching the failure. The corresponding shear stress and strain at point A denote the failure displacement δ_p and shear strength τ_p, respectively. Stage 2 is the falling stage (AB), which evidently shows the falling of shear stress after the soil fails. This falling of shear stress is attributed to the loss of cohesion and the tangent modulus of the τ-δ curve gives a negative value. Stage 3 is the plastic stage (BC) wherein the friction resistance starts to dominate after the loss of cohesion. In this stage, the tangent modulus of the curve approximates the initial elastic modulus at the beginning which subsequently decreases due to the occurrence of plastic shear stress. Stage 4 is the residual shear stress stage (CD) where the tangent modulus approaches zero and the shear stress is equal to residual shear stress. It should be noted that the τ-δ falling curve can be modified by ignoring Stages 2 and 3 (AB and BC), then it returns to the strain-hardening behavior. However, the shear strength in the falling curve refers to the shear stress at the end of Stage 1 which is different from its counterpart in the strain-hardening curve.

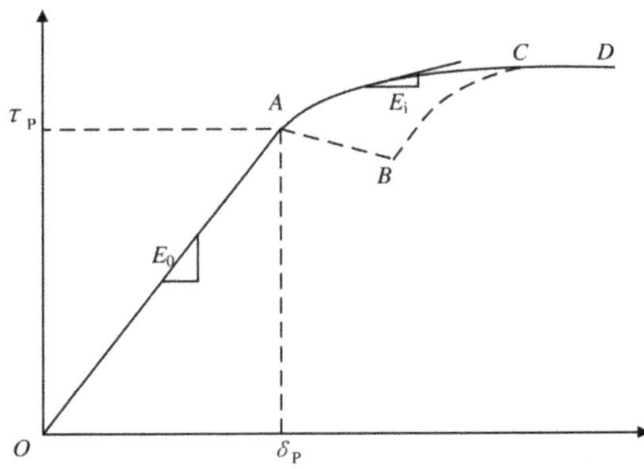

Figure 2. Falling τ-δ curve.

2.2.3. S-Type Falling τ-δ Behavior

Based on the τ-δ response of Nmcs in the scenario with high corrosion, the typical S-type falling curve is shown in Figure 3. As can be seen, the S-type falling curve consists of five stages. Stage 1 refers to the erosion softening stage (OA) where the surface of the sample is eroded by highly corrosive sulfuric acid and becomes relatively soft. Initially, the increment in the shear stress is relatively small as the displacement increases, giving rise to a relatively small initial modulus. As the shearing develops and enters the hard soil, which is less corroded, the tangent modulus tends to increase until stabilizing when accessing Stage 2, i.e., linear elastic stage (AB). The shear strength τ_p and corresponding displacement δ_p at failure occur at the end of Stage 2, i.e., point B. After reaching failure, the falling stage (BC), the frictional plastic stage (CD) and the residual shear stress stage (DE) emerge continuously. These three stages are similar to those in the above falling behavior. Thus, the S-type falling curve can be modified in a similar way as that for the above falling curve. However, the obtained modified curve is still unlike the strain-hardening curve which is convex to τ-axis; namely, an inflection point occurs on the curve, e.g., before this point, the curve is convex to δ-axis, while after this point, it is convex to τ-axis.

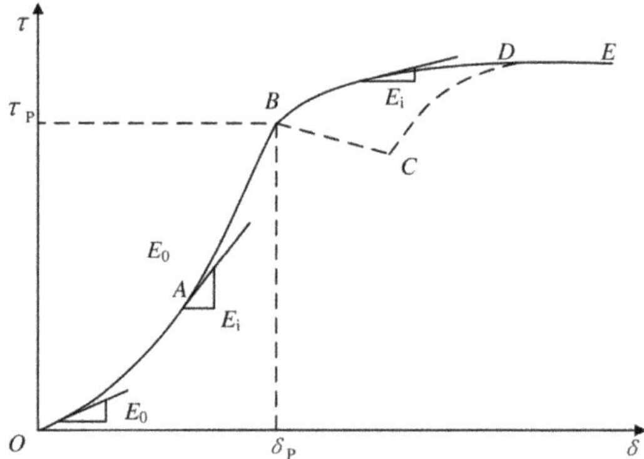

Figure 3. S-type Falling τ-δ curve.

2.3. Mathematical Characteristics of σ-δ Behavior

Based on the UCS (unconfined compression strength test) results, the stress–strain curve under uniaxial compression exhibits a strong softening and brittle failure behavior, as shown in Figure 4. As can be seen, the stress–strain curve is hump-shaped where the peak point of the curve gives the unconfined compressive strength σ_p corresponding to a strain of ε_p. This curve comprises three stages. In the hardening stage (OP), the stress initially increases linearly with strain; the gradient of which represents the initial elastic modulus E_0. This gradient starts to decrease in the later phase of the hardening stage until arriving at the peak, i.e., point P where the brittle failure takes place and the tangent modulus equals to zero. After that the stress reduces monotonically with strain in the softening stage (PM) and remains unchanged when entering the residual stress stage (MN). In sum, the mathematical features of the curve include: through the origin, having extreme value point, and converging at residual stress.

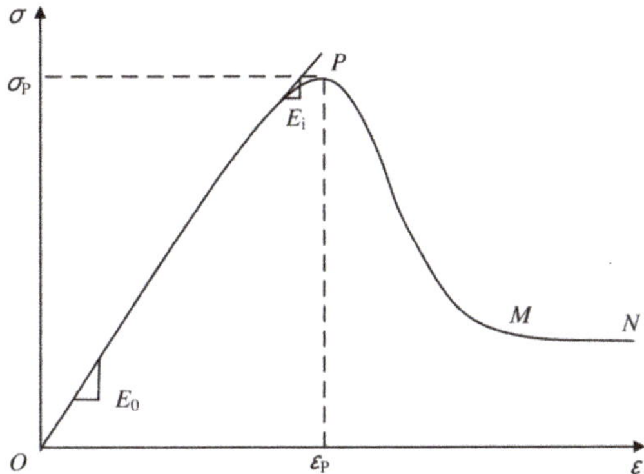

Figure 4. Axial stress–strain curve.

3. Established Mathematical Model

The characteristics of modified falling τ-δ curve are almost consistent with those of strain-hardening τ-δ curve. Therefore, this study attempts to establish a simple, mathematical constitutive model which is applicable for modeling both strain-hardening and modified falling τ-δ behaviors of Nmcs.

3.1. Conventional Shear Stress-Displacement (τ-δ) Models

The conventional shear stress-displacement models have three types: hyperbolic model, exponential model and power function model.

3.1.1. Hyperbolic Model

The most classical nonlinear model describing hardening curve is hyperbolic model, which is widely applied to the nonlinear modeling in various fields because of its simple expression, convenient simulation and easy determination of parameters. The hyperbolic model was proposed by Duncan and Chang [40] and its expression is

$$\tau = \frac{\delta}{1/E_0 + \delta/\tau_p},\tag{1}$$

where τ is the shear stress (kPa), δ is the shear displacement (mm), E_0 is the nominal elastic modulus (kPa/mm) and τ_p is the shear strength (kPa).

According to the τ-δ curves of shear test on the soil-structure contact surface, the hardening curve is rigid and plastic, and the hyperbolic model is difficult to meet this condition. In contrast, the strain in the τ-δ curve of Nmcs is mostly elastic before failure and only a small amount of plastic strain occurs after failure. In addition, the hyperbolic curve produces a large error in fitting the softening curve, which causes the deviation between the fitting results and the measured data due to its mathematical characteristics. Therefore, the hyperbolic model may be not suitable for modeling both hardening and softening behaviors.

3.1.2. Exponential Model

The exponential model converges faster, and it is suitable for the τ-δ curves with elastoplastic strain. Hence, it is superior to the hyperbolic model for hardening curves. The exponential model was firstly proposed based on the one-dimensional consolidation theory of Terzaghi, and was mainly applied to describe the stress–strain curve of soil. The mathematical constitutive function is

$$\tau = \tau_p\left(1 - e^{-E_0\delta/\tau_p}\right).\tag{2}$$

Cai et al. [41] investigated the mathematical defects of the exponential model using the half-value strength index. The results showed that the half-value strength index of both the exponential and the hyperbolic function is a fixed value which is only related to the peak strength and the initial elastic modulus. In this regard, the simulated stress–strain curve is only applicable for specific cases. In contrast, the half-value strength index of the power function is a parameter with a non-fixed value. The shape of its curve varies with the variation of the parameters, and hence it is widely applicable.

3.1.3. Power Function Model

Due to the mathematical characteristics of the power function model, its simulation effect is remarkable in modeling plastic curves. In addition, the power function model of the nonlinear constitutive model of clay has the parameter of tangent modulus index. The expression of the power function model is

$$\tau = \tau_p \left\{ 1 - \left[1 + \frac{(\theta - 1)E_0}{\tau_p} \delta \right]^{\frac{1}{1-\theta}} \right\}, \tag{3}$$

where the value of θ is larger than 1; when $\theta = 2$, the power function becomes the hyperbolic shear stress-displacement constitutive equation, so the hyperbolic function is only a special case of the power function. The first derivative of Equation (3) is given as

$$\frac{d\tau}{d\delta} = E_0 \left[1 + \frac{(\theta - 1)E_0}{\tau_p} \delta \right]^{\frac{\theta}{1-\theta}}. \tag{4}$$

Since $\theta > 1$, the first derivative of the power function is always greater than 0 in the domain of definition, and the curve increases monotonically. As the parameter θ changes, it is more applicable to the hardening curve and modified falling curve.

The second derivative of Equation (3) leads to

$$\frac{d^2\tau}{d\delta^2} = -\frac{E_0^2\theta}{\tau_p} \left[1 + \frac{(\theta - 1)E_0}{\tau_p} \delta \right]^{\frac{2\theta-1}{1-\theta}}. \tag{5}$$

It can be found that the second derivative of power function is always less than zero, meeting the characteristic of hardening and modified falling curves, i.e., convex to τ-axis. However, the inflection point occurs on the S-type τ-δ curve under a highly corrosive environment. Therefore, the second derivative of power function cannot be equal to zero.

3.2. Mathematical REP Model for τ-δ Behavior

Based on the above conventional mathematical models, a new REP (reinforced exponential and power function) mathematical model for modeling the shear stress-displacement behavior of Nmcs under acid erosion environment is proposed in this study. This model combines exponential and power functions, which is expressed as

$$\tau = a \left[1 - e^{-b\delta} (1 + k\delta)^{-\lambda} \right], \tag{6}$$

where a, b, λ, and k are parameters to be determined, and $a \geq 0, b > 0, \lambda > 0, k > 0$. When $k = 0$, Equation (6) degrades into exponential function.

Taking the limit and the first derivative of Equation (6) leads to

$$\begin{cases} \tau|_{\delta=+\infty} = a \\ \dfrac{d\tau}{d\delta} = abe^{-b\delta}(1 + k\delta)^{-\lambda} + a\lambda k e^{-b\delta}(1 + k\delta)^{-\lambda-1} \end{cases}. \tag{7}$$

The progressive limit of τ-δ curve is shear strength, i.e., τ_p. Hence,

$$a = \tau_p. \tag{8}$$

The first derivative of the new model with $\delta=0$ is the initial modulus of elasticity, i.e., E_0. Combining with Equation (8) gives

$$\tau_p b + \tau_p \lambda k = E_0, \tag{9}$$

where

$$b = \frac{E_0}{\tau_p} - \lambda k. \tag{10}$$

According to Equation (10), $(E_0/\tau_p - \lambda k) > 0$ if $b > 0$.

Based on the above, the REP model for the hardening τ-δ curve can be expressed as

$$\tau = \tau_p \left[1 - e^{-(E_0/\tau_p - \lambda k)\delta}(1 + k\delta)^{-\lambda} \right]. \tag{11}$$

In order to analyze whether REP model satisfies the mathematical characteristics of hardening curve, the zero point, the limit, the first derivative and the second derivative are discussed as below

$$
\begin{cases}
\tau|_{\delta=0} = 0 \\
\tau|_{\delta=+\infty} = \tau_p \\
\dfrac{d\tau}{d\delta} = E_0\left(E_0/\tau_p - \lambda k\right)e^{-(E_0/\tau_p - \lambda k)\delta}(1 + k\delta)^{-\lambda} + E_0\lambda k e^{-(E_0/\tau_p - \lambda k)\delta}(1 + k\delta)^{-\lambda-1} \\
\dfrac{d^2\tau}{d\delta^2} = -E_0\left(E_0/\tau_p - \lambda k\right)^2 e^{-(E_0/\tau_p - \lambda k)\delta}(1 + k\delta)^{-\lambda} \\
\qquad -2E_0\left(E_0/\tau_p - \lambda k\right)\lambda k e^{-(E_0/\tau_p - \lambda k)\delta}(1 + k\delta)^{-\lambda-1} - E_0\lambda(\lambda + 1)k^2(1 + k\delta)^{-\lambda-1}
\end{cases}
\tag{12}
$$

When $\delta = 0$, the shear stress equals to 0 so the curve passes through the origin, satisfying the first characteristic of the hardening curve. The coefficients in the first-order derivative equation are all positive, and the first-order derivative is always greater than 0 in the domain of definition, which satisfies the characteristic of monotonic increase. When the displacement δ approaches infinity, shear stress gradually gets close to shear strength of τ_p. Therefore, the new model goes through the origin and has both upper and lower bounds. The coefficients of the second derivative equation are all negative, and the second derivative of the new model is always less than 0. Hence, the REP model is theoretically suitable for the hardening and modified falling curves.

3.3. Mathematical Models for Stress-Displacement (σ-δ) Behavior

Likewise, the conventional mathematical model for the constitutive relation of stress- displacement (σ-δ) behavior also has many types such as hyperbolic, exponential function, power function, piecewise function and quadratic function. As aforementioned, the power function model has a better fitting effect than hyperbolic and exponential functions, while the piecewise function has some defects, e.g., it is troublesome to fit and has many parameters. Therefore, in the study of models for stress-displacement (σ-δ) behavior, only power function, quadratic function, and the proposed REP function are discussed.

For the power function of σ-δ behavior, it can be easily modified from Equation (3), that is

$$\sigma = \sigma_p \left\{ 1 - \left[1 + \frac{(\theta - 1)E_0}{\sigma_p} \varepsilon \right]^{\frac{1}{1-\theta}} \right\}, \tag{13}$$

where ε is the strain (%), σ is the stress (kPa) and σ_p is the peak value of the stress–strain curve (UCS).

The quadratic model is able to simulate the stress–strain curve of compacted cement soil with an obvious peak value; its expression is

$$\sigma = \sigma_{\mathrm{p}}\left[A\frac{\varepsilon}{\varepsilon_{\mathrm{p}}} - B\left(\frac{\varepsilon}{\varepsilon_{\mathrm{p}}}\right)^2\right],\tag{14}$$

where σ_{p} and ε_{p} are the maximum stress and the corresponding strain, respectively. A and B are the fitting parameters to be determined.

When $\varepsilon = 0$, the stress of power and quadratic models is 0, which satisfies the characteristic of the stress–strain curve passing through the origin. As aforementioned, in the process of power function fitting, it is unable to converge in the failure stage, while the strong softening stress–strain curve will soften immediately after reaching the peak failure. Therefore, the convergence of the power function may be not timely, which may lead to a large deviation from the measured curve. Although the quadratic model is able to converge in time after the peak value; when the strain approaches infinity, the stress is also infinite, which theoretically does not meet the characteristics of infinite convergence of the stress–strain curve.

The expression of REP model for the σ-δ behavior is slightly different from that for the τ-δ behavior, i.e., Equation (11)

$$\sigma = a\left[1 - e^{-b\varepsilon}(1 + k\varepsilon)^{-\lambda}\right],\tag{15}$$

where a, b, λ, and k are parameters to be determined, and the value range of each parameter should be suitable for the hump curve. When $\varepsilon = 0$, the σ value of the REP model also equals to zero which satisfies the characteristic of passing through the origin.

The first and second derivatives of the stress–strain curve are as follows

$$\begin{cases} \dfrac{d\sigma}{d\delta} = abe^{-b\varepsilon}(1 + k\varepsilon)^{-\lambda} + a\lambda k e^{-b\varepsilon}(1 + k\varepsilon)^{-\lambda-1} \\ \dfrac{d^2\sigma}{d\delta^2} = -ab^2 e^{-(E_0/\tau_{\mathrm{p}}-\lambda k)\delta}(1 + k\varepsilon)^{-\lambda} - 2ab\lambda k e^{-b\varepsilon}(1 + k\varepsilon)^{-\lambda-1} - a\lambda(\lambda + 1)k^2(1 + k\varepsilon)^{-\lambda-1} \end{cases}.\tag{16}$$

It can be observed that the first and second derivatives of REP model are affected by the value of each parameter, and the positive and negative signs are uncertain which enables its adaptability to complex strain-softening curves. The specific value range of each parameter and the judgment of the positive and negative sign of the first derivative and the second derivative need further study.

3.4. Application and Analysis

3.4.1. Hardening τ-δ Behavior

Based on the measured data of direct shear test, comparisons of using hyperbolic, exponential and power function models as well as the proposed REP model are made. Two cases with 5% and 7% cement mixture ratio were examined. To determine the aforementioned model parameters, e.g., λ and k, four tests with vertical pressures of 100, 200, 300, and 400 kPa were carried out respectively. The comparison results for the cases under a vertical pressure of 400 kPa are shown in Figures 5 and 6.

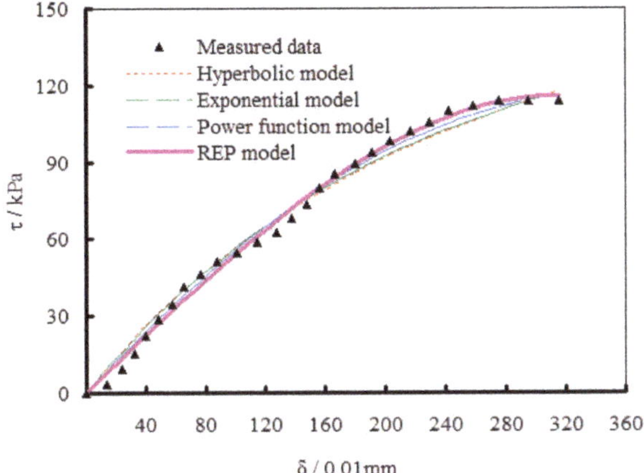

Figure 5. τ-δ curve for mix ratio with 5% cement content.

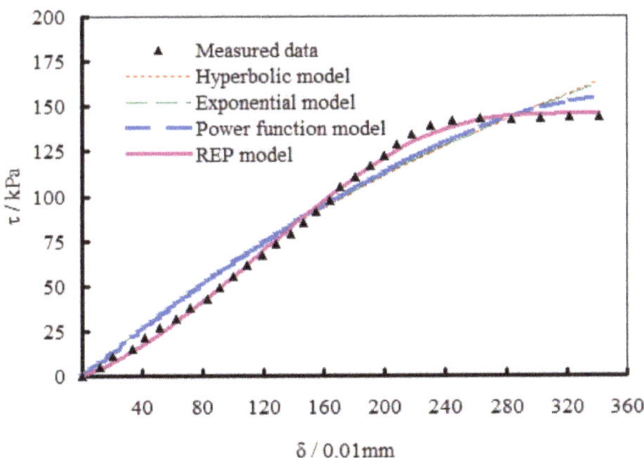

Figure 6. τ-δ curve for mix ratio with 7% cement content.

As shown in Figure 5, on the whole, the four models have a good fitting effect, but the elastic stage of the measured curve is not smooth, and there is a prominent inflection point locally. In this stage, the four models have a large deviation from the measured value. The tangent modulus of the measured curve decreases gradually in the plastic stage, and the hyperbolic, exponential and power functions converge slowly, all of which appear below the curve while the fitting effect of REP model is very evident. When entering failure stage, both the hyperbolic and exponential fittings cannot converge to τ_p, and lie in the upper part of the measured curve with big difference. In contrast, the power function converges well, and the fitting effect is better than the hyperbolic and exponential functions, although the end of the curve still appears above the measured curve. The REP model also has a good convergence effect and the whole fitted failure stage is very close to that of the measured curve.

As shown in Figure 6, the linearity of the measured curve in the elastic stage is not obvious while the hyperbolic, exponential, and power function curves are linear, deviating slightly from the measured curve. In the plastic deformation stage, the three functions appear below the measured curve, and the difference between the power function and the measured curve is the smallest. Similar to that observed in Figure 5, in the failure stage, the three models generally appear at the top of the curve. In contrast, the REP model is very close to the measured curve and shows an evidently favorable fitting effect.

Among the conventional models, the power function model is relatively superior in modeling τ-δ curves, particularly in the plastic stage where the computed results are closer to the measured curve and failure, and in the failure stage showing a faster convergence speed. Compared with the hyperbolic model, the exponential model has a better simulation effect. Moreover, the REP model shows superior fitting accuracy than the conventional mathematical models.

3.4.2. Modified Falling τ-δ Behavior

After adding Nm, the τ-δ curve exhibits modified falling type. In the comparison of different mathematical models, two cases with Nm mixing ratios of 10‰ and 20‰ under a vertical pressure of 400kPa were selected. The results are shown in Figures 7 and 8.

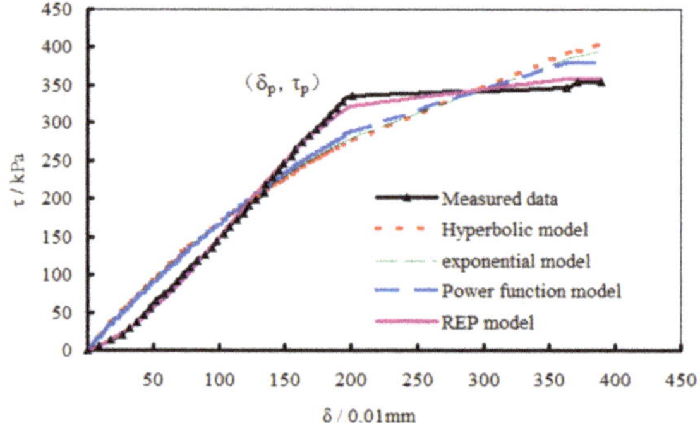

Figure 7. τ-δ curve for mix ratio with 10‰ Nm content.

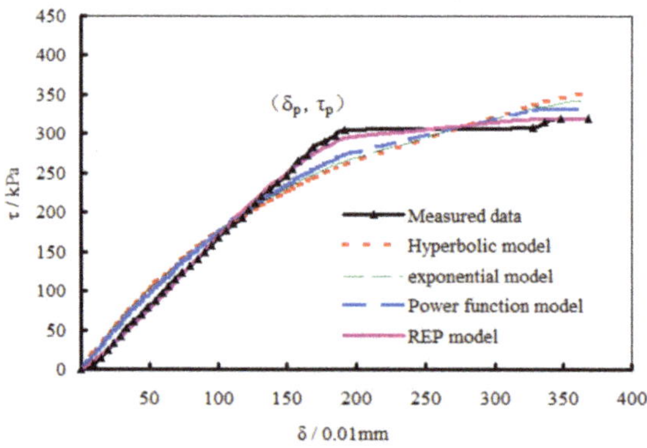

Figure 8. τ-δ curve for mix ratio with 20‰ Nm content.

As can be seen, the shape of the two measured curves is similar. Before failure, the modified curve is mostly elastic stage and there is no plastic transition stage showing significant linearity. In this stage, the conventional models produce non-linear curves, while the measured curve approximates a horizontal line in the failure stage. The curves of conventional models locate below the measured curve in the former part and above the measured curve in the later part. In the two stages, the conventional model intersects with the measured curve, exhibiting two "X" shapes. In general, the difference between conventional models and measured curve is relatively big while the REP model fits fairly well with the measured curve. Therefore, the proposed REP model has significant advantage in modeling the modified τ-δ curve. In contrast, it is easy for a double-"X" type discrepancy to appear in the conventional mathematical model simulation.

3.4.3. Strong Softening σ-δ Behavior

To compare the performance of the power function and quadratic function models as well as the REP model, UCS experimental results for cases with 0 mol/L and 0.08 mol/L H_2SO_4 erosion were adopted, Figure 9. As can be seen, the power function model behaves the worst in the modeling of the hump-shaped σ-δ curve. For instance, it is convex at the rising stage of the curve, which is contrary to the characteristic of concave in the measured curve. In addition, it cannot simulate the softening behavior as the measured curve although it converges to an almost constant value in the end. The quadratic function model performs better than the power function model. It is able to simulate the softening behavior and give a peak stress value. However, the shape of the quadratic curve is convex in all stages, which differs with the shape of measured curves. Worse still, there exists a large discrepancy in the maximum stress value and its corresponding strain value between the results of quadratic model and the measured results. This may raise an adverse effect in predicting UCS results. In contrast, the results of proposed REP model agree incredibly well with the measured results. The REP model not only shows consistent shapes, but also predicts a fairly close maximum stress value and a corresponding strain value. This is of great engineering value in predicting the compressive strength.

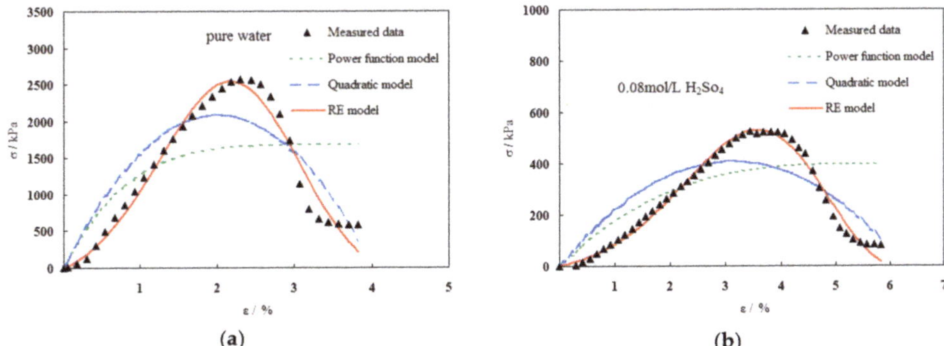

Figure 9. Comparison of stress–strain curves: (**a**) pure water condition (0 mol/L H_2SO_4); (**b**) 0.08 mol/L H_2SO_4.

To further examine the performance of the REP model, a series of comparisons for cases with varying erosion concentrations (i.e., 0.00 mol/L, 0.02 mol/L, 0.04 mol/L, 0.06 mol/L and 0.08 mol/L) is conducted, Figure 10. As can be seen, in the stress rising stage, the REP model fits fairly well with measured results; in the reducing stage, slight discrepancy occurs for the cases with 0.00 mol/L and 0.02 mol/L erosion concentrations. This could be attributed to the scattered data of the measured results. In the final residual stage, the REP model deviates from the measured data when the convergent residual stress was relatively large, such as the cases in pure water (i.e., 0.00 mol/L) and 0.02 mol/L acid erosion environment. When the residual stress was relatively small in the environment with a high concentration of acid erosion, the REP model gives better performance in fitting. Therefore, it can be found that the REP model can adapt to the measured curve under different acid erosion concentrations, with relatively high fitting accuracy. The fitting effect is more evident under a high concentration erosion environment. Table 1 provides the values of the four parameters in the REP model for these five cases.

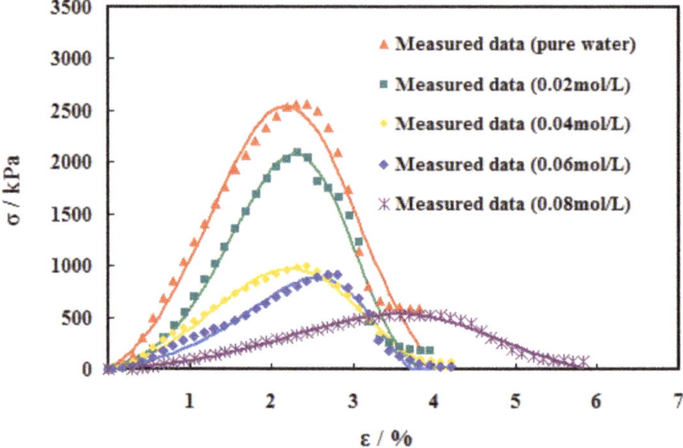

Figure 10. Comparison of stress–strain curves of different H_2SO_4 erosion concentrations with REP model fittings.

Table 1. Four parameter values of REP model under different erosion concentrations.

H_2SO_4 Erosion Concentrations	Parameter			
	a	b	k	λ
0.00 mol/L	223.23	16.62	0.08	131.20
0.02 mol/L	83.33	5.95	0.18	19.76
0.04 mol/L	100.70	7.53	0.12	48.53
0.06 mol/L	79.73	2.36	0.25	3.60
0.08 mol/L	40.00	2.93	0.12	14.04

In order to study the influence of erosion concentration on REP model parameters, an acid erosion factor was introduced

$$\omega(x, y, z, a_s) = \frac{\pi^2 x}{2\pi y + 4z a_s},\tag{17}$$

where x, y, and z are the parameters to be determined; a_s is the concentration of sulfuric acid solution.

In accordance with Table 1, substituting the acid erosion factor and acid erosion concentration a_s into the four parameters of REP model leads to

$$\begin{cases} a = 8\varphi \tan[\omega_a(a_s) + 1.84c] - 1.84c \\ b = -0.45c - 2.18\varphi \cos[\omega_b(a_s) + 2.18\varphi] \\ k = 0.11 \tan[\omega_k(a_s) + 0.16] - 0.16 \\ \lambda = -2.2c - a_s - 2.1c \sin[\omega_\lambda(a_s) + 2.2c] \end{cases},\tag{18}$$

where c and φ are the cohesion and internal friction angle of silty clay in the experiment, respectively.

Hence, the acid erosion factors of the four parameters can be obtained

$$\begin{cases} \omega_a(a_s) = \omega(-7.96\varphi, 0.881, 1.84c, a_s) \\ \omega_b(a_s) = \omega(2.18217\varphi, -0.100816, 0.44586c, a_s) \\ \omega_k(a_s) = \omega\left(0.1111, -6.592\varphi \times 10^{-6}, 0.158, a_s\right) \\ \omega_\lambda(a_s) = \omega(2.095c, 2.1927c, -105.507c, a_s) \end{cases}.\tag{19}$$

Combining with Equations (15) and (19), the damage model of sulfuric acid erosion can be expressed as

$$\sigma = a_{[\omega_a(a_s),A_s]}\left[1 - e^{-b_{[\omega_b(a_s),A_s]}\varepsilon}\left(1 + k_{[\omega_k(a_s),A_s]}\varepsilon\right)^{-\lambda_{[\omega_\lambda(a_s),A_s]}}\right]. \tag{20}$$

As shown in Equation (20), the damage model only contains parameters of sulfuric acid concentration (a_s), cohesion (c) and internal friction angle (φ), making the new model more suitable for acid erosion environment.

4. Mathematical CEL Model for Modified S-Type Falling τ-δ Behavior

As discussed above, the REP model has fairly good fitting effect which can be used for different types of stress–strain curves. However, for the S-type curve which has inflection points, it is time-consuming to calculate the zero point of the second derivative equation of REP model and determine the position of inflection points. To solve this, the REP is therefore simplified and a new CEL (coupled exponential and linear) model is put forth.

4.1. Mathematical CEL Model

The CEL model is actually a coupled exponential and linear function model.

Setting $\lambda = -1$, Equation (6) can be simplified as

$$\tau = a\left[1 - e^{-b\delta}(1 + k\delta)\right], \tag{21}$$

where $\alpha \geq 0$, $b > 0$ and $k > 0$; k is defined as the inflection factor.

Taking the limit of Equation (21) gives

$$\tau\big|_{\delta=+\infty} = a\left[1 - e^{-b\delta}(1 + k\delta)\right]\big|_{\delta=+\infty} = a(1 - 0) = a. \tag{22}$$

Since the S-type curve eventually converge to the shear strength τ_p so

$$a = \tau_p. \tag{23}$$

Taking the first derivative of Equation (21) leads to

$$\frac{d\tau}{d\delta} = abe^{-b\delta}(1 + k\delta) - ake^{-b\delta}. \tag{24}$$

If $\delta = 0$, then

$$\frac{d\tau}{d\delta}\big|_{\delta=0} = a(b - k). \tag{25}$$

The first derivative of the new model is the initial elastic modulus when the displacement is 0, i.e., $\delta = 0$. Combining with Equation (23) gives

$$\tau_p(b - k) = E_0. \tag{26}$$

Hence, the parameter b can be derived

$$b = \frac{E_0}{\tau_p} + k, \tag{27}$$

and the CEL model can be expressed as

$$\tau = E_0\left[1 - e^{-(\frac{E_0}{\tau_p} + k)\delta}(1 + k\delta)\right]. \tag{28}$$

In order to explore whether the CEL model satisfies the mathematical characteristics of the S-type curve, its zero point, limit, first derivative, and second derivative are discussed

$$\begin{cases} \tau|_{\delta=0} = E_0[1 - 1 \cdot (1-0)] = 0 \\ \tau|_{\delta=+\infty} = \tau_p \\ \dfrac{d\tau}{d\delta} = \tau_p\left(\dfrac{E_0}{\tau_p} + k\right)e^{-\left(\frac{E_0}{\tau_p}+k\right)\delta}(1+k\delta) - \tau_p k e^{-\left(\frac{E_0}{\tau_p}+k\right)\delta} \\ \dfrac{d^2\tau}{d\delta^2} = -\tau_p\left(\dfrac{E_0}{\tau_p}+k\right)^2 e^{-\left(\frac{E_0}{\tau_p}+k\right)\delta}(1+k\delta) + 2\tau_p\left(\dfrac{E_0}{\tau_p}+k\right)k e^{-\left(\frac{E_0}{\tau_p}+k\right)\delta} \end{cases} \tag{29}$$

According to Equation (29), the CEL model goes through the origin and converges to τ_p at infinity. Combining the similar items in the first derivative of Equation (29) gives

$$\frac{d\tau}{d\delta} = \left(E_0 + E_0 k\delta + \tau_p k^2 \delta\right)e^{-\left(\frac{E_0}{\tau_p}+k\right)\delta} > 0. \tag{30}$$

Hence, the first derivative of CEL model is always greater than 0 in the domain of definition, and the curve increases monotonically.

The displacement at zero point of the second derivative of Equation (29) is obtained by

$$\delta_c = \frac{\tau_p k - E_0}{\tau_p k^2 + E_0 k}. \tag{31}$$

To analyse the concavity of CEL model at zero point of second derivative, the relations are

$$\begin{cases} \dfrac{d^2\tau}{d\delta^2} > 0, \left(\delta < \dfrac{\tau_p k - E_0}{\tau_p k^2 + E_0 k}\right) \\ \dfrac{d^2\tau}{d\delta^2} < 0, \left(\delta > \dfrac{\tau_p k - E_0}{\tau_p k^2 + E_0 k}\right) \end{cases} \tag{32}$$

When $\delta < \delta_c$, $\tau'' > 0$ and the curve is convex to the δ-axis; when $\delta > \delta_c$, $\tau'' < 0$ and the curve is convex to the τ-axis.

4.2. Application and Analysis

Comparisons of hyperbolic, exponential and power function models and the above CEL model are carried out based on the experimental results of two cases with 0.09 mol/L H_2SO_4 erosion under vertical pressures of 300 kPa and 400 kPa, respectively. The results are shown in Figures 11 and 12.

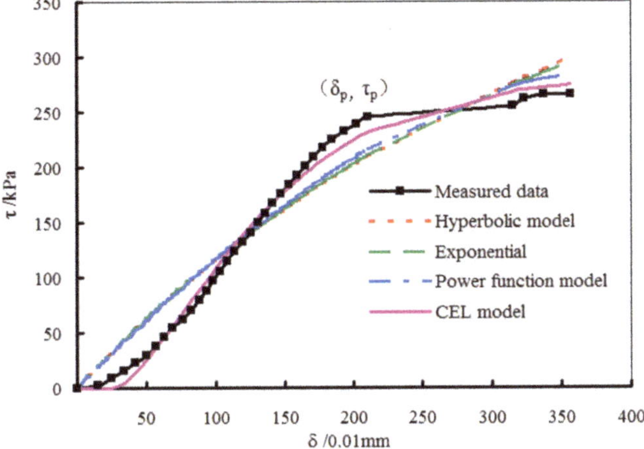

Figure 11. τ-δ curve under 300 kPa vertical pressure.

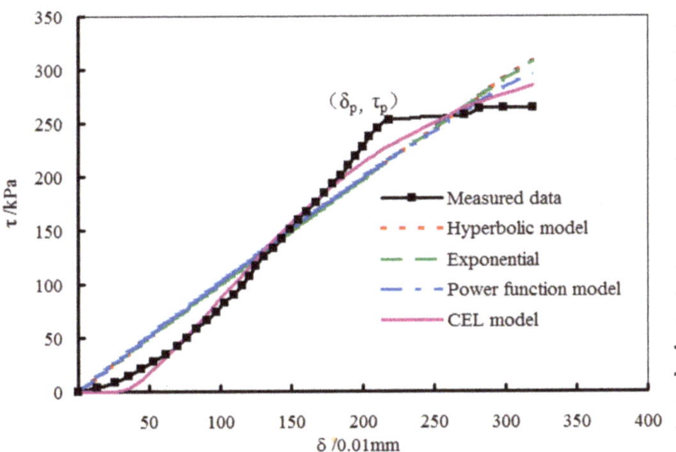

Figure 12. τ-δ curve under 400 kPa vertical pressure.

As can be seen, the curves of conventional models are well banded which are above the measured curve before the inflection point and below the measured curve after the inflection curve. In contrast, the fitting effect of CEL model is obviously better than that of the other conventional models. It is basically consistent with the measure curve in the S-shaped range and converges faster and closer to τ_p than the other models in the failure stage. However, at the initial stage, the CEL curve is slightly lower than the measure curve, while in the failure stage, the difference between the CEL curve and the measured curve is relatively evident.

5. Conclusions

This study analyses the experimental data of τ-δ and σ-δ constitutive relationships of Nmcs, and examines its characteristics under various conditions. Two mathematical models, i.e., REP (reinforced exponential and power function) and CEL (coupled exponential and linear) models, are proposed to overcome the shortcomings of conventional models in the modeling of hardening, modified falling,

S-type falling, and strong softening stress–strain behaviors. Some key findings are summarized as below:

- The τ-δ curves with varying cement mixing ratio, Nm mixing ratio and acid erosion concentrations exhibit different behaviors. For example, hardening behavior occurs with low cement mixing ratio; falling behavior takes place after adding Nm and S-type happens with acid erosion environment.
- The proposed REP model is able to satisfy the mathematical characteristics of the stress–strain curves with hardening and modified falling behaviors as well as strong softening behavior. Compared with conventional hyperbolic model, exponential and power function models which easily produce double-"X" discrepancies, the REP model has an evidently higher fitting accuracy in modeling both hardening and modified falling curves. In the modeling of strong softening behavior, the REP model also performs the best. In addition, by introducing an acid erosion factor, the REP model can be further expressed as a sulfuric acid erosion damage model which includes only three typical parameters, i.e., sulfuric acid concentration, cohesion of silty clay, and internal friction angle.
- In the modeling of S-type stress–strain behavior, the conventional models are unable to simulate the characteristic with inflection point and produce even bigger "X" discrepancy than that in the modeling of hardening and falling behavior. In contrast, the proposed CEL model performs much better especially in modeling the range around the inflection point at the early stage.

The applicability of the proposed method of this study in practice mainly include two parts. As introduced in the study, the seashore soft soil discussed herein was collected from coastal areas in Shaoxing, Zhejiang Province, China. Thus, for the regions with similar seashore soft soil, e.g., the Yangtze river delta region of China, the reported methods in this study could be used directly. On the other hand, for regions with different seashore soft soil (i.e., the soil specific gravity, LL, PL vary a lot), the proposed expressions are still applicable if the specific soil stress–strain behavior was exhibited, e.g., strong softening behavior. However, it should be noted that, for the latter scenario, the corresponding calculation parameters should be re-calibrated through laboratory tests.

Author Contributions: Investigation, C.Z.; writing-original draft, Y.F.; writing-review and editing, W.W.; validation, N.L. and A.Z. All authors have read and agreed to the published version of the manuscript.

Funding: This research was funded by the National Natural Science Foundation of China, grant number 41772311 and the Zhejiang Provincial Natural Science Foundation of China, grant number LY17E080016.

Conflicts of Interest: The authors declare no conflict of interest.

References

1. Al-Bared, M.A.M.; Marto, A.; Latifi, N.; Horpibulsuk, S. Sustainable improvement of marine clay using recycled blended tiles. *Geotech. Geol. Eng.* **2018**, *36*, 3135–3147. [CrossRef]
2. Yao, K.; An, D.L.; Wang, W.; Li, N.; Zhang, C.; Zhou, A.Z. Effect of nano-MgO on mechanical performance of cement stabilized silty clay. *Mar. Georesour. Geotechnol.* **2019**, *37*, 250–255. [CrossRef]
3. Yao, K.; Wang, W.; Li, N.; Zhang, C.; Wang, L. Investigation on strength and microstructure characteristics of nano-MgO admixed with cemented soft soil. *Constr. Build. Mater.* **2019**, *206*, 160–168. [CrossRef]
4. Puppala, A.J.; Griffin, J.A.; Hoyos, L.R.; Chomtid, S. Studies on sulfate-resistant cement stabilization methods to address sulfate-induced soil heave. *J. Geotech. Geoenviron. Eng.* **2004**, *130*, 391–402. [CrossRef]
5. Subramaniam, P.; Banerjee, S. Factors affecting shear modulus degradation of cement treated clay. *Soil Dyn. Earthq. Eng.* **2014**, *65*, 181–188. [CrossRef]
6. Xiao, H.W.; Lee, F.H.; Chin, K.G. Yielding of cement-treated marine clay. *Soils Found.* **2014**, *54*, 488–501. [CrossRef]
7. Pan, Y.T.; Liu, Y.; Chen, E. Probabilistic investigation on defective jet-grouted cut-off wall with random geometric imperfections. *Géotechnique* **2018**, *69*, 420–433. [CrossRef]
8. Pan, Y.T.; Liu, Y.; Lee, F.H.; Phoon, K.K. Analysis of cement-treated soil slab for deep excavation support–a rational approach. *Géotechnique* **2018**, *69*, 888–905. [CrossRef]

9. Du, H.J.; Pang, S.D. Value-added utilization of marine clay as cement replacement for sustainable concrete production. *J. Clean. Prod.* **2018**, *198*, 867–873. [CrossRef]

10. Park, C.G.; Yun, S.W.; Baveye, P.C.; Yu, C. Effect of Industrial By-Products on Unconfined Compressive Strength of Solidified Organic Marine Clayey Soils. *Materials* **2015**, *8*, 5098–5111. [CrossRef]

11. Du, H.J.; Pang, S.D. Dispersion and stability of graphene nanoplatelet in water and its influence on cement composites. *Constr. Build. Mater.* **2018**, *167*, 403–413. [CrossRef]

12. Kang, G.O.; Tsuchida, T.; Kim, Y.S. Strength and stiffness of cement-treated marine dredged clay at various curing stages. *Constr. Build. Mater.* **2017**, *132*, 71–84. [CrossRef]

13. Choobbasti, A.J.; Vafaei, A.; Kutanaei, S.S. Static and cyclic triaxial behavior of cemented sand with nanosilica. *J. Mater. Civ. Eng.* **2018**, *30*, 04018269. [CrossRef]

14. Choobbasti, A.J.; Kutanaei, S.S. Microstructure characteristics of cement-stabilized sandy soil using nanosilica. *J. Rock Mech. Geotech. Eng.* **2017**, *9*, 981–988. [CrossRef]

15. Choobbasti, A.J.; Vafaei, A.; Kutanaei, S.S. Mechanical properties of sandy soil improved with cement and nanosilica. *Open Eng.* **2015**, *5*, 111–116. [CrossRef]

16. Song, S.Q.; Jiang, L.H.; Jiang, S.H.; Yan, X.C.; Xu, N. The mechanical properties and electrochemical behavior of cement paste containing nano-MgO at different curing temperature. *Constr. Build. Mater.* **2018**, *164*, 663–671. [CrossRef]

17. Wang, W.; Zhang, C.; Li, N.; Tao, F.F.; Yao, K. Characterisation of nano magnesia–cement-reinforced seashore soft soil by direct-shear test. *Mar. Georesour. Geotechnol.* **2019**, *37*, 989–998. [CrossRef]

18. Wang, D.X.; Zhu, J.Y.; He, F.J. CO2 carbonation-induced improvement in strength and microstructure of reactive MgO-CaO-fly ash-solidified soils. *Constr. Build. Mater.* **2019**, *229*, 116914. [CrossRef]

19. Polat, R.; Demirboğa, R.; Karagöl, F. The effect of nano-MgO on the setting time, autogenous shrinkage, microstructure and mechanical properties of high performance cement paste and mortar. *Constr. Build. Mater.* **2017**, *156*, 208–218. [CrossRef]

20. Ghasabkolaei, N.; Choobbasti, A.J.; Roshan, N.; Ghasemi, S.E. Geotechnical properties of the soils modified with nanomaterials: A comprehensive review. *Arch. Civ. Mech. Eng.* **2017**, *17*, 639–650. [CrossRef]

21. Kutanaei, S.S.; Choobbasti, A.J. Experimental study of combined effects of fibers and nanosilica on mechanical properties of cemented sand. *J. Mater. Civ. Eng.* **2016**, *28*, 06016001. [CrossRef]

22. Kutanaei, S.S.; Choobbasti, A.J. Effects of nanosilica particles and randomly distributed fibers on the ultrasonic pulse velocity and mechanical properties of cemented sand. *J. Mater. Civ. Eng.* **2017**, *29*, 04016230. [CrossRef]

23. Deb, K. A mathematical model to study the soil arching effect in stone column-supported embankment resting on soft foundation soil. *Appl. Math. Model.* **2010**, *34*, 3871–3883. [CrossRef]

24. Ungureanu, N.; Biris, S.S.; Vladut, V.; Voicu, G.; Paraschiv, G. Studies on the mathematical modeling of artificial soil compaction. *Acta Tech. Corviniensis Bull. Eng.* **2015**, *8*, 85.

25. Meng, R.; Yin, D.; Zhou, C.; Wu, H. Fractional description of time-dependent mechanical property evolution in materials with strain softening behavior. *Appl. Math. Model.* **2016**, *40*, 398–406. [CrossRef]

26. Mei, G.X.; Chen, R.; Liu, J. New insight into developing mathematical models for predicting deformation-dependent lateral earth pressure. *Int. J. Geomech.* **2017**, *17*, 06017003. [CrossRef]

27. Ahmad, J.; Noor, M.; Jais, I.; Rosli, M.; Rahman, A.; Senin, S.; Ibrahim, A.; Hadi, B. Mathematical models for stress–strain curve prediction-a review. In *AIP Conference Proceedings*; AIP Publishing: Melville, NY, USA, 2018; p. 020005.

28. Tarasov, B.G.; Sadovskii, V.M.; Sadovskaya, O.V.; Cassidy, M.J.; Randolph, M.F. Modeling the static stress–strain state around the fan-structure in the shear rupture head. *Appl. Math. Model.* **2018**, *57*, 268–279. [CrossRef]

29. Ai, Z.Y.; Gui, J.C.; Mu, J.J. 3-D time-dependent analysis of multilayered cross-anisotropic saturated soils based on the fractional viscoelastic model. *Appl. Math. Model.* **2019**, *76*, 172–192. [CrossRef]

30. Cui, L.; Fall, M. Mathematical modeling of cemented tailings backfill: A review. *Int. J. Min. Reclam. Environ.* **2019**, *33*, 389–408. [CrossRef]

31. Wang, J.; Zhou, D.; Zhang, Y.Q.; Cai, W. Vertical impedance of a tapered pile in inhomogeneous saturated soil described by fractional viscoelastic model. *Appl. Math. Model.* **2019**, *75*, 88–100. [CrossRef]

32. Lee, K.; Chan, D.; Lam, K. Constitutive model for cement treated clay in a critical state frame work. *Soils Found.* **2004**, *44*, 69–77. [CrossRef]

33. Bergado, D.; Taechakumthorn, C.; Lorenzo, G.; Abuel-Naga, H.M. Stress-deformation behavior under anisotropic drained triaxial consolidation of cement-treated soft Bangkok clay. *Soils Found.* **2006**, *46*, 629–637. [CrossRef]

34. Chiu, C.F.; Zhu, W.; Zhang, C.L. Yielding and shear behavior of cement-treated dredged materials. *Eng. Geol.* **2009**, *103*, 1–12. [CrossRef]

35. Tremblay, H.; Leroueil, S.; Locat, J. Mechanical improvement and vertical yield stress prediction of clayey soils from eastern Canada treated with lime or cement. *Can. Geotech. J.* **2001**, *38*, 567–579. [CrossRef]

36. Du, Y.J.; Horpibulsuk, S.; Wei, M.L.; Suksiripattanapong, C.; Liu, M.D. Modeling compression behavior of cement-treated zinc-contaminated clayey soils. *Soils Found.* **2014**, *54*, 1018–1026. [CrossRef]

37. Subramaniam, P.; Sreenadh, M.M.; Banerjee, S. Critical state parameters of dredged Chennai marine clay treated with low cement content. *Mar. Georesour. Geotechnol.* **2016**, *34*, 603–616. [CrossRef]

38. Yu, B.W.; Du, Y.J.; Jin, F.; Liu, C.Y. Multiscale study of sodium sulfate soaking durability of low plastic clay stabilized by reactive magnesia-activated ground granulated blast-furnace slag. *J. Mater. Civ. Eng.* **2016**, *28*, 04016016. [CrossRef]

39. Wang, W.; Li, Y.; Yao, K.; Li, N.; Zhou, A.Z.; Zhang, C. Strength properties of nano-MgO and cement stabilized coastal silty clay subjected to sulfuric acid attack. *Mar. Georesour. Geotechnol.* **2019**. [CrossRef]

40. Duncan, J.M.; Chang, C.Y. Nonlinear analysis of stress and strain in soils. *J. Soil Mech. Found. Div.* **1970**, *96*, 1629–1653.

41. Cai, Y.Y.; Zheng, C.T.; Qi, Z.B.; Mu, K.; Yu, J. Nonlinear constitutive model of clay based on power function. *J. Lanzhou Univ. Technol.* **2012**, *38*, 126–130.

Article

Detecting and Handling Cyber-Attacks in Model Predictive Control of Chemical Processes

Zhe Wu [1], Fahad Albalawi [2], Junfeng Zhang [1], Zhihao Zhang [1] and Helen Durand [3] and Panagiotis D. Christofides [1,4,*]

1   Department of Chemical and Biomolecular Engineering, University of California,
    Los Angeles, CA 90095-1592, USA; zhewu2008@gmail.com (Z.W.);
    jf-zhang13@mails.tsinghua.edu.cn (J.Z.); zhihaozhang@ucla.edu (Z.Z.)
2   Department of Electrical and Computer Engineering, Taif University, Taif 21974, Saudi Arabia;
    eng.fahad19@gmail.com
3   Department of Chemical Engineering and Materials Science, Wayne State University,
    Detroit, MI 48202, USA; helen.durand@wayne.edu
4   Department of Electrical and Computer Engineering, University of California,
    Los Angeles, CA 90095-1592, USA
*   Correspondence: pdc@seas.ucla.edu

Received: 24 August 2018; Accepted: 21 September 2018; Published: 25 September 2018

Abstract: Since industrial control systems are usually integrated with numerous physical devices, the security of control systems plays an important role in safe operation of industrial chemical processes. However, due to the use of a large number of control actuators and measurement sensors and the increasing use of wireless communication, control systems are becoming increasingly vulnerable to cyber-attacks, which may spread rapidly and may cause severe industrial incidents. To mitigate the impact of cyber-attacks in chemical processes, this work integrates a neural network (NN)-based detection method and a Lyapunov-based model predictive controller for a class of nonlinear systems. A chemical process example is used to illustrate the application of the proposed NN-based detection and LMPC methods to handle cyber-attacks.

Keywords: industrial cyber-physical systems; cyber-attacks; neural network; model predictive control; nonlinear chemical processes

---

1. Introduction

Recently, the security of process control systems has become crucially important since control systems are vulnerable to cyber-attacks, which are a series of computer actions to compromise the security of control systems (e.g., integrity, stability and safety) [1,2]. Since cyber-physical systems (CPS) or supervisory control and data acquisition (SCADA) systems are usually large-scale, geographically dispersed and life-critical systems where embedded sensors and actuators are connected into a network to sense and control the physical devices [3], the failure of cybersecurity can lead to unsafe process operation, and potentially to catastrophic consequences in the chemical process industries, causing environmental damage, capital loss and human injuries. Among cyber-attacks, targeted attacks are severe threats for control systems because of their specific designs with the aim of modifying the control actions applied to a chemical process (for example, the Stuxnet worm aims to modify the data sent to a Programmable Logic Controller [4]). Additionally, targeted attacks are usually stealthy and difficult to detect using classical detection methods since they are designed based on some known information of control systems (e.g., the process state measurement). Therefore, designing an advanced detection system (e.g., machine learning-based detection methods [5,6]) and a suitable optimal control scheme for nonlinear processes in the presence of targeted cyber-attacks is an important open issue.

Due to the rapid development of computer networks of CPS in the past two to three decades, the components (e.g., sensors, actuators, and controllers) in a large-scale process control system are now connected through wired/wireless networks, which makes these systems more vulnerable to cyber-attacks that can damage the operation of physical layers besides cyber layers. Additionally, since the development of most of the existing detection methods still depends partly on human analysis, the increased use of data and the designs of stealthy cyber-attacks pose challenges to the development of timely detection methods with high detection accuracy. In this direction, the design of cyber-attacks, the anomaly detection methods focusing on physical layers, and the corresponding resilient control methods have received a lot of attention. A typical method of detection [4] is using a model of the process and comparing the model output predictions with the actual measured outputs. In [7], a dynamic watermarking method was proposed to detect cyber-attacks via a technique of injecting private excitation into the system. Moreover, four representative detection methods were summarized in [3] as Bayesian detection with binary hypothesis, weighted least squares, χ^2-detector based on Kalman filters and quasi-fault detection and isolation methods.

Besides the detection of cyber-attacks, the design of resilient control schemes also plays an important role in operating a chemical process reliably under cyber-attacks. To guarantee the process performance (e.g., robustness, stability, safety, etc.) and mitigate the impact of cyber-attacks, resilient state estimation and resilient control strategies have attracted considerable research interest. In [2,8], resilient estimators were designed to reconstruct the system states accurately. An event-triggered control system was proposed in [9] to tolerate Denial-of-service (DoS) attacks without jeopardizing the stability of the closed-loop system.

On the other hand, as a widely-used advanced control methodology in industrial chemical plants, model predictive control (MPC) achieves optimal performance of multiple-input multiple-output processes while accounting for state and input constraints [10]. Based on Lyapunov methods (e.g., a Lyapunov-based control law), the Lyapunov-based model predictive control (LMPC) method was developed to ensure stability and feasibility in an explicitly-defined subset of the region of attraction of the closed-loop system [11,12]. Additionally, process operational safety can also be guaranteed via control Lyapunov-barrier function-based constraints in the framework of LMPC [13]. At this stage, however, the potential safety/stability problem in MPC caused by cyber-attacks has not been studied with the exception of a recent work that provides a quantitative framework for the evaluation of resilience of control systems with respect to various types of cyber-attacks [14].

Motivated by this, we develop an integrated data-based cyber-attack detection and model predictive control method for nonlinear systems subject to cyber-attacks. Specifically, a cyber-attack (e.g., a min-max cyber-attack) that aims to destabilize the closed-loop system via a sensor tamper is considered and applied to the closed-loop process. Under such a cyber-attack, the closed-loop system under the MPC without accounting for the cyber-attack cannot ensure closed-loop stability. To detect potential cyber-attacks, we take advantage of machine learning methods, which are widely-used in clustering, regression, and other applications such as model order reduction [15–17], to build a neural network (NN)-based detection system. First, the NN training dataset was obtained for three conditions: (1) The system without disturbances and cyber-attacks (i.e., nominal system); (2) The system with only process disturbances considered; (3) The system with only cyber-attacks considered. Then, a NN detection method is trained off-line to derive a model that can be used on-line to predict cyber-attacks. In addition, considering the classification accuracy of the NN, a sliding detection window is employed to reduce false cyber-attack alarms. Finally, a Lyapunov-based model predictive control (LMPC) method that utilizes the state measurement from secure, redundant sensors is developed to reduce the impact of cyber-attacks and re-stabilize the closed-loop system in finite time.

The rest of the paper is organized as follows: in Section 2, the class of nonlinear systems considered and the stabilizability assumptions are given. In Section 3, we introduce the min-max cyber-attack, develop a NN-based detection system and a Lyapunov-based model predictive controller (LMPC) that guarantees recursive feasibility and closed-loop stability under sample-and-hold implementation within an explicitly characterized set of initial conditions. In Section 4, a nonlinear chemical process example is used to demonstrate the applicability of the proposed cyber-attack detection and control method.

2. Preliminaries

2.1. Notation

Throughout the paper, the notation $|\cdot|$ is used to denote the Euclidean norm of a vector, the notation $|\cdot|_Q$ denotes a weighted Euclidean norm of a vector (i.e., $|x|_Q^2 = x^T Q x$ where Q is a positive definite matrix). x^T denotes the transpose of x. $\mathbf{R}_+$ denotes the set $[0, \infty)$. The notation $L_f V(x)$ denotes the standard Lie derivative $L_f V(x) := \frac{\partial V(x)}{\partial x} f(x)$. For given positive real numbers β and ϵ, $\mathcal{B}_\beta(\epsilon) := \{x \in \mathbf{R}^n \mid |x - \epsilon| < \beta\}$ is an open ball around ϵ with a radius of β. Set subtraction is denoted by "\", i.e., $A \backslash B := \{x \in \mathbf{R}^n \mid x \in A, x \notin B\}$. $\lceil x \rceil$ maps x to the least integer greater than or equal to x and $\lfloor x \rfloor$ maps x to the greatest integer less than or equal to x. The function $f(\cdot)$ is of class $\mathcal{C}^1$ if it is continuously differentiable in its domain. A continuous function $\alpha : [0, a) \to [0, \infty)$ is said to belong to class $\mathcal{K}$ if it is strictly increasing and is zero only when evaluated at zero.

2.2. Class of Systems

The class of continuous-time nonlinear systems considered is described by the following state-space form:

$$\dot{x} = f(x) + g(x)u + d(x)w, \ x(t_0) = x_0 \tag{1}$$

where $x \in \mathbf{R}^n$ is the state vector, $u \in \mathbf{R}^m$ is the manipulated input vector, and $w \in W$ is the disturbance vector, where $W := \{w \in \mathbf{R}^q \mid |w| \leq \theta, \ \theta \geq 0\}$. The control action constraint is defined by $u \in U = \{u_{min} \leq u \leq u_{max}\} \subset \mathbf{R}^m$, where u_{min} and u_{max} represent the minimum and the maximum value vectors of inputs allowed, respectively. $f(\cdot)$, $g(\cdot)$ and $d(\cdot)$ are sufficiently smooth vector and matrix functions of dimensions $n \times 1$, $n \times m$ and $n \times q$, respectively. Without loss of generality, the initial time t_0 is taken to be zero ($t_0 = 0$), and it is assumed that $f(0) = 0$, and thus, the origin is a steady-state of the system of Equation (1) with $w(t) \equiv 0$, (i.e., $(x_s^*, u_s^*) = (0, 0)$). In the manuscript, we assume that every measured state is measured by multiple sensors that are isolated from one another such that if one sensor measurement is tampered by cyber-attacks, a secure network or some secure way can still be used to send the correct sensor measurements of $x(t)$ to the controller. This can also be viewed as secure, redundant sensors or just having an alternative, secure network to send the sensor measurements to the controller. However, if this assumption does not hold, i.e., no secure sensors are available, then the system has to be shut down after the detection of cyber-attacks, or to be operated in an open-loop manner thereafter with an accurate process model.

2.3. Stabilizability Assumptions and Lyapunov-Based Control

Consider the nominal system of Equation (1) with $w(t) \equiv 0$. We first assume that there exists a stabilizing feedback control law $u = \Phi(x) \in U$ such that the origin of the nominal system of Equation (1) can be rendered asymptotically stable for all $x \in D_1 \subset \mathbf{R}^n$, where D_1 is an open neighborhood of the origin, in the sense that there exists a positive definite $\mathcal{C}^1$ control Lyapunov function V that satisfies the small control property and the following inequalities:

$$\alpha_1(|x|) \leq V(x) \leq \alpha_2(|x|), \tag{2a}$$

$$\frac{\partial V(x)}{\partial x} F(x, \Phi(x), 0) \leq -\alpha_3(|x|), \tag{2b}$$

$$\left| \frac{\partial V(x)}{\partial x} \right| \leq \alpha_4(|x|) \tag{2c}$$

where $\alpha_j(\cdot), j = 1, 2, 3, 4$ are class $\mathcal{K}$ functions. $F(x, u, w)$ is used to represent the system of Equation (1) (i.e., $F(x, u, w) = f(x) + g(x)u + d(x)w$).

An example of a feedback control law that is continuous for all x in a neighborhood of the origin and renders the origin asymptotically stable is the following control law [18]:

$$\varphi_i(x) = \begin{cases} -\dfrac{p + \sqrt{p^2 + |q|^4}}{|q|^2} q, & \text{if} \quad q \neq 0 \\ 0, & \text{if} \quad q = 0 \end{cases} \tag{3a}$$

$$\Phi_i(x) = \begin{cases} u_i^{min}, & \text{if} \quad \varphi_i(x) < u_i^{min} \\ \varphi_i(x), & \text{if} \quad u_i^{min} \leq \varphi_i(x) \leq u_i^{max} \\ u_i^{max}, & \text{if} \quad \varphi_i(x) > u_i^{max} \end{cases} \tag{3b}$$

where p denotes $L_f V(x)$ and q denotes $(L_g V(x))^T = [L_{g_1} V(x) \cdots L_{g_m} V(x)]^T$. $\varphi_i(x)$ of Equation (3a) represents the ith component of the control law $\Phi(x)$ before considering saturation of the control action at the input bounds. $\Phi_i(x)$ of Equation (3b) represents the ith component of the saturated control law $\Phi(x)$ that accounts for the input constraints $u \in U$. Based on the controller $\Phi(x)$ that satisfies Equation (2), the set of initial conditions from which the controller $\Phi(x)$ can stabilize the origin of the input-constrained system of Equation (1) is characterized as: $\phi_n = \{x \in \mathbf{R}^n \mid \dot{V} + \kappa V(x) \leq 0, u = \Phi(x) \in U, \kappa > 0\}$. Additionally, we define a level set of $V(x)$ inside ϕ_n as $\Omega_\rho := \{x \in \phi_n \mid V(x) \leq \rho\}$, which represents a stability region of the closed-loop system of Equation (1).

3. Cyber-Attack and Detection Methodology

From the perspective of process control systems, cyber-attacks are malicious signals that can compromise actuators, sensors or their communication networks. Specifically, among sensor cyber-attacks, DoS attacks, replay attacks and deception attacks are the three most common and easily implementable ones by attackers [5]. On the other hand, since stealthy cyber-attacks are designed to damage the performance of CPS (e.g., stability and safety), developing more reliable detection and control methods that can detect, locate and mitigate cyber-attacks in a timely fashion and control the damage within a tolerable limit is imperative.

In this section, the min-max cyber-attack designed to damage closed-loop stability of the system of Equation (1) is first introduced. Subsequently, a general model-based detection method [4] and the corresponding stealthy cyber-attacks that can evade such detection are presented. Therefore, to better detect different types of cyber-attacks, the data-based detection scheme that utilizes machine learning methods is finally developed with a sliding detection window.

3.1. Min-Max Cyber-Attack

In this subsection, we first consider a deception sensor cyber-attack, in which the minimum or maximum allowable sensor measurement values are fed into process control systems (e.g., a Lyapunov-based control system with a stability region Ω_ρ defined by a level set of Lyapunov function $V(x)$) to drive the closed-loop states away from their expected values and finally ruin the stability of the closed-loop system. Since $\forall x \in \Omega_\rho$, there exists a feasible control action $u = \Phi(x)$ such that $\dot{V} < 0$, closed-loop stability is maintained within the stability region Ω_ρ under $\Phi(x)$. Assuming that attackers know the stability region of the system of Equation (1) in advance and have access to some of the sensors (but not all), to remain undetectable by a simple stability region-based detection method (i.e., the cyber-attack is detected if the state is out of the stability region), the min-max cyber-attack is designed with the following form such that the fake sensor measurements are still inside Ω_ρ:

$$\bar{x} = \arg \max_{x \in \mathbf{R}} \{V(x) \leq \rho\} \tag{4}$$

where $\tilde{x}$ is the tampered sensor measurement. Since the controller needs to get access to true state measurements to maintain closed-loop stability in a state feedback control system, wrong state measurements under cyber-attacks can affect control actions and eventually drive the state away from its set-point. In the section "Application to a chemical process example", it is shown that if attackers apply a min-max cyber-attack to safety-critical sensors (e.g., temperature or pressure sensors in a chemical reactor) in process control systems, closed-loop stability may not be maintained (i.e., the closed-loop state goes out of Ω_ρ) and the system may have to be shut down.

3.2. Model-Based Detection and Stealthy Cyber-Attack

Based on the known process model of Equation (1), a cumulative sum (CUSUM) statistic detection method [4] can be developed to minimize the detection time when a cyber-attack occurs. Specifically, the CUSUM statistic method detects cyber-attacks by calculating the cumulative sum of the deviation between expected and measured states. The method is developed by the following equations:

$$S(k) = (S(k-1) + z(k))^+, \; S(0) = 0 \tag{5a}$$

$$D(S(k)) = \begin{cases} 1, & \text{if } S(k) > S_{TH} \\ 0, & \text{otherwise} \end{cases} \tag{5b}$$

where $S(k)$ is the nonparametric CUSUM statistic and S_{TH} is the threshold of the detection of cyber-attacks. $(S)^+ = S$, if $S \geq 0$ and $(S)^+ = 0$ otherwise. D is the detection indicator where $D = 1$ indicates that the cyber-attack is confirmed or there is no cyber-attack if $D = 0$. $z(k)$ is the deviation between expected states $\tilde{x}(t_k)$ and measured states $x(t_k)$ at time $t = t_k$: $z(k) := |\tilde{x}(t_k) - x(t_k)| - b$ where $\tilde{x}(t_k)$ is derived using the known process model, the state and the control action at $t = t_{k-1}$, and b is a small positive constant to reduce the false alarm rate due to disturbances.

With a carefully selected S_{TH}, the model-based detection method can detect many sensor cyber-attacks efficiently. However, the above model-based method may be evaded and becomes invalid for stealthy cyber-attacks if attackers know more about the system (e.g., the system model and the principles of the detection method). For example, three advanced stealthy cyber-attacks were proposed in [4] to damage the system without triggering the threshold of the model-based detection method. Specifically, a surge cyber-attack is designed to maximize the damage for the first few steps (similar to min-max cyber-attacks) and switch to cyber-attacks with small perturbations for the rest of time when $S(k)$ reaches S_{TH}. The form of a surge cyber-attack is given by the following equations:

$$x(t_k) = \begin{cases} x(t_k)^{min}, & \text{if } S(k) \leq S_{TH} \\ \tilde{x}(t_k) - |S_{TH} + b - S(k-1)|, & \text{otherwise} \end{cases} \tag{6}$$

The above surge cyber-attack is able to maintain $S(k)$ within its threshold and therefore is undetectable by the above detection method. In this case, the defenders should either develop more advanced detection methods for stealthy cyber-attacks (i.e., it becomes an interactive decision-making process between an attacker and a defender [19]), or develop a detection method from another perspective, for example, a data-based method. Since the purpose of any type of stealthy cyber-attack is to change the normal operation and destroy the performance of the system of Equation (1), the dynamic operation of the system of Equation (1) (e.g., dynamic trajectories in state-space) under cyber-attacks becomes different from that of the nominal system of Equation (1). The deviation of the data can be regarded as an intrinsic indicator for detection of cyber-attacks. In this direction, a data-based detection system is developed via machine learning methods in the next subsection.

3.3. Detection via Machine Learning Techniques

Machine learning has a wide range of applications in classification, regression, and clustering problems. To detect cyber-attacks, classification methods can be utilized to determine whether there is a cyber-attack on the system of Equation (1) or not. The data-based learning problems are usually categorized into unsupervised learning and supervised learning.

Unsupervised learning (e.g., k-means clustering) uses unlabeled data to derive a model that can split the data into different categories. On the other hand, supervised learning aims to develop a function that maps an input to an output based on labeled dataset (input-output pairs). There are two types of supervised learning tools, (1) classification tools (e.g., k-nearest neighbor (k-NN), support vector machine (SVM), random forest, neural networks) are used to develop a function based on labeled training datasets to predict the class of a new set of data that was not used in the training stage; (2) regression tools (e.g., linear regression, support vector regression, etc.) aim to predict the outcome of an event based on the relationship between variables obtained from the training datasets (labeled input-output pairs) [20]. Since supervised learning concerns labeled training data, we utilize a neural network (NN) algorithm to predict whether the system of Equation (1) is nominally operating, under disturbances or under cyber-attacks. Subsequently, a Lyapunov-based model predictive controller is proposed to stabilize the closed-loop system during the absence and presence of cyber-attacks.

3.4. NN-Based Detection System

Since the evolution of the closed-loop state from the initial condition $x(0) = x_0 \in \Omega_\rho$ is determined by both the nonlinear system model of Equation (1) and the design of process control systems, it is difficult to distinguish normal operation from the operation under cyber-attacks. Moreover, even if a detection method is developed for a specific cyber-attack (e.g., min-max cyber-attack), the detection strategy is not guaranteed to identify a different type of cyber-attack. Motivated by these concerns, this work proposes a data-based detection system for different types of cyber-attacks by using machine learning methods.

As a widely-used machine learning method, neural networks build a general class of nonlinear functions from input variables to output variables. The basic structure of a feed-forward multiple-input-single-output neural network with one hidden layer is given in Figure 1, where N_{uj}, $j = 1, 2, \ldots, n$ denotes the input variables in the input layer, θ_{1i}, $i = 1, 2, \ldots, h$ denotes the neurons in the hidden layer and N_y denotes the output in the output layer. Specifically, the hidden neurons θ_{1i} and the output N_y (i.e., the classification result) are obtained by the following equations, respectively [21]:

$$\theta_{1i} = \sigma_1 \left(\sum_{j=1}^{n} N_{wij}^{(1)} N_{uj} + N_{wi0}^{(1)} \right) \tag{7}$$

$$N_y = \sigma_2 \left(\sum_{j=1}^{h} N_{wj}^{(2)} \theta_{1j} + N_{w0}^{(2)} \right) \tag{8}$$

where σ_1, σ_2 are nonlinear activation functions, $N_{wij}^{(1)}$ and $N_{wj}^{(2)}$ are weights, and $N_{wi0}^{(1)}$, $N_{w0}^{(2)}$ are biases. For simplicity, the input vector $\mathbf{N_u}$ will be used to denote all the inputs N_{uj}, and the weight matrix $\mathbf{N_w}$ will be used to represent all the weights and biases in Equations (7) and (8). The neurons in the hidden layer receive the weighted sum of inputs and use activation functions σ_1 (e.g., ReLu function $\sigma(x) = \max(0, x)$ or sigmoid function $\sigma(x) = 1/(1 + e^{-x})$) to bring in the nonlinearity such that the NN is not a simple linear combination of the inputs. The output neuron generates the class label via a linear combination of hidden neurons and an activation function σ_2 (e.g., sigmoid function for two classes or softmax function $\sigma_i(x) = e^{x_i} / \sum_{k=1}^{K} e^{x_k}$ for multiple classes where K is the number of classes).

Given a set of training data including the input vectors $\mathbf{N_u^i}$, $i = 1, 2, \ldots N_T$ and the corresponding classified labels (i.e., target vectors $\mathbf{N_t^i}$), the NN model is trained by minimizing the following error function (i.e., loss function):

$$\mathbf{E(N_w)} = \frac{1}{2} \sum_{i=1}^{N_T} |\mathbf{N_y^i}(\mathbf{N_u^i}, \mathbf{N_w}) - \mathbf{N_t^i}|^2 \tag{9}$$

where $\mathbf{N_y^i}(\mathbf{N_u^i}, \mathbf{N_w})$ is the predicted class for the input $\mathbf{N_u^i}$ under $\mathbf{N_w}$. The above nonlinear optimization problem is solved using the stochastic gradient descent (SGD) method, in which the backpropagation method is utilized to calculate the gradient of $\mathbf{E(N_w)}$. Meanwhile, the weight matrix $\mathbf{N_w}$ is updated by the following equation:

$$\mathbf{N_w} := \mathbf{N_w} - \eta \nabla \mathbf{E(N_w)} \tag{10}$$

where η is the learning rate to control the speed of convergence. Additionally, to avoid over-fitting during the training process, k-fold cross-validation is employed to randomly partition the original dataset into $k-1$ subsets of training data and 1 subset of validation data, and early-stopping is activated once the error on the validation set stops decreasing.

Finally, the classification accuracy of the validation dataset is utilized to demonstrate the performance of the neural network since the validation dataset is independent of the training dataset and is not used in training the NN model. Specifically, the classification accuracy (i.e., the test accuracy) of the trained NN model is obtained by the following equation:

$$N_{acc} = \frac{n_c}{n_{val}} \tag{11}$$

where n_c is the number of data samples with correct predicted classes, and n_{val} is the total number of data samples in the validation dataset. In general, the NN performance depends on many factors, e.g., the size of dataset, the number of hidden layers and nodes, and the intensity and the amount of disturbance applied [22–24]. In Remark 1, the method of determining the number of layers and nodes is introduced.

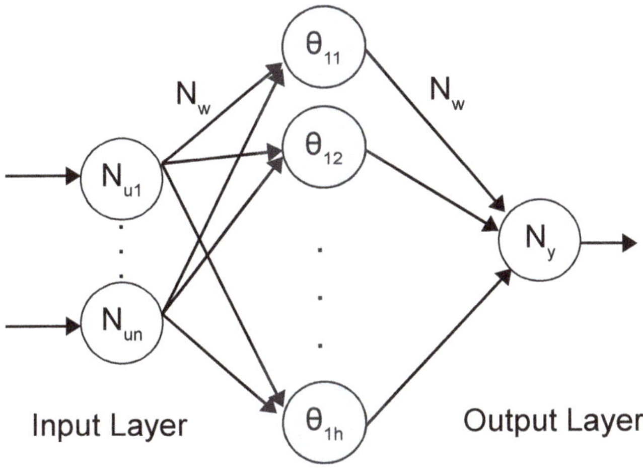

Figure 1. Basic structure of a feed-forward neural network used for cyber-attack detection.

In this paper, the NN is developed to derive a model M to classify three classes: the nominal closed-loop system, the closed-loop system with disturbances, and the closed-loop system under cyber-attacks. A large dataset of time-varying states for various initial conditions (i.e., dynamic trajectories) of the above three cases is used as the input to the neural network. The output of the neural network is the classified class. Since the feed-forward NN is a static model with a fixed input dimension (i.e., fixed time length) but the detection method should be applied during the dynamic operation of the system of Equation (1), multiple NN models with various sizes of input datasets (i.e., various time lengths) are used for the detection of cyber-attacks in real time until the time length corresponding to the available data since the beginning of the time of operation becomes equal to the time length that is preferred to be utilized for the remainder of the operating time. Specifically, given a training dataset of time-series state vectors (i.e., closed-loop trajectories): $N_u \in \mathbf{R}^{n \times T}$ where n is the number of states and T is the number of sampling steps of each trajectory, the NN model is obtained and applied as follows: (1) the NN is trained with data corresponding to time lengths from the initial time to T sampling steps in intervals of N_a sampling steps, i.e., the ith NN model M_i is trained using data from $t = 0$ to $t = iN_a$, where $i = 1, 2, \ldots, T/N_a$ and T is a multiple integer of N_a; (2) when incorporating the NN-based detection system in MPC, real-time state measurement data can be readily utilized in the corresponding NN model M_i to check if there is a cyber-attack so far.

Remark 1. *With an appropriate structure (i.e., number of layers and hidden neurons) of the neural network, the weight matrix* $\mathbf{N_w}$ *is calculated by Equation (10) and will be utilized to derive the classification accuracy of Equation (11). However, in general, there is no systematic method to determine the structure of a neural network since it highly depends on the number of training data samples and also the complexity of the model needed for classification. Therefore, in practice, the neural network is initiated with one hidden layer with a few hidden neurons. If the classification result is unsatisfactory, we increase the hidden neurons number and further layers with appropriate regularization are added to improve the performance.*

Remark 2. *It is noted that the above classification accuracy of the NN model represents the ratio of the number of correct predictions to the total number of predictions for all classes. If we only consider the case of binary classification (i.e., whether the system is under cyber-attacks or not), sensitivity (also called recall or true positive rate) and specificity (also called true negative rate) are also useful measures. Specifically, sensitivity measures the proportion of actual cyber-attacks that are correctly identified as such, while specificity measures the proportion of actual non-cyber-attacks that are correctly identified as such. Therefore, in the presence of multiple types of cyber-attacks or disturbances, it becomes straightforward to learn the performance of the NN-based method to detect true cyber-attacks via sensitivity and specificity.*

3.5. Sliding Detection Window

Since the classification accuracy of a NN is not perfect, false alarms may be triggered based on a one-time detection (i.e., non-cyber-attack case may be identified as cyber-attack). In order to reduce the false alarm rates, a detection indicator D_i generated by each sub-model M_i and a sliding detection window with length N_s are proposed as follows:

$$D_i = \begin{cases} 1, & \text{if attack is detected by } M_i \\ 0, & \text{if no attack is detected by } M_i \end{cases} \tag{12}$$

Based on the detection indicator D_i at every N_a sampling steps, the weighted sum of detection indicators within the sliding detection window D_I shown in Figure 2 at $t = t_k = k\Delta$ is calculated as follows:

$$D_I = \sum_{j=\lceil (k-N_s+1)/N_a \rceil}^{\lfloor k/N_a \rfloor} \gamma^{\lfloor \frac{k}{N_a} \rfloor - j} D_j \tag{13}$$

where γ is a detection factor that gives more weight to recent detections within the sliding window because the classification accuracy of the NN increases as more data is used for training. If $D_I \geq D_{TH}$, where D_{TH} is a threshold that indicates a real cyber-attack in the closed-loop system, then the cyber-attack is confirmed and reported by the NN-based detection system; otherwise, the detection system remains silent and the sliding window will be rolled one sampling time. To balance false alarms and missed detections, the threshold D_{TH} is determined via extensive closed-loop simulations under cyber-attacks to derive a desired detection rate.

Additionally, since there is no guaranteed feasible control action that can drive the state back towards the origin once the state of the system of Equation (1) is outside the stability region Ω_ρ due to the way of characterizing ϕ_n and Ω_ρ, it is also necessary to check whether the state is in Ω_ρ, especially when cyber-attacks occur but have not been detected yet. Therefore, to prevent the system state from entering a region in state-space where closed-loop stability is not guaranteed, the boundedness of the state vector within the stability region is also checked using the state measurement from redundant, secure sensors at the time when $D_i = 1$. If the state x has already left Ω_ρ, closed-loop stability is no longer guaranteed and in this case further safety system components (e.g., physical safety devices) need to be activated to avoid dangerous operations [25]. However, if $x \in \Omega_\rho$, the state measurement will be read from redundant, secure sensors instead of the original sensors to avoid deterioration of stability under the potential cyber-attack indicated by $D_i = 1$.

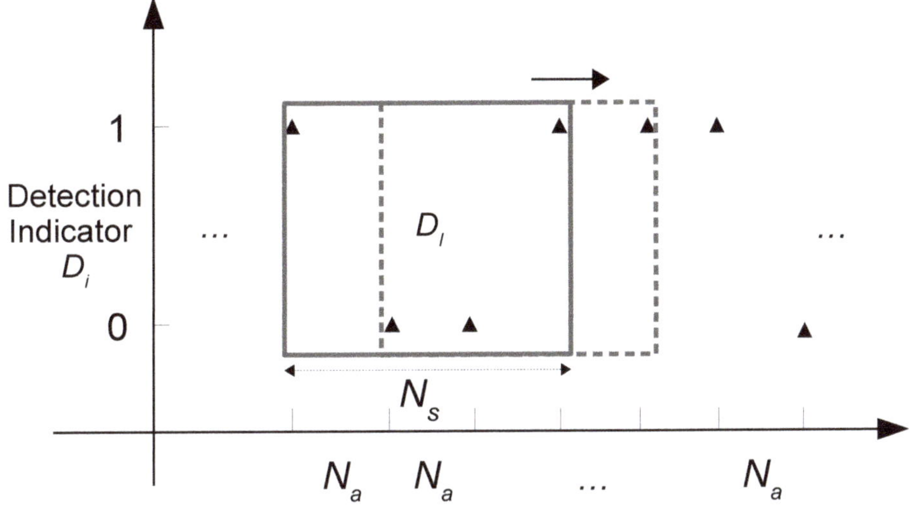

Figure 2. The sliding detection window with detection activated every N_a sampling steps, where triangles represent the detection indicator D_i and the box with length N_s represents the sliding detection window.

Remark 3. *The sliding window with length N_s is employed to reduce false alarm rates. Considering that the classification accuracy derived is not perfect, the idea behind the sliding detection window is that a cyber-attack is confirmed only if it has been detected for a few times continuously instead of a one-time detection. The length of sliding window N_s will balance the efficiency of detection and false alarm rates. Specifically, a larger N_s and a higher detection threshold D_{TH} ($D_I \geq D_{TH}$ within the sliding detection window represents the confirmation of a cyber-attack) lead to longer detection time but a lower false alarm rate, while a smaller N_s and a lower D_{TH} have the opposite effect. Therefore, N_s and D_{TH} should be determined well to achieve a balanced performance between detection efficiency and false alarm rate.*

Remark 4. *The above supervised learning-based cyber-attack detection method is able to distinguish the normal operation of the system of Equation (1) from the abnormal operation under cyber-attacks, provided that there is a large amount of labeled data available for training. However, for those unknown cyber-attacks which are never used for training, the detection is not guaranteed. Specifically, if there exists an unknown cyber-attack that is distinct from the trained cyber-attacks, the NN-based detection method may not be able to identify it as a cyber-attack. In this case, an unsupervised learning-based detection method may achieve better performance by clustering unknown cyber-attack data into a new class. However, if the unknown cyber-attack shares similar* *properties (e.g., similar attack mechanism) with a trained cyber-attack, the NN method may still be able to detect it and classify it as one of the available classes. For example, it is demonstrated in the section "Application to a chemical process example" that the unknown surge cyber-attack can still be detected by the NN-based detection system that is trained for min-max cyber-attacks because of the similarity between these two cyber-attacks.*

Remark 5. *Since different types of cyber-attacks may have various purposes, targeted sensors and attack duration, the dynamic behavior of a closed-loop system varies with different cyber-attacks, which can be eventually reflected by the data of states. Besides the detection of cyber-attacks, the above NN-based detection method is also able to recognize the types of cyber-attacks by training the NN model with data of various types of cyber-attacks labeled as different classes. As a result, the NN model can not only detect the occurrence of cyber-attacks, but also can identify the type of a cyber-attack if the data of that particular cyber-attack has been utilized for training.*

4. Lyapunov-Based MPC (LMPC)

To cope with the threats of the above sensor cyber-attacks, a feedback control method that accounts for the corruption of some sensor measurements should be designed by defenders to mitigate the impact of cyber-attacks and still stabilize the system of Equation (1) at its steady-state. Based on the assumption of the existence of a Lyapunov function $V(x)$ and a controller $u = \Phi(x)$ that satisfy Equation (2), the LMPC that utilizes the accurate measurement from redundant, secure sensors is proposed as the following optimization problem:

$$\mathcal{J} = \min_{u \in S(\Delta)} \int_{t_k}^{t_{k+N}} L_t(\tilde{x}(t), u(t)) dt \tag{14a}$$

$$\text{s.t} \quad \dot{\tilde{x}}(t) = f(\tilde{x}(t)) + g(\tilde{x}(t))u(t) \tag{14b}$$

$$\tilde{x}(t_k) = x(t_k) \tag{14c}$$

$$u(t) \in U, \ \forall \, t \in [t_k, t_{k+N}) \tag{14d}$$

$$\dot{V}(x(t_k), u(t_k)) \leq \dot{V}(x(t_k), \Phi(x(t_k))),$$

$$\text{if } V(x(t_k)) > \rho_{min}, \tag{14e}$$

$$V(\tilde{x}(t)) \leq \rho_{min}, \ \forall \, t \in [t_k, t_{k+N})$$

$$\text{if } V(x(t_k)) \leq \rho_{min} \tag{14f}$$

where $\tilde{x}(t)$ is the predicted state trajectory, $S(\Delta)$ is the set of piecewise constant functions with period Δ, and N is the number of sampling periods in the prediction horizon. $\dot{V}(x(t_k), u(t_k))$ represents the time derivative of $V(x)$, i.e., $\frac{\partial V}{\partial x}(f(\tilde{x}(t)) + g(\tilde{x}(t))u(t))$. We assume that the states of the closed-loop system are measured at each sampling time instance, and will be used as the initial condition in the optimization problem of LMPC in the next sampling step. Specifically, based on the measured state $x(t_k)$ at $t = t_k$, the above optimization problem is solved to obtain the optimal solution $u^*(t)$ over the prediction horizon $t \in [t_k, t_{k+N})$. The first control action of $u^*(t)$, i.e., $u^*(t_k)$, is sent to the control actuators to be applied over the next sampling period. Then, at the next sampling time $t_{k+1} := t_k + \Delta$, the optimization problem is solved again, and the horizon will be rolled one sampling time.

In the optimization problem of Equation (14), the objective function of Equation (14a) that is minimized is the integral of $L_t(\tilde{x}(t), u(t))$ over the prediction horizon, where the function $L_t(x, u)$ is usually in a quadratic form (i.e., $L_t(x, u) = x^T R x + u^T Q u$, where R and Q are positive definite matrices). The constraint of Equation (14b) is the nominal system of Equation (1) (i.e., $w(t) \equiv 0$) to predict the evolution of the closed-loop state. Equation (14c) defines the initial condition of the nominal process system of Equation (14b,14d) defines the input constraints over the prediction horizon. The constraint of Equation (14e) requires that $V(\tilde{x})$ for the system decreases at least at the rate under $\Phi(x)$ at t_k when $V(x(t_k)) > \rho_{min}$. However, if $x(t_k)$ enters a small neighborhood around the origin $\Omega_{\rho_{min}} := \{x \in \phi_n \,|\, V(x) \leq \rho_{min}\}$, in which $\dot{V}$ is not required to be negative due to the sample-and-hold implementation of the LMPC, the constraint of Equation (14f) is activated to maintain the state inside $\Omega_{\rho_{min}}$ afterwards.

When the cyber-attack is detected by $D_i = 1$ but not confirmed by $D_I \geq D_{TH}$ yet, the optimization problem of the LMPC of Equation (14) uses the state measurement from redundant, secure sensors instead of the original sensors as the initial condition $x(t_k)$ for the optimization problem of Equation (14) until the next instance of detection. However, if the cyber-attack is finally confirmed by $D_I \geq D_{TH}$, the misbehaving sensor will be isolated, and the optimization problem of the LMPC of Equation (14) starts to use the state measurement from secure sensors instead of the compromised state measurement as the initial condition $x(t_k)$ for the optimization problem of Equation (14) for the remaining time of process operation. The structure of the entire cyber-attack-detection-control system is shown in Figure 3.

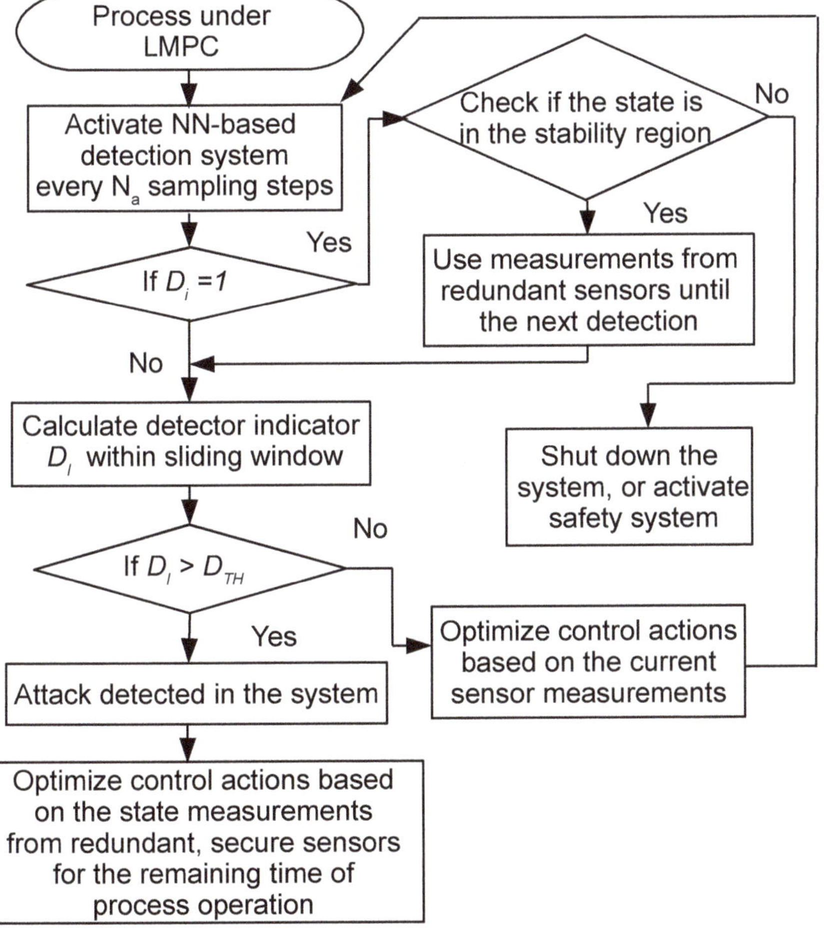

Figure 3. Basic structure of the proposed integrated NN-based detection and LMPC control method.

If the cyber-attack is detected and confirmed before the closed-loop state is driven out of the stability region, it follows that the closed-loop state is always bounded in the stability region Ω_ρ thereafter and ultimately converges to a small neighborhood $\Omega_{\rho_{min}}$ around the origin for any $x_0 \in \Omega_\rho$

under the LMPC of Equation (14). The detailed proof can be found in [11]. An example trajectory is shown in Figure 4.

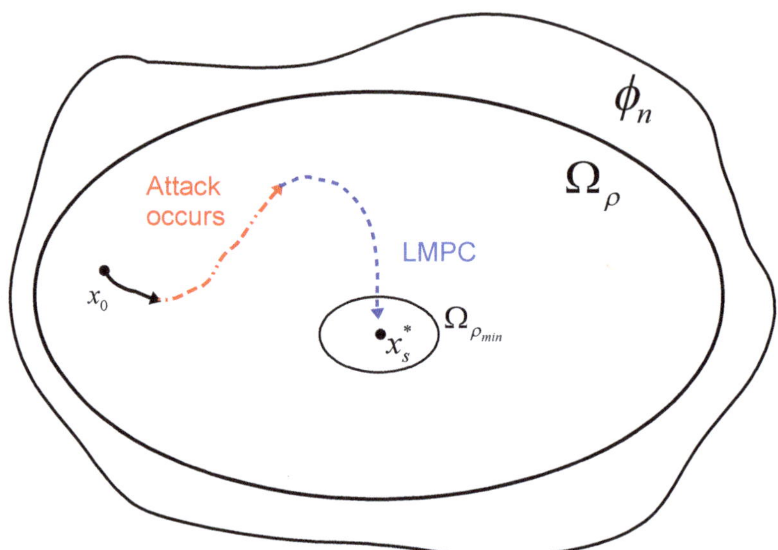

Figure 4. A schematic representing the stability region Ω_ρ and the small neighborhood $\Omega_{\rho_{min}}$ around the origin. The trajectory first moves away from the origin due to the cyber-attack and finally re-converges to $\Omega_{\rho_{min}}$ under the LMPC of Equation (14) after the detection of the cyber-attack by the proposed detection scheme.

Remark 6. *It is noted that the speed of detection (which depends heavily on the size of the input data to the NN, the number of hidden layers and the type of activation functions) plays an important role in stabilizing the closed-loop system of Equation (1) since the operation of the closed-loop system under the LMPC of Equation (14) becomes unreliable after cyber-attacks occur. In other words, if we can detect cyber-attacks in a short time, the LMPC can switch to redundant, secure sensors and still be able to stabilize the system at the origin before it leaves the stability region Ω_ρ. Additionally, the probability of closed-loop stability can be derived based on the classification accuracy of the NN-based detection method and its activation frequency N_a. Specifically, given the classification accuracy $p_{nn} \in [0,1]$, if the NN-based detection system is activated every $N_a = 1$ sampling step, the probability of the cyber-attack being detected at each sampling step (i.e., $D_i = 1$) is equal to p_{nn}, which implies that the probability of closed-loop stability $\forall x_0 \in \Omega_\rho$ is no less than p_{nn}. Moreover, for safety reasons, the region of initial conditions can be chosen as a conservative sub-region (i.e., $\Omega_{\rho_e} := \{x \in \phi_n \mid V(x) \leq \rho_e\}$, where $\rho_e < \rho$) inside the stability region to avoid the rapid divergence of states under cyber-attacks and improve closed-loop stability. For example, let $\rho_e = max\{V(x(t)) \mid V(x(t + \Delta)) \leq \rho, u \in U\}$ such that $\forall x(t_k) \in \Omega_{\rho_e}$, $x(t_{k+1})$ still stays in Ω_ρ despite a miss of detection of cyber-attacks. Therefore, the probability of closed-loop stability $\forall x_0 \in \Omega_{\rho_e}$ under the LMPC of Equation (14) reaches $1 - (1 - p_{nn})^2$ (i.e., the probability of cyber-attacks being detected within two sampling periods).*

Remark 7. *It is demonstrated in [11] that in the presence of sufficiently small bounded disturbances (i.e., $|w(t)| \leq \theta$), closed-loop stability is still guaranteed for the system of Equation (1) under the sample-and-hold implementation of the LMPC of Equation (14) with a sufficiently small sampling period Δ. In this case, it is undesirable to treat the disturbance as a cyber-attack and trigger the false alarm. Therefore, the detection system should account for the disturbance case and have the capability to distinguish cyber-attacks from disturbances (i.e., the system with disturbances should be classified as a distinct class or treated as the nominal system).*

5. Application to a Chemical Process Example

In this section, we utilize a chemical process example to illustrate the application of the proposed detection and control methods for potential cyber-attacks. Consider a well-mixed, non-isothermal continuous stirred tank reactor (CSTR) where an irreversible first-order exothermic reaction takes place. The reaction converts the reactant A to the product B via the chemical reaction $A \rightarrow B$. A heating jacket that supplies or removes heat from the reactor is used. The CSTR dynamic model derived from material and energy balances is given below:

$$\frac{dC_A}{dt} = \frac{F}{V_L}(C_{A0} - C_A) - k_0 e^{-E/RT} C_A \tag{15a}$$

$$\frac{dT}{dt} = \frac{F}{V_L}(T_0 - T) - \frac{\Delta H k_0}{\rho C_p} e^{-E/RT} C_A + \frac{Q}{\rho C_p V_L} \tag{15b}$$

where C_A is the concentration of reactant A in the reactor, T is the temperature of the reactor, Q denotes the heat supply/removal rate, and V_L is the volume of the reacting liquid in the reactor. The feed to the reactor contains the reactant A at a concentration C_{A0}, temperature T_0, and volumetric flow rate F. The liquid has a constant density of ρ and a heat capacity of C_p. k_0, E and ΔH are the reaction pre-exponential factor, activation energy and the enthalpy of the reaction, respectively. Process parameter values are listed in Table 1. The control objective is to operate the CSTR at the equilibrium point $(C_{As}, T_s) = (0.57 \text{ kmol/m}^3, 395.3 \text{ K})$ by manipulating the heat input rate $\Delta Q = Q - Q_s$, and the inlet concentration of species A, $\Delta C_{A0} = C_{A0} - C_{A0_s}$. The input constraints for ΔQ and ΔC_{A0} are $|\Delta Q| \leq 0.0167$ kJ/min and $|\Delta C_{A0}| \leq 1$ kmol/m^3, respectively.

Table 1. Parameter values of the CSTR.

$T_0 = 310$ K	$F = 100 \times 10^{-3}$ m^3/min
$V_L = 0.1$ m^3	$E = 8.314 \times 10^4$ kJ/kmol
$k_0 = 72 \times 10^9$ min^{-1}	$\Delta H = -4.78 \times 10^4$ kJ/kmol
$C_p = 0.239$ kJ/(kg K)	$R = 8.314$ kJ/(kmol K)
$\rho = 1000$ kg/m^3	$C_{A0_s} = 1.0$ kmol/m^3
$Q_s = 0.0$ kJ/min	$C_{A_s} = 0.57$ kmol/m^3
$T_s = 395.3$ K	

To place Equation (15) in the form of the class of nonlinear systems of Equation (1), deviation variables are used in this example, such that the equilibrium point of the system is at the origin of the state-space. $x^T = [C_A - C_{As} \; T - T_s]$ represents the state vector in deviation variable form, and $u^T = [\Delta C_{A0} \; \Delta Q]$ represents the manipulated input vector in deviation variable form.

The explicit Euler method with an integration time step of $h_c = 10^{-5}$ min is applied to numerically simulate the dynamic model of Equation (15). The nonlinear optimization problem of the LMPC of Equation (14) is solved using the IPOPT software package [26] with the sampling period $\Delta = 10^{-3}$ min.

We construct a Control Lyapunov Function using the standard quadratic form $V(x) = x^T P x$, with the following positive definite P matrix:

$$P = \begin{bmatrix} 9.35 & 0.41 \\ 0.41 & 0.02 \end{bmatrix} \tag{16}$$

Under the LMPC of Equation (14) without cyber-attacks, closed-loop stability is achieved for the nominal system of Equation (15) in the sense that the closed-loop state is always bounded in the stability region Ω_ρ with $\rho = 0.2$ and ultimately converges to $\Omega_{\rho_{min}}$ with $\rho_{min} = 0.002$ around the origin. However, if a min-max cyber-attack is added to tamper the sensor measurement of temperature of the system of Equation (15), closed-loop stability is no longer guaranteed. Specifically, the min-max cyber-attack is designed to be of the following form:

$$\bar{x}_1 = x_1 \tag{17a}$$

$$\bar{x}_2 = \min\{\underset{x_2 \in \mathbf{R}}{\arg\max}\{x^T P x \le \rho\}\} \tag{17b}$$

where $x_1 = C_A - C_{As}$, $x_2 = T - T_s$, and $\bar{x}_1$, $\bar{x}_2$ are the corresponding state measurements under min-max cyber-attacks. In this example, the min-max cyber-attack of Equation (17) is designed such that the measurement of concentration remains unchanged, and the measurement of temperature is tampered to be the minimum value that keeps the state at the boundary of the stability region Ω_ρ.

In Figures 5 and 6, the temperature sensor measurement is intruded by a min-max cyber-attack at time $t = 0.067$ min. Without any cyber-attack detection system, it is shown in Figure 5 that the LMPC of Equation (14) keeps operating the system of Equation (15) using false sensor measurements blindly and finally drives the closed-loop state out of the stability region Ω_ρ.

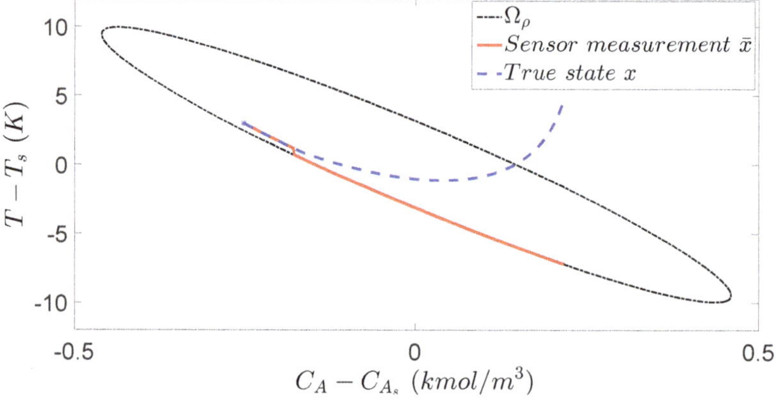

Figure 5. The state-space profile for the CSTR of Equation (15) under the LMPC of Equation (14) and under a min-max cyber-attack for the initial condition $(-0.25, 3)$.

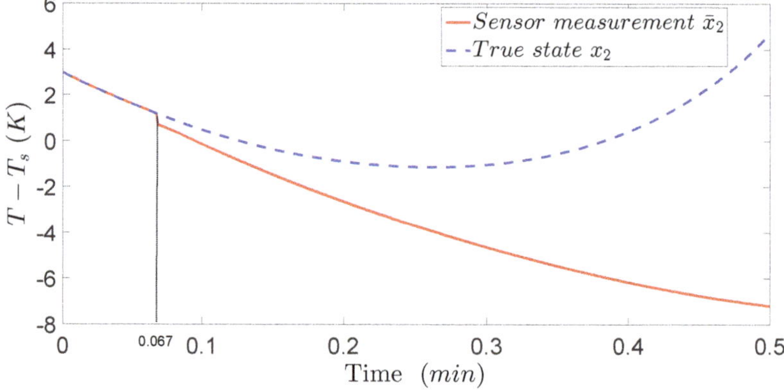

Figure 6. The true state profile ($x_2 = T - T_s$) and the sensor measurements ($\bar{x}_2 = \bar{T} - T_s$) of the closed-loop system under the LMPC of Equation (14) and under a min-max cyber-attack for the initial condition $(-0.25, 3)$, where the vertical dotted line shows the time the cyber-attack is added.

To handle the min-max cyber-attack, the model-based detection system of Equation (5) and the NN-based detection method are applied to the system of Equation (15). The simulation results are shown in Figures 7–13. Subsequently, the application of the NN-based detection method to the system under other cyber-attacks and the presence of disturbances is demonstrated in Figures 14–16. Specifically, we first demonstrate the application of the model-based detection system of Equation (5) and of the LMPC of Equation (14), where $S_{TH} = 1$ and $b = -0.5$ are chosen through closed-loop simulations. In Figure 7, the min-max cyber-attack of Equation (17) is added at 0.06 min and is detected at 0.1 min before the closed-loop state comes out of Ω_ρ. The variation of the CUSUM statistic $S(k)$ is shown in Figure 8, in which $S(k)$ remains at b when there is no cyber-attack and exceeds S_{TH} at 0.1 min. After the min-max cyber-attack is detected, the true states are obtained from redundant, secure sensors and the LMPC of Equation (14) drives the closed-loop state into $\Omega_{\rho_{min}}$.

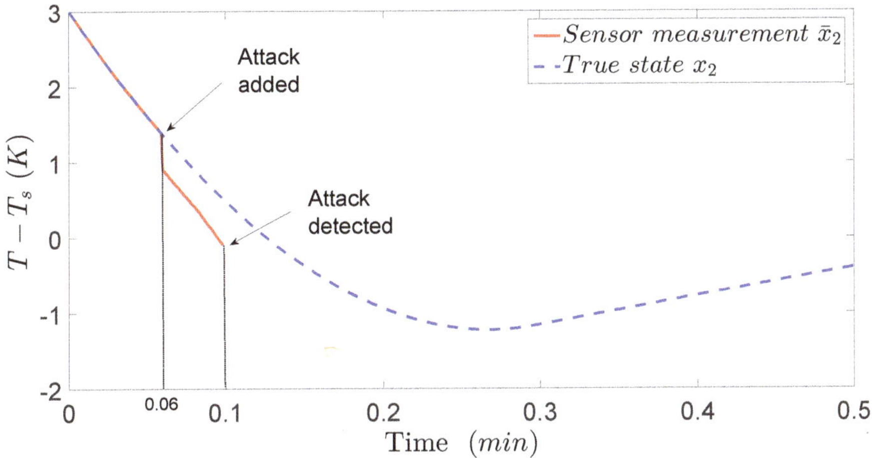

Figure 7. Closed-loop state profiles ($x_2 = T - T_s$, $\bar{x}_2 = \bar{T} - T_s$) for the initial condition $(-0.25, 3)$ under the LMPC of Equation (14) and the model-based detection system.

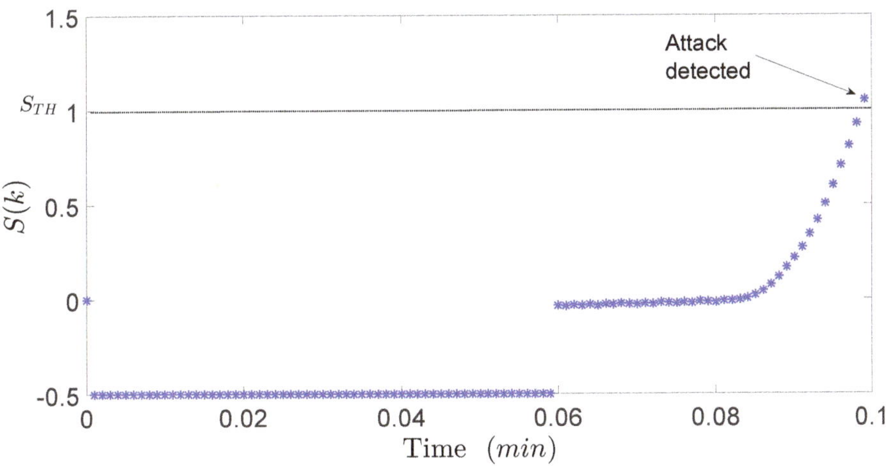

Figure 8. The variation of $S(k)$ for the initial condition $(-0.25, 3)$ under the LMPC of Equation (14) and the model-based detection system.

Next, the NN-based detection system and the LMPC of Equation (14) are implemented to mitigate the impact of cyber-attacks. The feed-forward NN model with two hidden layers is built in Python using the Keras library. Specifically, 3000 time-series balanced data samples of the closed-loop states of the nominal system, the system with disturbances, and the system under min-max cyber-attacks from

$t = 0$ to $t = 1$ min are used to train the neural network to generate the classification of three classes, where class 0, 1, and 2 stand for the system under min-max cyber-attacks, the nominal system and the system with disturbances, respectively. It is demonstrated that 3000 time-series data is sufficient to build the NN for the CSTR example because dataset size smaller than 3000 leads to lower classification accuracy while the increase of dataset size over 3000 does not significantly improve the classification accuracy but brings more computation time as found in our calculations. 3000 data samples are split into 2000 training data, 500 validation data and 500 test data, respectively. $V(x) = x^T P x$ is utilized as the input vector to the NN model. The structure of the NN model is listed in Table 2. Additionally, to improve the performance of the NN model, batch normalization is utilized after each hidden layer to improve the performance of the NN algorithm.

Table 2. Feed-forward NN model.

	Neurons	Activation Functions
First Hidden Layer	120	ReLu
Second Hidden Layer	100	ReLu
Output Layer	1	Softmax

To apply the NN-based detection method, we first investigate the relationship of the classification accuracy of the NN with respect to the size of the dataset. Specifically, assuming that the min-max cyber-attack occurs at a random sampling step before 0.1 min, the first NN model $M_{0.1}$ is trained at $t = 0.1$ min using the data of states from $t = 0$ to 0.1 min. As shown in Figure 9, early-stopping is activated at the 8th iteration (epoch) of training when validation accuracy ceases to increase. The averaged classification accuracy at $t = 0.1$ min is obtained by training the same model $M_{t=0.1}$ for 10 times independently. The above process is repeated by increasing the size of the dataset by 0.02 min every time to derive the models for different time instances (i.e., $M_{t=0.12}$, $M_{t=0.14}$, ...). The minimum, the maximum and the averaged classification accuracy at each detection time instance are shown in Figure 10.

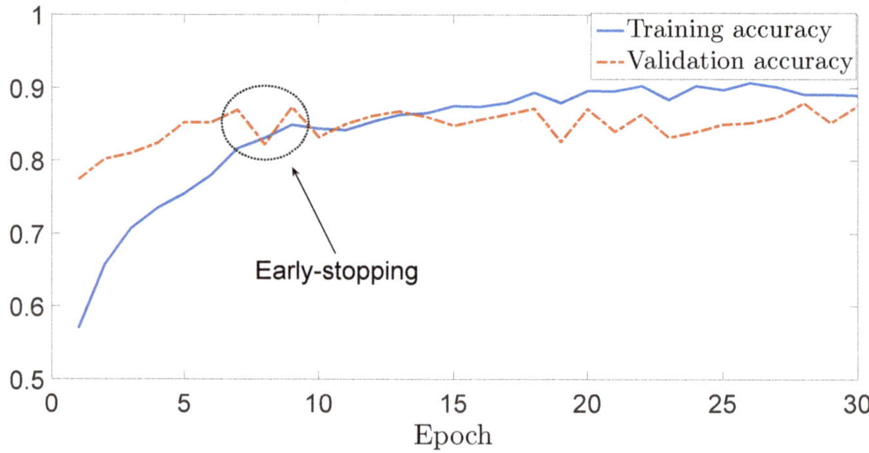

Figure 9. The variation of training accuracy and validation accuracy for the NN model $M_{0.1}$, where early-stopping is activated at the 8th epoch of training.

Figure 10 shows that the averaged test accuracy increases as more state measurements are collected after the cyber-attack occurs, and is up to 95% with state measurements for a long period of time. This suggests that the detection based on recent models is more reliable and deserves higher weights in the sliding window. The confusion matrix of the above NN for three classes: the system under min-max cyber-attack, the nominal system, and the system with disturbances is given in Table 3. Additionally, besides the NN method, other supervised learning-based classification methods including k-NN, SVM and random forests are also applied to the same dataset and obtained the averaged test accuracies, sensitivities and specificities within 0.28 min as listed in Table 4.

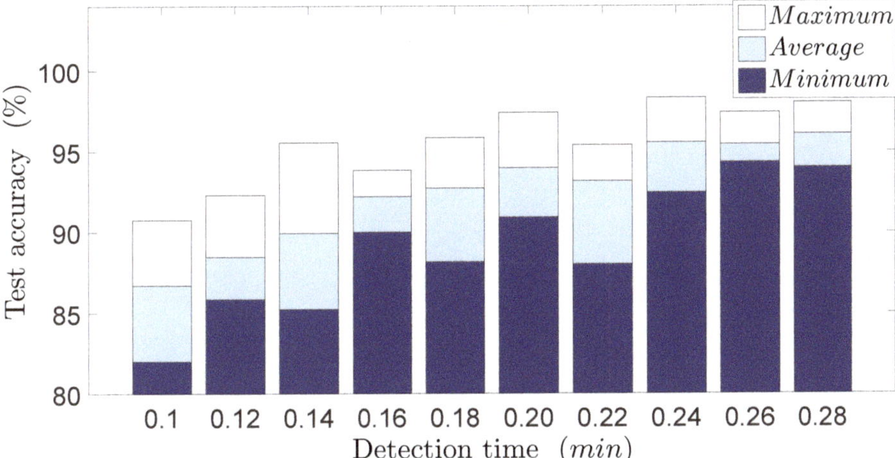

Figure 10. The test accuracy of neural network with respect to the size of training and test data.

Table 3. Confusion matrix of the neural network.

	Actual Class 0: Min-Max Cyber-Attack	Actual Class 1: Nominal System	Actual Class 2: The System with Disturbances
Predicted Class 0:	198	1	3
Predicted Class 1:	0	140	10
Predicted Class 2:	0	0	148

Table 4. Comparison of the performance of different detection models.

Models	Test Accuracy	Sensitivity	Specificity
k-NN	71.1%	90.9%	99.5%
SVM	83.0%	93.0%	87.8%
Random Forest	96.2%	100.0%	96.2%
Neural Network	95.8%	98.0%	98.6%

When the detection of cyber-attacks is incorporated into the closed-loop system of Equation (15) under the LMPC of Equation (14), the detection system is called every $N_a = 5$ sampling periods. The sliding window length is $N_s = 15$ sampling periods and the threshold for the detection indicator is $D_{TH} = 1.6$. The detection system is activated from $t = 0.1$ min such that a desired test accuracy is achieved with enough data. The closed-loop state-space profiles under the NN-based detection system with the stability region Ω_ρ check and the detection system without the Ω_ρ check are shown in Figures 11 and 12.

Specifically, in Figure 11, it is demonstrated that without the stability region check, the closed-loop state leaves Ω_ρ before the cyber-attack is confirmed. However, under the detection system with the boundedness check of Ω_ρ, the closed-loop state is always bounded in Ω_ρ by switching to redundant sensors at the first detection of min-max cyber-attacks. In Figure 12, it is shown that after the min-max cyber-attack is confirmed at $t = 0.115$ min, the misbehaving sensor is isolated and the LMPC of Equation (14) starts using the measurement of temperature from redundant sensors and re-stabilizes the system at the origin. The simulations demonstrate that it takes around 0.8 min for the closed-loop state trajectory to enter and remain in $\Omega_{\rho_{min}}$ under the LMPC of Equation (14) once the min-max cyber-attack is detected. The corresponding input profiles for the closed-loop system of Equation (1) under the NN-based detection system with the Ω_ρ check are shown in Figure 13, where it is observed that a sharp change of ΔC_{A0} occurs from $t = 0.095$ min to $t = 0.115$ min due to the min-max cyber-attack.

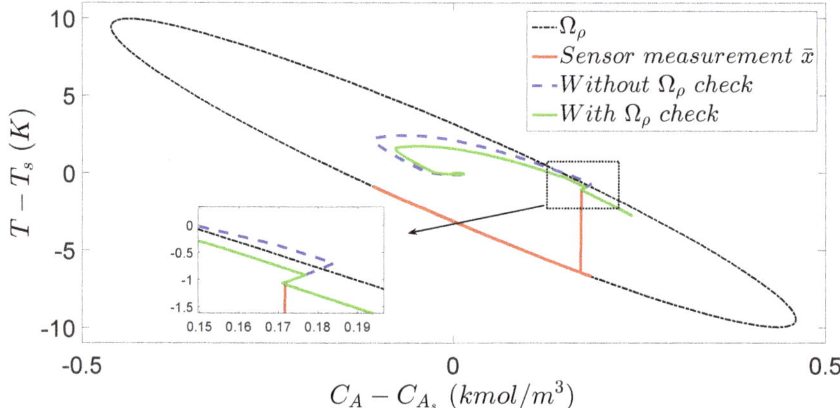

Figure 11. The state-space profile for the closed-loop CSTR with the initial condition $(0.24, -2.78)$, where a min-max cyber-attack is detected by the NN-based detection system and mitigated by the LMPC of Equation (14).

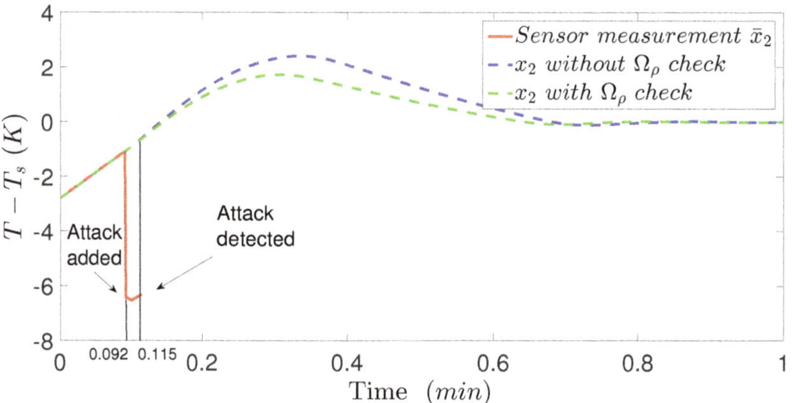

Figure 12. Closed-loop state profiles ($x_2 = T - T_s$, $\bar{x}_2 = \bar{T} - T_s$) for the initial condition $(0.24, -2.78)$ under the LMPC of Equation (14) and the NN-based detection system.

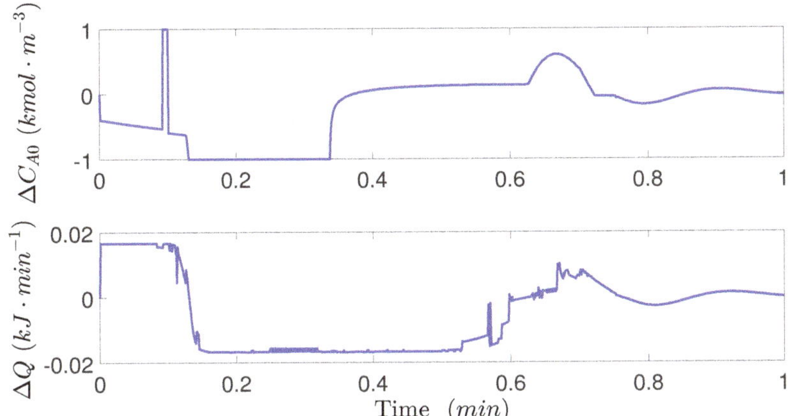

Figure 13. Manipulated input profiles ($u_1 = \Delta C_{A0}$, $u_2 = \Delta Q$) for the initial condition $(0.24, -2.78)$ under the LMPC of Equation (14) and the NN-based detection system.

Additionally, when both disturbances and min-max cyber-attacks are present, it is demonstrated that the NN-based detection system is still able to detect the min-max cyber-attack and re-stabilize the closed-loop system of Equation (15) in the presence of disturbances by following the same steps as in the pure-cyber-attack case. The bounded disturbances w_1 and w_2 are added in Equation (15a,15b) as standard Gaussian white noise with zero mean and variances $\sigma_1 = 0.1$ kmol/(m^3 min) and $\sigma_2 = 2$ K/min, respectively. Also, the disturbance terms are bounded as follows: $|w_1| \leq 0.1$ kmol/(m^3 min), and $|w_2| \leq 2$ K/min, respectively. The closed-loop state and input profiles are shown in Figures 14–16. Specifically, in Figure 15, it is demonstrated that the min-max cyber-attack occurs at 0.08 min and is confirmed at 0.115 min before the closed-loop state leaves Ω_ρ. In the presence of disturbances, the misbehaving sensor is isolated and the closed-loop states are driven to a neighborhood around the origin under the LMPC of Equation (14). In Figure 16, it is demonstrated that the manipulated inputs show variation around the steady-state values $(0, 0)$ when the closed-loop system reaches a neighborhood of the steady-state due to the bounded disturbances.

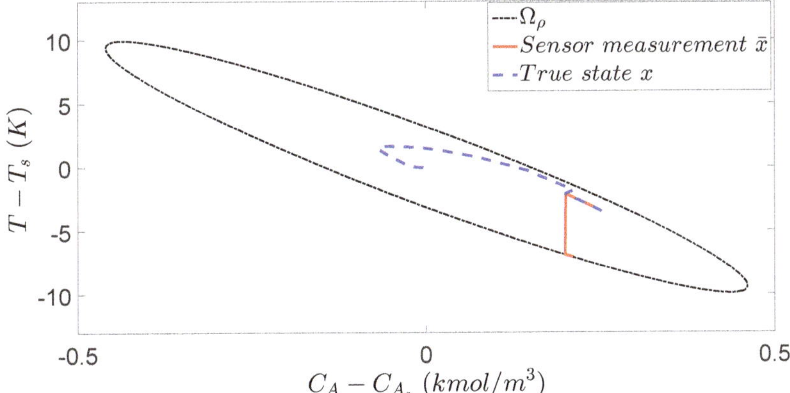

Figure 14. The state-space profiles for the closed-loop CSTR with bounded disturbances and the initial condition $(0.25, -3)$, where a min-max attack is detected by the NN-based detection system and mitigated by the LMPC of Equation (14).

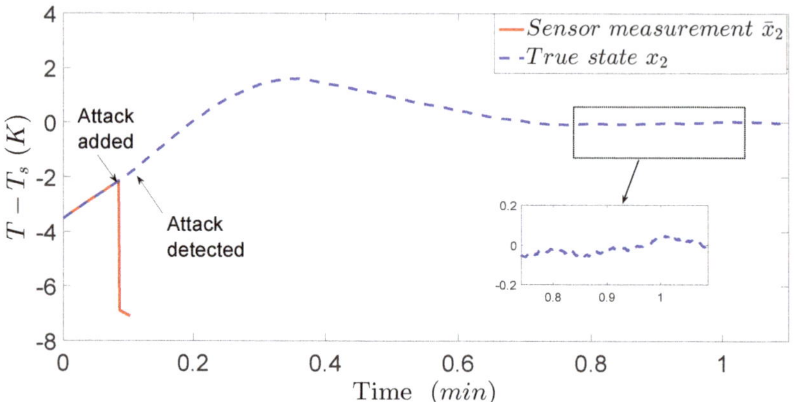

Figure 15. State profiles ($x_2 = T - T_s$, $\bar{x}_2 = \bar{T} - T_s$) for the closed-loop CSTR with bounded disturbances and the initial condition $(0.25, -3)$ under the LMPC of Equation (14) and the NN-based detection system.

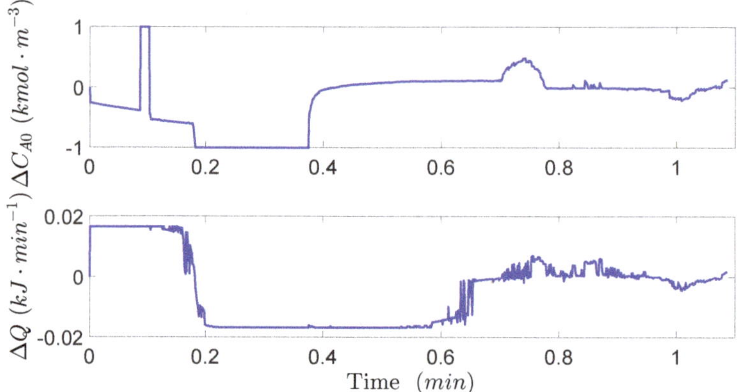

Figure 16. Manipulated input profiles ($u_1 = \Delta C_{A0}$, $u_2 = \Delta Q$) for the closed-loop CSTR with bounded disturbances and the initial condition $(0.25, -3)$ under the LMPC of Equation (14).

Lastly, since the surge cyber-attack of Equation (6) is undetectable by the model-based detection method, we also test the performance of the NN-based detection on the surge cyber-attack due to the similarity between surge cyber-attacks and min-max cyber-attacks (i.e., the surge cyber-attack works as a min-max attack for the first few sampling steps). It is demonstrated in simulations that 89% of surge cyber-attacks can be detected by the NN-based detection system that is trained for min-max cyber-attacks only, which implies that the NN-based detection method can be applied to many other cyber-attacks with similar properties.

Moreover, when cyber-attacks with different properties are taken into account, for example, the replay attack (i.e., $\bar{x} = X$, where X is the set of past measurements of states), the NN-based detection system can still efficiently distinguish the type of cyber-attacks and disturbances by re-training the NN model. The new NN model is built with labeled training data for the case of min-max, replay, nominal and with disturbances, for which the classification accuracy within 0.28 min is up to 85%. As a result, the NN-based detection model can be readily updated with the data of new cyber-attacks without changing the entire structure of detection or control systems.

6. Conclusions

In this work, we proposed an integrated NN-based detection and model predictive control method for nonlinear process systems to account for potential cyber-attacks. The NN-based detection system was first developed with the sliding detection window to detect cyber-attacks. Based on that, the Lyapunov-based MPC was developed with the stability region check triggered by the detection indicator to achieve closed-loop stability in the sense that the closed-loop state remained within a well-characterized stability region and was ultimately driven to a small neighborhood around the origin. Finally, the proposed integrated NN-based detection and LMPC method was applied to a nonlinear chemical process example. The simulation results demonstrated that the min-max cyber-attack was successfully detected before the state exited the stability region, and the closed-loop system was stabilized under the LMPC by using the measurements from redundant secure sensors. The good performance of the proposed approach with respect to surge and replay cyber-attacks was also demonstrated. The value and importance of the NN-based detection method is twofold. First, the NN-based detection method is able to detect cyber-attacks without having to know the process model if a large amount of past data is available. This is very important as nowadays most SCADA systems are large-scale networks with complicated process models, while the big data processing becoming available in both storage and computation. Second, compared to other detection methods, the NN-based detection is easy to implement. The proposed detection and control method can improve the safeness of processes by effectively detecting known (or similar to known) cyber-attacks and also can be readily updated to handle new, unknown cyber-attacks. However, NN-based detection method also has its limitations. Although it achieves desired performance for a trained, known cyber-attack, it is not guaranteed to work for an unknown, new cyber-attack unless it shares similar properties with known cyber-attacks.

Author Contributions: Investigation, Z.W., F.A., J.Z., Z.Z. and H.D.; Methodology Z.W., F.A., J.Z., Z.Z. and H.D.; Writing, Z.W. and H.D.; Supervision, P.D.C.

Funding: Financial support from the National Science Foundation and the Department of Energy is gratefully acknowledged.

Conflicts of Interest: The authors declare that they have no conflict of interest regarding the publication of the research article.

References

1. Ye, N.; Zhang, Y.; Borror, C.M. Robustness of the Markov-chain model for cyber-attack detection. *IEEE Trans. Reliab.* **2004**, *53*, 116–123. [CrossRef]
2. Fawzi, H.; Tabuada, P.; Diggavi, S. Secure estimation and control for cyber-physical systems under adversarial attacks. *IEEE Trans. Autom. Control* **2014**, *59*, 1454–1467. [CrossRef]
3. Ding, D.; Han, Q.L.; Xiang, Y.; Ge, X.; Zhang, X.M. A survey on security control and attack detection for industrial cyber-physical systems. *Neurocomputing* **2018**, *275*, 1674–1683. [CrossRef]
4. Cárdenas, A.A.; Amin, S.; Lin, Z.S.; Huang, Y.L.; Huang, C.Y.; Sastry, S. Attacks against process control systems: Risk assessment, detection, and response. In Proceedings of the 6th ACM Symposium on Information, Computer And Communications Security, Hong Kong, China, 22–24 March 2011; pp. 355–366.
5. Singh, J.; Nene, M.J. A survey on machine learning techniques for intrusion detection systems. *Int. J. Adv. Res. Comput. Commun. Eng.* **2013**, *2*, 4349–4355.
6. Ozay, M.; Esnaola, I.; Vural, F.T.Y.; Kulkarni, S.R.; Poor, H.V. Machine learning methods for attack detection in the smart grid. *IEEE Trans. Neural Netw. Learn. Syst.* **2016**, *27*, 1773–1786. [CrossRef] [PubMed]
7. Satchidanandan, B.; Kumar, P.R. Dynamic watermarking: Active defense of networked cyber–physical systems. *Proc. IEEE* **2017**, *105*, 219–240. [CrossRef]
8. Pajic, M.; Weimer, J.; Bezzo, N.; Sokolsky, O.; Pappas, G.J.; Lee, I. Design and implementation of attack-resilient cyberphysical systems: With a focus on attack-resilient state estimators. *IEEE Control Syst.* **2017**, *37*, 66–81.
9. Dolk, V.S.; Tesi, P.; De Persis, C.; Heemels, W.P.M.H. Event-triggered control systems under denial-of-service attacks. *IEEE Trans. Control Netw. Syst.* **2017**, *4*, 93–105. [CrossRef]

10. Rawlings, J.B.; Mayne, D.Q. *Model Predictive Control: Theory and Design*; Nob Hill Pub.: San Francisco, CA, USA, 2009.
11. Mhaskar, P.; El-Farra, N.H.; Christofides, P.D. Stabilization of nonlinear systems with state and control constraints using Lyapunov-based predictive control. *Syst. Control Lett.* **2006**, *55*, 650–659. [CrossRef]
12. Muñoz de la Peña, D.; Christofides, P.D. Lyapunov-based model predictive control of nonlinear systems subject to data losses. *IEEE Trans. Autom. Control* **2008**, *53*, 2076–2089. [CrossRef]
13. Wu, Z.; Albalawi, F.; Zhang, Z.; Zhang, J.; Durand, H.; Christofides, P.D. Control Lyapunov-barrier function-based model, predictive control of nonlinear systems. In Proceedings of the American Control Conference, Milwaukee, WI, USA, 27–29 June 2018; pp. 5920–5926.
14. Durand, H. A Nonlinear Systems Framework for Cyberattack Prevention for Chemical Process Control Systems. *Mathematics* **2018**, *6*, 169. [CrossRef]
15. Narasingam, A.; Kwon, J.S.I. Data-driven identification of interpretable reduced-order models using sparse regression. *Comput. Chem. Eng.* **2018**. [CrossRef]
16. Narasingam, A.; Siddhamshetty, P.; Kwon, J.S.I. Temporal clustering for order reduction of nonlinear parabolic PDE systems with time-dependent spatial domains: Application to a hydraulic fracturing process. *AIChE J.* **2017**, *63*, 3818–3831. [CrossRef]
17. Sidhu, H.S.; Narasingam, A.; Siddhamshetty, P.; Kwon, J.S.I. Model order reduction of nonlinear parabolic PDE systems with moving boundaries using sparse proper orthogonal decomposition: Application to hydraulic fracturing. *Comput. Chem. Eng.* **2018**, *112*, 92–100. [CrossRef]
18. Lin, Y.; Sontag, E.D. A universal formula for stabilization with bounded controls. *Syst. Control Lett.* **1991**, *16*, 393–397. [CrossRef]
19. Li, Y.; Shi, L.; Cheng, P.; Chen, J.; Quevedo, D.E. Jamming attacks on remote state estimation in cyber-physical systems: A game-theoretic approach. *IEEE Trans. Autom. Control* **2015**, *60*, 2831–2836. [CrossRef]
20. Tsai, C.F.; Hsu, Y.F.; Lin, C.Y.; Lin, W.Y. Intrusion detection by machine learning: A review. *Expert Syst. Appl.* **2009**, *36*, 11994–12000. [CrossRef]
21. Bishop, C.M. *Pattern Recognition and Machine Learning (Information Science and Statistics)*; Springer: New York, NY, USA, 2006; p. 049901.
22. Alexandridis, K.; Maru, Y. Collapse and reorganization patterns of social knowledge representation in evolving semantic networks. *Inf. Sci.* **2012**, *200*, 1–21. [CrossRef]
23. Daqi, G.; Yan, J. Classification methodologies of multilayer perceptrons with sigmoid activation functions. *Pattern Recognit.* **2005**, *38*, 1469–1482. [CrossRef]
24. Xu, B.; Liu, X.; Liao, X. Global exponential stability of high order Hopfield type neural networks. *Appl. Math. Comput.* **2006**, *174*, 98–116. [CrossRef]
25. Zhang, Z.; Wu, Z.; Durand, H.; Albalawi, F.; Christofides, P.D. On integration of feedback control and safety systems: Analyzing two chemical process applications. *Chem. Eng. Res. Des.* **2018**, *132*, 616–626. [CrossRef]
26. Wächter, A.; Biegler, L.T. On the implementation of an interior-point filter line-search algorithm for large-scale nonlinear programming. *Math. Programm.* **2006**, *106*, 25–57. [CrossRef]

Article

Real-Time Optimization and Control of Nonlinear Processes Using Machine Learning

Zhihao Zhang [1], Zhe Wu [1], David Rincon [1] and Panagiotis D. Christofides [1,2,*]

[1] Department of Chemical and Biomolecular Engineering, University of California, Los Angeles, CA 90095-1592, USA; zhihaozhang@ucla.edu (Z.Z.); wuzhe@ucla.edu (Z.W.); fdrinconc@g.ucla.edu (D.R.)

[2] Department of Electrical and Computer Engineering, University of California, Los Angeles, CA 90095-1592, USA

* Correspondence: pdc@seas.ucla.edu

Received: 3 September 2019; Accepted: 20 September 2019; Published: 24 September 2019

Abstract: Machine learning has attracted extensive interest in the process engineering field, due to the capability of modeling complex nonlinear process behavior. This work presents a method for combining neural network models with first-principles models in real-time optimization (RTO) and model predictive control (MPC) and demonstrates the application to two chemical process examples. First, the proposed methodology that integrates a neural network model and a first-principles model in the optimization problems of RTO and MPC is discussed. Then, two chemical process examples are presented. In the first example, a continuous stirred tank reactor (CSTR) with a reversible exothermic reaction is studied. A feed-forward neural network model is used to approximate the nonlinear reaction rate and is combined with a first-principles model in RTO and MPC. An RTO is designed to find the optimal reactor operating condition balancing energy cost and reactant conversion, and an MPC is designed to drive the process to the optimal operating condition. A variation in energy price is introduced to demonstrate that the developed RTO scheme is able to minimize operation cost and yields a closed-loop performance that is very close to the one attained by RTO/MPC using the first-principles model. In the second example, a distillation column is used to demonstrate an industrial application of the use of machine learning to model nonlinearities in RTO. A feed-forward neural network is first built to obtain the phase equilibrium properties and then combined with a first-principles model in RTO, which is designed to maximize the operation profit and calculate optimal set-points for the controllers. A variation in feed concentration is introduced to demonstrate that the developed RTO scheme can increase operation profit for all considered conditions.

Keywords: real-time optimization; nonlinear processes; process control; model predictive control; chemical reactor control; distillation column control

---

1. Introduction

In the last few decades, chemical processes have been studied and represented with different models for real-time optimization (RTO) and model predictive control (MPC) in order to improve the process steady-state and dynamic performance. The available models range from linear to nonlinear and from first-principles models to neural network models, among others [1]. For many applications, first-principles models are the preferable choice, especially when applied with process systems methodologies [2]. However, first-principles models are difficult to maintain due to the variation of some parameters. Furthermore, it could be difficult or impractical to obtain first-principles models for large-scale applications [3]. As a well-tested alternative, machine learning method, especially neural network models are able to represent complicated nonlinear systems [4,5]. Neural networks fit the data in an input-output fashion using fully-connected layers within the hidden output layers [6]. However, due to their general structures, neural networks lack physical knowledge in their formulation. To alleviate the above problem, this work integrates neural network models with first-principles models. Specifically, first-principles models are used to represent the well-known part of the process and embedding physical knowledge in the formulation, while the complex nonlinear part of the process is represented with neural networks. This proposed hybrid formulation is then applied in the context of real-time optimization and model predictive control in two chemical processes.

The machine learning method has been part of process system engineering for at least 30 years in which the feed-forward neural network is the most classical structure found in the literature [7]. For instance, neural networks have been proposed as an alternative to first-principles models for the classical problems of process engineering [7], such as modeling, fault diagnosis, product design, state estimation, and process control. The neural network model has also gained much interest in the chemical engineering field, and more comprehensive reviews with detailed information on neural networks in chemical processes are available in [7,8]. For example, an artificial neural networks was applied to approximate pressure-volume-temperature data in refrigerant fluids [9]. Complex reaction kinetic data have been fitted using a large experimental dataset with neural networks to approximate the reaction rate and compared with standard kinetics methods, showing that neural networks can represent kinetic data at a faster pace [10]. Reliable predictions of the vapor-liquid equilibrium has been developed by means of neural networks in binary ethanol mixtures [11]. Studies on mass transfer have shown good agreements between neural network predictions and experimental data in the absorption performance of packed columns [12].

Since the applications with standard neural networks rely on fully-connected networks, the physical interpretation of the obtained model can be a difficult task. One solution is to integrate physical knowledge into the neural network model. For example, the work in [13] proposed a learning technique in which the neural network can be physically interpretable depending on the specifications. Similarly, the work in [14] designed a neural network with physical-based knowledge using hidden layers as intermediate outputs and prioritized the connection between inputs and hidden layers based on the effect of each input with the corresponding intermediate variables. Another method to add more physical knowledge into neural networks is to combine first-principles models with neural networks as hybrid modeling [15]. For instance, biochemical processes have been represented with mass balances for modeling the bioreactor system and with artificial neural networks for representing the cell population system [16]. Similarly, an experimental study for a bio-process showed the benefits of the hybrid approach in which the kinetic models of the reaction rates were identified with neural networks [17]. In crystallization, growth rate, nucleation kinetics, and agglomeration phenomena have been represented by neural networks, while mass, energy, and population balances have been used as a complement to the system's behavior [18]. In industry, hybrid modeling using rigorous models and neural networks has also been tested in product development and process design [19]. However, most of the applications with hybrid modeling are limited to the open-loop case.

Real-time optimization (RTO) and model predictive control (MPC) are vital tools for chemical process performance in industry in which the process model plays a key role in their formulations [20,21]. RTO and MPC have been primarily implemented based on first-principles models, while the difference is that RTO is based on steady-steady models and MPC is based on dynamical models [20,21]. In both RTO and MPC, the performance depends highly on the accuracy of the process model. To obtain a more accurate model, machine learning methods have been employed within MPC [6] and within RTO [22], as well. In practice, it is common to use process measurements to construct neural network models for chemical processes. However, the obtained model from process operations may lack robustness and accuracy for parameter identification, as was shown in [23]. As a consequence, there has been significant effort to include hybrid models in process analysis, MPC, and process optimization [24–30] in order to reduce the dependency on data and infuse physical knowledge. At this stage, little attention has been paid to utilizing the full benefit of employing hybrid modeling in both the RTO and MPC layers.

Motivated by the above, this work demonstrates the implementation of a hybrid approach of combining a first-principles model and a neural network model in the RTO and MPC optimization problems. Specifically, the nonlinear part of the first-principles model is replaced by a neural network model to represent the complex, nonlinear term in a nonlinear process. We note that in our previous works, we developed recurrent neural network models from process data for use in MPC without using any information from a first-principles model or process structure in the recurrent neural network model formulation [4,5,31]. Furthermore, the previous works did not consider the use of neural network models to describe nonlinearities in the RTO layer and focused exclusively on model predictive control. In the present work, we use neural networks to describe nonlinearities arising in chemical processes and embed these neural network models in first-principles process models used in both RTO (nonlinear steady-state process model) and MPC (nonlinear dynamic process model), resulting in the use of hybrid model formulations in both layers. The rest of the paper is organized as follows: in Section 2, the proposed method that combines neural network with the first-principles model is discussed. In Section 3, a continuous stirred tank reactor (CSTR) example is utilized to illustrate the combination of neural network models and first-principles models in RTO and Lyapunov-based MPC, where the reaction rate equation is represented by a neural network model. In Section 4, an industrial distillation column is co-simulated in Aspen Plus Dynamics and MATLAB. A first-principles steady-state model of the distillation column is first developed, and a neural network model is constructed for phase equilibrium properties. The combined model is then used in RTO to investigate the performance of the proposed methodology.

2. Neural Network Model and Application

2.1. Neural Network Model

The neural network model is a nonlinear function $y = f_{NN}(x)$ with input vector $x = [x_1, x_2, ..., x_n]$ and output vector $y = [y_1, y_2, ..., y_m]$. Mathematically, a neural network function is defined as a series of functional transformations. The structure of a two-layer (one hidden-layer) feed-forward neural network is shown in Figure 1, where $h_1, h_2, ..., h_p$ are hidden neurons [32,33]. Specifically, the hidden neurons h_j and the outputs y_k are obtained by Equation (1):

$$h_j = \sigma_1(\sum_{i=1}^{n} w_{ji}^{(1)} x_i + w_{j0}^{(1)}), \quad j = 1, 2, ..., p \tag{1a}$$

$$y_k = \sigma_2(\sum_{i=1}^{p} w_{ki}^{(2)} h_i + w_{k0}^{(2)}), \quad k = 1, 2, ..., m \tag{1b}$$

where parameters $w_{ji}^{(1)}$ and $w_{ki}^{(2)}$ are weights in the first and the second layer and parameters $w_{j0}^{(1)}$ and $w_{k0}^{(2)}$ are biases. σ_1 and σ_2 are nonlinear element-wise transformations $\sigma : R^1 \to R^1$, which are generally chosen to be sigmoid functions such as the logistic sigmoid $S(x) = 1/(1 + e^{-x})$ or hyperbolic tangent function $tanh(x) = 2/(1 + e^{-2x}) - 1$. Each hidden neuron h_j is calculated by an activation function σ_1 with a linear combination of input variables x_i. Each output variable y_k is also calculated by an activation function σ_2 with a linear combination of hidden neurons h_i. Since the neural network models in this work are developed to solve regression problems, no additional output unit activation functions are needed. All the neural network models in this work will follow the structure discussed in this section.

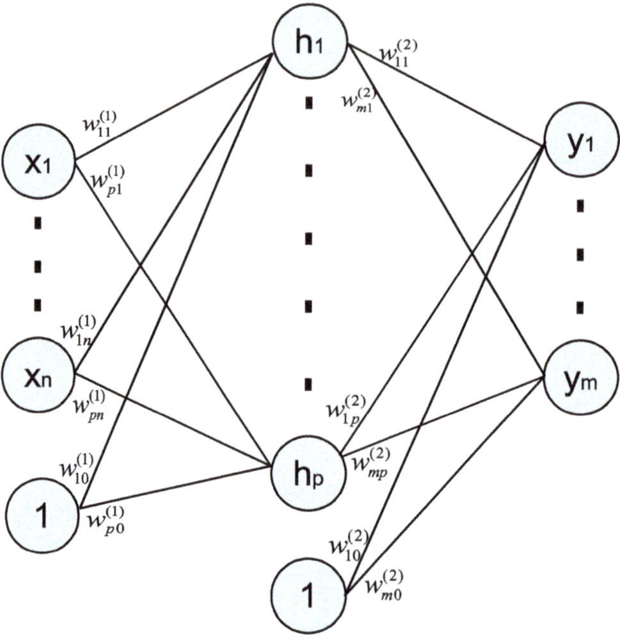

Figure 1. A feed-forward neural network with input $x_1, ..., x_n$, hidden neurons $h_1, h_2, ..., h_p$, and outputs $y_1, y_2, ..., y_m$. Each weight $w_{ji}^{(k)}$ is marked on the structure. Neuron "1" is used to represent the biases.

Given a set of input vectors $\{x^n\}$ together with a corresponding set of target output vectors $\{\hat{y}^n\}$ as a training set of N data points, the neural network model is trained by minimizing the following sum-of-squares error function [33]:

$$E(w) = \frac{1}{2} \sum_{n=1}^{N} \|y(x^n, w) - \hat{y}^n\|^2 \tag{2}$$

The proper weight vectors w are obtained by minimizing the above cost function via the gradient descent optimization method:

$$w^{\tau+1} = w^\tau - \eta \nabla E(w^\tau) \tag{3}$$

where τ labels the iteration, $\eta > 0$ is known as the learning rate, and $\nabla E(w^\tau)$ is the derivative of the cost function with respect to weight w. The weight vectors are optimized by moving through weight space in a succession of Equation (3) with some initial value $w(0)$. The gradient of an error function $\nabla E(w)$ is evaluated by back propagation method. Additionally, data are first normalized, and then, k-fold cross-validation is used to separate the dataset into the training and validation set in order to avoid model overfitting.

2.2. Application of Neural Network Models in RTO and MPC

In the chemical engineering field, model fitting is a popular technique in both academia and industry. In most applications, a certain model formulation needs to be assumed first, and then, the model is fitted with experiment data. However, a good approximation is not guaranteed since the assumed model formulation may be developed based on deficient assumptions and uncertain mechanism, which lead to an inaccurate model. Alternatively, neural network model can be employed to model complex, nonlinear systems since neural networks do not require any a priori knowledge about the process and are able to fit any nonlinearity with a sufficient number of layers and neurons according to the universal approximation theorem [34]. The obtained neural network model can be used together with existing first-principles models. Specifically, the combination of the neural network model and first-principles model can be used in optimization problems, such as real-time optimization (RTO) and model predictive control (MPC).

2.2.1. RTO with the Neural Network Model

Real-time optimization (RTO) maximizes the economic productivity of the process subject to operational constraints via the continuous re-evaluation and alteration of operating conditions of a process [35]. The economically-optimal plant operating conditions are determined by RTO and sent to the controllers to operate the process at the optimal set-points [36].

Since RTO is an optimization problem, an explicit steady-state model is required in order to obtain optimal steady-states. First-principles models are commonly used in RTO; however, first-principles models may not represent the real process well due to model mismatch, and thus lead to non-optimal steady-states or even infeasible steady-states. In these cases, the machine learning method becomes a good solution to improve model accuracy. Specifically, a neural network model can be used to replace the complicated nonlinear part of the steady-state model to increase the accuracy of the first-principles model.

In general, the RTO problem is formulated as the optimization problem of Equation (4), where $x \in R^n$ is the state and $\hat{x} \in R^m$ is part of the state. $g(\hat{x})$ is a nonlinear function of $\hat{x}$, which is a part of the steady-state model.

$$
\begin{aligned}
\min_{x} \quad & cost\ function(x) \\
\text{s.t.} \quad & F(x, g(\hat{x})) = 0 \\
& other\ constraints
\end{aligned}
\tag{4}
$$

Since it is difficult to obtain an accurate functional form of $g(\hat{x})$, a neural network $F_{NN}(\hat{x})$ is developed using simulation data to replace $g(\hat{x})$ in Equation (4). Therefore, the RTO based on the integration of first-principles model and neural network model is developed as follows:

$$
\begin{aligned}
\min_{x} \quad & cost\ function(x) \\
\text{s.t.} \quad & F(x, F_{NN}(\hat{x})) = 0 \\
& other\ constraints
\end{aligned}
\tag{5}
$$

2.2.2. MPC with Neural Network Models

Model predictive control (MPC) is an advanced control technique that uses a dynamic process model to predict future states over a finite-time horizon to calculate the optimal input trajectory. Since MPC is able to account for multi-variable interactions and process constraints, it has been widely used to control constrained multiple-input multiple-output nonlinear systems [37]. Since MPC is an optimization problem, an explicit dynamic model is required to predict future states and make optimal decisions. First-principles models can be developed and used as the prediction model in MPC; however, first-principles models suffer from model mismatch, which might lead to offsets and other issues. Therefore, machine learning methods can be used to reduce model mismatch by replacing the complicated nonlinear part of the dynamic model with a neural network model.

In general, MPC can be formulated as the optimization problem of Equation (6), where the notations follow those in Equation (4) and $\dot{x} = F(x, g(\hat{x}))$ is the first-principles dynamic process model.

$$
\begin{aligned}
\min_{u} \quad & cost\ function(x, u) \\
\text{s.t.} \quad & \dot{x} = F(x, g(\hat{x}), u) \\
& other\ constraints
\end{aligned}
\tag{6}
$$

Similar to Equation (5), a neural network $F_{NN}(\hat{x})$ is developed using simulation data to replace $g(\hat{x})$ in Equation (6). As a result, the MPC based on the integration of the first-principles model and neural network model is developed as follows:

$$
\begin{aligned}
&\min_{u} \quad cost\ function(x, u) \\
&\text{s.t.} \quad \dot{x} = F(x, F_{NN}(\hat{x}), u) \\
&\qquad\ other\ constraints
\end{aligned}
\tag{7}
$$

Remark 1. *To derive stability properties for the closed-loop system under MPC, additional stabilizing constraints can be employed within the MPC of Equation (7) (e.g., terminal constraints [38] and Lyapunov-based constraints [39]). In this work, a Lyapunov-based MPC (LMPC) is developed to achieve closed-loop stability in the sense that the close-loop state is bounded in a stability region for all times and is ultimately driven to the origin. The discussion and the proof of closed-loop stability under LMPC using machine learning-based models can be found in [4,31].*

Remark 2. *All the optimization problems of MPC and RTO in this manuscript are solved using IPOPT, which is an interior point optimizer for large-scale nonlinear programs. The IPOPT solver was run on the OPTI Toolbox in MATLAB. It is noted that the global optimum of the nonlinear optimization problem is not required in our case, since the control objective of MPC is to stabilize the system at its set-point, rather than to find the globally-optimal trajectory. The Lyapunov-based constraints can guarantee closed-loop stability in terms of convergence to the set-point for the nonlinear system provided that a feasible solution (could be a locally-optimal solution) to the LMPC optimization problem exists.*

Remark 3. *In the manuscript, the MPC is implemented in a sample-and-hold fashion, under which the control action remains the same over one sampling period, i.e., $u(t) = u(x(t_k))$, $\forall t \in [t_k, t_{k+1})$, where t_{k+1} represents $t_k + \Delta$ and Δ is the sampling period. Additionally, one possible way to solve the optimization problems of Equations (6) and (7) is to use continuous-time optimization schemes. This method has recently gained researchers attention and can be found in [40,41].*

Remark 4. *In this work, the neural network is used to replace the nonlinear term in the first-principles model, for which it is generally difficult to obtain an accurate functional form from first-principles calculations. It should be noted that the neural network $F_{NN}(\hat{x})$ was developed as an input-output function to replace only a part (static nonlinearities) of the first-principles model, and thus does not replace the entire steady-state model or dynamic model.*

3. Application to a Chemical Reactor Example

3.1. Process Description and Simulation

The first example considers a continuous stirred tank reactor (CSTR), where a reversible exothermic reaction $A \leftrightarrow B$ takes place [42,43]. After applying mass and energy balances, the following dynamic model is achieved to describe the process:

$$
\begin{aligned}
\frac{dC_A}{dt} &= \frac{1}{\tau}(C_{A_0} - C_A) - k_A e^{\frac{-E_A}{RT}} C_A + k_B e^{\frac{-E_B}{RT}} C_B \\
\frac{dC_B}{dt} &= -\frac{1}{\tau} C_B + k_A e^{\frac{-E_A}{RT}} C_A - k_B e^{\frac{-E_B}{RT}} C_B \\
\frac{dT}{dt} &= \frac{-\Delta H}{\rho C_P}(k_A e^{\frac{-E_A}{RT}} C_A - k_B e^{\frac{-E_B}{RT}} C_B) + \frac{1}{\tau}(T_0 - T) + \frac{Q}{\rho C_P V}
\end{aligned}
\tag{8}
$$

In the model of Equation (8), C_A, C_B are the concentrations of A and B in the reactor, and T is the temperature of the reactor. The feed temperature and concentration are denoted by T_0 and C_{A_0}, respectively. k_A and k_B are the pre-exponential factor for the forward reaction and reverse reaction, respectively. E_A and E_B are the activation energy for the forward reaction and reverse reaction, respectively. τ is the residence time in the reactor; ΔH is the enthalpy of the reaction; and C_P is the heat capacity of the mixture liquid. The CSTR is equipped with a jacket to provide heat to the reactor at rate Q. All process parameter values and steady-state values are listed in Table 1. Additionally, it is noted that the second equation of Equation (8) for C_B is unnecessary if C_{A_0} is fixed due to $C_B = C_{A_0} - C_A$. This does not hold when C_{A_0} is varying, and thus, the full model is used in this work for generality.

Table 1. Parameter values and steady-state values for the continuous stirred tank reactor (CSTR) case study.

$T_0 = 400$ K	$\tau = 60$ s
$k_A = 5000$ /s	$k_B = 10^6$ /s
$E_A = 1 \times 10^4$ cal/mol	$E_B = 1.5 \times 10^4$ cal/mol
$R = 1.987$ cal/(mol K)	$\Delta H = -5000$ cal/mol
$\rho = 1$ kg/L	$C_P = 1000$ cal/(kg K)
$C_{A_0} = 1$ mol/L	$V = 100$ L
$C_{A_s} = 0.4977$ mol/L	$C_{B_s} = 0.5023$ mol/L
$T_{A_s} = 426.743$ K	$Q_s = 40386$ cal/s

When the tank temperature T is too low, the reaction rate is maintained as slow such that the reactant A does not totally reacted during the residence time, and thus, the reactant conversion $(1 - C_A/C_{A_0})$ is low. When the tank temperature T is too high, the reversible exothermic reaction equilibrium turns backwards so that the reactant conversion $(1 - C_A/C_{A_0})$ also drops. As a result, there exists a best tank temperature to maximize the reactant conversion. Figure 2 shows the variation of the CSTR steady-state (i.e., concentration C_A and temperature T) under varying heat input rate Q, where Q is not explicitly shown in Figure 2. Specifically, the minimum point of C_A represents the steady-state of C_A and T, under which the highest conversion rate (conversion rate $= 1 - C_A/C_{A_0}$) is achieved. Therefore, the CSTR process should be operated at this steady-state for economic optimality if no other cost is accounted for.

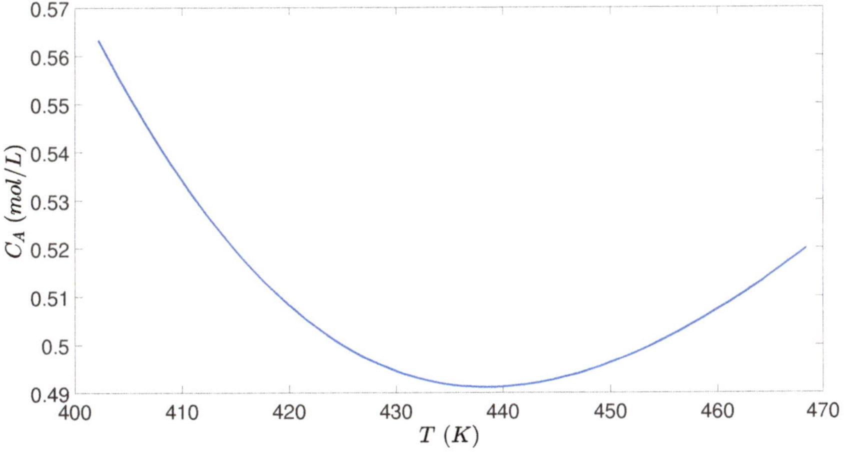

Figure 2. Steady-state profiles (C_A and T) for the CSTR of Equation (8) under varying heat input rate Q, where the minimum of C_A is achieved at Q= 59,983 cal/s.

3.2. Neural Network Model

In the CSTR model of Equation (8), the reaction rate $r = k_A e^{\frac{-E_A}{RT}} C_A - k_B e^{\frac{-E_B}{RT}} C_B$ is a nonlinear function of C_A, C_B, and T. To obtain this reaction rate from experiment data, an assumption of the reaction rate mechanism and reaction rate function formulation is required. In practice, it could be challenging to obtain an accurate reaction rate expression using the above method if the reaction mechanism is unknown and the rate expression is very complicated.

In this work, a neural network model is built to represent the reaction rate r as a function of C_A, C_B, and T (i.e., $r = F_{NN}(C_A, C_B, T)$), and then, the neural network model replaces the first-principles rate equation in the process model. Specifically, around eight million data were generated by the original reaction rate expression $r = k_A e^{\frac{-E_A}{RT}} C_A - k_B e^{\frac{-E_B}{RT}} C_B$ with different values of C_A, C_B, and T. The dataset was generated such that various reaction rates under different operating conditions (i.e., temperature, concentrations of A and B) were covered. The operating conditions were discretized equidistantly. Specifically, we tried the activation functions such as tanh, sigmoid, and ReLU for hidden layers and a linear unit and softmax function for the output layer. It is demonstrated that the choice of activation functions for the output layer significantly affected the performance of the neural network in a regression problem, while those for the hidden layers achieved similar results. $tanh(x) = 2/(1 + e^{-2x}) - 1$ was ultimately chosen as the activation function for the hidden layers, and a linear unit was used for the output layer since they achieved the best training performance with the mean squared error less than 10^{-7}. Data were first normalized and then fed to the MATLAB Deep Learning toolbox to train the model. The neural network model had one hidden layer with 10 neurons. The parameters were trained using Levenberg–Marquardt optimization algorithm. In terms of the accuracy of the neural network model, the coefficient of determination R^2 was 1, and the error histogram of Figure 3 demonstrates that the neural network represented the reaction rate with a high accuracy, as can be seen from the error distribution (we note that error metrics used in classification problems like the confusion matrix, precision, recall, and f1-score were not applicable to the regression problems considered in this work). In the process model of Equation (8), the first-principles reaction rate term $k_A e^{\frac{-E_A}{RT}} C_A - k_B e^{\frac{-E_B}{RT}} C_B$ was replaced with the obtained neural network $F_{NN}(C_A, C_B, T)$. The integration of the first-principles model and the neural network model that was used in RTO and MPC will be discussed in the following sections.

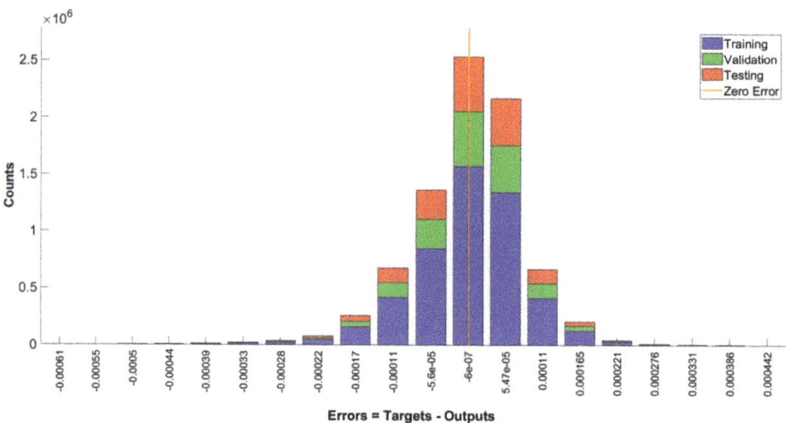

Figure 3. Error distribution histogram for training, validation, and testing data.

Remark 5. *The activation function plays an important role in the neural network training process and may affect its prediction performance significantly. Specifically, in the CSTR example, since it is known that the* *reaction rate is generally in the form of exponential functions, we tried tanh and sigmoid activation functions. It is demonstrated that both achieved the desired performance with mean squared error less than 10^{-7}.*

3.3. RTO and Controller Design

3.3.1. RTO Design

It is generally accepted that energy costs vary significantly compared to capital, labor, and other expenses in an actual plant. Therefore, in addition to productivity, it is important to account for energy cost in the real-time optimization of plant operation. Specifically, in this example, the heating cost was regarded as the entire energy cost since other energy costs may be lumped into the heating energy cost. The overall cost function is defined as follows:

$$total\ cost = \frac{C_A}{C_{A_0}} + heat\ price \times Q \tag{9}$$

Equation (9) attempts to find the balance between the reactant conversion and heat cost. A simple linear form was taken between Q and C_A in this case study since it was sufficient to illustrate the relationship between energy cost and reactant conversion. The above total cost was optimized in real time to minimize the cost of the CSTR process, by solving the optimization problem of Equation (10).

$$\min_{C_A, C_B, T, Q} \quad total\ cost = \frac{C_A}{C_{A_0}} + heat\ price \times Q \tag{10a}$$

$$\text{s.t.} \quad 0 = \frac{1}{\tau}(C_{A_0} - C_A) - F_{NN}(C_A, C_B, T) \tag{10b}$$

$$0 = -\frac{1}{\tau}C_B + F_{NN}(C_A, C_B, T) \tag{10c}$$

$$0 = \frac{-\Delta H}{\rho C_P}F_{NN}(C_A, C_B, T) + \frac{1}{\tau}(T_0 - T) + \frac{Q}{\rho C_P V} \tag{10d}$$

$$C_A \in [0, 1] \tag{10e}$$

$$C_B \in [0, 1] \tag{10f}$$

$$T \in [400, 500] \tag{10g}$$

$$Q \in [0, 10^5] \tag{10h}$$

The constraints of Equation (10b), Equation (10c), and Equation (10d) are the steady-state models of the CSTR process, which set the time derivative of Equation (8) to zero and replace the reaction rate term by the neural network model built in Section 3.2. Since the feed concentration C_{A_0} is 1 mol/L, C_A and C_B must be between 0 and 1 mol/L. The temperature constraint [400, 500] and energy constraint $[0, 10^5]$ are the desired operating conditions. At the initial steady-state, the heat price is 7×10^{-7}, and the CSTR operates at $T = 426.7$ K, $C_A = 0.4977$ mol/L and $Q = 40{,}386$ cal/s. The performance is not compromised too much since $C_A = 0.4977$ mol/L is close to the optimum value $C_A = 0.4912$ mol/L, while the energy saving is considerable when $Q = 40{,}386$ cal/s is compared to the optimum value $Q = 59{,}983$ cal/s. In the presence of variation in process variables or heat price, RTO recalculates the optimal operating condition, given that the variation is measurable every RTO period. The RTO of Equation (10) is solved every RTO period, and then sends steady-state values to controllers as the optimal set-points for the next 1000 s. Since the CSTR process has a relatively fast dynamics, a small RTO period of 1000 s is chosen to illustrate the performance of RTO.

3.3.2. Controller Design

In order to drive the process to the optimal steady-state, a Lyapunov-based model predictive controller (LMPC) is developed in this section. The controlled variables are C_A, C_B, and T, and the manipulated variable is heat rate Q. The CSTR is initially operated at the steady-state $[C_{A_s} C_{B_s}\ T_s] =$

[0.4977 mol/L   0.5023 mol/L   426.743 K], with steady-state $Q_{j_s} = 40,386$ cal/s. At the beginning of each RTO period, a new set of steady-states are calculated, and then, the input and the states are represented in their deviation variable form as $u = Q - Q_s$ and $x^T = [C_A - C_{A_s}\ C_B - C_{B_s}\ T - T_s]$, such that the systems of Equation (8) together with $F_{NN}(C_A, C_B, T)$ can be written in the form of $\dot{x} = f(x) + g(x)u$. A Lyapunov function is designed using the standard quadratic form $V(x) = 100{,}000x_1^2 + 100{,}000x_2^2 + x_3^2$, and the parameters are chosen to ensure that all terms are of similar order of magnitude since temperature is varying in a much larger range compared to concentration. We characterize the stability region Ω_ρ as a level set of Lyapunov function, i.e., $\Omega_\rho = \{x \in R^3 \mid V(x) \leq \rho\}$. For the system of Equation (8), the stability region Ω_ρ with $\rho = 1000$ is found based on the above Lyapunov function V and the following controller $h(x)$ [44]:

$$
h(x) = \begin{cases} -\dfrac{L_f V + \sqrt{L_f V^2 + L_g V^4}}{L_g V^2} L_g V & \text{if} \quad L_g V \neq 0 \\ 0 & \text{if} \quad L_g V = 0 \end{cases}
\tag{11}
$$

where $L_f V(x)$ denotes the standard Lie derivative $L_f V(x) := \frac{\partial V(x)}{\partial x} f(x)$. The control objective is to stabilize C_A, C_B, and T in the reactor at its steady-state by manipulating the heat rate Q. A Lyapunov-based model predictive controller (LMPC) is designed to bring the process to the steady-state calculated by the RTO. Specifically, the LMPC is presented by the following optimization problem:

$$
\min_{u \in S(\Delta)} \int_{t_k}^{t_{k+N}} \left(\|\tilde{x}(\tau)\|_{Q_c}^2 + \|u(\tau)\|_{R_c}^2 \right) d\tau \tag{12a}
$$

$$
\text{s.t.} \quad \dot{\tilde{x}}(t) = f(\tilde{x}(t)) + g(\tilde{x}(t))u(t) \tag{12b}
$$

$$
\tilde{x}(t_k) = x(t_k) \tag{12c}
$$

$$
u(t) \in U, \ \forall t \in [t_k, t_{k+N}) \tag{12d}
$$

$$
\frac{\partial V(x(t_k))}{\partial x}(f(x(t_k)) + g(x(t_k))u(t_k)) \leq \frac{\partial V(x(t_k))}{\partial x}(f(x(t_k)) + g(x(t_k))h(x(t_k))) \tag{12e}
$$

where $\tilde{x}$ is the predicted state, N is the number of sampling periods within the prediction horizon, and $S(\Delta)$ is the set of piece-wise constant functions with period Δ. The LMPC optimization problem calculates the optimal input trajectory over the entire prediction horizon $t \in [t_k, t_{k+N})$, but only applies the control action for the first sampling period, i.e., $u(t) = u(x(t_k))$, $\forall t \in [t_k, t_{k+1})$. In the optimization problem of Equation (12), Equation (12a) is the objective function minimizing the time integral of $\|\tilde{x}(\tau)\|_{Q_c}^2 + \|u(\tau)\|_{R_c}^2$ over the prediction horizon. Equation (12b) is the process model of Equation (8) in its deviation form and is used to predict the future states. A neural network $F_{NN}(x_1, x_2, x_3)$ is used to replace $k_A e^{\frac{-E_A}{RT}} C_A - k_B e^{\frac{-E_B}{RT}} C_B$ in Equation (8). Equation (12c) uses the state measurement $x(t_k)$ at $t = t_k$ as the initial condition $\tilde{x}(t_k)$ of the optimization problem. Equation (12d) defines the input constraints over the entire prediction horizon, where $U = [0 - Q_s\ 10^5 - Q_s]$. The constraint of Equation (12e) is used to decrease $V(x)$ such that the state $x(t)$ is forced to move towards the origin. It guarantees that the origin of the closed-loop system is rendered asymptotically stable under LMPC for any initial conditions inside the stability region Ω_ρ. The detailed proof of closed-loop stability can be found in [39].

To simulate the dynamic model of Equation (8) numerically under the LMPC of Equation (12), we used the explicit Euler method with an integration time step of $h_c = 10^{-2}$ s. Additionally, the optimization problem of the LMPC of Equation (12) is solved using the solver IPOPT in the OPTI Toolbox in MATLAB with the following parameters: sampling period $\Delta = 5$ s; prediction horizon $N = 10$. $Q_c = [1\ 0\ 0;\ 0\ 1\ 0;\ 0\ 0\ 5 \times 10^{-5}]$ and $R_c = 10^{-11}$ were chosen such that the magnitudes of the states and of the input in $\|\tilde{x}(\tau)\|_{Q_c}^2$ and $\|u(\tau)\|_{R_c}^2$ have the similar order.

3.4. Simulation Results

In the simulation, a variation of heat price is introduced to demonstrate the performance of the designed RTO and MPC. Since the heat price is changing as shown in Figure 4, the initial steady-state is no longer the optimal operating condition. The RTO of Equation (10) is solved at the beginning of each RTO period to achieve a set of improved set-points, which will be tracked by the MPC of Equation (12). With the updated set-points, the CSTR process keeps adjusting operating conditions accounting for varying heat price. After the controller receives the set-points, the MPC of Equation (12) calculates input u to bring x to the new set-point, and finally, both state x and input u are maintained at their new steady-states. The concentration profiles, temperature profile, and heat rate profile are shown in Figures 5–7.

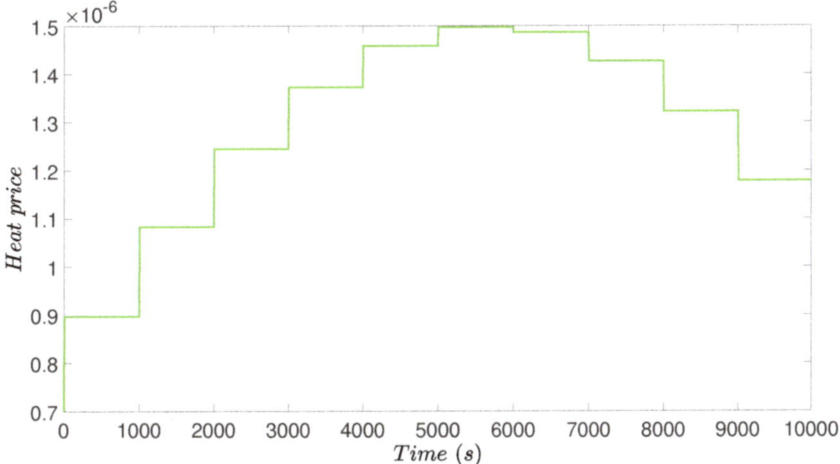

Figure 4. Heat price profile during the simulation, where the heat price first increases and then decreases to simulate heat rate price changing.

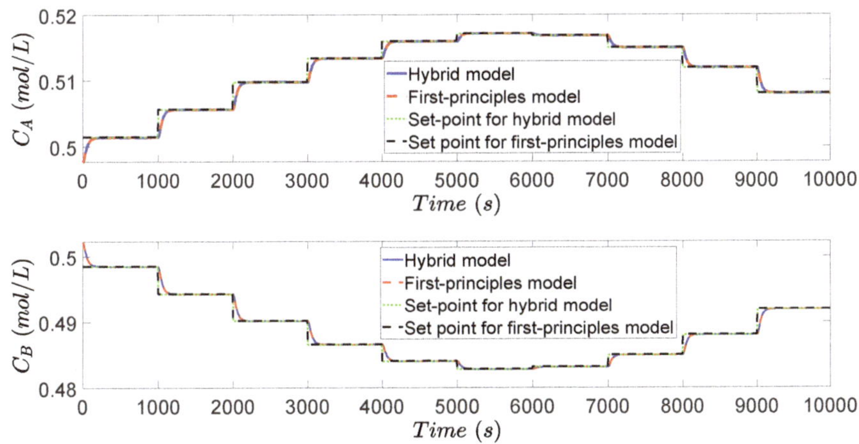

Figure 5. Evolution of the concentration of A and B for the CSTR case study under the proposed real-time optimization (RTO) and MPC.

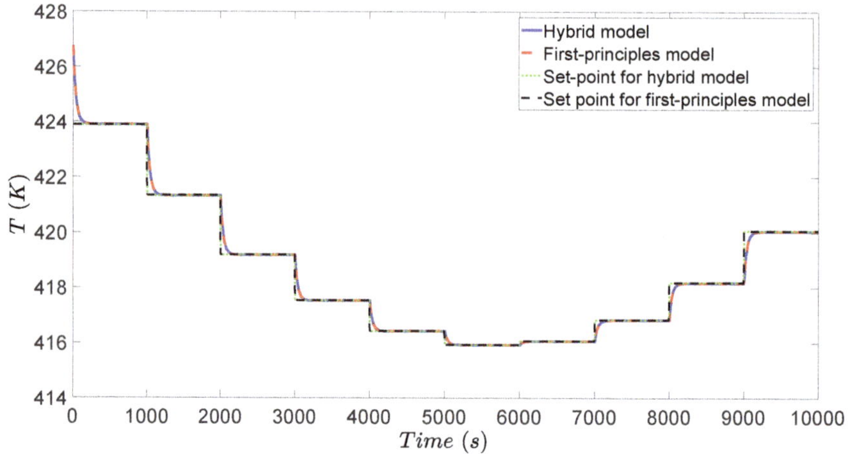

Figure 6. Evolution of the reactor temperature T for the CSTR case study under the proposed RTO and MPC scheme.

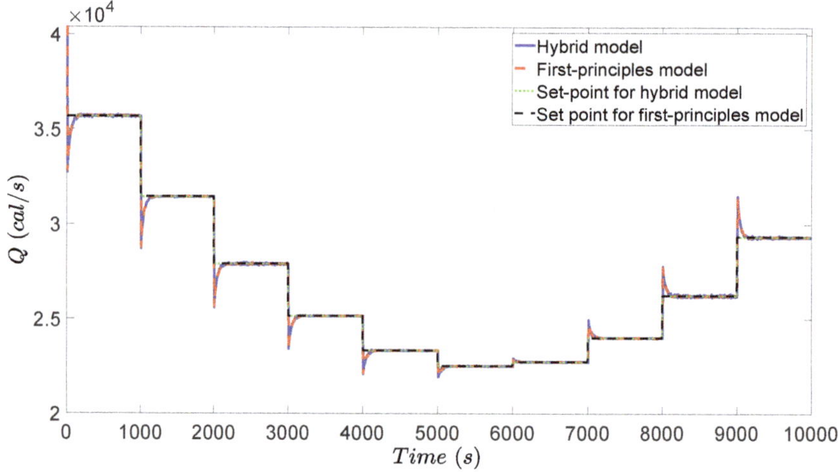

Figure 7. Evolution of the manipulated input, the heating rate Q, for the CSTR example under the proposed RTO and MPC scheme.

During the first half of the simulation, heat price rises up to a doubled value. Considering the increasing heat price, the operation tends to decrease the heat rate to reduce the energy cost, while compromising the reactant conversion. Therefore, the energy cost and reactant conversion will be balanced by RTO to reach a new optimum. As demonstrated in Figure 5, C_A increases and C_B decreases during the first half of simulation, which implies that less reactant A is converted to product B in the tank. The reactor temperature also drops as shown in Figure 6, which corresponds to the reducing heat rate as shown in Figure 7.

Total cost is calculated by Equation (9) using state measurements of C_A and Q from the closed-loop simulation and is plotted in Figure 8. The total cost with fixed steady-state is also calculated and plotted for comparison. After the heat price starts to increase, both total costs inevitably increase. Since RTO keeps calculating better steady-states compared to the initial steady-state, the total cost under RTO increases less than the simulation without RTO. The total cost is integrated with time to demonstrate the difference in cost increment, using Equation (13).

$$cost\ increase = \int_0^{t_{final}} \|total\ cost - initial\ cost\| \, dt \tag{13}$$

where *initial cost* $= 0.526$ and $t_{final} = 10{,}000$ s. The ratio of cost increment between simulations with RTO and without RTO is $195 : 241$. Although the operating cost increases because of rising heat price, RTO reduces the cost increment by approximately a factor of $1/5$, when compared to the fixed operating condition without RTO.

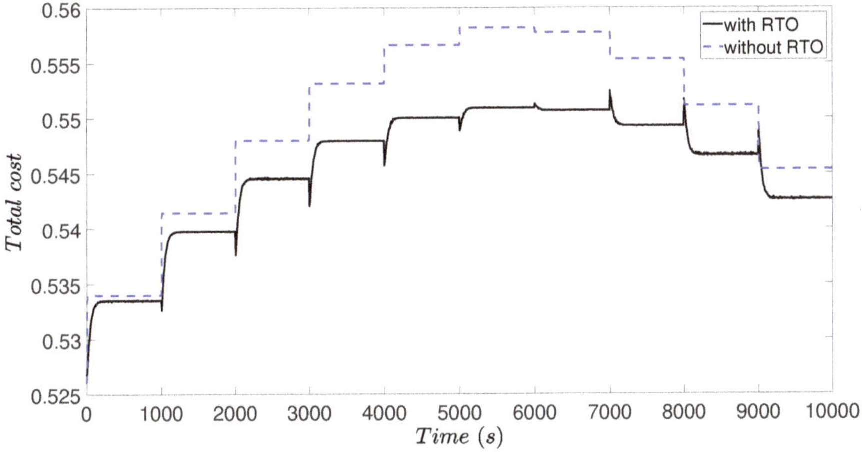

Figure 8. Comparison of the total operation cost for the CSTR example for simulations with and without RTO adapting to the heat rate price changing.

The combination of neural network models and first-principles models works well in both RTO and MPC. Additionally, it is shown in Figures 5–7 that the RTO with the combined first-principles/neural-network model calculates the same steady-state when compared to the RTO with a pure first-principles model. Moreover, the MPC also drives all the states to the set-points without offset when the MPC uses the combination of a neural network model with a first-principles model. In this case study, the neural network model is accurate such that the combination of neural network and first-principles model attains almost the same closed-loop result as the pure first-principles model (curves overlap when plotted in the same figure as is done in Figures 5–7, where the blue curve denotes the solution under MPC with the combined first-principles/neural network model, the red curve denotes the solution under MPC with the first-principles model, the green curve denotes the set-points calculated by RTO with the hybrid model, and the black curve denotes the set-points calculated by RTO with the first-principles model). Additionally, we calculated the accumulated relative error (i.e., $E = \frac{\int_{t=0}^{t=10,000s} |T_f - T_h| dt}{\int_{t=0}^{t=10,000s} T_f dt}$) between the temperature curves (Figure 6) under the first-principles model (i.e., T_f) and under the hybrid model (i.e., T_h) over the entire operating period from $t = 0$ to $t = 10{,}000$ s. It was obtained that $E = 4.98 \times 10^{-6}$, which is sufficiently small. This implies that the neural network successfully approximated the nonlinear term of reaction rate. In practice, neural network could be more effective when the reaction rate is very complicated and depends on more variables and the reaction mechanism is unknown.

4. Application to a Distillation Column

4.1. Process Description, Simulation, and Model

4.1.1. Process Description

A simple binary separation of propane from isobutane in a distillation column was used for the second case study [45]. Aspen Plus (Aspen Technology, Inc., Bedford, MA, USA) and Aspen Plus Dynamics V10.0 were utilized to perform high-fidelity dynamic simulation for the distillation column. Specifically, Aspen Plus uses the mass and energy balances to calculate the steady-state of the process based on a process flowsheet design and carefully-chosen thermodynamic models. After the steady-state model is solved in Aspen Plus, it can be exported to a dynamic model in Aspen Plus Dynamics, which runs dynamic simulations based on the obtained steady-state models and detailed process parameters [46,47].

A schematic of the distillation process is shown in Figure 9. The feed to the separation process was at 20 atm, 322 K and 1 kmol/s, with a propane mole fraction of 0.4 and an isobutane mole fraction of 0.6. After a valve controlling the feed flow rate, the feed enters the distillation column at Tray 14. The feed tray is carefully chosen to achieve the best separation performance and minimum energy cost, as discussed in [45]. The column has 30 trays with a tray spacing of 0.61 m, and the diameter of the tray is 3.85 m and 4.89 m for the rectifying section and stripping section, respectively. At the initial steady-state, the distillate product has a propane mole fraction 0.98 and a flow rate 0.39 kmol, while the bottom product has a propane mole fraction 0.019 and a flow rate 0.61 kmol. The reflux ratio is 3.33, together with condenser heat duty -2.17×10^7 W and reboiler heat duty 2.61×10^7 W. The pressure at the top and bottom is 16.8 atm and 17 atm. Both the top and bottom products are followed by a pump and a control valve. All the parameters are summarized in Table 2.

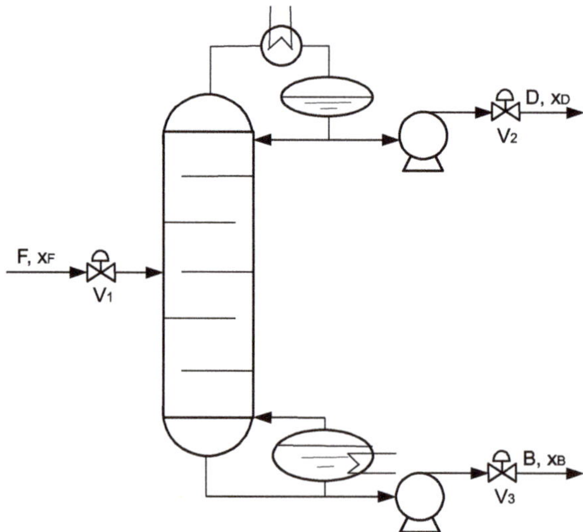

Figure 9. A schematic diagram of the distillation column implemented in Aspen Plus Dynamics.

In our simulation, the involved components of propane and isobutane were carefully chosen, and the CHAO-SEA model was selected for the thermodynamic property calculation. The steady-state model was first built in Aspen Plus using the detailed information as discussed above and the parameters in Table 2. Then, the achieved steady-state simulation was exported to the dynamic model as a pressure-driven model, based on additional parameters such as reboiler size and drum size. After checking the open-loop response of the dynamic model, controllers will be designed in Section 4.3.2.

Table 2. Parameter values and steady-state values for the distillation column case study.

$F = 1$ kmol	$x_F = 0.4$
$T_F = 322$ K	$P_F = 20$ atm
$q = 1.24$	$N_F = 14$
$N_T = 30$	$Diameter_{reboiler} = 5.08$ m
$Length_{reboiler} = 10.16$ m	$Diameter_{reflux\ drum} = 4.08$ m
$Length_{reflux\ drum} = 8.16$ m	
steady-state condition:	$R = 3.33$
$x_B = 0.019$	$x_D = 0.98$
$P_{bottom} = 17$ atm	$P_{top} = 16.8$ atm
$B = 0.61$ kmol/L	$D = 0.39$ kmol/L
$Q_{top} = -2.17 \times 10^7$ W	$Q_{bottom} = 2.61 \times 10^7$ W

4.1.2. Process Model

In order to calculate the steady-state of the distillation process, an analytic steady-state model is developed in this section. Since the Aspen model cannot be used in the optimization problem explicitly, this analytic steady-state model will be used in the RTO.

The analytic steady-state model consists of five variables, which are the reflux ratio R, the distillate mole flow rate D, the bottom mole flow rate B, the distillate mole fraction x_D, and the bottom mole fraction x_B. For clarification, x is denoted as the mole fraction for the light component propane. Other parameters include feed conditions: feed molar flow rate F, feed mole fraction x_F, feed heat condition q; column parameters: total number of trays N_T, feed tray N_F; component property: relative volatility α. Three equations were developed for the steady-state model.

The first equation $F_1(D, B) = 0$ is the overall molar balance between feed and products, as shown in Equation (14).

$$F = D + B \tag{14}$$

The second equation $F_2(D, B, x_D, x_B) = 0$ is the overall component balance of light component propane, as shown in Equation (15):

$$Fx_F = Dx_D + Bx_B \tag{15}$$

The third equation applies the binary McCabe–Thiele method. The constant molar overflow assumptions of the McCabe–Thiele method were held in this case study: the liquid and vapor flow rates were constant in a given section of the column. Equilibrium was also assumed to be reached on each tray. The top tray was defined as the first tray. To apply the McCabe–Thiele method, the rectifying operating line (ROL), stripping operating line (SOL), and phase equilibrium were developed as follows:

Rectifying operating line (ROL):

$$y_{n+1} = \frac{R}{R+1}x_n + \frac{x_D}{R+1} \tag{16}$$

Stripping operating line (SOL):

$$y_{n+1} = \frac{RD + qF}{(R+1)D - (1-q)F}x_n + \frac{F - D}{(R+1) - (1-q)F}x_B \tag{17}$$

Phase equilibrium:

$$x_n = \frac{y_n}{\alpha - (\alpha - 1)y_n} \tag{18}$$

where $\alpha = \frac{y_{C3}/x_{C3}}{y_{C4}/x_{C4}} = 1.79$ is the approximate relative volatility between propane and isobutane at a pressure 16.9 atm, which is the mean of the top and bottom pressure.

The third equation $F_3(R, D, x_D, x_B) = 0$ is expressed in Equation (19) below:

$$y_1 = x_D \tag{19a}$$

$$x_n = \frac{y_n}{\alpha - (\alpha - 1)y_n} \qquad n = 1, 2 ... N_T \tag{19b}$$

$$y_{n+1} = \frac{R}{R+1}x_n + \frac{x_D}{R+1} \qquad n = 1, 2 ... N_F - 1 \tag{19c}$$

$$y_{n+1} = \frac{RD + qF}{(R+1)D - (1-q)F}x_n + \frac{F - D}{(R+1) - (1-q)F}x_B \qquad n = N_F, N_F + 1 ... N_T - 1 \tag{19d}$$

$$x_{N_T} = x_B \tag{19e}$$

The third equation $F_3(R, D, x_D, x_B) = 0$ ties the distillate mole fraction x_D to the bottom mole fraction x_B by calculating both liquid and vapor mole fractions through all trays from top to bottom. Equation (19a) defines the vapor mole fraction y_1 on the first tray as the distillate mole fraction x_D. Then, the liquid mole fraction x_1 on the first tray can be calculated by the phase equilibrium of Equation (19b). Subsequently, the vapor mole fraction y_2 on the second tray is calculated by the ROL of Equation (19c). The calculation is repeated until x_{14} and y_{14} are obtained. Then, y_{15} is calculated by the SOL of Equation (19d), instead of ROL. Then, x_{15} can be calculated again by the phase equilibrium of Equation (19b). The above calculations are repeated until x_{30} and y_{30} are obtained, and $x_{30} = x_B$ since the liquid on the last tray is the bottom product. In this way, all the variables (i.e., R, D, x_D, x_B) have values that satisfy $F_3(R, D, x_D, x_B) = 0$.

There are five variables R, D, B, x_D, x_B and three equations F_1, F_2, F_3, which implies that there are two degrees of freedom. In order to determine the whole process operating condition, two more states need to be fixed, potentially by RTO. It is necessary to point out that the concentrations x_i and y_i on each tray can be calculated by Equation (19) if all five variables R, D, B, x_D, x_B are determined. Additionally, if the equilibrium temperature-component curve $T = f_e(x)$ (bubble point curve) or $T = f_e(y)$ (dew point curve) are provided, then the temperature on each tray T_i can also be calculated by simply using $T_i = f_e(x_i)$ or $T_i = f_e(y_i)$.

4.2. Neural Network Model

Phase equilibrium properties are usually nonlinear, and the first-principles models are often found to be inaccurate and demand modifications. In the above steady-state model, the phase equilibrium $x_n = \frac{y_n}{\alpha - (\alpha - 1)y_n}$ of Equation (19b) assumes that relative volatility α is constant; however, the relative volatility α does not hold constant with varying concentration and pressure. Therefore, a more accurate model for phase equilibrium $x \sim y$ can improve the model performance. Similarly, dew point curve $T \sim y$ can be built from first-principles formulation upon Raoult's Law and the Antoine equation. However, the Antoine equation is an empirical equation, and it is hard to relate saturated pressure with temperature accurately, especially for a mixture. As a result, the machine learning method can be used to achieve a better model to represent the phase equilibrium properties.

In this case study, a neural network $(x, T) = F_{NN}(y)$ was built, with one input (vapor phase mole fraction y) and two outputs (equilibrium temperature T and liquid phase mole fraction x). One thousand five hundred data of T, x, and y were generated by the Aspen property library and were then normalized and fed into the MATLAB Deep Learning toolbox. $tanh(x) = 2/(1 + e^{-2x}) - 1$ was chosen as the activation function. The neural network model had one hidden layer with five neurons. The parameters were trained according to Levenberg–Marquardt optimization, and the mean squared error for the test dataset was around 10^{-7}. It is demonstrated in Figure 10 that the neural network model fits the data from the Aspen property library very well, where the blue solid curve is the neural network model prediction and the red curve denotes the Aspen model. Additionally, we calculated the accumulated relative error (i.e., $E = \frac{\int_{y=0}^{y=1} |T_f - T_h| dy}{\int_{y=0}^{y=1} T_f dy}$) between the temperature curves

(Figure 10) under the Aspen model (i.e., T_f) and under the neural network model (i.e., T_h) and $E = 2.32 \times 10^{-6}$; the result was similar for the liquid mole fraction curves. This sufficiently small error implies that the neural network model successfully approximated the nonlinear behavior of the thermodynamic properties. Additionally, the coefficient of determination R^2 was 1, and the error histogram of Figure 11 demonstrated that the neural network model represented the thermodynamic properties with great accuracy.

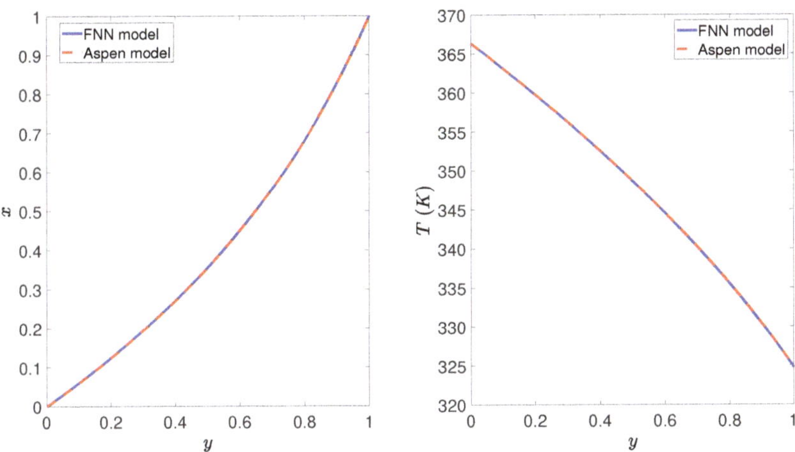

Figure 10. Comparison of the neural network model and the Aspen model.

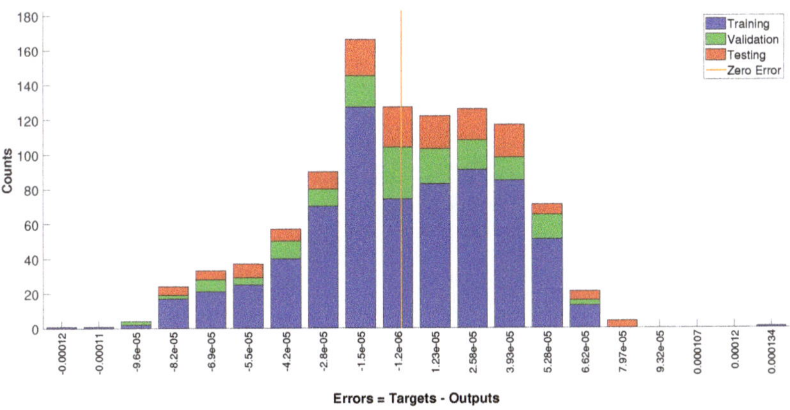

Figure 11. Error distribution histogram for training, validation, and testing data.

After training the neural network model, the first-principles phase equilibrium expression $x_n = \frac{y_n}{\alpha - (\alpha - 1)y_n}$ in Equation (19b) is replaced by the neural network phase equilibrium expression $x_n = F_{NN,1}(y_n)$, and then, the integrated model of first-principles model and neural network model is used in RTO as discussed in the following sections. In addition, the second output of the neural network model $T_n = F_{NN,2}(y_n)$ can be combined together with Equation (19) to calculate the temperature on each tray, which will be used later to calculate the set-points for the controllers.

4.3. RTO and Controller Design

4.3.1. RTO Design

Since the process has two degrees of freedom, the operating condition has not been determined. An RTO was designed for the distillation process to obtain the optimal operating condition. Since RTO needs an objective function, a profit was developed to represent the operation profit. According to the products, feed, and energy price in [45], the profit is defined by Equation (20).

$$
\begin{aligned}
Profit &= price_{top}D + price_{bottom}B - price_{feed}F - price_{energy}Q \\
&= price_{top}D + price_{bottom}B - price_{feed}F - price_{energy}(L(R+1)F) \\
&= Profit(R, D, B, x_D, x_B)
\end{aligned}
\tag{20}
$$

The profit equals the profit of product subtracting the cost of feed and energy. The profit that will be used in RTO is represented as a function of R, D, B, x_D, x_B. As a result, heat duty Q of both the condenser and reboiler is approximated by $Q = L(R+1)F$, where $L = 1.29 \times 10^7$ J/kmol is the molar latent heat of the mixture. Moreover, mass-based prices are changed to mole-based prices because all flow rates are mole-based. The price of the top distillate rises linearly as the mole fraction x_D increases in order to demonstrate that the higher purity product has a higher price.

$$
\begin{aligned}
price_{top} &= (0.528 + (x_D - 0.97))\$/kg \times 44.1 kg/kmol = 23.29 + 44.1(x_D - 0.97)\ \$/kmol \\
price_{bottom} &= 0.264\$/kg \times 58.1 kg/kmol = 15.34\ \$/kmol \\
price_{feed} &= 0.264\$/kg \times 52.5 kg/kmol = 13.86\ \$/kmol \\
price_{energy} &= 6.11 \times 10^{-8}\ \$/J
\end{aligned}
\tag{21}
$$

To maximize the operation profit, the RTO problem is formulated as Equation (22).

$$
\min_{R,D,B,x_D,x_B} \quad -Profit(R, D, B, x_D, x_B)
\tag{22a}
$$

$$
\text{s.t.} \quad F_1(D, B) = 0
\tag{22b}
$$

$$
F_2(D, B, x_D, x_B) = 0
\tag{22c}
$$

$$
F_3(D, x_D, x_B, R) = 0
\tag{22d}
$$

$$
R \in [0, \infty]
\tag{22e}
$$

$$
D \in [0, 1]
\tag{22f}
$$

$$
B \in [0, 1]
\tag{22g}
$$

$$
x_D \in [0, 1]
\tag{22h}
$$

$$
x_B \in [0, 1]
\tag{22i}
$$

Equation (22a) minimizes the negative profit with respective to five optimization variables R, D, B, x_D, x_B. The first three constraint Equation (22b), Equation (22c), and Equation (22d) are the steady-state model of Equation (14), Equation (15) and Equation (19), as discussed in Section 4.1.2. The neural network model $x_n = F_{NN,1}(y_n)$ replaces $x_n = \frac{y_n}{\alpha - (\alpha-1)y_n}$ in Equation (19). Constraints on the optimization variables are determined based on process parameters. Specifically, reflux ratio R can be any positive number; D and B should be between 0 and 1 because the feed had only 1 kmol/s; x_D and x_B should be also between zero and one because they are mole fractions. Since there are two degrees of freedom in the optimization problem, two steady-state values are sent to the controllers as set-points.

4.3.2. Controller Design

Six controllers were added in the distillation column, four of which had fixed set-points and two of which received set-points from RTO. The control scheme is shown in Figure 12.

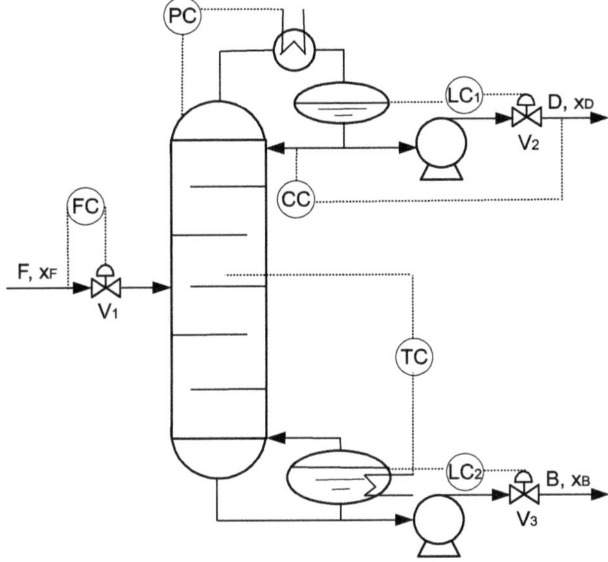

Figure 12. A schematic diagram of the control structure implemented in the distillation column. Flow rate controller FC, pressure controller PC, and both level controllers LC_1 and LC_2 have fixed set-points, and concentration controller CC and temperature controller TC receive set-points from the RTO.

(1) A flow rate controller FC is controlling the feed mole flow rate at 1 kmol/s by manipulating feed valve V_1. A fixed feed flow rate helps to fix the parameters in the first-principles steady-state model.

(2) A pressure controller PC is controlling the column top pressure at 16.8 atm by manipulating condenser heat duty Q_{top}. A fixed column pressure helps to operate the process with fixed thermodynamic properties.

(3) A level controller LC_1 is controlling the reflux drum liquid level at 5.1 m by manipulating the distillate outlet valve V_2. A certain liquid level in the condenser is required to avoid flooding or drying.

(4) A level controller LC_2 is controlling the reboiler liquid level at 6.35 m by manipulating the bottom outlet valve V_3. A certain liquid level in the reboiler is required to avoid flooding or drying.

(5) A concentration controller CC is controlling the distillate C_3 mole fraction by manipulating the reflux mole flow rate. A time delay of 5 min was added to simulate the concentration measurement delay. At the beginning of each RTO period, RTO sends the optimized distillate C_3 mole fraction x_D to concentration controller CC as the set-point. Then, controller CC adjusts the reflux flow to track the mole fraction to its set-point.

(6) A temperature controller TC is controlling temperature T_7 on Tray 7, by manipulating reboiler heat duty Q_{bottom}. A time delay of 1 min was added to simulate the temperature measurement delay. Tray temperature control is common in industry, and two methods were carried out to determine the best tray temperature to be controlled. A steady-state simulation was used to obtain the temperature profile along the tube to find out that the temperature changes among Tray 6, Tray 7, and Tray 8 were greater than those among other trays. One more simulation was performed to get the gain of tray temperature as a response to a small change in the reboiler heat duty. It was also found that the temperature on Tray 7 had a greater gain than those on other trays. As a result, Tray 7 was chosen as the controlled variable.

At the beginning of the RTO period, RTO optimizes the profit and calculates a set of steady-states. Given the optimum value of R, D, B, x_D, x_B, the steady-state model of $F_1 = 0$, $F_2 = 0$, and $F_3 = 0$ were used again to obtain the concentration profile in the distillation column. Then, the neural network model $T_n = F_{NN,2}(y_n)$ was used to calculate the temperature on Tray 7. After that, the tray temperature T_7 was sent to the controller TC and will be tracked to its set-point by manipulating the reboiler heat duty.

Flow rate controller FC, pressure controller PC, and both level controllers LC_1 and LC_2 had fixed set-points, which stabilized the process to operate at fixed operation parameters. Concentration controller CC and temperature controller TC received set-points from RTO at the beginning of RTO period and drove the process to more profitable steady-state. All the PI parameters were tuned by the Ziegler–Nichols method and are shown in Table 3.

Table 3. Proportional gain and integral time constant of all the PI controllers in the distillation case study.

	K_C	τ_I/min
FC	0.5	0.3
PC	15	12
LC_1	2	150
LC_2	4	150
CC	0.1	20
TC	0.6	8

4.4. Simulation Results

To demonstrate the effectiveness of RTO, a variation in feed mole fraction x_F was introduced to the process, as shown in Figure 13. At the beginning of each RTO period (20 h), one measurement of feed mole fraction x_F was sent to RTO to optimize the profit. Then, a set of steady-states was achieved from RTO and was sent to the controllers as set-points.

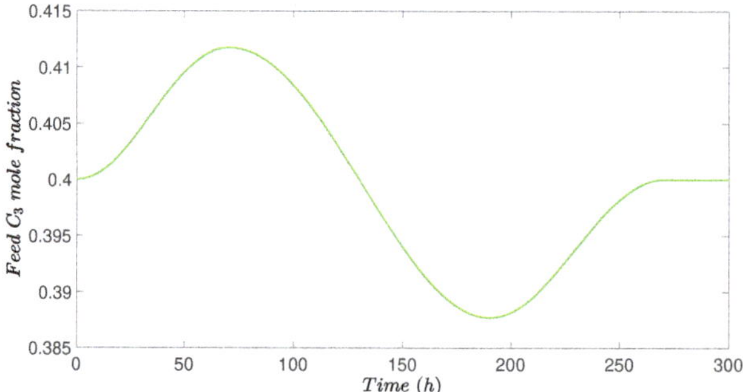

Figure 13. The feed concentration profile of the distillation column, which is changing with respect to time.

The simulation results are shown in Figures 14 and 15. In Figure 14, the set-point of x_D increases as feed concentration x_F increases at the beginning of simulation, because higher distillate concentration is more profitable and more feed concentration x_F allows further separation to achieve a higher concentration in the distillate. The set-point for x_D also decreased later when feed concentration x_F decreased. At the beginning of the simulation, reflux flow increased to reach higher x_D set-points, and reflux flow never reached a steady-state during the whole simulation because the feed component kept changing as shown in Figure 13. In some cases, the mole fraction x_D did not track exactly the set-point because of the ever-changing feed, too small set-point change, and coupled effect with other variables and controllers.

Figure 15 illustrates the performance of temperature controller TC. When the feed x_F increased, the set-point for Tray 7 temperature T_7 decreased according to RTO. The controller then manipulated the reboiler heat duty to track the tray temperature with a good performance as shown in Figure 15. It is noted in Figure 15 that the reboiler heat duty increased as tray temperature decreased at the beginning of the simulation. The reason is that the reboiler heat duty mainly dependent on the liquid flow into the reboiler and the vapor flow leaving the reboiler. Since the reflux flow was increased by the concentration controller CC at the beginning of simulation, both the liquid flow into the reboiler and vapor leaving the reboiler increased, thus increasing reboiler heat duty.

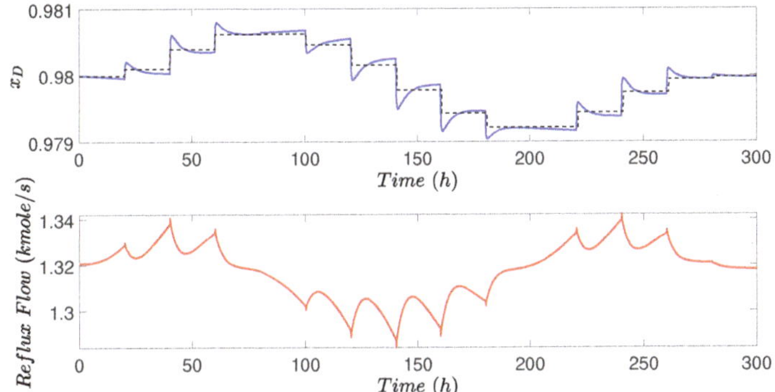

Figure 14. Controlled output x_D and manipulated input *reflux flow* for the concentration controller CC in the distillation process under the proposed RTO scheme.

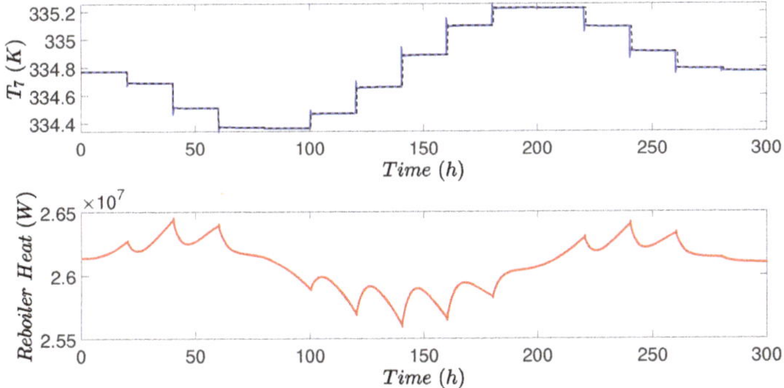

Figure 15. Controlled output T_7 and manipulated input *reboiler heat* for the temperature controller TC in the distillation process under the proposed RTO scheme.

Other controllers stayed at the fixed set-points throughout the simulation by adjusting their manipulated inputs. Therefore, we are not showing the plots for other controllers. It is demonstrated in Figure 16 that the RTO increased the operation profit when distillation column had a varying feed concentration. The profit in Figure 16 was calculated by the profit definition of Equation (20), using the closed-loop simulation data for variables D, B, F, and R. The black line is the operation profit calculated by the closed-loop simulation where the four controllers (FC, PC, LC_1, and LC_2) had fixed set-points and the two controllers (CC and TC) had varying set-points from RTO. The blue line is the simulation where the set-points of all controllers were fixed at the initial steady-state and the controlled variables stayed at the initial set-point by adjusting manipulated variables in the presence of the same feed variation in Figure 13. Although the feed concentration kept changing each second and RTO updated the steady-state only each 20 h, the profit was still improved significantly by RTO, as shown in Figure 16.

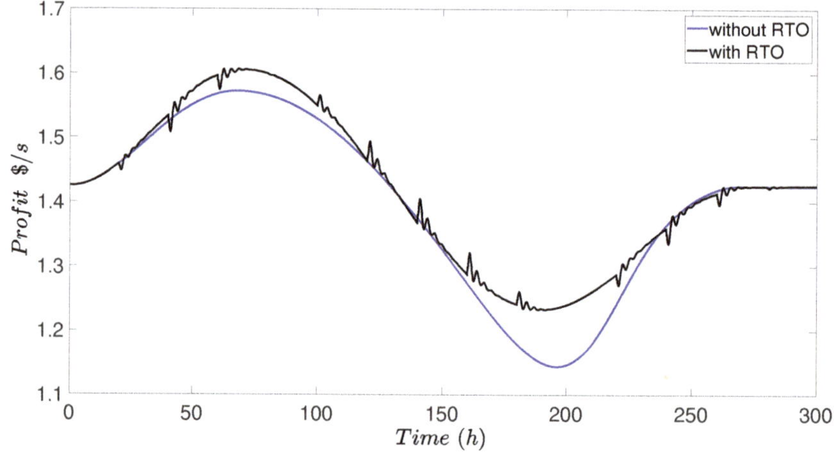

Figure 16. Comparison of the operation profit for the distillation process for closed-loop simulations with and without RTO adapting for change in the feed concentration.

In this case study, a neural network model was combined only with the steady-state first-principles model, not the dynamic model. Additionally, it was demonstrated that the steady-states calculated by RTO using a combination of models were very close to the steady-state values in the Aspen simulator, which means that the combination of the neural network model and first-principles model was of high accuracy. The neural network model was used to represent the phase equilibrium properties for RTO to calculate the optimal steady-state in this work. Neural network models can be useful when the phase equilibrium is highly nonlinear such that the first-principles model is inaccurate. Additionally, it can be used when a large number of states are included in thermodynamic equations, such as pressure or more concentrations for the multi-component case.

5. Conclusions

In this work, we presented a method for integrating neural network modeling with first-principles modeling in the model used in RTO and MPC. First, a general framework that integrates neural network models with first-principle models in the optimization problems of RTO and MPC was discussed. Then, two chemical process examples were studied in this work. In the first case study, a CSTR with reversible exothermic reaction was utilized to analyze the performance of integrating the neural network model and first-principles model in RTO and MPC. Specifically, a neural network was first built to represent the nonlinear reaction rate. An RTO was designed to find the operating steady-state providing the optimal balance between the energy cost and reactant conversion. Then, an LMPC was designed to stabilize the process to the optimal operating condition. A variation in energy price was introduced, and the simulation results demonstrated that RTO minimized the operation cost and yielded a closed-loop performance that was very close to the one attained by RTO/MPC using the first-principles model. In the second case study, a distillation column was studied to demonstrate an application to a large-scale chemical process. A neural network was first trained to obtain the phase equilibrium properties. An RTO scheme was designed to maximize the operation profit and calculate the optimal set-points for the controllers using a neural network model with a first-principles model. A variation in the feed concentration was introduced to demonstrate that RTO increased operation profit for all considered conditions. In closing, it is important to note that the two simulation studies only demonstrated how the proposed approach can be applied and provided some type of "proof of concept" on the use of hybrid models in RTO and MPC, but certainly, both examples yield limited conclusions and cannot substitute for an industrial/experimental implementation to evaluate the proposed approach, which would be the subject of future work.

Author Contributions: Z.Z. developed the idea of incorporating machine learning in real-time optimization and model predictive control, performed the simulation studies, and prepared the initial draft of the paper. Z.W. and D.R. revised this manuscript. P.D.C. oversaw all aspects of the research and revised this manuscript.

Acknowledgments: Financial support from the National Science Foundation and the Department of Energy is gratefully acknowledged.

Conflicts of Interest: The authors declare that they have no conflict of interest regarding the publication of the research article.

References

1. Bhutani, N.; Rangaiah, G.; Ray, A. First-principles, data-based, and hybrid modeling and optimization of an industrial hydrocracking unit. *Ind. Eng. Chem. Res.* **2006**, *45*, 7807–7816. [CrossRef]

2. Pantelides, C.C.; Renfro, J. The online use of first-principles models in process operations: Review, current status and future needs. *Comput. Chem. Eng.* **2013**, *51*, 136–148. [CrossRef]

3. Quelhas, A.D.; de Jesus, N.J.C.; Pinto, J.C. Common vulnerabilities of RTO implementations in real chemical processes. *Can. J. Chem. Eng.* **2013**, *91*, 652–668. [CrossRef]

4. Wu, Z.; Tran, A.; Rincon, D.; Christofides, P.D. Machine Learning-Based Predictive Control of Nonlinear Processes. Part I: Theory. *AIChE J.* **2019**, *65*, e16729. [CrossRef]

5. Wu, Z.; Tran, A.; Rincon, D.; Christofides, P.D. Machine Learning-Based Predictive Control of Nonlinear Processes. Part II: Computational Implementation. *AIChE J.* **2019**, *65*, e16734. [CrossRef]

6. Lee, M.; Park, S. A new scheme combining neural feedforward control with model-predictive control. *AIChE J.* **1992**, *38*, 193–200. [CrossRef]

7. Venkatasubramanian, V. The promise of artificial intelligence in chemical engineering: Is it here, finally? *AIChE J.* **2019**, *65*, 466–478. [CrossRef]

8. Himmelblau, D. Applications of artificial neural networks in chemical engineering. *Korean J. Chem. Eng.* **2000**, *17*, 373–392. [CrossRef]

9. Chouai, A.; Laugier, S.; Richon, D. Modeling of thermodynamic properties using neural networks: Application to refrigerants. *Fluid Phase Equilib.* **2002**, *199*, 53–62. [CrossRef]

10. Galván, I.M.; Zaldívar, J.M.; Hernandez, H.; Molga, E. The use of neural networks for fitting complex kinetic data. *Comput. Chem. Eng.* **1996**, *20*, 1451–1465. [CrossRef]

11. Faúndez, C.A.; Quiero, F.A.; Valderrama, J. Phase equilibrium modeling in ethanol+ congener mixtures using an artificial neural network. *Fluid Phase Equilib.* **2010**, *292*, 29–35. [CrossRef]

12. Fu, K.; Chen, G.; Sema, T.; Zhang, X.; Liang, Z.; Idem, R.; Tontiwachwuthikul, P. Experimental study on mass transfer and prediction using artificial neural network for CO_2 absorption into aqueous DETA. *Chem. Eng. Sci.* **2013**, *100*, 195–202. [CrossRef]

13. Bakshi, B.; Koulouris, A.; Stephanopoulos, G. Wave-Nets: Novel learning techniques, and the induction of physically interpretable models. In *Wavelet Applications*; International Society for Optics and Photonics: Orlando, FL, USA, 1994; Volume 2242, pp. 637–648.

14. Lu, Y.; Rajora, M.; Zou, P.; Liang, S. Physics-embedded machine learning: Case study with electrochemical micro-machining. *Machines* **2017**, *5*, 4. [CrossRef]

15. Psichogios, D.C.; Ungar, L.H. A hybrid neural network-first principles approach to process modeling. *AIChE J.* **1992**, *38*, 1499–1511. [CrossRef]

16. Oliveira, R. Combining first principles modelling and artificial neural networks: A general framework. *Comput. Chem. Eng.* **2004**, *28*, 755–766. [CrossRef]

17. Chen, L.; Bernard, O.; Bastin, G.; Angelov, P. Hybrid modelling of biotechnological processes using neural networks. *Control Eng. Pract.* **2000**, *8*, 821–827. [CrossRef]

18. Georgieva, P.; Meireles, M.; de Azevedo, S. Knowledge-based hybrid modelling of a batch crystallisation when accounting for nucleation, growth and agglomeration phenomena. *Chem. Eng. Sci.* **2003**, *58*, 3699–3713. [CrossRef]

19. Schuppert, A.; Mrziglod, T. Hybrid Model Identification and Discrimination with Practical Examples from the Chemical Industry. In *Hybrid Modeling in Process Industries*; CRC Press: Boca Raton, FL, USA, 2018; pp. 63–88.

20. Qin, S.; Badgwell, T. A survey of industrial model predictive control technology. *Control Eng. Pract.* **2003**, *11*, 733–764. [CrossRef]

21. Câmara, M.; Quelhas, A.; Pinto, J. Performance evaluation of real industrial RTO systems. *Processes* **2016**, *4*, 44–64. [CrossRef]

22. Lee, W.J.; Na, J.; Kim, K.; Lee, C.; Lee, Y.; Lee, J.M. NARX modeling for real-time optimization of air and gas compression systems in chemical processes. *Comput. Chem. Eng.* **2018**, *115*, 262–274. [CrossRef]

23. Agbi, C.; Song, Z.; Krogh, B. Parameter identifiability for multi-zone building models. In Proceedings of the 2012 IEEE 51st IEEE Conference on Decision and Control (CDC), Maui, HI, USA, 10–13 December 2012; pp. 6951–6956.

24. Yen-Di Tsen, A.; Jang, S.S.; Wong, D.S.H.; Joseph, B. Predictive control of quality in batch polymerization using hybrid ANN models. *AIChE J.* **1996**, *42*, 455–465. [CrossRef]

25. Klimasauskas, C.C. Hybrid modeling for robust nonlinear multivariable control. *ISA Trans.* **1998**, *37*, 291–297. [CrossRef]

26. Chang, J.; Lu, S.; Chiu, Y. Dynamic modeling of batch polymerization reactors via the hybrid neural-network rate-function approach. *Chem. Eng. J.* **2007**, *130*, 19–28. [CrossRef]

27. Noor, R.M.; Ahmad, Z.; Don, M.M.; Uzir, M.H. Modelling and control of different types of polymerization processes using neural networks technique: A review. *Can. J. Chem. Eng.* **2010**, *88*, 1065–1084. [CrossRef]

28. Wang, J.; Cao, L.L.; Wu, H.Y.; Li, X.G.; Jin, Q.B. Dynamic modeling and optimal control of batch reactors, based on structure approaching hybrid neural networks. *Ind. Eng. Chem. Res.* **2011**, *50*, 6174–6186. [CrossRef]

29. Chaffart, D.; Ricardez-Sandoval, L.A. Optimization and control of a thin film growth process: A hybrid first principles/artificial neural network based multiscale modelling approach. *Comput. Chem. Eng.* **2018**, *119*, 465–479. [CrossRef]

30. Schweidtmann, A.M.; Mitsos, A. Deterministic global optimization with artificial neural networks embedded. *J. Opt. Theory Appl.* **2019**, *180*, 925–948. [CrossRef]

31. Wu, Z.; Christofides, P.D. Economic Machine-Learning-Based Predictive Control of Nonlinear Systems. *Mathematics* **2019**, *7*, 494. [CrossRef]

32. Goodfellow, I.; Bengio, Y.; Courville, A. *Deep Learning*; MIT Press: Cambridge, MA, USA, 2016.

33. Bishop, C. *Pattern Recognition and Machine Learning*; Springer: New York, NY, USA, 2006.

34. Haykin, S. *Neural Networks: A Comprehensive Foundation*; Prentice Hall PTR: Upper Saddle River, NJ, USA, 1994.

35. Naysmith, M.; Douglas, P. Review of real time optimization in the chemical process industries. *Dev. Chem. Eng. Miner. Process.* **1995**, *3*, 67–87. [CrossRef]

36. Rawlings, J.; Amrit, R. Optimizing process economic performance using model predictive control. In *Nonlinear Model Predictive Control*; Springer: Berlin, Germany, 2009; pp. 19–138.

37. Ellis, M.; Durand, H.; Christofides, P. A tutorial review of economic model predictive control methods. *J. Process Control* **2014**, *24*, 1156–1178. [CrossRef]

38. Rawlings, J.B.; Bonné, D.; Jørgensen, J.B.; Venkat, A.N.; Jørgensen, S.B. Unreachable setpoints in model predictive control. *IEEE Transa. Autom. Control* **2008**, *53*, 2209–2215. [CrossRef]

39. Mhaskar, P.; El-Farra, N.H.; Christofides, P.D. Stabilization of nonlinear systems with state and control constraints using Lyapunov-based predictive control. *Syst. Control Lett.* **2006**, *55*, 650–659. [CrossRef]

40. Wang, L. Continuous time model predictive control design using orthonormal functions. *Int. J. Control* **2001**, *74*, 1588–1600. [CrossRef]

41. Hosseinzadeh, M.; Cotorruelo, A.; Limon, D.; Garone, E. Constrained Control of Linear Systems Subject to Combinations of Intersections and Unions of Concave Constraints. *IEEE Control Syst. Lett.* **2019**, *3*, 571–576. [CrossRef]

42. Daoutidis, P.; Kravaris, C. Dynamic output feedback control of minimum-phase multivariable nonlinear processes. *Chem. Eng. Sci.* **1994**, *49*, 433–447. [CrossRef]
43. Economou, C.; Morari, M.; Palsson, B. Internal model control: Extension to nonlinear system. *Ind. Eng. Chem. Process Des. Dev.* **1986**, *25*, 403–411. [CrossRef]
44. Lin, Y.; Sontag, E. A universal formula for stabilization with bounded controls. *Syst. Control Lett.* **1991**, *16*, 393–397. [CrossRef]
45. Luyben, W.L. *Distillation Design and Control Using Aspen Simulation*; John Wiley & Sons: Hoboken, NJ, USA, 2013.
46. Al-Malah, K.I. *Aspen Plus: Chemical Engineering Applications*; John Wiley & Sons: Hoboken, NJ, USA, 2016.
47. Aspen Technology, Inc. *Aspen Plus User Guide*; Aspen Technology, Inc.: Cambridge, MA, USA, 2003.

Article

Target Fusion Detection of LiDAR and Camera Based on the Improved YOLO Algorithm

Jian Han, Yaping Liao, Junyou Zhang *, Shufeng Wang and Sixian Li

College of Transportation, Shandong University of Science and Technology, Huangdao District,
Qingdao 266590, China; hanjianzrx@163.com (J.H.); liaoyapingsk@163.com (Y.L.);
shufengwang@sdust.edu.cn (S.W.); gsypsbyw666@163.com (S.L.)
* Correspondence: junyouzhang@sdust.edu.cn; Tel.: +86-139-0532-3314

Received: 7 September 2018; Accepted: 17 October 2018; Published: 19 October 2018

Abstract: Target detection plays a key role in the safe driving of autonomous vehicles. At present, most studies use single sensor to collect obstacle information, but single sensor cannot deal with the complex urban road environment, and the rate of missed detection is high. Therefore, this paper presents a detection fusion system with integrating LiDAR and color camera. Based on the original You Only Look Once (YOLO) algorithm, the second detection scheme is proposed to improve the YOLO algorithm for dim targets such as non-motorized vehicles and pedestrians. Many image samples are used to train the YOLO algorithm to obtain the relevant parameters and establish the target detection model. Then, the decision level fusion of sensors is introduced to fuse the color image and the depth image to improve the accuracy of the target detection. Finally, the test samples are used to verify the decision level fusion. The results show that the improved YOLO algorithm and decision level fusion have high accuracy of target detection, can meet the need of real-time, and can reduce the rate of missed detection of dim targets such as non-motor vehicles and pedestrians. Thus, the method in this paper, under the premise of considering accuracy and real-time, has better performance and larger application prospect.

Keywords: autonomous vehicle; target detection; multi-sensors; fusion; YOLO

1. Introduction

To improve road traffic safety, autonomous vehicles have become the mainstream of future traffic development in the world. Target recognition is one of the fundamental parts to ensure the safe driving of autonomous vehicles, which needs the help of various sensors. In recent years, the most popular sensors include LiDAR and color camera, due to their excellent performance in the field of obstacle detection and modeling.

The color cameras can capture images of real-time traffic scenes and use target detection to find where the target is located. Compared with the traditional target detection methods, the deep learning-based detection method can provide more accurate information, and therefore has gradually become a research trend. In deep learning, convolutional neural networks combine artificial neural networks and convolutional algorithms to identify a variety of targets. It has good robustness to a certain degree of distortion and deformation [1] and You only look once (YOLO) is a target real-time detection model based on convolutional neural network. For the ability to learn massive data, capability to extract point-to-point feature and good real-time recognition effect [2], YOLO has become a benchmark in the field of target detection. Gao et al. [3] clustered the selected initial candidate boxes, reorganized the feature maps, and expanded the number of horizontal candidate boxes to construct the YOLO-based pedestrian (YOLO-P) detector, which reduced the missed rate for pedestrians. However, the YOLO model was limited to static image detection, making a greater limitation in the detection of pedestrian dynamic changes. Thus, based on the original YOLO, Yang et al. [4] merged it with the detection algorithm DPM (Deformable Part Model) and R-FCN (Region-based Fully Convolutional Network), designed an extraction algorithm that could reduce the loss of feature information, and then used this algorithm to identify situations involving privacy in the smart home environment. However, this algorithm divides the grid of the recognition image into 14×14. Although dim objects can be extracted, the workload does not meet the requirement of real-time. Nguyen et al. [5] extracted the information features of grayscale image and used them as the input layer of YOLO model. However, the process of extracting information using the alternating direction multiplier method to form the input layer takes much more time, and the application can be greatly limited.

LiDAR can obtain three-dimensional information of the driving environment, which has unique advantages in detecting and tracking obstacle detection, measuring speed, navigating and positioning vehicle. Dynamic obstacle detection and tracking is the research hotspot in the field of LiDAR. Many scholars have conducted a lot of research on it. Azim et al. [6] proposed the ratio characteristics method to distinguish moving obstacles. However, it is only uses numerical values to judge the type of object, which might result in the high missed rate when the regional point cloud data are sparse, or the detection region is blocked. Zhou et al. [7] used a distance-based vehicle clustering algorithm to identify vehicles based on multi-feature information fusion after confirming the feature information, and used a deterministic method to perform the target correlation. However, the multi-feature information fusion is cumbersome, the rules are not clear, and the correlated methods cannot handle the appearance and disappearance of goals. Asvadi et al. [8] proposed a 3D voxel-based representation method, and used a discriminative analysis method to model obstacles. This method is relatively novel, and can be used to merge the color information from images in the future to provide more robust static/moving obstacle detection.

All of these above studies use a single sensor for target detection. The image information of color camera will be affected by the ambient light, and LiDAR cannot give full play to its advantages in foggy and hot weather. Thus, the performance and recognition accuracy of the single sensor is low in the complex urban traffic environment, which cannot meet the security needs of autonomous vehicles.

To adapt to the complexity and variability of the traffic environment, some studies use color camera and LiDAR to detect the target simultaneously on the autonomous vehicle, and then provide sufficient environmental information for the vehicle through the fusion method. Asvadi et al. [9] uses a convolutional neural network method to extract the obstacle information based on three detectors designed by combining the dense depth map and dense reflection map output from the 3D LiDAR and the color images output from the camera. Xue et al. [10] proposed a vision-centered multi-sensor fusion framework for autonomous driving in traffic environment perception and integrated sensor information of LiDAR to achieve efficient autonomous positioning and obstacle perception through geometric and semantic constraints, but the process and algorithm of multiple sensor fusion are too complex to meet the requirements of real-time. In addition, references [9,10] did not consider the existence of dimmer targets such as pedestrians and non-motor vehicle.

Based on the above analysis, this paper presents a multi-sensor (color camera and LiDAR) and multi-modality (color image and LiDAR depth image) real-time target detection system. Firstly, color image and depth image of the obstacle are obtained using color camera and LiDAR, respectively, and are input into the improved YOLO detection model frame. Then, after the convolution and pooling processing, the detection bounding box for each mode is output. Finally, the two types of detection bounding boxes are fused on the decision-level to obtain the accurate detection target.

In particular, the contributions of this article are as follows:

(1)   By incorporating the proposed secondary detection scheme into the algorithm, the YOLO target detection model is improved to detect the targets effectively. Then, decision level fusion is introduced to fuse the image information of LiDAR and color camera output from the YOLO model. Thus, it can improve the target detection accuracy.

(2)   The proposed fusion system has been built in related environments, and the optimal parameter configuration of the algorithm has been obtained through training with many samples.

2. System Method Overview

2.1. LiDAR and Color Camera

The sensors used in this paper include color camera and LiDAR, as shown in Figure 1.

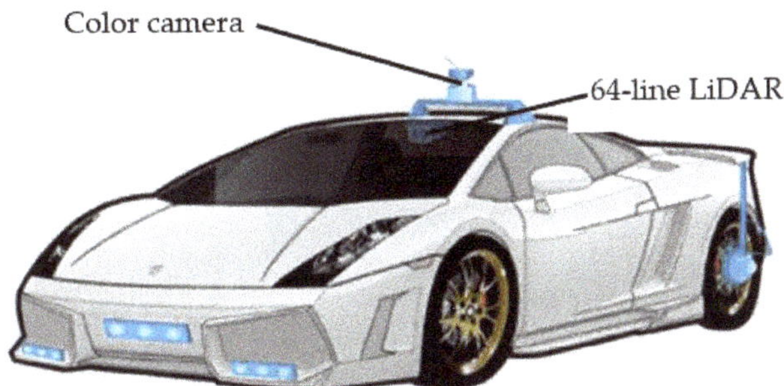

Figure 1. Installation layout of two sensors.

The LiDAR is a Velodyne (Velodyne LiDAR, San Jose, CA, USA) 64-line three-dimensional radar system which can send a detection signal (laser beam) to a target, and then compare the received signal reflected from the target (the echo of the target) with the transmitted signal. After proper processing, the relevant information of the target can be obtained. The LiDAR is installed at the top center of a vehicle and capable of detecting environmental information through high-speed rotation scanning [11]. The LiDAR can emit 64 laser beams at the head. These laser beams are divided into four groups and each group has 16 laser emitters [12]. The head rotation angle is $360°$ and the detectable distance is 120 m [13]. The 64-line LiDAR has 64 fixed laser transmitters. Through a fixed pitch angle, it can get surrounding environmental information for each Δt and output a series of three-dimensional coordinate points. Then, the 64 points $(p_1, p_2, \ldots, p_{64})$ acquired by the transmitter are marked, and the distance from each point in the scene to the LiDAR is used as the pixel value to obtain a depth image. The color camera is installed under the top LiDAR. The position of the camera is adjusted according to the axis of the transverse and longitudinal center of the camera image and the transverse and longitudinal orthogonal plane formed with the laser projector, so that the camcorder angle and the yaw angle are approximated to 0, and the pitch angle is approximately to 0. Color images can be obtained directly from color cameras, but images output from LiDAR and camera must be matched in time and space to realize the information synchronization of the two.

2.2. Image Calibration and Synchronization

To integrate information in the vehicle environment perceptual system, information calibration and synchronization need to be completed.

2.2.1. Information Calibration

(1) The installation calibration of LiDAR: The midpoints of the front bumper and windshield can be measured with a tape measure, and, according to these two midpoints, the straight line of central axis of the test vehicle can be marked by the laser thrower. Then, on the central axis, a straight line perpendicular to the central axis is marked at a distance of 10 m from the rear axle of the test vehicle; the longitudinal axis of the radar center can be measured by a ruler, and corrected by the longitudinal beam perpendicular to the ground with a laser thrower, to make the longitudinal axis and the beam coincide, and the lateral shift of the radar is approximately 0 m. The horizontal beam of the laser thrower is coincided with the transverse axis of the radar, then the lateral shift of the radar is approximately 0 m.

(2) The installation calibration of camera: The position of the camera is adjusted according to the axis of the transverse and longitudinal center of the camera image and the transverse and longitudinal orthogonal plane formed with the laser projector, so that the camcorder angle and the yaw angle are approximated to 0. Then, the plumb line is used to adjust the pitch angle of the camera to approximately 0.

2.2.2. Information Synchronization

(1) Space matching

Space matching requires the space alignment of vehicle sensors. Assuming that the Velodyne coordinate system is $O_v - X_v Y_v Z_v$ and the color camera coordinate system is $O_p - X_p Y_p Z_p$, the coordinate system is in translational relationship with respect to the Velodyne coordinate system. The fixing angle between the sensors is adjusted to unify the camera coordinates to the Velodyne coordinate system. Assuming that the vertical height of the LiDAR and color camera is Δh, the conversion relationship of a point "M" in space is as follows:

$$
\begin{bmatrix} X_V^m \\ Y_V^m \\ Z_V^m \end{bmatrix} = \begin{bmatrix} X_P^m \\ Y_P^m \\ Z_P^m \end{bmatrix} + \begin{bmatrix} 0 \\ 0 \\ \Delta h \end{bmatrix}
\tag{1}
$$

(2) Time matching

The method of matching in time is to create a data collection thread for the LiDAR and the camera, respectively. By setting the same acquisition frames rate of 30 fps, the data matching on the time is achieved.

2.3. The Process of Target Detection

The target detection process based on sensor fusion is shown in Figure 2. After collecting information from the traffic scene, the LiDAR and the color camera output the depth image and the color image, respectively, and input them into the improved YOLO algorithm (the algorithm has been trained by many images collected by LiDAR and color camera) to construct target detection Models 1 and 2. Then, the decision-level fusion is performed to obtain the final target recognition model, which realizes the multi-sensor information fusion.

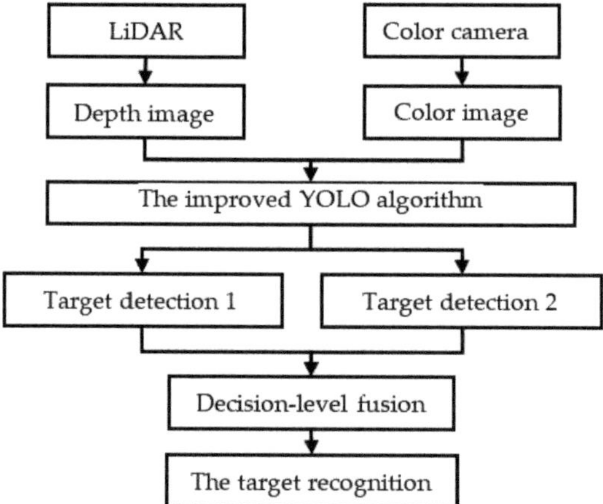

Figure 2. The flow chart of multi-modal target detection.

3. Obstacle Detection Method

3.1. The Original YOLO Algorithm

You Only Look Once (YOLO) is a single convolution neural network to predict the bounding boxes and the target categories from full images, which divides the input image into $S \times S$ cells and predicts multiple bounding boxes with their class probabilities for each cell. The architecture of YOLO is composed of input layer, convolution layer, pooling layer, fully connected layer and output layer. The convolution layer is used to extract the image features, the full connection layer is used to predict the position of image and the estimated probability values of target categories, and the pooling layer is responsible for reducing the pixels of the slice.

The YOLO network architecture is shown in Figure 3 [14].

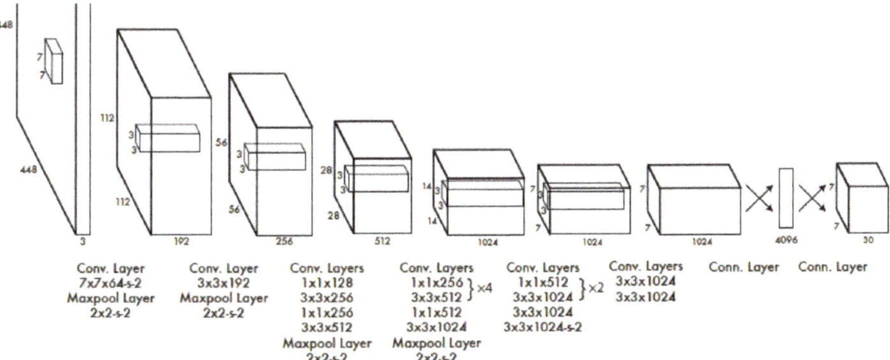

Figure 3. The YOLO network architecture. The detection network has 24 convolutional layers followed by two fully connected layers. Alternating 1×1 convolutional layers reduce the features space from preceding layers. We pre-train the convolutional layers on the ImageNet classification task at half the resolution (224×224 input images) and then double the resolution for detection.

Assume that B is the number of sliding windows used for each cell to predict objects and C is the total number of categories, then the dimensions of output layer is $S \times S \times (B \times 5 + C)$.

The output model of each detected border is as follows:

$$T = (x, y, w, h, c) \tag{2}$$

where (x, y) represents the center coordinates of the bounding box and (w, h) represents the height and width of the detection bounding box. The above four indexes have been normalized with respect to the width and height of the image. c is the confidence score, which reflects the probability value of the current window containing the accuracy of the detection object, and the formula is as follows:

$$c = P_o \times P_{\text{IOU}} \tag{3}$$

where P_o indicates the probability of including the detection object in the sliding window, P_{IOU} indicates the overlapping area ratio of the sliding window and the real detected object.

$$P_{\text{IOU}} = \frac{\text{Area}\left(BB_i \cap BB_g\right)}{\text{Area}\left(BB_i \cup BB_g\right)} \tag{4}$$

In the formula, BB_g is the detection bounding box, and BB_g is the reference standard box based on the training label.

For the regression method in the YOLO, the loss function can be calculated as follows:

$$F(\text{loss}) = \lambda coord \sum_{i=0}^{S^2} \sum_{j=0}^{B} 1_{ij}^{obj} [(x_i - \hat{x}_i)^2 + (y_i - \hat{y}_i)^2] + \lambda coord \sum_{i=0}^{S^2} \sum_{j=0}^{B} 1_{ij}^{obj} [(\sqrt{\omega_i} - \sqrt{\hat{\omega}_i})^2 + (\sqrt{h_i} - \sqrt{\hat{h}_i})^2]$$
$$+ \sum_{i=0}^{S^2} \sum_{j=0}^{B} 1_{ij}^{obj} (C_i - \hat{C}_i)^2 + \lambda noobj \sum_{i=0}^{S^2} \sum_{j=0}^{B} 1_{ij}^{noobj} (C_i - \hat{C}_i)^2 + \sum_{i=0}^{S^2} 1_i^{obj} \sum_{c \in classes} (p_i(c) - \hat{p}_i(c))^2 \tag{5}$$

1_i^{obj} denotes that the grid cell i contains part of the traffic objects. 1_{ij}^{obj} represents the j bounding box in grid cell i. Conversely, 1_i^{noobj} represents the j bounding box in grid cell i which does not contain any part of traffic objects. The time complexity of Formula (5) is $O((k + c) \times S^2)$, which is calculated for one image.

3.2. The Improved YOLO Algorithm

In the application process of the original YOLO algorithm, the following issues are found:

(1)   YOLO imposes strong spatial constraints on bounding box predictions since each grid cell only predicts two boxes and can only have one class. This spatial constraint limits the number of nearby objects that our model can predict.

(2)   The cell division of the image is set as 7×7 in the original YOLO model, which can only detect large traffic objects such as buses, cars and trucks, but does not meet the requirements of cell division of the picture for dim objects such as non-motor vehicles and pedestrians. When the target is close to the safe distance from the autonomous vehicle and the confidence score of the detection target is low, it is easy to ignore the existence of the target to cause security risk.

Based on the above deficiencies, this paper improves the original YOLO algorithm as follows: (1) To eliminate the problem of redundant time caused by the identification of undesired targets, and according to the size and driving characteristics of common targets in traffic scenes, the total number of categories is set to six types, including {bus, car, truck, non-motor vehicle, pedestrian and others}. (2) For the issue of non-motor vehicle and pedestrian detection, this paper proposes a secondary image detection scheme. Then, the cell division of the image is kept as 7×7, the sliding window convolution kernel is set as 3×3.

The whole target detection process of the improved YOLO algorithm is shown in Figure 4, and the steps are as follows:

(1)   When the target is identified, the confidence score c is higher than the maximum threshold τ_1, indicating that the recognition accuracy is high, and the frame model of target detection is directly output.

(2)   When the recognition categories are {bus, car and truck}, and the confidence score is $\tau_0 \leq c < \tau_1$ (τ_0 is the minimum threshold), indicating such targets are large in size and easy to detect, and they can be recognized at the next moment, the current border detection model can be directly output.

(3)   When the recognition categories are {non-motor vehicle and pedestrian}, the confidence score is $\tau_0 \leq c < \tau_1$. Due to the dim size and mobility of such targets, it is impossible to accurately predict the position of the next moment. At this time, this target is marked as {others}, indicating that it is required to be detected further. Then, the next steps need to be performed:

(3a)   When the distance l between the target marked as {others} and the autonomous vehicle is less than the safety distance l_0 (the distance that does not affect decision making; if the distance exceeds it, the target can be ignored), i.e., $l \leq l_0$, the slider region divided as {other} is marked, and the region is subdivided into 9×9 cells. The secondary convolution operation is performed again. When the confidence score c of the secondary detection is higher than the threshold τ_1, the border model of {others} is output, and the category is changed from {others} to {non-motor vehicle} or {pedestrian}. When the confidence score c of the secondary detection is lower than the threshold τ_1, it is determined that the target does not belong to the classification item, and the target is eliminated.

(3b) When $l > l_0$, this target is kept as {others}. It does not require a secondary convolution operation.

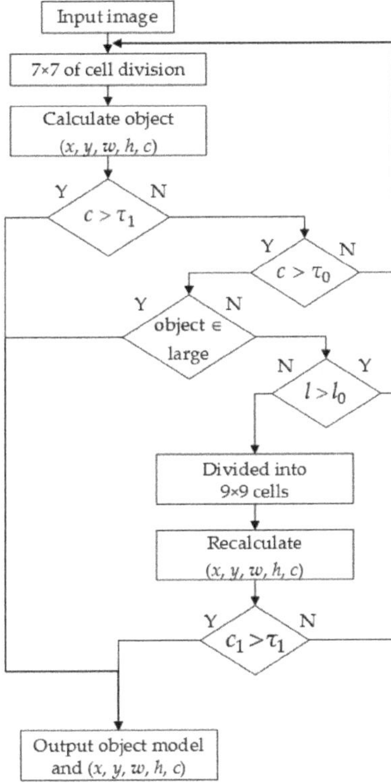

Figure 4. The flow chart of secondary image detection program. Object ∈ large means that targets are {bus, car, truck}.

The original YOLO algorithm fails to distinguish and recognize the targets according to their characteristics, and may lose some targets. The improved YOLO algorithm can try to detect the target twice in a certain distance according to the characteristic of dim of pedestrians and non-motor vehicles. Thus, it is can reduce the missing rate of the target and output a more comprehensive scene model and ensure the safe driving of vehicles.

4. Decision-Level Fusion of the Detection Information

After inputting the depth image and color image into the improved YOLO model algorithm, the detected target frame and confidence score are output, and then the final target model is output based on the fusion distance measurement matrix for decision level fusion.

4.1. Theory of Data Fusion

It is assumed that multiple sensors measure the same parameter, and the data measured by the i sensor and the j sensor are X_i and X_j, and both obey the Gaussian distribution, and their pdf (probability distribution function) curve is used as the characteristic function of the sensor and is denoted as $p_i(x)$, $p_j(x)$. x_i and x_j are the observations of X_i and X_j, respectively. To reflect the deviation between x_i and x_j, the confidence distance measure is introduced [15]:

$$d_{ij} = 2 \int_{x_i}^{x_j} p_i(x/x_i)dx \tag{6}$$

$$d_{ji} = 2 \int_{x_j}^{x_i} p_j(x/x_j)dx \tag{7}$$

Among them:

$$p_i(x/x_i) = \frac{1}{\sqrt{2\pi}\sigma_i} \exp\{-\frac{1}{2}[\frac{x - x_i}{\sigma_i}]^2\} \tag{8}$$

$$p_j(x/x_j) = \frac{1}{\sqrt{2\pi}\sigma_j} \exp\{-\frac{1}{2}[\frac{x - x_j}{\sigma_j}]^2\} \tag{9}$$

The value of d_{ij} is called the confidence distance measure of the i sensor and the j sensor observation, and its value can be directly obtained by means of the error function erf (θ), namely:

$$d_{ij} = \text{erf}[\frac{x_j - x_i}{\sqrt{2}\sigma_i}] \tag{10}$$

$$d_{ji} = \text{erf}[\frac{x_i - x_j}{\sqrt{2}\sigma_j}] \tag{11}$$

If there are n sensors measuring the same indicator parameter, the confidence distance measure d_{ij} $(i, j = 1, 2, ..., n)$ constitutes the confidence distance matrix D_n of the multi-sensor data:

$$D_n = \begin{bmatrix} d_{11} & d_{12} & \cdots & d_{1n} \\ d_{21} & d_{22} & \cdots & d_{2n} \\ \vdots & \vdots & & \vdots \\ d_{n1} & d_{n2} & \cdots & d_{nn} \end{bmatrix} \tag{12}$$

The general fusion method is to use experience to give an upper bound β_{ij} of fusion, and then the degree of fusion between sensors is:

$$r_{ij} = \begin{cases} 1, & d_{ij} \leq \beta_{ij} \\ 0, & d_{ij} > \beta_{ij} \end{cases} \tag{13}$$

In this paper, there are two sensors, i.e., LiDAR and color camera, so $i, j = 1, 2$. Then, taking $\beta_{ij} = 0.5$ [16], r_{12} is set as the degree of fusion between the two sensors. Figure 5 explains the fusion process.

(1) When $r_{12} = 0$, it means that the two sets of border models (green and blue areas) do not completely overlap. At this time, the overlapping area is taken as the final detection model (red area). The fusion process is shown in Figure 5a,b.
(2) When $r_{12} = 1$, it indicates that the two border models (green and blue areas) basically coincide with each other. At this time, all border model areas are valid and expanded to the standard border model (red area). The fusion process is shown in Figure 5b,c.

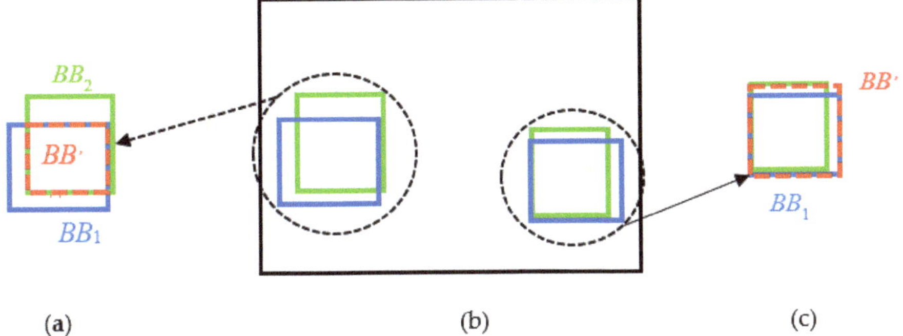

Figure 5. Decision-level fusion diagram of detection model. Blue area (BB_1) is the model output from the depth image. Green area (BB_2) is the model output from the color image. Red area (BB') is the final detection model. When $r_{12} = 0$, the fusion process is shown in (**a**). The models not to be fused are shown in (**b**). When $r_{12} = 1$, the fusion process is shown in (**c**).

Simple average rules between scores are applied in confidence scores. The formula is as follows:

$$c = \frac{c_1 + c_2}{2} \tag{14}$$

where c_1 is the confidence score of target Model 1, and c_2 is the confidence score of target Model 2. In addition, it should be noted that, when there is only one bounding box, to reduce the missed detection rate, this bounding box information is retained as the final output result. The final target detection model can be output through decision-level fusion and confidence scores.

4.2. The Case of the Target Fusion Process

An example of the target fusion process is shown in Figure 6, and the confidence scores obtained using different sensors can be seen in Table 1.

(1)　Figure 6A is a processed depth image. It can be seen that the improved YOLO algorithm identifies two targets, a and b, and gives the confidence scores of 0.78 and 0.55, respectively.
(2)　Figure 6B is a color image. It can be seen that three targets, a, b, and c, are identified and the confidence scores are given as 0.86, 0.53 and 0.51, respectively.
(3)　The red box in Figure 6C is the final target model after fusion:

    (1)　For target a, according to the decision-level fusion scheme, the result $r_{12} \leq 0$ is obtained; then, the overlapping area is taken as the final detection model, and the confidence score after fusion is 0.82, as shown in Figure 6C (a′).
    (2)　For target b, according to the decision-level fusion scheme, the result $r_{12} \geq 0$ is obtained; then, the union of all regions is taken as the final detection model, and the confidence score after fusion is 0.54, as shown in Figure 6C (b′).
    (3)　For target c, since there is no such information in Figure 6A, and Figure 6B identifies the pedestrian information on the right, according to the fusion rule, the bounding box information of c in Figure 6B is retained as the final output result, and the confidence score is kept as 0.51, as shown in Figure 6C (c′).

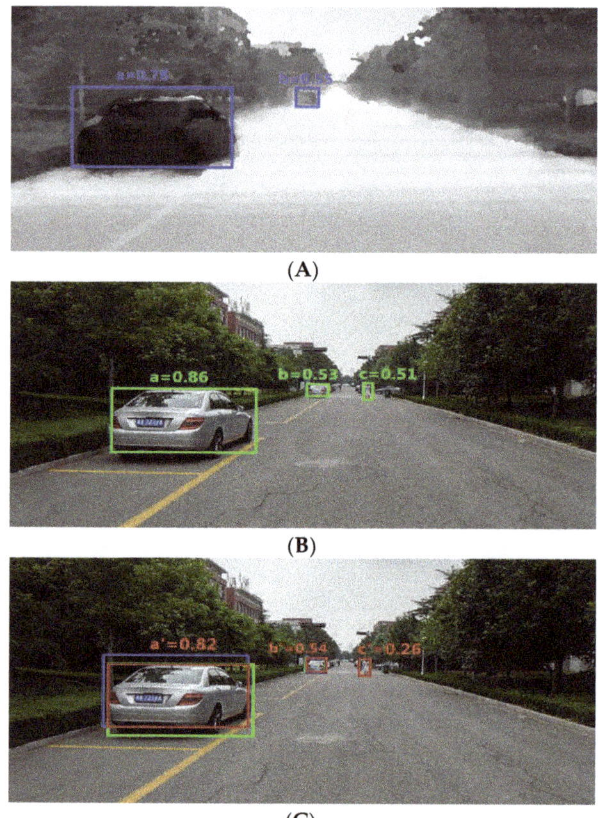

(A)

(B)

(C)

Figure 6. An example of target detection fusion process. (**A**) is a processed depth image. The models detected a and b are shown with blue. (**B**) is a color image. The models detected a, b and c are shown with green. (**C**) is the final target model after fusion. The models fused a', b' and c' are shown with red.

Table 1. Confidence scores obtained using different sensors.

Sensor	Confidence Score (Detected Object from Left to Right)		
	a (a')	b (b')	c (c')
LiDAR	0.78	0.55	–
Color camera	0.86	0.53	0.51
The fusion of both	0.82	0.54	0.26

5. Results and Discussion

5.1. Conditional Configuration

The target detection training dataset included 3000-frame resolution images of 1500 × 630 and was divided into six different categories: bus, car, truck, non-motor vehicle, pedestrian and others. The dataset was partitioned into three subsets: 60% as training set (1800 observations), 20% as validation set (600 observations), and 20% as testing set (600 observations).

The autonomous vehicles collected data on and off campus. The shooting equipment included a color camera and a Velodyne 64-line LiDAR. The camera was synchronized with a 10 Hz spining LiDAR. The Velodyne has 64-layer vertical resolution, 0.09 angular resolutions, 2 cm of distance accuracy, and captures 100 k points per cycle [9]. The processing platform was completed in the PC segment, including the i5 processor (Intel Corporation, Santa Clara, CA, USA) and GPU (NVIDIA, Santa Clara, CA, USA). The improved YOLO algorithm was accomplished by building a Darket framework and using Python (Python 3.6.0, JetBrains, Prague, The Czech Republic) for programming.

5.2. Time Performance Testing

The whole process included the extraction of depth image and color image, and they were, respectively, substituted into the improved YOLO algorithm and the proposed decision-level fusion scheme as the input layer. The improved YOLO algorithm involved the image grid's secondary detection process and is therefore slightly slower than the normal recognition process. The amount of computation to implement the different steps of the environment and algorithm is shown in Figure 7. In the figure, it can be seen that the average time to process each frame is 81 ms (about 13 fps). Considering that the operating frequency of the camera and Velodyne LiDAR is about 10 Hz, it can meet the real-time requirements of traffic scenes.

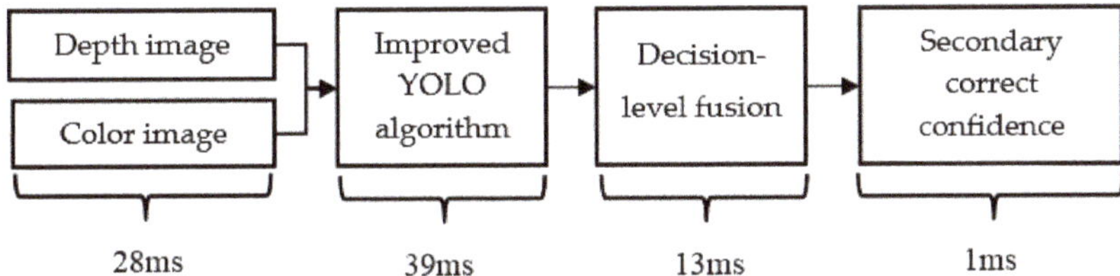

Figure 7. Processing time for each step of the inspection system (in ms).

5.3. Training Model Parameters Analysis

The training of the model takes more time, so the setting of related parameters in the model has a great impact on performance and accuracy. Because the YOLO model involved in this article has been modified from the initial model, the relevant parameters in the original model need to be reconfigured through training tests.

The training step will affect the training time and the setting of other parameters. For this purpose, eight steps of training scale were designed. Under the learning rate of 0.001 given by YOLO, the confidence prediction score, actual score, and recognition time of the model are statistically analyzed. Table 1 shows the performance of the BB_2 model, and Figure 7 shows the example results of the BB_2 model under D1 (green solid line), D3 (violet solid line), D7 (yellow solid line) and D8 (red solid line).

Table 2 shows that, with the increase of training steps, the confidence score for the BB_2 model is constantly increasing, and the actual confidence level is also in a rising trend. When the training step reaches 10,000, the actual confidence score arrives at the highest value of 0.947. However, when the training step reaches 20,000, the actual confidence score begins to fall, and the recognition time also slightly increases, which is related to the configuration of model and the selection of learning rate.

Table 2. Performance of BB_2 model under different steps.

Mark	Number of Steps	Estimated Confidence	Actual Confidence	Recognition Time (ms)
D1	4000	0.718	0.739	38.42
D2	5000	0.740	0.771	38.40
D3	6000	0.781	0.800	38.33
D4	7000	0.825	0.842	38.27
D5	8000	0.862	0.885	38.20
D6	9000	0.899	0.923	38.12
D7	10,000	0.923	0.947	38.37
D8	20,000	0.940	0.885	38.50

Figure 8 shows the vehicle identification with the training steps of 4000, 6000, 10,000, and 20,000. The yellow dotted box indicates the recognition rate when the learning rate is 10,000. Clearly, the model box basically covers the entire goal and almost no redundant area. Based on the above analysis, the number of steps set in this paper is 10,000.

Figure 8. Performance comparison of BB_2 model under 4 kinds of training steps.

The learning rate determines the speed at which the parameters are moved to the optimal value. To find the optimal learning rate, the model performances with the learning rate of 10^{-7}, 10^{-6}, 10^{-5}, 10^{-4}, 10^{-3}, 10^{-2}, 10^{-1} and 1 are estimated, respectively, when the training step is set to 10,000.

Table 3 shows the estimated confidence scores and final scores of the output detection models BB_1 and BB_2 under different learning rates. Figure 9 shows the change trend of the confidence score. After analyzing Table 3 and Figure 9, we can see that, with the decrease of learning rate, all of the confidence prediction score and actual score of model experienced a rising trend firstly and then decreasing. When the learning rate reaches D3 (10^{-2}), the confidence score reaches a maximum value, and the confidence level remains within a stable range with the change of learning rate. Based on the above analysis, when the learning rate is 10^{-2}, the proposed model can obtain a more accurate recognition rate.

Table 3. Model performance under different learning rates.

Mark	Learning Rate	Estimated Confidence		Actual Confidence	
		BB_1	BB_2	BB_1	BB_2
D1	1	0.772	0.73	0.827	0.853
D2	10^{-1}	0.881	0.864	0.911	0.938
D3	10^{-2}	0.894	0.912	0.932	0.959
D4	10^{-3}	0.846	0.85	0.894	0.928
D5	10^{-4}	0.76	0.773	0.889	0.911
D6	10^{-5}	0.665	0.68	0.874	0.892
D7	10^{-6}	0.619	0.62	0.833	0.851
D8	10^{-7}	0.548	0.557	0.802	0.822

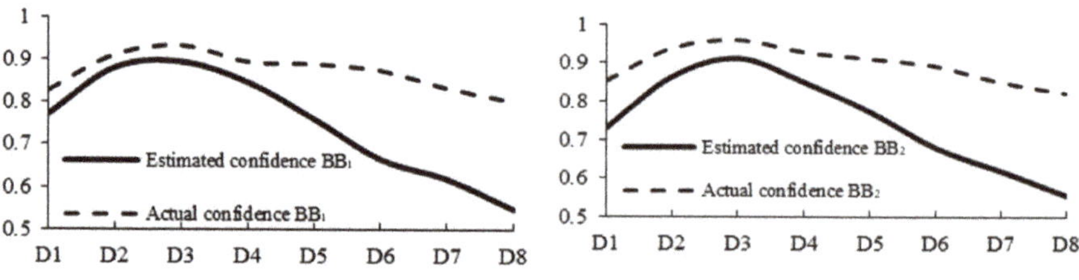

Figure 9. Performance trends under different learning rates.

5.4. Evaluation of Experiment Results

The paper takes the IOU as the evaluation criteria of recognition accuracy obtained by comparing the BB_i ($i = 1, 2$) of output model and the BB_g of actual target model, and defines three evaluation grades:

(1) Low precision: Vehicle targets can be identified within the overlap area, and the identified effective area accounts for 60% of the model total area.
(2) Medium precision: Vehicle targets are more accurately identified in overlapping areas, and the identified effective area accounts for 80% of the model's total area.
(3) High precision: The vehicle is accurately identified in the overlapping area, and the identified effective area accounts for 90% of the model total area. Figure 10 is used to describe the definition of evaluation grade. The red dotted frame area is the target actual area and the black frame area is the area BB_i output from the model.

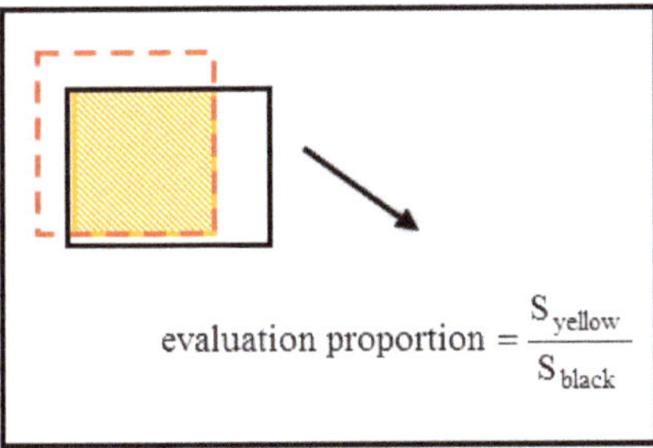

Figure 10. The definition of evaluation grade. The yellow area is the identified effective area. The black frame area is model's total area. The above proportion is the ratio between yellow area and black area.

To avoid the influence caused by the imbalance of all kinds of samples, the precision and recall were introduced to evaluate the box model under the above three levels:

$$\text{Precision} = \frac{TP}{TP + FP} \tag{15}$$

$$\text{Recall} = \frac{TP}{TP + FN} \tag{16}$$

In the formula, TP, FP, and FN indicate the correctly defined examples, wrongly defined examples and wrongly negative examples, respectively. The Precision–Recall diagram for each model BB_i ($i = 1, 2$) is calculated, as shown in Figure 11a,b.

When the recall is less than 0.4, all the accuracy under the three levels is high; when the recall reaches around 0.6, only the accuracy of the level hard decreases sharply and tends to zero, while the accuracy of the other two levels is basically maintained at a relatively high level. Therefore, when the requirements of level for target detection is not very high, the method proposed in this paper can fully satisfy the needs of vehicle detection under real road conditions.

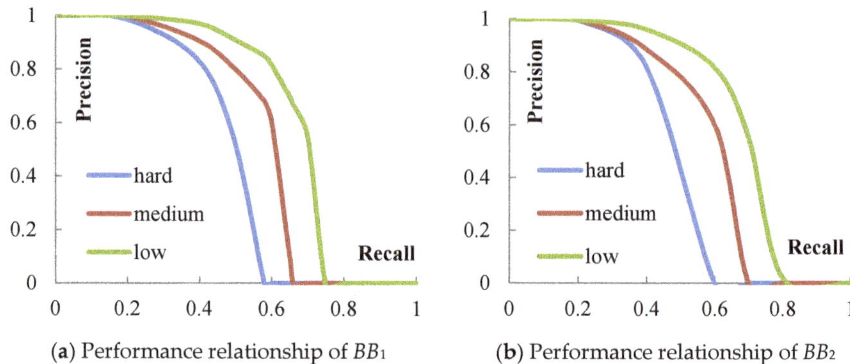

(**a**) Performance relationship of BB_1 (**b**) Performance relationship of BB_2

Figure 11. Detection performance of the target. (**A**) is the performance relationship of model BB_1. (**B**) is the performance relationship of model BB_2.

5.5. Method Comparison

The method proposed in this paper is compared with the current more advanced algorithms. The indicators are mainly mAP (mean average precision) and FPS (frames per second). The results obtained are shown in Table 4.

Table 4. Comparison of the training results of all algorithms.

Algorithms	mAP	FPS
YOLO [17]	63.4	45
Fast R-CNN [18]	70.0	0.5
Faster R-CNN [19]	73.2	7
Projection [20]	96.2	8
3D FCN [21]	64.2	0.2
Vote3D [22]	47.9	2
the improved YOLO algorithm	82.9	13

In Table 4, the recognition accuracy of the improved algorithm proposed in this paper is better than that of the original YOLO algorithm. This is related to the fusion decision of the two images and the proposed secondary image detection scheme. To ensure the accuracy, the detection frame number of the improved YOLO dropped from 45 to 13, and the running time increased, but it can fully meet the normal target detection requirements and ensure the normal driving of autonomous vehicles.

6. Conclusions

This paper presents a detection fusion system with integrating LiDAR and color camera. Based on the original YOLO algorithm, the second detection scheme is proposed to improve the YOLO algorithm for dim targets such as non-motorized vehicles and pedestrians. Then, the decision level fusion of sensors is introduced to fuse the color image of color camera and the depth image of LiDAR to improve the accuracy of the target detection. The final experimental results show that, when the training step is set to 10,000 and the learning rate is 0.01, the performance of the model proposed in this paper is optimal and the Precision–Recall performance relationship could satisfy the target detection in most cases. In addition, in the aspect of algorithm comparison, under the requirement of both accuracy and real-time, the method of this paper has better performance and a relatively large research prospect.

Since the samples needed in this paper are collected from several traffic scenes, the coverage of the traffic scenes is relatively narrow. In the future research work, we will gradually expand the complexity of the scenario and make further improvements to the YOLO algorithm. In the next experimental session, the influence of environmental factors will be considered, because the image-based identification method is greatly affected by light. At different distances (0–20 m, 20–50 m, 50–100 m, and >100 m), the intensity level of light is different, so how to deal with the problem of light intensity and image resolution is the primary basis for target detection.

Author Contributions: Conceptualization, J.Z.; Data curation, J.H. and Y.L.; Formal analysis, J.H.; Funding acquisition, S.W.; Investigation, J.H. and J.Z.; Methodology, Y.L.; Project administration, Y.L.; Resources, J.H.; Software, J.Z.; Supervision, J.Z. and S.W.; Validation, S.W.; Visualization, Y.L.; Writing—original draft, J.H.; and Writing—review and editing, J.Z., S.W. and S.L.

Funding: This work was financially supported by the National Natural Science Foundation of China under Grant No. 71801144, and the Science & Technology Innovation Fund for Graduate Students of Shandong University of Science and Technology under Grant No. SDKDYC180373.

Conflicts of Interest: The authors declare no conflict of interest.

References

1. Batch Re-normalization of Real-Time Object Detection Algorithm YOLO. Available online: http://www.arocmag.com/article/02-2018-11-055.html (accessed on 10 November 2017).

2. Liu, Y.; Zhang, Y.; Zhang, X. Adaptive spatial pooling for image classification. *Pattern Recognit.* **2016**, *55*, 58–67. [CrossRef]

3. Gao, Z.; Li, S.B.; Chen, J.N.; Li, Z.J. Pedestrian detection method based on YOLO network. *Comput. Eng.* **2018**, *44*, 215–219, 226.

4. Improved YOLO Feature Extraction Algorithm and Its Application to Privacy Situation Detection of Social Robots. Available online: http://kns.cnki.net/kcms/detail/11.2109.TP.20171212.0908.023.html (accessed on 12 December 2017).

5. Nguyen, V.T.; Nguyen, T.B.; Chung, S.T. ConvNets and AGMM Based Real-time Human Detection under Fisheye Camera for Embedded Surveillance. In Proceedings of the 2016 International Conference on Information and Communication Technology Convergence (ICTC), Jeju, South Korea, 19–21 October 2016; pp. 840–845.

6. Azim, A.; Aycard, O. Detection, Classification and Tracking of Moving Objects in a 3D Environment. In Proceedings of the IEEE Intelligent Vehicles Symposium, Alcala de Henares, Spain, 3–7 June 2012; pp. 802–807.

7. Zhou, J.J.; Duan, J.M.; Yang, G.Z. A vehicle identification and tracking method based on radar ranging. *Automot. Eng.* **2014**, *36*, 1415–1420, 1414.

8. Asvadi, A.; Premebida, C.; Peixoto, P.; Nunes, U. 3D Lidar-based static and moving obstacle detection in driving environments: An approach based on voxels and multi-region ground planes. *Robot. Auton. Syst.* **2016**, *83*, 299–311. [CrossRef]

9. Asvadi, A.; Garrote, L.; Premebida, C.; Peixoto, P.; Nunes, U.J. Multimodal Vehicle Detection: Fusing 3D-LIDAR and Color Camera Data. *Pattern Recognit. Lett.* **2017**. [CrossRef]

10. Xue, J.R.; Wang, D.; Du, S.Y. A vision-centered multi-sensor fusing approach to self-localization and obstacle perception for robotic cars. *Front. Inf. Technol. Electron. Eng.* **2017**, *18*, 122–138. [CrossRef]

11. Wang, X.Z.; Li, J.; Li, H.J.; Shang, B.X. Obstacle detection based on 3d laser scanner and range image for intelligent vehicle. *J. Jilin Univ. (Eng. Technol. Ed.)* **2016**, *46*, 360–365.

12. Glennie, C.; Lichti, D.D. Static calibration and analysis of the Velodyne HDL-64E S2 for high accuracy mobile scanning. *Remote Sens.* **2010**, *2*, 1610–1624. [CrossRef]

13. Yang, F.; Zhu, Z.; Gong, X.J.; Liu, J.L. Real-time dynamic obstacle detection and tracking using 3D lidar. *J. Zhejiang Univ.* **2012**, *46*, 1565–1571.

14. Zhang, J.M.; Huang, M.T.; Jin, X.K.; Li, X.D. A real-time Chinese traffic sign detection algorithm based on modified YOLOv2. *Algorithms* **2017**, *10*, 127. [CrossRef]

15. Han, F.; Yang, W.H.; Yuan, X.G. Multi-sensor Data Fusion Based on Correlation Function and Fuzzy Clingy Degree. *J. Proj. Rocket. Missiles Guid.* **2009**, *29*, 227–229, 234.

16. Chen, F.Z. Multi-sensor data fusion mathematics. *Math. Pract. Theory* **1995**, *25*, 11–15.

17. Redmon, J.; Divvala, S.; Girshick, R.; Farhadi, A. You Only Look Once: Unified, Real-Time Object Detection. In Proceedings of the IEEE Conference on Computer Vision and Pattern Recognition, Las Vegas, NV, USA, 27–30 June 2016; pp. 779–788.

18. Girshick, R. Fast R-CNN. In Proceedings of the 2015 IEEE International Conference on Computer Vision (ICCV), Santiago, Chile, 7–13 December 2015; pp. 1440–1448.
19. Ren, S.; He, K.; Girshick, R. Faster R-CNN: Towards real-time object detection with region proposal networks. *Int. Conf. Neural Inf. Process. Syst.* **2015**, *39*, 91–99. [CrossRef] [PubMed]
20. Wei, P.; Cagle, L.; Reza, T.; Ball, J.; Gafford, J. LiDAR and camera detection fusion in a real-time industrial multi-sensor collision avoidance system. *Electronics* **2018**, *7*, 84. [CrossRef]
21. Li, B. 3D fully convolutional network for vehicle detection in point cloud. *arXiv*, **2017**; arXiv:1611.08069.
22. Voting for Voting in Online Point Cloud Object Detection. Available online: Https://www.researchgate. net/publication/314582192_Voting_for_Voting_in_Online_Point_Cloud_Object_Detection (accessed on 13 July 2015).

Article

Energy-Based Control and LMI-Based Control for a Quadrotor Transporting a Payload

María-Eusebia Guerrero-Sánchez [1,†], Omar Hernández-González [1,†], Rogelio Lozano [2], Carlos-D. García-Beltrán [3], Guillermo Valencia-Palomo [4,*] and Francisco-R. López-Estrada [5]

1 Tecnológico Nacional de México/Instituto Tecnológico Superior de Coatzacoalcos, Carretera Antigua Mina-Coatza, km. 16.5, Col. Reserva Territorial, Coatzacoalcos 96536, Veracruz, Mexico; maguerreros@itesco.edu.mx (M.-E.G.-S.); ohernandezg@itesco.edu.mx (O.H.-G.)
2 Sorbonne Universités, UTC CNRS UMR 7253 Heudiasyc, 60203 Compiègne, France; rlozano@utc.fr
3 Tecnológico Nacional de México/Centro Nacional de Investigación y Desarrollo Tecnológico, Interior Internado Palmira S/N, Col. Palmira, Cuernavaca 62490, Morelos, Mexico; cgarcia@cenidet.edu.mx
4 Tecnológico Nacional de México/Instituto Tecnológico de Hermosillo, Ave. Tecnológico y Periférico Poniente S/N, Hermosillo Sonora 83170, Mexico
5 Tecnológico Nacional de México/Instituto Tecnológico de Tuxtla Gutiérrez, TURIX-Dynamics Diagnosis and Control Group, Carr. Panam. km 1080, A.P. 599, Tuxtla Gutierrez 29050, Mexico; frlopez@ittg.edu.mx
* Correspondence: gvalencia@ith.mx
† These authors contributed equally to this work.

Received: 8 October 2019; Accepted: 6 November 2019; Published: 11 November 2019

Abstract: This paper presents the control of a quadrotor with a cable-suspended payload. The proposed control structure is a hierarchical scheme consisting of an energy-based control (EBC) to stabilize the vehicle translational dynamics and to attenuate the payload oscillation, together with a nonlinear state feedback controller based on an linear matrix inequality (LMI) to control the quadrotor rotational dynamics. The payload swing control is based on an energy approach and the passivity properties of the system's translational dynamics. The main advantage of the proposed EBC strategy is that it does not require excessive computations and complex partial differential equations (PDEs) for implementing the control algorithm. We present a new methodology for using an LMI to synthesize the controller gains for Lipschitz nonlinear systems with larger Lipschitz constants than other classical techniques based on LMIs. This theoretical approach is applied to the quadrotor rotational dynamics. Stability proofs based on the Lyapunov theory for the controller design are presented. The designed control scheme allows for the stabilization of the system in all its states for the three-dimensional case. Numerical simulations demonstrating the effectiveness of the controller are provided.

Keywords: energy-based control; payload swing attenuation; linear matrix inequalities; quadrotor; larger Lipschitz constants

---

1. Introduction

In recent years, systems for the transportation of suspended-payload using unmanned aerial vehicles (UAVs) have attracted research interest. Some important applications are described in [1]. Quadrotor vehicles exhibit complex dynamic behavior, and if a cable-suspended payload is added to a quadrotor, it increases the complexity of the system because additional degrees of freedom (DOFs) due to the payload oscillation are introduced. Moreover, if uncontrolled, a cable-suspended payload changes the dynamics of the flying vehicle and it can result in an unstable system.

Recently, some control methodologies have been developed to attenuate the payload swing and to solve the general problem of a quadrotor carrying a cable-suspended payload. For example, a control algorithm based on a backstepping strategy is obtained in [2], that ensures trajectory tracking of the quadrotor regardless of the payload swing. However, attitude control of the UAV is not considered. Additionally, a control design for a two-dimensional quadrotor with a cable-suspended payload that enables tracking of the vehicle rotation, the payload rotation, or the payload position is presented in [3], and it was extended to the three-dimensional case in [4].

In [5], the authors develop a nested saturation controller capable of driving the vehicle to a specified position while simultaneously limiting the payload dynamic effect. For this work, Nicotra et al. considered the design for the two-dimensional case only and the attitude control of the vehicle was not considered. Moreover, a feed-forward control algorithm for reducing or canceling the payload's oscillation is introduced in [6]. This controller was designed by implementing the input shaping theory. In [7], the authors present a geometric controller to exponentially stabilize the aerial robot position while aligning the links vertically below the vehicle. Similarly, a tracking control law for a UAV with a load attached by a cable represented as successively-attached inflexible links was designed in [8]. A fixed-gain geometric PD control strategy is developed to reach the desired goal for a nominal load. An adaptive control law for an aerial robot carrying a payload attached using a cable was presented in [9]. In [10], an algorithm for parameterizing aerial vehicles transporting a payload employing a complementary constraint is presented. A nonlinear attitude controller is developed in [11] to stabilize the altitude, and the translational dynamics control law is introduced by converting the vehicle velocity and position error into rotation control. Rego and Raffo [12] address trajectory tracking for a two-dimensional aerial robot transporting a payload. A discrete-time mixed $H2/H\infty$ linear control strategy is presented. In [13], an active-model-based linear controller is designed for a UAV transporting a payload. A linear model is obtained considering the vehicle in hover flight mode. In [14], a path tracking controller is developed based on existing Lyapunov-based path tracking control laws for free-flying aerial vehicles, which are further backstepped through the vehicle rotation dynamics.

A passivity-based control technique is used in [15] to control the UAV such that cable-suspended payload swing reduction is achieved for the planar case. Here, the attitude control of the vehicle is not considered. Also, interconnection and damping assignment passivity-based control (IDA-PBC) without total energy-shaping for a UAV transporting a cable-suspended payload for planar maneuvers is developed in [16,17]. Two control laws with total energy-shaping are presented in [18], where the closed-loop inertia matrices are modified. These works compute PDEs for synthesizing the control law. For this reason, the control strategy is only designed within the longitudinal plane. The control design for the three-dimensional case yields complex partial differential equations (PDEs).

In the literature, unmanned aerial vehicles have been controlled using energy-based controllers. In [19], the design of two nonlinear controllers to stabilize an aerial vehicle characterized with quaternions and their axis-angle depiction is studied. Also, [20] introduces a PBC for a vertical take-off and landing (VToL) vehicle. An estimator of unmodelled dynamics and external wrench acting on the UAV and based on the momentum of the system is used to compensate such disturbance effects. Moreover, [21] develops an IDA-PBC methodology that is able to change the apparent vehicle dynamical parameters, while [22] proposes a robust control of underactuated aerial manipulators via IDA-PBC.

On the other hand, some works apply control strategies based on linear matrix inequalities to UAVs. In [23], a method for using LMIs to synthesize controller gains for a UAV system is presented. In [24], a nonlinear state feedback controller based on LMIs, and a technique with pole placement constraint (PDC) is synthesized. The requirements of stability and pole placement in the linear matrix inequality (LMI) region are formulated based on the Lyapunov direct method.

In this work, the control approach is based on a hierarchical scheme considering the well-known time-scale separation between rotational and translational dynamics of the quadrotor. On one hand, the objective of this paper was to design an energy-based control law for the outer-loop (i.e., for the underactuated dynamics of the system). This control law is proposed for the three-dimensional case, and it is based on the translation dynamics, which is able to lead the vehicle to a desired position while simultaneously reducing the payload swing. Compared with similar works that present control laws based on passivity and energy, particularly for underactuated systems, the proposed controller avoids solving complex PDEs to obtain the control law. On the other hand, a feedback controller based on an LMI for the inner loop which is fully actuated is presented. The controller based on an LMI for the rotational dynamics results in a control algorithm with relative simplicity and with an easy analysis to demonstrate its stability.

The contribution of this paper is the synthesis of a new controller for a class of Lipschitz nonlinear systems. An important limitation of the classic results for Lipschitz nonlinear systems is that they perform well only for small values of the Lipschitz constant. In the case when the Lipschitz constant is large, most of the existing controller design approaches fail to contribute a solution to the LMI. This article introduces an algorithm that operates for larger Lipschitz constants compared with classical results in the literature.

This paper is organized as follows. Section 2 describes the dynamical model for a three-dimensional aerial vehicle carrying a payload. Section 3 presents an approach for LMI-based Lipschitz nonlinear systems. This theoretical approach is applied to stabilize the quadrotor rotational dynamics. Section 4 proposes an energy-based control to stabilize the vehicle translational dynamics and to attenuate the payload swing. Section 5 presents numerical simulations and results. Finally, Section 6 gives conclusions and perspectives.

2. Dynamic Model

In this section, we present the mathematical model of a quadrotor transporting a payload connected by a cable. The aim is to present a dynamic model that mathematically describes the relationship between the quadrotor, the cable, and the payload. For this purpose, consider a rigid body with mass m, being transported by a quadrotor as shown in Figure 1. Note that in addition to the six DOFs of the UAV, the payload adds another two DOFs, resulting in a system with eight DOFs and four inputs.

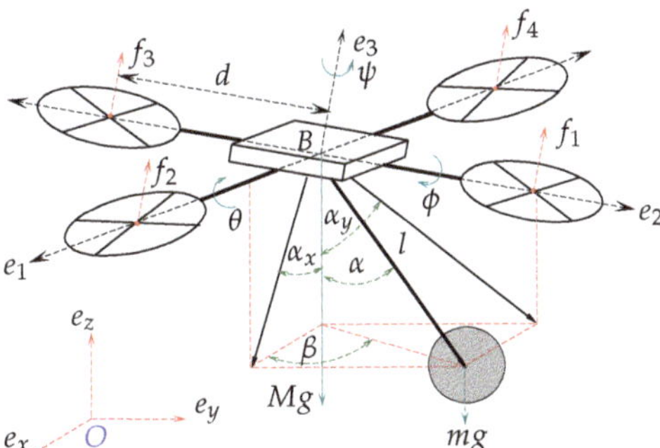

Figure 1. Three-dimensional quadrotor with a cable-suspended payload.

The following assumptions were made for modeling the quadrotor with a cable-suspended payload:

(a) The cable is attached to the center of mass of the quadrotor and the air drag is negligible.
(b) The cable connecting the payload and the quadrotor fuselage is considered rigid, massless, and inelastic.
(c) The payload can be considered as a mass point.
(d) Mass distribution of the quadrotor is symmetrical in the x-y plane.

As shown in Figure 1, the body-fixed frame is defined by $B = \{e_1, e_2, e_3\}$ and the inertial frame by $O = \{e_x, e_y, e_z\}$. The location of the mass center of the vehicle relative to O is represented by

$\zeta = \begin{bmatrix} x & y & z \end{bmatrix}^T$, the attitude of the quadrotor is denoted by $\eta = \begin{bmatrix} \psi & \theta & \phi \end{bmatrix}^T$ (i.e., yaw, pitch, and roll, respectively). α defines the payload swing angle in space and has two components α_x and α_y, $\mu = \begin{bmatrix} \alpha_x & \alpha_y \end{bmatrix}^T$. α_x is the swing angle projected on the XZ plane and α_y is the swing angle projected on the YZ plane, β represents the angle of the X axis and the projected line of the cable to the Y plane. Thus, the state vector is denoted by $q = \begin{bmatrix} x & y & z & \psi & \theta & \phi & \alpha_x & \alpha_y \end{bmatrix}^T \in \mathbb{R}^8$. The control input is represented by $u = \begin{bmatrix} f & \tau \end{bmatrix}^T \in \mathbb{R}^4$, where $f = f_1 + f_2 + f_3 + f_4$ defines the total thrust magnitude and $\tau = \begin{bmatrix} \tau_\psi & \tau_\theta & \tau_\phi \end{bmatrix}^T$ denotes the input torques.

The mass of the quadrotor and the payload are defined by M and m, respectively. The length of the cable is represented by l, the gravitational acceleration by g. Finally, the distance between the motors and the gravity center is equal to d.

In this work, the system is modeled using the Euler–Lagrange formulation.

2.1. Euler–Lagrange Formulation

Let $\zeta_p = \begin{bmatrix} x_p & y_p & z_p \end{bmatrix}^T \in \mathbb{R}^3$ be the payload position in the inertial frame. Thus, the quadrotor and payload positions are related by

$$\zeta_p = \zeta + lp,$$

where

$$p = \begin{bmatrix} \sin(\alpha_x)\cos(\alpha_y) & \sin(\alpha_y) & -\cos(\alpha_x)\cos(\alpha_y) \end{bmatrix}^T.$$

The equations of motion are obtained using the Euler–Lagrange method. The kinetic energy of the payload-quadrotor system is given by

$$K_{Q-P} = \frac{1}{2}M\dot{\zeta}^T\dot{\zeta} + \frac{1}{2}m\dot{\zeta}_p^T\dot{\zeta}_p + \frac{1}{2}\dot{\eta}^T J\dot{\eta}, \tag{1}$$

where $J = diag[I_{xx}, I_{yy}, I_{zz}]_{3\times3}$ is a symmetric positive definite constant inertia matrix of the quadrotor with respect to B.

The total potential energy is defined by

$$V_{Q-P} = (M + m)gz - mgl\cos(\alpha_x)\cos(\alpha_y). \tag{2}$$

From (1) and (2), the Lagrangian is given by

$$L = \frac{1}{2}M\dot{\zeta}^T\dot{\zeta} + \frac{1}{2}m\dot{\zeta}_p^T\dot{\zeta}_p + \frac{1}{2}\dot{\eta}^T J\dot{\eta} - (M + m)gz + mgl\cos(\alpha_x)\cos(\alpha_y). \tag{3}$$

Using the Lagrangian and the derived formula for the equations of motion:

$$\mathcal{M}(q)\ddot{q} + C(q,\dot{q})\dot{q} + G(q) = \bar{B}u, \tag{4}$$

where $\mathcal{M}(q) \in \mathbb{R}^{8\times8}$ denotes the inertia matrix, which is symmetric and positive definite, $C(q,\dot{q}) \in \mathbb{R}^{8\times8}$ the Coriolis and centrifugal matrix, $G(q) \in \mathbb{R}^8$ the gravitational vector, and the matrix $\bar{B} \in \mathbb{R}^{8\times4}$

is determined by the manner in which the control $u \in \mathbb{R}^4$ is the input of the system. These matrices are defined as

$$\mathcal{M}(q) = \begin{bmatrix} m+M & 0 & 0 & 0 & 0 & 0 & mlc_{\alpha_x}c_{\alpha_y} & -mls_{\alpha_x}s_{\alpha_y} \\ 0 & m+M & 0 & 0 & 0 & 0 & 0 & mlc_{\alpha_y} \\ 0 & 0 & m+M & 0 & 0 & 0 & mls_{\alpha_x}c_{\alpha_y} & mlc_{\alpha_x}s_{\alpha_y} \\ 0 & 0 & 0 & I_{xx} & 0 & 0 & 0 & 0 \\ 0 & 0 & 0 & 0 & I_{yy} & 0 & 0 & 0 \\ 0 & 0 & 0 & 0 & 0 & I_{zz} & 0 & 0 \\ mlc_{\alpha_x}c_{\alpha_y} & 0 & mls_{\alpha_x}c_{\alpha_y} & 0 & 0 & 0 & ml^2c_{\alpha_y}^2 & 0 \\ -mls_{\alpha_x}s_{\alpha_y} & mlc_{\alpha_y} & mlc_{\alpha_x}s_{\alpha_y} & 0 & 0 & 0 & 0 & ml^2 \end{bmatrix}$$

$$C(q) = \begin{bmatrix} 0 & 0 & 0 & 0 & 0 & 0 \\ 0 & 0 & 0 & 0 & 0 & 0 \\ 0 & 0 & 0 & 0 & 0 & 0 \\ 0 & 0 & 0 & 0 & 0 & (I_{yy}-I_{zz})\dot{\theta} \\ 0 & 0 & 0 & 0 & 0 & J_r\Omega_r + (I_{zz}-I_{xx})\dot{\psi} \\ 0 & 0 & 0 & 0 & (I_{xx}-I_{yy})\dot{\psi} - J_r\Omega_r & 0 \\ 0 & 0 & 0 & 0 & 0 & 0 \\ 0 & 0 & 0 & 0 & 0 & 0 \end{bmatrix}$$

$$\begin{matrix} -ml(s_{\alpha_x}c_{\alpha_y}\dot{\alpha}_x + c_{\alpha_x}s_{\alpha_y}\dot{\alpha}_y) & -ml(s_{\alpha_x}c_{\alpha_y}\dot{\alpha}_y + c_{\alpha_x}s_{\alpha_y}\dot{\alpha}_x) \\ 0 & -mls_{\alpha_y}\dot{\alpha}_y \\ ml(-s_{\alpha_x}s_{\alpha_y}\dot{\alpha}_y + c_{\alpha_x}c_{\alpha_y}\dot{\alpha}_x) & ml(-s_{\alpha_x}s_{\alpha_y}\dot{\alpha}_x + c_{\alpha_y}c_{\alpha_x}\dot{\alpha}_x) \\ 0 & 0 \\ 0 & 0 \\ 0 & 0 \\ -ml^2s_{\alpha_y}c_{\alpha_y}\dot{\alpha}_y & -ml^2s_{\alpha_y}c_{\alpha_y}\dot{\alpha}_x \\ ml^2s_{\alpha_y}c_{\alpha_y}\dot{\alpha}_x & 0 \end{matrix}$$

$$G(q) = \begin{bmatrix} 0 & 0 & (M+m)g & 0 & 0 & 0 & mgls_{\alpha_x}c_{\alpha_y} & mglc_{\alpha_x}s_{\alpha_y} \end{bmatrix}^T$$

$$\bar{B}(q) = \begin{bmatrix} s_\phi s_\psi + c_\phi c_\psi s_\theta & c_\phi s_\theta s_\psi - c_\psi s_\phi & c_\theta c_\phi & 0 & 0 & 0 & 0 & 0 \\ 0 & 0 & 0 & 1 & 0 & 0 & 0 & 0 \\ 0 & 0 & 0 & 0 & 1 & 0 & 0 & 0 \\ 0 & 0 & 0 & 0 & 0 & 1 & 0 & 0 \end{bmatrix}^T$$

where $s_\theta = \sin(\theta)$, $c_\theta = \cos(\theta)$.

We can see in the above expressions that the cable-suspended payload affects the translational motion, but the rotational motion is not affected. Then, the vehicle attitude dynamics is decoupled from the payload–quadrotor translational dynamics. Thus, the quadrotor with a cable-suspended payload model can be divided into payload–quadrotor translational and quadrotor rotational dynamics. The following subsections show the translational and rotational dynamics.

2.2. Translational Dynamics

From (1) the model considering the translational motion is

$$\tilde{\mathcal{M}}(\tilde{q})\ddot{\tilde{q}} + \tilde{C}(\tilde{q},\dot{\tilde{q}})\dot{\tilde{q}} + \tilde{G}(\tilde{q}) = \bar{B}\tilde{u}, \tag{5}$$

where $\tilde{q} = \begin{bmatrix} \xi & \mu \end{bmatrix}^T$, $\tilde{u} = \begin{bmatrix} F_x & F_y & F_z \end{bmatrix}^T = \begin{bmatrix} f(s_\phi s_\psi + c_\phi c_\psi s_\theta) & f(c_\phi s_\theta s_\psi - c_\psi s_\phi) & fc_\theta c_\phi \end{bmatrix}^T,$

the matrices are given by

$$\tilde{M}(\tilde{q}) = \begin{bmatrix} m+M & 0 & 0 & mlc_{\alpha_x}c_{\alpha_y} & -mls_{\alpha_x}s_{\alpha_y} \\ 0 & m+M & 0 & 0 & mlc_{\alpha_y} \\ 0 & 0 & m+M & mls_{\alpha_x}c_{\alpha_y} & mlc_{\alpha_x}s_{\alpha_y} \\ mlc_{\alpha_x}c_{\alpha_y} & 0 & mls_{\alpha_x}c_{\alpha_y} & ml^2c_{\alpha_y}^2 & 0 \\ -mls_{\alpha_x}s_{\alpha_y} & mlc_{\alpha_y} & mlc_{\alpha_x}s_{\alpha_y} & 0 & ml^2 \end{bmatrix} \tag{6}$$

$$\tilde{C}(\tilde{q},\dot{\tilde{q}}) = \begin{bmatrix} 0 & 0 & 0 & -ml(s_{\alpha_x}c_{\alpha_y}\dot{\alpha}_x + c_{\alpha_x}s_{\alpha_y}\dot{\alpha}_y) & -ml(s_{\alpha_x}c_{\alpha_y}\dot{\alpha}_y + c_{\alpha_x}s_{\alpha_y}\dot{\alpha}_x) \\ 0 & 0 & 0 & 0 & -mls_{\alpha_y}\dot{\alpha}_y \\ 0 & 0 & 0 & ml(-s_{\alpha_x}s_{\alpha_y}\dot{\alpha}_y + c_{\alpha_x}c_{\alpha_y}\dot{\alpha}_x) & ml(-s_{\alpha_x}s_{\alpha_y}\dot{\alpha}_x + c_{\alpha_y}c_{\alpha_x}\dot{\alpha}_y) \\ 0 & 0 & 0 & -ml^2s_{\alpha_y}c_{\alpha_y}\dot{\alpha}_y & -ml^2s_{\alpha_y}c_{\alpha_y}\dot{\alpha}_x \\ 0 & 0 & 0 & ml^2s_{\alpha_y}c_{\alpha_y}\dot{\alpha}_x & 0 \end{bmatrix} \tag{7}$$

$$\tilde{G}(\tilde{q}) = \begin{bmatrix} 0 & 0 & (M+m)g & mgls_{\alpha_x}c_{\alpha_y} & mglc_{\alpha_x}s_{\alpha_y} \end{bmatrix}^T \tag{8}$$

$$\tilde{B}\tilde{u} = \begin{bmatrix} F_x & F_y & F_z & 0 & 0 \end{bmatrix}^T \tag{9}$$

where $s_\theta = \sin(\theta)$, $c_\theta = \cos(\theta)$.

2.3. Rotational Dynamics

From (1) the model considering the rotational motion is

$$\mathcal{M}^*(\eta)\,\ddot{\eta} + C^*(\eta,\dot{\eta})\,\dot{\eta} = B^*\tau,$$

where

$$\mathcal{M}^*(\eta) = dig\begin{bmatrix} I_{xx}, I_{yy}, I_{zz} \end{bmatrix}_{3\times3}, B^* = I_{3\times3}$$

$$C^*(\eta) = \begin{bmatrix} 0 & 0 & (I_{yy} - I_{zz})\dot{\theta} \\ 0 & 0 & J_r\Omega_r + (I_{zz} - I_{xx})\dot{\psi} \\ 0 & (I_{xx} - I_{yy})\dot{\psi} - J_r\Omega_r & 0 \end{bmatrix},$$

where J_r is the rotational moment of inertia.

3. LMI-Based Approach for Lipschitz Nonlinear Systems

In this section, we propose a nonlinear state feedback controller based on a linear matrix inequality to stabilize the quadrotor rotational dynamics.

Taking the state vector of the attitude dynamics as $\delta(t) = \begin{bmatrix} \psi & \dot{\psi} & \theta & \dot{\theta} & \phi & \dot{\phi} \end{bmatrix}^T$, one obtains

$$\dot{\delta} = \begin{bmatrix} 0 & 1 & 0 & 0 & 0 & 0 \\ 0 & 0 & 0 & 0 & 0 & 0 \\ 0 & 0 & 0 & 1 & 0 & 0 \\ 0 & 0 & 0 & 0 & 0 & -a_1\Omega_r \\ 0 & 0 & 0 & 0 & 0 & 1 \\ 0 & 0 & 0 & a_2\Omega_r & 0 & 0 \end{bmatrix}\delta + \begin{bmatrix} 0 & 0 & 0 \\ b_1 & 0 & 0 \\ 0 & 0 & 0 \\ 0 & b_2 & 0 \\ 0 & 0 & 0 \\ 0 & 0 & b_3 \end{bmatrix}\tau + \begin{bmatrix} 0 \\ I_1\dot{\theta}\dot{\phi} \\ 0 \\ I_2\dot{\psi}\dot{\phi} \\ 0 \\ I_3\dot{\psi}\dot{\theta} \end{bmatrix}. \tag{10}$$

Therefore, we can represent the rotational dynamics as a particular nonlinear system of the form

$$\dot{\delta} = A\delta + B\tau + \varphi(\delta,t),$$

where A and B are constant matrices, B is chosen such that (A, B) is controllable, and $\varphi(\delta, t)$ denotes the nonlinearities of the system. The parameters are given by:

$$I_1 = \frac{I_{zz} - I_{yy}}{I_{xx}}, \quad I_2 = \frac{I_{xx} - I_{zz}}{I_{yy}}, \quad I_3 = \frac{I_{yy} - I_{xx}}{I_{zz}}, \quad a_1 = \frac{J_r}{I_{yy}}, \quad a_2 = \frac{J_r}{I_{zz}}, \quad b_1 = \frac{1}{I_{xx}}, \quad b_2 = \frac{d}{I_{yy}}, \quad b_3 = \frac{d}{I_{zz}}.$$

Now, we consider that the following assumptions and lemmas are accomplished.

Assumption 1. *The function $\varphi(\delta, t)$ is Lipschitz w.r.t. δ, with a Lipschitz constant γ, if:*

$$\|\varphi(\delta, t) - \varphi(\tilde{\delta}, t)\| \leq \gamma \|\delta - \tilde{\delta}\|. \tag{11}$$

Lemma 1. *Let $\varphi : [\bar{a}, \bar{b}] \times \mathcal{D} \to \mathbb{R}^n$ be continuous for some domain $\mathcal{D} \subset \mathbb{R}^n$. Suppose that $\left[\partial\varphi/\partial\delta\right](\delta, t)$ exists and is continuous on $[\bar{a}, \bar{b}] \times \mathcal{D}$. If, for a convex subset $W \subset \mathcal{D}$, there is a constant $\gamma \geqslant 0$ such that*

$$\left\| \frac{\partial\varphi}{\partial\delta}(t, \delta) \right\| \leqslant \gamma$$

on $[\bar{a}, \bar{b}] \times W$, then

$$\|\varphi(\delta, t) - \varphi(\tilde{\delta}, t)\| \leq \gamma \|\delta - \tilde{\delta}\|$$

for all $t \in [\bar{a}, \bar{b}], \delta \in W$, and $\tilde{\delta} \in W$.

Lemma 2. *If $\varphi(\delta, t)$ and $\left[\partial\varphi/\partial\delta\right](\delta, t)$ are continuous on $[\bar{a}, \bar{b}] \times \mathbb{R}^n$, then φ is globally Lipschitz in δ on $[\bar{a}, \bar{b}] \times \mathbb{R}^n$ if and only if $\left[\partial\varphi/\partial\delta\right](\delta, t)$ is uniformly bounded on $[\bar{a}, \bar{b}] \times \mathbb{R}^n$.*

Lemma 3. *If $\varphi(\delta, t)$ and $\left[\partial\varphi/\partial\delta\right](\delta, t)$ are continuous on $[\bar{a}, \bar{b}] \times \mathcal{D}$, for some domain $\mathcal{D} \subset \mathbb{R}^n$, then φ is locally Lipschitz in δ on $[\bar{a}, \bar{b}] \times \mathcal{D}$ [25].*

Assumption 2. *For a and $b \in \mathbb{R}^n$ and $\varepsilon > 0$ we have*

$$2a^T b \leq \varepsilon^{-1} a^T a + \varepsilon b^T b. \tag{12}$$

3.1. Classical LMIs for the Quadrotor's Orientation

We propose a controller $u = -K\delta$ based on an LMI. Firstly, one introduces a classical LMI approach, which states the following lemma.

Lemma 4. *Consider system (10) and Assumption 1. Assume also that the nonlinearity φ is locally Lipschitz (Lemma 3) with Lipschitz constant γ (Lemma 1). Then, there exist a constant $\vartheta > 0$, matrices $X = X^T > 0$, and W with appropriate dimensions such that the following LMI is satisfied:*

$$\begin{bmatrix} AX + XA^T - BW^T - WB^T + 2(\vartheta - \gamma)X & \gamma I_n + X \\ \gamma I_n + X & -I_n \end{bmatrix} < 0, \tag{13}$$

with the feedback gain $K = W^T X^{-1}$. The system (10) is exponentially stable when the control input is $u = -K\delta$.

We now need to obtain the value of the Lipschitz constant γ of the nonlinearities of the rotation system function $\varphi(\delta, t)$. From the rotational dynamics model (10), the function $\varphi(\delta, t)$ is given by

$$\varphi(\delta, t) = \begin{bmatrix} 0 & I_1 \dot{\theta}\dot{\phi} & 0 & I_2 \dot{\psi}\dot{\phi} & 0 & I_3 \dot{\psi}\dot{\theta} \end{bmatrix}^T.$$

We are interested in calculating a Lipschitz constant for $\varphi(\delta, t)$ over the convex set

$$\mathcal{W} = \left\{ (\psi, \dot{\psi}, \theta, \dot{\theta}, \phi, \dot{\phi}) \mid (|\psi| \leq l_\psi, |\dot{\psi}| \leq l_{\dot{\psi}}, |\theta| \leq l_\theta, |\dot{\theta}| \leq l_{\dot{\theta}}, |\phi| \leq l_\phi, |\dot{\phi}| \leq l_{\dot{\phi}}) \right\}.$$

Applying Lemma 1, one gets:

$$
\begin{aligned}
\sup_{\delta \in \mathcal{W}} \left| \frac{\partial \varphi(\delta, t)}{\partial \delta} \right| &= \sup_{x \in \mathcal{W}} \left\| \begin{bmatrix} 0 & 0 & 0 & 0 & 0 & 0 \\ 0 & 0 & 0 & I_1\dot{\phi} & 0 & I_1\dot{\theta} \\ 0 & 0 & 0 & 0 & 0 & 0 \\ 0 & I_2\dot{\psi} & 0 & 0 & 0 & I_2\dot{\phi} \\ 0 & 0 & 0 & 0 & 0 & 0 \\ 0 & I_3\dot{\theta} & 0 & I_3\dot{\psi} & 0 & 0 \end{bmatrix} \right\| \leq \gamma \\
&= \max\left\{ |I_1\dot{\phi}| + |I_1\dot{\theta}|, |I_2\dot{\psi}| + |I_2\dot{\phi}|, |I_3\dot{\theta}| + |I_3\dot{\psi}| \right\} \\
&= I_2\left\{ |\dot{\psi}| + |\dot{\phi}| \right\} = I_3\left\{ |\dot{\theta}| + |\dot{\psi}| \right\} \\
&\leq I_2\left(l_{\dot{\psi}} + l_{\dot{\phi}} \right) = I_3\left(l_{\dot{\theta}} + l_{\dot{\psi}} \right).
\end{aligned}
$$

Then,

$$\gamma = I_2\left(l_{\dot{\psi}} + l_{\dot{\phi}} \right) = I_3\left(l_{\dot{\theta}} + l_{\dot{\psi}} \right).$$

The velocities are bounded as $l_{\dot{\psi}} = l_{\dot{\theta}} = l_{\dot{\phi}} = \frac{13\pi}{9}$ rad/s because the rotors are driven by DC permanent magnet motors, which support a maximum voltage of 9 V. This implies that the torque input vector τ is bounded and that the rotation speed capability of the motors has a maximum. When 9 V is applied to the motor the angular speed reaches $l_{\dot{\psi}} = l_{\dot{\theta}} = l_{\dot{\phi}} = \frac{13\pi}{9}$ rad/s. Considering these bounded values and the values of the Table 1, we can compute the Lipschitz constant as $\gamma = \frac{26\pi}{9}$. Then, φ is locally Lipschitz.

Table 1. Parameters.

Parameter	Value	[Units]
d	0.25	[m]
I_{xx}	$7.6566e^{-3}$	[kgm^2]
I_{yy}	$3.8278e^{-3}$	[kgm^2]
I_{zz}	$3.8278e^{-3}$	[kgm^2]
J_r	$28.385e^{-6}$	[kgm^2]

With $\gamma = \frac{26\pi}{9}$, we try to solve the LMI (13) using the LMI Toolbox® in MATLAB® software. However, the LMI is infeasible. It may happen that the classical LMI controller can deal with the problem only for smaller values of the Lipschitz constant. Thus, we propose an LMI following some ideas from [26]. The LMI presented in [26] is conceived for observer design. The new LMI design techniques are significantly less conservative than the classical LMI design technique.

3.2. Rotational Subsystem Control

Based on the linear-state-feedback approach, the control law for the attitude dynamics is $u(t) = -K\tilde{\delta}(t)$. Therefore, the orientation dynamics can be constructed as follows:

$$
\begin{aligned}
\dot{\tilde{\delta}} &= A\tilde{\delta}(t) + Bu(t) + \varphi(\tilde{\delta}(t), t) \\
&= A\tilde{\delta}(t) - BK\tilde{\delta}(t) + \varphi(\tilde{\delta}(t), t).
\end{aligned} \tag{14}
$$

The goal is to find a suitable B and K, such that $\delta \to \tilde{\delta}$. Thus, the attitude error dynamics $e(t)$ is represented as:

$$\dot{e}(t) = \dot{\delta} - \dot{\tilde{\delta}}$$
$$= (A - BK)e(t) + \varphi(\delta(t)) - \varphi(\tilde{\delta}(t)). \tag{15}$$

The following assumptions are needed for the derivation of the control law.

Assumption 3. *There exists a matrix G with appropriate dimensions such that:*

$$\|\varphi(\delta, t) - \varphi(\tilde{\delta}, t)\| \le \|G(\delta - \tilde{\delta})\|. \tag{16}$$

This matrix G is a sparsely populated matrix. $\|G(\delta - \tilde{\delta})\|$ can be much smaller than the constant $\gamma\|(\delta - \tilde{\delta})\|$ used earlier in Equation (11) for the same nonlinear function.

Let us now consider a larger Lipschitz constant of the nonlinear system. We can achieve a state feedback controller that is able to bring the state of the nonlinear system $\delta(t)$ to the desired state $\tilde{\delta}(t)$. This controller is given in the following statement:

Theorem 1. *For attitude error dynamics (15), assume that Assumption 1, Lemmas 1 and 3 are satisfied and there exist a constant $\varepsilon > 0$, matrices $X = X^T > 0$, and W with suitable dimensions, such that the following LMI holds:*

$$\begin{bmatrix} A\,X + XA^T - BW^T - WB^T + 2\vartheta X + XG^T + GX & \varepsilon I_n + XG^T \\ \varepsilon I_n + GX & -\varepsilon I_n \end{bmatrix} < 0, \tag{17}$$

where I_n denotes an identity matrix with appropriate dimensions, $\vartheta > 0$ is a constant, and $W = (K^T X)^{-1}$, the matrix K is a suitable feedback gain. Then, system (15) is exponentially stable, implying that the systems (10) and (14) are exponentially stable, then $\delta(t) \to \tilde{\delta}(t)$.

Proof. Define the Lyapunov function $V = e^T P e$. From the trajectory error (15), one gets:

$$\dot{V} = e^T P \dot{e} + \dot{e}^T P e$$
$$= e^T \left(P(A - BK) + (A - BK)^T P \right) e + 2e^T P \left(\varphi(\delta(t)) - \varphi(\tilde{\delta}(t)) \right). \tag{18}$$

From Assumption 3, one gets:

$$\|2e^T(t)P(\varphi(u(t), \delta(t)) - \varphi(u(t), \tilde{\delta}(t)))\| \le 2\|e^T(t)P\|\|\varphi(\delta(t)) - \varphi(\tilde{\delta}(t))\|$$
$$\le 2\|e^T(t)P\|\|Ge\|. \tag{19}$$

According to Assumption 2, $a = \|Pe(t)\|$ and $b = \|Ge\|$, one can rewrite (19) as follows:

$$\|2e^T(t)P(\varphi(\delta(t)) - \varphi(\tilde{\delta}(t)))\| \le \epsilon_1^{-1} e^T PPe + \epsilon_1 e^T G^T Ge. \tag{20}$$

Now, replacing (20) into (18), one gets:

$$\dot{V} \le e^T(t) \left(P(A - BK) + (A - BK)^T P + \epsilon_1^{-1} PP + \epsilon_1 G^T G \right) e(t). \tag{21}$$

If $\dot{V} \le -2\vartheta e^T(t)Pe(t) < 0$, where $\vartheta > 0$, one can rewrite (21) as:

$$\dot{V} \le e^T(t) \left(P(A - BK) + (A - BK)^T P + \epsilon_1^{-1} PP + \epsilon_1 G^T G + 2\vartheta P \right) e(t). \tag{22}$$

Indeed, the attitude error dynamics (15) is exponentially stable, and hence the two coupled systems (10) and (14) are exponentially stable. Using the Schur complement, Equation (22) can be easily represented in an LMI as:

$$\begin{bmatrix} P(A-BK)+(A-BK)^T P+2\vartheta P+G^T P+PG & P+\epsilon_1 G^T \\ P+\epsilon_1 G^T & -\epsilon_1 I_n \end{bmatrix} < 0. \tag{23}$$

Multiplying the above inequality by $\begin{bmatrix} P^{-1} & 0 \\ 0 & \epsilon_1^{-1} I_n \end{bmatrix}$ from the left-hand and right-hand sides, respectively, and letting $X = P^{-1}$, $\varepsilon = \epsilon_1^{-1}$, and $W = (KP^{-1})^T$, then the above inequality is further transformed into the following LMI:

$$\begin{bmatrix} A\,X+XA^T-BW^T-WB^T+2\vartheta X+XG^T+GX & \varepsilon I_n+XG^T \\ \varepsilon I_n+GX & -\varepsilon I_n \end{bmatrix} < 0. \tag{24}$$

If suitable $X > 0$ matrix and W are selected such that the LMI (17) is satisfied, then the attitude error dynamics (15) with the feedback gain $K = W^T X^{-1}$ is exponentially stable, implying that the coupled systems (10) and (14) are exponentially synchronized. $\square$

Now, we apply the main result of this paper (Theorem 1) to system (10). Then, we compute a controller for the rotational dynamics of the vehicle with guaranteed stability. Using Theorem 1, the LMI is then solved to obtain the control gain matrix K with $\varepsilon = 1$, $\vartheta = 25$, and $\gamma = \frac{26\pi}{9}$. Therefore, one easily obtains K from (24) by using the MATLAB® LMI Toolbox®:

$$K = \begin{bmatrix} 3.281 \times 10^8 & 2.235 \times 10^6 & 0 & 0 & 0 & 0 \\ 0 & 0 & 38.75 & 1.16 & -0.0114 & -2.27 \times 10^{-4} \\ 0 & 0 & 0.0114 & 2.27 \times 10^{-4} & 38.75 & 1.16 \end{bmatrix}. \tag{25}$$

The results of the state feedback controller with the gain matrix (25) are shown in Figures 2 and 3. Figure 2 displays the quadrotor attitude (roll, pitch, and yaw) with $\delta_0 = \begin{bmatrix} 0° & 0° & 1° & 0° & 2° & 0° \end{bmatrix}$. Note that the stabilization time is about 0.2 s, thus the state feedback controller with the gain matrix K calculated by LMI provides good transient performance. Figure 3 shows the input torques. Here we can observe that they are smooth.

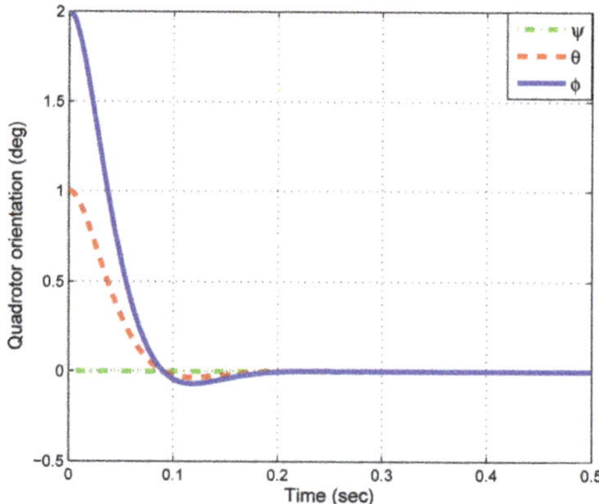

Figure 2. Quadrotor attitude.

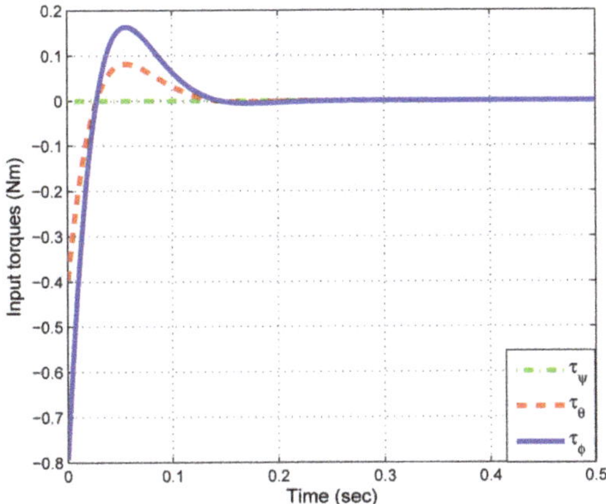

Figure 3. Control inputs.

4. Energy-Based Control

In this section, we propose an energy-based strategy to control the quadrotor's translation movements and to attenuate the payload oscillation.

4.1. Planar Case

In order to show the energy-based control law design in a simple manner, let us consider the system as a planar system operating in the XZ axis, as illustrated in Figure 4. The controller synthesis for the three-dimensional case is presented in Section 4.2.

In this case, $y = \psi = \phi = \alpha_y = 0$, then $\tilde{q} = \begin{bmatrix} x & z & \alpha \end{bmatrix}^T$, $\tilde{u} = \begin{bmatrix} F_x & F_z \end{bmatrix}^T = \begin{bmatrix} f\sin\theta & f\cos\theta \end{bmatrix}^T$ and the matrices of the model (5) for the translational dynamics are defined as

$$
\tilde{M}(\tilde{q}) = \begin{bmatrix} M+m & 0 & mlc_\alpha \\ 0 & M+m & mls_\alpha \\ mlc_\alpha & mls_\alpha & ml^2 \end{bmatrix}, \quad \tilde{C}(\tilde{q},\dot{\tilde{q}}) = \begin{bmatrix} 0 & 0 & -mls_\alpha\dot{\alpha} \\ 0 & 0 & mlc_\alpha\dot{\alpha} \\ 0 & 0 & 0 \end{bmatrix}
\tag{26}
$$

$$
\tilde{G}(\tilde{q}) = \begin{bmatrix} 0 \\ (M+m)g \\ mlgs_\alpha \end{bmatrix}, \quad \tilde{B}\tilde{u} = \begin{bmatrix} F_x \\ F_z \\ 0 \end{bmatrix}.
\tag{27}
$$

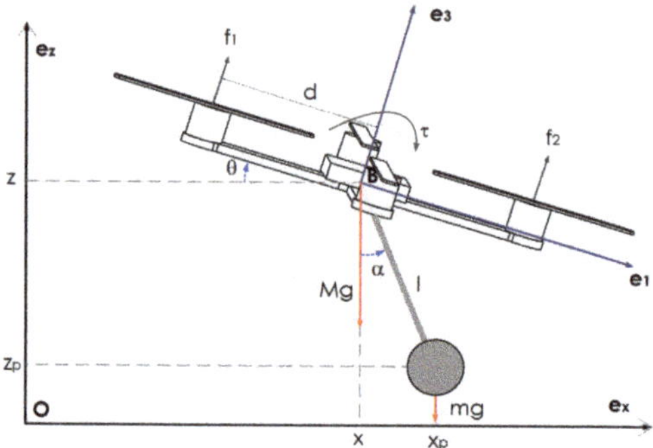

Figure 4. Two-dimensional quadrotor with a cable-suspended payload.

The total energy of the translational dynamics can be described by

$$
\begin{aligned}
H(\tilde{q}, \mathring{q}) &= \frac{1}{2}\mathring{q}^T \tilde{\mathcal{M}}\mathring{q} + V(\tilde{q}) \\
&= \frac{1}{2}\mathring{q}^T \tilde{\mathcal{M}}\mathring{q} + (M+m)gz - mgl\cos(\alpha).
\end{aligned}
\tag{28}
$$

Differentiating (28) along the trajectories of the system, we obtain

$$
\dot{H}(\tilde{q}, \mathring{q}) = \mathring{q}^T \tilde{\mathcal{M}}(\tilde{q})\ddot{\mathring{q}} + \frac{1}{2}\mathring{q}^T \dot{\tilde{\mathcal{M}}}(\tilde{q})\mathring{q} + \mathring{q}^T \tilde{G}(\tilde{q}).
$$

Substituting (5) into the above yields

$$
\dot{H}(\tilde{q}, \mathring{q}) = \mathring{q}^T \left(\tilde{u} - \tilde{C}(\tilde{q}, \mathring{q})\mathring{q} - \tilde{G}(\tilde{q}) \right) + \frac{1}{2}\mathring{q}^T \dot{\tilde{\mathcal{M}}}(\tilde{q})\mathring{q} + \mathring{q}^T \tilde{G}(\tilde{q}).
$$

Taking into account that the skew-symmetric relationship $\mathring{q}^T \left(\frac{1}{2}\dot{\tilde{\mathcal{M}}}(\tilde{q})\mathring{q} - \tilde{C}(\tilde{q}, \mathring{q}) \right)\mathring{q} = 0$ is satisfied, we obtain

$$
\begin{aligned}
\dot{H}(\tilde{q}, \mathring{q}) &= \mathring{q}^T \tilde{u} \\
&= \dot{x}F_x + \dot{z}F_z.
\end{aligned}
$$

Considering $\bar{x} = x - \tilde{x}$ and $\bar{z} = z - \tilde{z}$, the total energy in terms of the error is given by

$$
\dot{H} = \dot{\bar{x}}F_x + \dot{\bar{z}}F_z.
\tag{29}
$$

We propose the following Lyapunov candidate function:

$$
E = \frac{1}{2}\bar{H}^2 + \frac{k_{vx}}{2}\dot{\bar{x}}^2 + \frac{k_{vz}}{2}\dot{\bar{z}}^2 + \frac{k_{px}}{2}\bar{x}^2 + \frac{k_{pz}}{2}\bar{z}^2,
\tag{30}
$$

where k_{px}, k_{pz} are proportional constant gains and the k_{vx}, k_{vz} constants inject damping into the system. Differentiating (30) with respect to time, we have

$$
\begin{aligned}
\dot{E} &= \bar{H}\dot{\bar{H}} + k_{vx}\dot{\bar{x}}\ddot{\bar{x}} + k_{vz}\dot{\bar{z}}\ddot{\bar{z}} + k_{px}\bar{x}\dot{\bar{x}} + k_{pz}\bar{z}\dot{\bar{z}} \\
&= \dot{\bar{x}}\left(\bar{H}F_x + k_{vx}\ddot{\bar{x}} + k_{px}\bar{x} \right) + \dot{\bar{z}}\left(\bar{H}F_z + k_{vz}\ddot{\bar{z}} + k_{vz}\bar{z} \right).
\end{aligned}
\tag{31}
$$

We can obtain $\ddot{x}$ and $\ddot{z}$ from (5), (26), and (27):

$$\ddot{x} = \frac{M\left(lm\sin\bar\alpha\,\dot{\bar\alpha}^2 + F_x\right) + F_x m\cos^2\bar\alpha + F_z m\cos\bar\alpha\sin\bar\alpha}{M^2 + Mm},$$

$$\ddot{z} = -g + \frac{F_z m + M\left(F_z - lm\dot{\bar\alpha}^2\cos\bar\alpha\right) - F_z m\cos^2\bar\alpha + F_x m\cos\bar\alpha\sin\bar\alpha}{M^2 + Mm}.$$

Introducing the above into (31), we get

$$\dot{E} = \dot{x}\left[\bar{H}F_x + \frac{k_{vx}M\left(lm\sin\bar\alpha\,\dot{\bar\alpha}^2 + F_x\right) + k_{vx}mF_x\cos^2\bar\alpha + k_{vx}mF_z\cos\bar\alpha\sin\bar\alpha}{M(M+m)} + k_{xx}\bar{x}\right] +$$
$$\dot{z}\left[\bar{H}F_z + \frac{k_{vz}M\left(-lm\cos\bar\alpha\,\dot{\bar\alpha}^2 + F_z\right) + k_{vz}mF_z - k_{vz}mF_z\cos^2\bar\alpha + k_{vz}mF_x\cos\bar\alpha\sin\bar\alpha}{M(M+m)} + k_{xz}\bar{z} - k_{vz}g\right].$$

We propose a control law such that

$$\begin{bmatrix} E + \frac{k_{vx}}{M+m} + \frac{k_{vx}m\cos^2\alpha}{M(M+m)} & \frac{k_{vx}m\cos\alpha\sin\alpha}{M(M+m)} \\[2mm] \frac{k_{vz}m\cos\alpha\sin\alpha}{M(M+m)} & E + \frac{k_{vz}}{M} - \frac{k_{vz}m\cos^2\alpha}{M(M+m)} \end{bmatrix}\begin{bmatrix} F_x \\ F_z \end{bmatrix} = \begin{bmatrix} -\frac{k_{vx}lm\sin\alpha\dot\alpha^2}{M+m} - k_{xx}x - k_{ix}\dot{x} \\[2mm] \frac{k_{vz}lm\cos\alpha\dot\alpha^2}{M+m} - k_{xz}z + k_{vz}g - k_{iz}\dot{z} \end{bmatrix}. \quad (32)$$

The matrix that multiplies the vector $\begin{bmatrix} F_x & F_z \end{bmatrix}^T$ is a nonsingular matrix. Which leads to

$$\dot{E} = -k_{ix}\dot{\bar{x}}^2 - k_{iz}\dot{\bar{z}}^2.$$

From (32) we can obtain

$$F_x = \frac{-k_{ix}\dot{\bar{x}} - k_{xx}\bar{x} - \frac{k_{vx}lm\sin\bar\alpha\,\dot{\bar\alpha}^2}{M+m} - \dfrac{k_{vx}m\cos\bar\alpha\sin\bar\alpha\left(\frac{k_{vz}lm\cos\bar\alpha\,\dot{\bar\alpha}^2}{M+m} - k_{xz}\bar{z} + k_{vz}g - k_{iz}\dot{\bar{z}}\right)}{M(M+m)\bar{H} + k_{vz}(M+m) - k_{vz}m\cos^2\bar\alpha}}{\bar{H} + \frac{k_{vx}}{M+m} + \frac{k_{vx}m\cos^2\bar\alpha}{M(M+m)} - \dfrac{k_{vx}k_{vz}m^2\cos^2\bar\alpha\sin^2\bar\alpha}{M(M+m)(M(M+m)\bar{H} + k_{vz}(M+m) - k_{vz}m\cos^2\bar\alpha)}},$$

$$F_z = \frac{-k_{iz}\dot{\bar{z}} - k_{xz}\bar{z} + \frac{k_{vz}lm\cos\bar\alpha\,\dot{\bar\alpha}^2}{M+m} + k_{vz}g + \dfrac{k_{vz}m\cos\bar\alpha\sin\bar\alpha\left(\frac{k_{vx}lm\sin\bar\alpha\,\dot{\bar\alpha}^2}{M+m} + k_{xx}\bar{x} + k_{ix}\dot{\bar{x}}\right)}{M(M+m)\bar{H} + k_{vx}M + k_{vx}m\cos^2\bar\alpha}}{\bar{H} + \frac{k_{vz}}{M} - \frac{k_{vz}m\cos^2\bar\alpha}{M(M+m)} - \dfrac{k_{vx}k_{vz}m^2\cos^2\bar\alpha\sin^2\bar\alpha}{M(M+m)(M(M+m)\bar{H} + k_{vx}M + k_{vx}m\cos^2\bar\alpha)}}.$$

4.2. Three-Dimensional Case

The mathematical model for the three-dimensional case is presented in (5)–(9). The total energy of the system is

$$H(\tilde{q}, \dot{\tilde{q}}) = \frac{1}{2}\dot{\tilde{q}}^T\tilde{M}\dot{\tilde{q}} + V(\tilde{q})$$
$$= \frac{1}{2}\dot{\tilde{q}}^T\tilde{M}\dot{\tilde{q}} + (M+m)gz - mgl\cos(\alpha_x)\cos(\alpha_y). \quad (33)$$

In a similar way to the planar case, differentiating (33) along the trajectories of the system, we obtain

$$\dot{H}(\tilde{q}, \dot{\tilde{q}}) = \dot{\tilde{q}}^T\tilde{u}$$
$$= \dot{x}F_x + \dot{y}F_y + \dot{z}F_z. \quad (34)$$

The total energy in terms of the error can be rewritten as

$$\dot{H} = \dot{\tilde{x}} F_x + \dot{\tilde{y}} F_y + \dot{\tilde{z}} F_z. \tag{35}$$

Consider the following Lyapunov candidate function:

$$E(\tilde{q}, \dot{\tilde{q}}) = \frac{1}{2}\bar{H}^2 + \frac{k_{vx}}{2}\dot{\tilde{x}}^2 + \frac{k_{vy}}{2}\dot{\tilde{y}}^2 + \frac{k_{vz}}{2}\dot{\tilde{z}}^2 + \frac{k_{px}}{2}\tilde{x}^2 + \frac{k_{py}}{2}\tilde{y}^2 + \frac{k_{pz}}{2}\tilde{z}^2. \tag{36}$$

Differentiating (36) along the trajectories of the system, it follows that

$$
\begin{aligned}
\dot{E} &= \bar{H}\dot{\bar{H}} + k_{vx}\dot{\tilde{x}}\ddot{\tilde{x}} + k_{vy}\dot{\tilde{y}}\ddot{\tilde{y}} + k_{vz}\dot{\tilde{z}}\ddot{\tilde{z}} + k_{px}\tilde{x}\dot{\tilde{x}} + k_{py}\tilde{y}\dot{\tilde{y}} + k_{pz}\tilde{z}\dot{\tilde{z}} \\
&= \dot{\tilde{x}}\left(\bar{H}F_x + k_{vx}\ddot{\tilde{x}} + k_{px}\tilde{x}\right) + \dot{\tilde{y}}\left(\bar{H}F_y + k_{vy}\ddot{\tilde{y}} + k_{py}\tilde{y}\right) + \dot{\tilde{z}}\left(\bar{H}F_z + k_{vz}\ddot{\tilde{z}} + k_{pz}\tilde{z}\right).
\end{aligned} \tag{37}
$$

From (5)–(9) we can obtain $\ddot{\tilde{x}}$, $\ddot{\tilde{y}}$ and $\ddot{\tilde{z}}$. These expressions are defined as:

$$\ddot{\tilde{x}} = \frac{F_x\left((M+m)\left(s_{\tilde{\alpha}_y}^2 c_{\tilde{\alpha}_x}^4 + c_{\tilde{\alpha}_x}^2 c_{\tilde{\alpha}_y}^2\right) + Mc_{\tilde{\alpha}_y}^4 s_{\tilde{\alpha}_x}^2\right) - F_y mc_{\tilde{\alpha}_x}^2 c_{\tilde{\alpha}_y}^2 s_{\tilde{\alpha}_x} s_{\tilde{\alpha}_y} + F_z ms_{\tilde{\alpha}_x} c_{\tilde{\alpha}_x} c_{\tilde{\alpha}_y}^3 + \lambda(\mu,\dot{\mu})}{(M^2 + Mm)\left(s_{\tilde{\alpha}_x}^4 s_{\tilde{\alpha}_y}^2 + s_{\tilde{\alpha}_x}^2 s_{\tilde{\alpha}_y}^4 - 3s_{\tilde{\alpha}_x}^2 s_{\tilde{\alpha}_y}^2 + 1\right)},$$

$$\ddot{\tilde{y}} = \frac{F_y\left((M+m)\left(s_{\tilde{\alpha}_x}^2 c_{\tilde{\alpha}_y}^4 + c_{\tilde{\alpha}_x}^2 c_{\tilde{\alpha}_y}^2\right) + Mc_{\tilde{\alpha}_x}^4 s_{\tilde{\alpha}_y}^2\right) - F_x mc_{\tilde{\alpha}_x}^2 c_{\tilde{\alpha}_y}^2 s_{\tilde{\alpha}_x} s_{\tilde{\alpha}_y} + F_z ms_{\tilde{\alpha}_y} c_{\tilde{\alpha}_y} c_{\tilde{\alpha}_x}^3 + \kappa(\mu,\dot{\mu})}{(M^2 + Mm)\left(s_{\tilde{\alpha}_x}^4 s_{\tilde{\alpha}_y}^2 + s_{\tilde{\alpha}_x}^2 s_{\tilde{\alpha}_y}^4 - 3s_{\tilde{\alpha}_x}^2 s_{\tilde{\alpha}_y}^2 + 1\right)},$$

$$\ddot{\tilde{z}} = \frac{F_z\left((M+m)\left(s_{\tilde{\alpha}_y}^2 c_{\tilde{\alpha}_x}^4 - s_{\tilde{\alpha}_x}^2 c_{\tilde{\alpha}_y}^4\right) - Mc_{\tilde{\alpha}_x}^2 c_{\tilde{\alpha}_y}^2\right) - F_x mc_{\tilde{\alpha}_y}^3 c_{\tilde{\alpha}_x} s_{\tilde{\alpha}_x} - F_y ms_{\tilde{\alpha}_y} c_{\tilde{\alpha}_y} c_{\tilde{\alpha}_x}^3 + (M^2 + Mm)\,go(\mu) + \chi(\mu,\dot{\mu})}{(M^2 + Mm)\left(s_{\tilde{\alpha}_x}^4 s_{\tilde{\alpha}_y}^2 + s_{\tilde{\alpha}_x}^2 s_{\tilde{\alpha}_y}^4 - 3s_{\tilde{\alpha}_x}^2 s_{\tilde{\alpha}_y}^2 + 1\right)},$$

where

$$
\begin{aligned}
\lambda(\tilde{\alpha}_x, \tilde{\alpha}_y) &= Mlm\left(s_{\tilde{\alpha}_x} c_{\tilde{\alpha}_x}^2 c_{\tilde{\alpha}_y}^4 \left(\dot{\tilde{\alpha}}_x^2 + \dot{\tilde{\alpha}}_y^2\right) + c_{\tilde{\alpha}_x}^2 c_{\tilde{\alpha}_y}^2 s_{\tilde{\alpha}_x} s_{\tilde{\alpha}_y}^2 \dot{\tilde{\alpha}}_y^2 + 2c_{\tilde{\alpha}_x} c_{\tilde{\alpha}_y}^3 s_{\tilde{\alpha}_x}^2 s_{\tilde{\alpha}_y} \dot{\tilde{\alpha}}_x \dot{\tilde{\alpha}}_y + c_{\tilde{\alpha}_y}^4 s_{\tilde{\alpha}_x}^3 \dot{\tilde{\alpha}}_x^2\right), \\
\kappa(\bar{\mu}, \dot{\bar{\mu}}) &= Mlm\left(s_{\tilde{\alpha}_y} c_{\tilde{\alpha}_y}^2 c_{\tilde{\alpha}_x}^4 \left(\dot{\tilde{\alpha}}_x^2 + \dot{\tilde{\alpha}}_y^2\right) + c_{\tilde{\alpha}_x}^2 c_{\tilde{\alpha}_y}^2 s_{\tilde{\alpha}_y} s_{\tilde{\alpha}_x}^2 \dot{\tilde{\alpha}}_x^2 + 2c_{\tilde{\alpha}_y} c_{\tilde{\alpha}_x}^3 s_{\tilde{\alpha}_y}^2 s_{\tilde{\alpha}_x} \dot{\tilde{\alpha}}_x \dot{\tilde{\alpha}}_y + c_{\tilde{\alpha}_x}^4 s_{\tilde{\alpha}_y}^3 \dot{\tilde{\alpha}}_y^2\right), \\
\chi(\bar{\mu}, \dot{\bar{\mu}}) &= Mlm\left(c_{\tilde{\alpha}_y}^3 c_{\tilde{\alpha}_x}^3 \left(\dot{\tilde{\alpha}}_x^2 + \dot{\tilde{\alpha}}_y^2\right) + c_{\tilde{\alpha}_x} c_{\tilde{\alpha}_y}^3 s_{\tilde{\alpha}_x}^2 \dot{\tilde{\alpha}}_x^2 - 2s_{\tilde{\alpha}_y} c_{\tilde{\alpha}_x}^2 c_{\tilde{\alpha}_y}^2 s_{\tilde{\alpha}_x} \dot{\tilde{\alpha}}_x \dot{\tilde{\alpha}}_y - c_{\tilde{\alpha}_x}^3 s_{\tilde{\alpha}_y}^2 c_{\tilde{\alpha}_y} \dot{\tilde{\alpha}}_y^2\right), \\
o(\bar{\mu}) &= \left(c_{\tilde{\alpha}_y}^4 s_{\tilde{\alpha}_x}^2 + c_{\tilde{\alpha}_x}^2 c_{\tilde{\alpha}_y}^2 - s_{\tilde{\alpha}_y}^2 c_{\tilde{\alpha}_x}^4\right).
\end{aligned}
$$

Substituting $\ddot{\tilde{x}}$, $\ddot{\tilde{y}}$, and $\ddot{\tilde{z}}$ into (37) yields

$$
\begin{aligned}
\dot{E} = \dot{\tilde{x}}&\left[\bar{H}F_x + k_{vx}\frac{F_x\left((M+m)\left(s_{\tilde{\alpha}_y}^2 c_{\tilde{\alpha}_x}^4 + c_{\tilde{\alpha}_x}^2 c_{\tilde{\alpha}_y}^2\right) Mc_{\tilde{\alpha}_y}^4 s_{\tilde{\alpha}_x}^2\right) - F_y mc_{\tilde{\alpha}_x}^2 c_{\tilde{\alpha}_y}^2 s_{\tilde{\alpha}_x} s_{\tilde{\alpha}_y} + F_z ms_{\tilde{\alpha}_x} c_{\tilde{\alpha}_x} c_{\tilde{\alpha}_y}^3 + \lambda(\bar{\mu},\dot{\bar{\mu}})}{(M^2 + Mm)\left(s_{\tilde{\alpha}_x}^4 s_{\tilde{\alpha}_y}^2 + s_{\tilde{\alpha}_x}^2 s_{\tilde{\alpha}_y}^4 - 3s_{\tilde{\alpha}_x}^2 s_{\tilde{\alpha}_y}^2 + 1\right)} + k_{px}\tilde{x}\right] + \\[2ex]
\dot{\tilde{y}}&\left[\bar{H}F_y + k_{vy}\frac{F_y\left((M+m)\left(s_{\tilde{\alpha}_x}^2 c_{\tilde{\alpha}_y}^4 + c_{\tilde{\alpha}_x}^2 c_{\tilde{\alpha}_y}^2\right) + Mc_{\tilde{\alpha}_x}^4 s_{\tilde{\alpha}_y}^2\right) - F_x mc_{\tilde{\alpha}_x}^2 c_{\tilde{\alpha}_y}^2 s_{\tilde{\alpha}_x} s_{\tilde{\alpha}_y} + F_z ms_{\tilde{\alpha}_y} c_{\tilde{\alpha}_y} c_{\tilde{\alpha}_x}^3 + \kappa(\bar{\mu},\dot{\bar{\mu}})}{(M^2 + Mm)\left(s_{\tilde{\alpha}_x}^4 s_{\tilde{\alpha}_y}^2 + s_{\tilde{\alpha}_x}^2 s_{\tilde{\alpha}_y}^4 - 3s_{\tilde{\alpha}_x}^2 s_{\tilde{\alpha}_y}^2 + 1\right)} + k_{py}\tilde{y}\right] + \\[2ex]
\dot{\tilde{z}}&\left[\bar{H}F_z + k_{vz}\frac{F_z\left((M+m)\left(s_{\tilde{\alpha}_y}^2 c_{\tilde{\alpha}_x}^4 - s_{\tilde{\alpha}_x}^2 c_{\tilde{\alpha}_y}^4\right) - Mc_{\tilde{\alpha}_x}^2 c_{\tilde{\alpha}_y}^2\right) - F_x mc_{\tilde{\alpha}_y}^3 c_{\tilde{\alpha}_x} s_{\tilde{\alpha}_x} - F_y ms_{\tilde{\alpha}_y} c_{\tilde{\alpha}_y} c_{\tilde{\alpha}_x}^3}{(M^2 + Mm)\left(s_{\tilde{\alpha}_x}^4 s_{\tilde{\alpha}_y}^2 + s_{\tilde{\alpha}_x}^2 s_{\tilde{\alpha}_y}^4 - 3s_{\tilde{\alpha}_x}^2 s_{\tilde{\alpha}_y}^2 + 1\right)}\right. \\[2ex]
&\left. + k_{vz}\frac{(M^2 + Mm)\,go(\bar{\mu}) + \chi(\bar{\mu},\dot{\bar{\mu}})}{(M^2 + Mm)\left(s_{\tilde{\alpha}_x}^4 s_{\tilde{\alpha}_y}^2 + s_{\tilde{\alpha}_x}^2 s_{\tilde{\alpha}_y}^4 - 3s_{\tilde{\alpha}_x}^2 s_{\tilde{\alpha}_y}^2 + 1\right)} + k_{pz}\tilde{z}\right].
\end{aligned}
$$

We propose a control law such that

$$\dot{E} = -k_{ix}\dot{\tilde{x}}^2 - k_{iy}\dot{\tilde{y}}^2 - k_{iz}\dot{\tilde{z}}^2.$$

Finally, we solve the following system of equations to obtain F_x, F_y, and F_z.

$$F_x \left(\bar{H} + \frac{k_{vx}\left((M+m)\left(s_{\bar{\alpha}_y}^2 c_{\bar{\alpha}_x}^4 + c_{\bar{\alpha}_x}^2 c_{\bar{\alpha}_y}^2\right) + Mc_{\bar{\alpha}_y}^4 s_{\bar{\alpha}_x}^2\right)}{\delta(\bar{\mu})} \right) - F_y \frac{k_{vx}mc_{\bar{\alpha}_x}^2 c_{\bar{\alpha}_y}^2 s_{\bar{\alpha}_x} s_{\bar{\alpha}_y}}{\delta(\bar{\mu})} + F_z \frac{k_{vx}ms_{\bar{\alpha}_x} c_{\bar{\alpha}_x} c_{\bar{\alpha}_y}^3}{\delta(\bar{\mu})}$$
$$+ \frac{\lambda(\bar{\mu},\dot{\bar{\mu}})}{\delta(\bar{\mu})} + k_{px}\bar{x} = -k_{ix}\dot{\bar{x}}, \tag{38}$$

$$F_y \left(\bar{H} + \frac{k_{vy}\left((M+m)\left(s_{\bar{\alpha}_x}^2 c_{\bar{\alpha}_y}^4 + c_{\bar{\alpha}_x}^2 c_{\bar{\alpha}_y}^2\right) + Mc_{\bar{\alpha}_x}^4 s_{\bar{\alpha}_y}^2\right)}{\delta(\bar{\mu})} \right) - F_x \frac{k_{vy}mc_{\bar{\alpha}_x}^2 c_{\bar{\alpha}_y}^2 s_{\bar{\alpha}_x} s_{\bar{\alpha}_y}}{\delta(\bar{\mu})} + F_z \frac{k_{vy}ms_{\bar{\alpha}_y} c_{\bar{\alpha}_y} c_{\bar{\alpha}_x}^3}{\delta(\bar{\mu})}$$
$$+ \frac{\kappa(\bar{\mu},\dot{\bar{\mu}})}{\delta(\bar{\mu})} + k_{py}\bar{y} = -k_{iy}\dot{\bar{y}}, \tag{39}$$

$$F_z \left(\bar{H} + \frac{k_{vz}\left((M+m)\left(s_{\bar{\alpha}_y}^2 c_{\bar{\alpha}_x}^4 - s_{\bar{\alpha}_x}^2 c_{\bar{\alpha}_y}^4\right) - Mc_{\bar{\alpha}_x}^2 c_{\bar{\alpha}_y}^2\right)}{\delta(\bar{\mu})} \right) - F_x \frac{k_{vz}mc_{\bar{\alpha}_y}^3 c_{\bar{\alpha}_x} s_{\bar{\alpha}_x}}{\delta(\bar{\mu})} - F_y \frac{k_{vz}ms_{\bar{\alpha}_y} c_{\bar{\alpha}_y} c_{\bar{\alpha}_x}^3}{\delta(\bar{\mu})}$$
$$+ \frac{(M^2+Mm)\,go(\bar{\mu})}{\delta(\bar{\mu})} + \frac{\chi(\bar{\mu},\dot{\bar{\mu}})}{\delta(\bar{\mu})} + k_{pz}\bar{z} = -k_{iz}\dot{\bar{z}}, \tag{40}$$

where

$$\delta(\bar{\mu}) = \left(M^2 + Mm\right)\left(s_{\bar{\alpha}_x}^4 s_{\bar{\alpha}_y}^2 + s_{\bar{\alpha}_x}^2 s_{\bar{\alpha}_y}^4 - 3s_{\bar{\alpha}_x}^2 s_{\bar{\alpha}_y}^2 + 1\right).$$

5. Numerical Simulations and Results

In order to check the performance of the designed control scheme, some simulations were carried out. The objective was to move the vehicle transporting a payload to the desired position of a square of 1 m length at 1 m height. The desired trajectory is then defined by,

$$\bar{\varsigma}_d = \begin{cases} \begin{bmatrix} 0 & 0 & 1 \end{bmatrix}^T, & t < 2 \\ \begin{bmatrix} 0 & 1 & 1 \end{bmatrix}^T, & 2 <= t < 14 \\ \begin{bmatrix} 1 & 1 & 1 \end{bmatrix}^T, & 14 <= t < 26 \\ \begin{bmatrix} 1 & 0 & 1 \end{bmatrix}^T, & 26 <= t < 38 \\ \begin{bmatrix} 0 & 0 & 1 \end{bmatrix}^T, & 38 <= t <= 50 \end{cases}$$

The values of the model parameters used for the simulation were the following: $M = 0.4$ kg, $m = 0.03$ kg, $l = 0.35$ m, $g = 9.81$ m/s^2.

These parameters are close to real aerial platforms. The corresponding simulation results are presented in the following figures.

Figure 5 illustrates the x, y, and z positions of the vehicle during the validation. We can see that the position for each axis was stabilized according to the desired reference points. The position z was regulated in less than 2 s, the performance of the x and y positions dynamics were similar and were regulated in less than 5 s.

Figure 6 displays the payload swing angles α and β. It is clear that the proposed control law exhibited good performance, since the payload swing angles were regulated to $0°$ at around 5 s and the maximum overshoot was $\pm 2.2°$.

The simulation results of the feedback controller based on LMI (inner-loop of the system) show the quadrotor's orientation dynamics in Figure 7. We can see that the attitude converged to the desired points with a null steady-state error. The evolution of the control inputs f, τ_ψ, τ_θ, and τ_ϕ are presented in Figure 8. Finally, a three-dimensional view of the path followed by the vehicle is depicted in Figure 9.

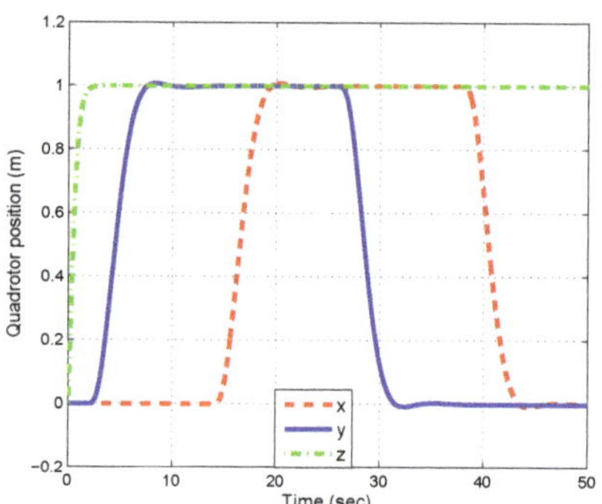

Figure 5. Quadrotor position.

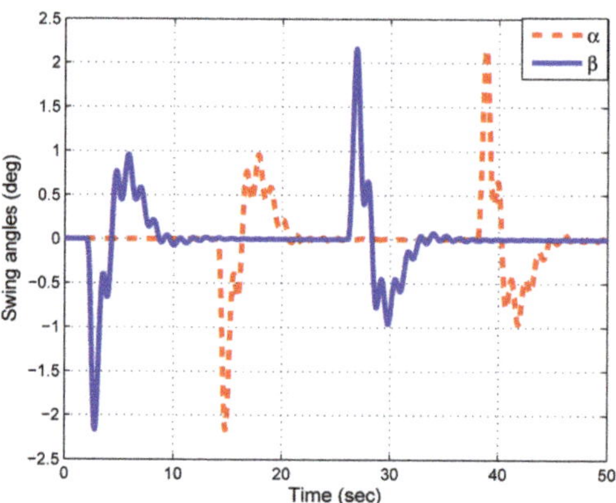

Figure 6. Swing angles.

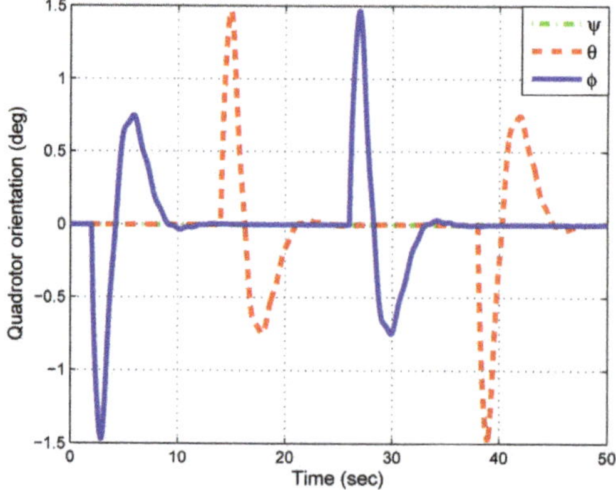

Figure 7. Quadrotor attitude.

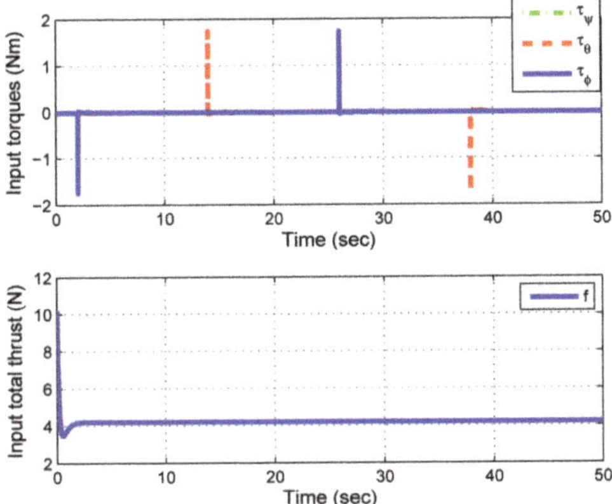

Figure 8. Control inputs.

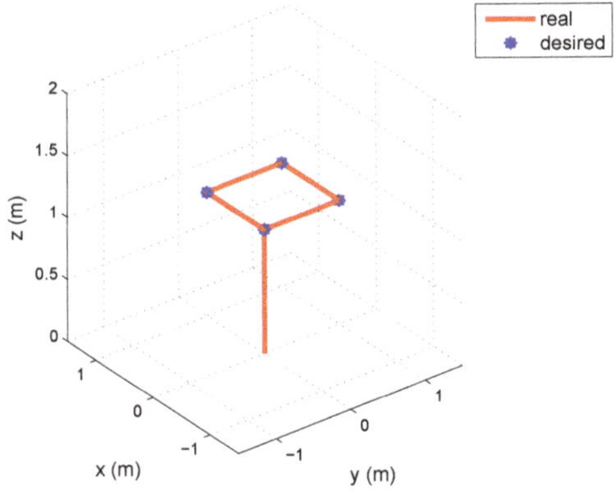

Figure 9. Three-dimensional trajectory.

One more numerical experiment was carried out. Figures 10–13 illustrate the tracking of an ascending circular trajectory to prove the efficiency of the proposed controller in a scenario that involves simultaneous variations of both α and β angles. These figures show that the proposed control strategy was capable of achieving accurate trajectory tracking since the aircraft converged to the reference trajectory while attenuating the swing angles of the payload.

In summary, these numerical experiments show that the proposed control scheme presented a satisfactory performance in position control and the attenuation of cable-suspended payload swing. It succeeded in transporting the payload to a desired position with attenuation of the oscillation angles. In contrast, the algorithm of [18] involves solving complicated partial differential equations for obtaining the control law, and they cannot be solved for the 3D case.

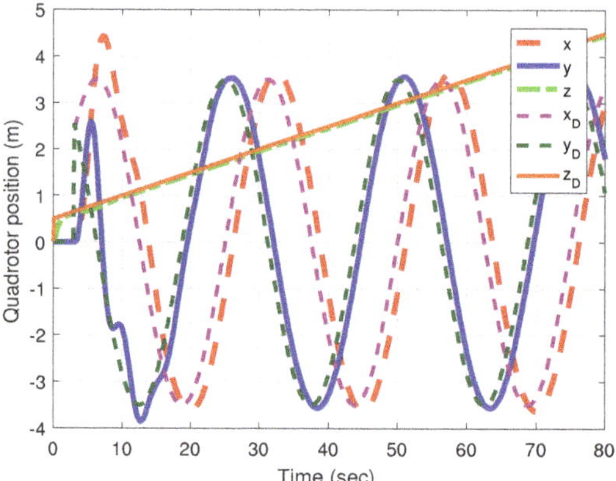

Figure 10. Quadrotor position.

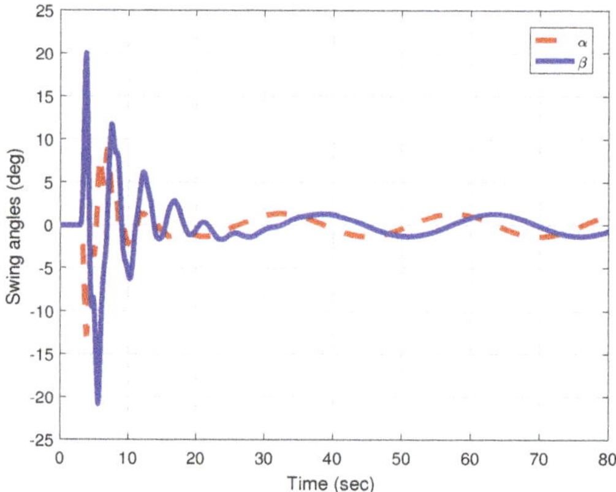

Figure 11. Swing angles.

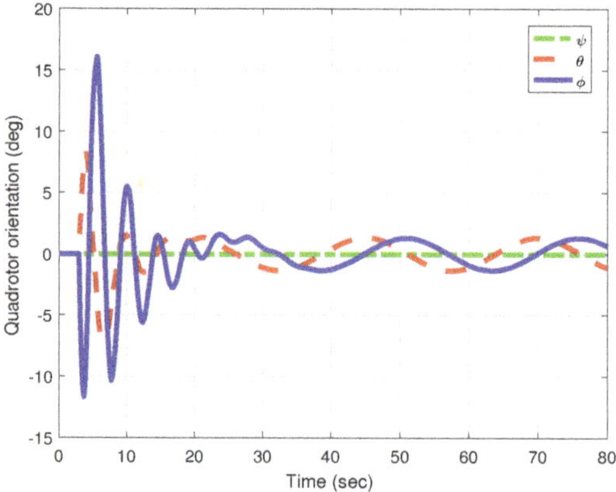

Figure 12. Quadrotor attitude.

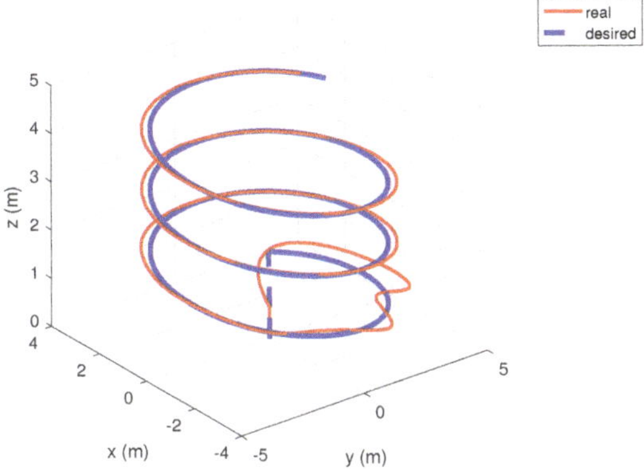

Figure 13. Three-dimensional trajectory.

6. Conclusions

This work presents an energy-based control strategy and a nonlinear state feedback controller based on a linear matrix inequality to solve the problem of transporting a cable-suspended payload by an unmanned aerial vehicle. On one hand, a new methodology based on an LMI for stabilizing the orientation dynamics is proposed. The main contribution is that we can employ the proposed methodology for Lipschitz nonlinear systems with larger Lipschitz constants than other classical techniques based on LMIs. Moreover, the LMI-based controller results in a control algorithm with relative simplicity and guaranteed stability. In addition, the design of the LMI-based control takes into account physical limits of the system such as the maximum motor voltage or its rotation speed capability through the velocity and torque bounds which are used to calculate the Lipschitz constant, while with the energy-based control these limits are not part of the controller design. On the other hand, an energy-based control to stabilize the quadrotor's translational dynamics and to attenuate the cable-suspended payload swing is designed in this work. This strategy is based on an energy approach and the passivity properties of the translational dynamics. Passivity-based control is employed, as this part of the system is underactuated. The main contribution is that the computation of excessive and complex partial differential equations is not needed to obtain the control law. The results showed an excellent performance of the proposed control scheme. Thus, the new approach achieves precise payload positioning with rapid oscillation attenuation.

Future work will extend the energy-based control method in order to consider variations in the payload. In addition, the methodology will be extended so that the linear matrix inequality can be replaced by an algebraic Riccati equation.

Author Contributions: Conceptualization, M.-E.G.-S. and O.H.-G.; methodology, M.-E.G.-S., O.H.-G., R.L., C.-D.G.-B., G.V.-P., and F.-R.L.-E.; formal analysis, M.-E.G.-S. and O.H.-G.; writing—original draft preparation, M.-E.G.-S. and O.H.-G.; writing—review and editing, G.V.-P. and F.-R.L.-E.; supervision, R.L. and C.-D.G.-B.

Funding: This work has been partially supported by PRODEP Mexican Program grant number 511-6/18-10333.

Conflicts of Interest: The authors declare no conflict of interest.

References

1. Bernard, M.; Kondak, K.; Maza, I.; Ollero, A. Autonomous transportation and deployment with aerial robots for search and rescue missions. *J. Field Robot.* **2011**, *28*, 914–931. [CrossRef]
2. Klausen, K.; Fossen, T.I.; Johansen, T.A. Nonlinear control of a multirotor UAV with suspended load. In Proceedings of the IEEE International Conference on Unmanned Aircraft Systems (ICUAS), Denver, CO, USA, 9–12 June 2015; pp. 176–184.

3.   Sreenath, K.; Michael, N.; Kumar, V. Trajectory generation and control of a quadrotor with a cable-suspended load-a differentially-flat hybrid system. In Proceedings of the IEEE International Conference on Robotics and Automation (ICRA), Karlsruhe, Germany, 6–10 May 2013; pp. 4888–4895.

4.   Sreenath, K.; Lee, T.; Kumar, V. Geometric control and differential flatness of a quadrotor UAV with a cable-suspended load. In Proceedings of the IEEE Conference on Decision and Control, Florence, Italy, 10–13 December 2013; pp. 2269–2274.

5.   Nicotra, M.M.; Garone, E.; Naldi, R.; Marconi, L. Nested saturation control of an uav carrying a suspended load. In Proceedings of the IEEE American Control Conference, Portland, OR, USA, 4–6 June 2014; pp. 3585–3590.

6.   Sadr, S.; Moosavian, S.A.A.; Zarafshan, P. Dynamics modeling and control of a quadrotor with swing load. *J. Robot.* **2014**, *2014*.10.1155/2014/265897. [CrossRef]

7.   Goodarzi, F.A.; Lee, D.; Lee, T. Geometric control of a quadrotor UAV transporting a payload connected via flexible cable. *Int. J. Control. Autom. Syst.* **2015**, *13*, 1486–1498. [CrossRef]

8.   Dai, S.; Lee, T.; Bernstein, D.S. Adaptive control of a quadrotor UAV transporting a cable-suspended load with unknown mass. In Proceedings of the IEEE Conference on Decision and Control, Los Angeles, CA, USA, 15–17 December 2014; pp. 6149–6154.

9.   Palunko, I.; Cruz, P.; Fierro, R. Agile load transportation: Safe and efficient load manipulation with aerial robots. *IEEE Robot. Autom. Mag.* **2012**, *19*, 69–79. [CrossRef]

10. Foehn, P.; Falanga, D.; Kuppuswamy, N.; Tedrake, R.; Scaramuzza, D. Fast Trajectory Optimization for Agile Quadrotor Maneuvers with a Cable-Suspended Payload. In Proceedings of the Robotics XIII: Science and Systems, Cambridge, MA, USA, 12–16 July 2017; pp. 1–10.

11. Qian, L.; Liu, H.H. Dynamics and control of a quadrotor with a cable suspended payload. In Proceedings of the IEEE Canadian Conference on Electrical and Computer Engineering (CCECE), Windsor, ON, Canada, 30 April–3 May 2017; pp. 1–4.

12. Rego, B.S.; Raffo, G.V. Suspended load path tracking control using a tilt-rotor UAV based on zonotopic state estimation. *J. Frankl. Inst.* **2019**, *356*, 1695–1729. [CrossRef]

13. Yi, K.; Liang, X.; He, Y.; Yang, L.; Han, J. Active-Model-Based Control for the Quadrotor Carrying a Changed Slung Load. *Electronics* **2019**, *8*, 461. [CrossRef]

14. Cabecinhas, D.; Cunha, R.; Silvestre, C. A trajectory tracking control law for a quadrotor with slung load. *Automatica* **2019**, *106*, 384–389. [CrossRef]

15. Weijers, M.; Carloni, R. Minimum swing control of a uav with a cable suspended load. In *Robotics and Mechatronics*; Technical Report; University of Twente: Enschede, The Netherlands, 2015.

16. Guerrero, M.; Mercado, D.; Lozano, R.; García, C. IDA-PBC methodology for a quadrotor uav transporting a cable-suspended payload. In Proceedings of the IEEE International Conference on Unmanned Aircraft Systems (ICUAS), Denver, CO, USA, 9–12 June 2015; pp. 470–476.

17. Guerrero, M.E.; Mercado, D.; Lozano, R.; García, C. Passivity based control for a quadrotor UAV transporting a cable-suspended payload with minimum swing. In Proceedings of the IEEE Conference on Decision and Control (CDC), Osaka, Japan, 15–18 December 2015; pp. 6718–6723.

18. Guerrero-Sánchez, M.E.; Mercado-Ravell, D.A.; Lozano, R.; García-Beltrán, C.D. Swing-attenuation for a quadrotor transporting a cable-suspended payload. *ISA Trans.* **2017**, *68*, 433–449.

19. Guerrero-Sánchez, M.E.; Abaunza, H.; Castillo, P.; Lozano, R.; García-Beltrán, C.D. Quadrotor Energy-Based Control Laws: a Unit-Quaternion Approach. *J. Intell. Robot. Syst.* **2017**, *88*, 347–377.

20. Ruggiero, F.; Cacace, J.; Sadeghian, H.; Lippiello, V. Passivity-based control of VToL UAVs with a momentum-based estimator of external wrench and unmodeled dynamics. *Robot. Auton. Syst.* **2015**, *72*, 139–151. [CrossRef]

21. Yüksel, B.; Secchi, C.; Bülthoff, H.H.; Franchi, A. Reshaping the physical properties of a quadrotor through IDA-PBC and its application to aerial physical interaction. In Proceedings of the IEEE International Conference on Robotics and Automation (ICRA), Hong Kong, China, 31 May–7 June 2014; pp. 6258–6265.

22. Acosta, J.Á.; Sanchez, M.; Ollero, A. Robust control of underactuated aerial manipulators via IDA-PBC. In Proceedings of the IEEE Conference on Decision and Control, Los Angeles, CA, USA, 15–17 December 2014; pp. 673–678.

23. Ryan, T.; Kim, H.J. LMI-based gain synthesis for simple robust quadrotor control. *IEEE Trans. Autom. Sci. Eng.* **2013**, *10*, 1173–1178. [CrossRef]

24. Yacef, F.; Bouhali, O.; Khebbache, H.; Boudjema, F. Takagi-sugeno model for quadrotor modelling and control using nonlinear state feedback controller. *Int. J. Control. Theory Comput. Model.* **2012**, *2*, 9–24. [CrossRef]

25. Khalil, H.K. *Nonlinear Systems*; Macmillan: New York, NY, USA, 2002.

26. Phanomchoeng, G.; Rajamani, R. Observer design for Lipschitz nonlinear systems using Riccati equations. In Proceedings of the IEEE American Control Conference, Baltimore, MD, USA, 30 June–2 July 2010; pp. 6060–6065.

Article

A Theoretical Rigid Body Model of Vibrating Screen for Spring Failure Diagnosis

Yue Liu [1,*], Shuangfu Suo [1,*], Guoying Meng [2,*], Deyong Shang [2], Long Bai [3] and Jianwen Shi [1]

[1] Department of Mechanical Engineering, State Key Laboratory of Tribology, Tsinghua University, Beijing 100084, China; shjw6222@mail.tsinghua.edu.cn

[2] School of Mechanical Electronic & Information Engineering, China University of Mining & Technology-Beijing, Beijing 10083, China; shangdeyong@cumtb.edu.cn

[3] Mechanical & Electrical Engineering School, Beijing Information Science & Technology University, Beijing 100192, China; bailong0316jn@126.com

* Correspondence: liuyue2018@mail.tsinghua.edu.cn (Y.L.); sfsuo@tsinghua.edu.cn (S.S.); mgy@cumtb.edu.cn (G.M.)

Received: 7 January 2019; Accepted: 5 March 2019; Published: 9 March 2019

Abstract: Springs are critical components in mining vibrating screen elastic supports. However, long-term alternating loads are likely to lead to spring failures, likely resulting in structural damages to the vibrating screen and resulting in a lower separation efficiency. Proper dynamic models provide a basis for spring failure diagnosis. In this paper, a six-degree-of-freedom theoretical rigid body model of a mining vibrating screen is proposed, and a dynamic equation is established in order to explore the dynamic characteristics. Numerical simulations, based on the Newmark-β algorithm, are carried out, and the results indicate that the model proposed is suitable for revealing the dynamic characteristics of the mining vibrating screen. Meanwhile, the mining vibrating screen amplitudes change with the spring failures. Therefore, six types of spring failure are selected for simulations, and the results indicate that the spring failures lead to an amplitude change for the four elastic support points in the x, y, and z directions, where the changes depend on certain spring failures. Hence, the key to spring failure diagnosis lies in obtaining the amplitude change rules, which can reveal particular spring failures. The conclusions provide a theoretical basis for further study and experiments in spring failure diagnosis for a mining vibrating screen.

Keywords: mining vibrating screen; theoretical rigid body model; spring failures diagnosis; amplitudes change

1. Introduction

Mining vibrating screens are important equipment for mine washing and processing, and are widely used for mine grading, dehydration, and desliming in China [1,2], working as a forced vibration system under alternating loads [3–5]. The SLK3661W double-deck linear mining vibrating screen is shown in Figure 1, and its main structures include a screen box and four elastic supports, designed using principles of symmetry. As shown in Figure 2, the screen box is assembled from an exciter, a lateral plate, an exciting beam, reinforcing beams, upper-bearing beams, under-bearing beams, an upper-screen deck, and an under-screen deck. Additionally, each elastic support is composed of several metal helical springs. These springs are critical components in a mining vibrating screen's elastic supports, which directly affect the working performance of the mining vibrating screen [6,7]. However, long-term alternating loads are highly likely to lead to spring failure through spring stiffness decrease [8], causing a negative influence on the mining vibrating screen. On one hand, spring failures could lead to structural damages, such as beam fracture or lateral plate cracks [9–11]. On the other hand, spring failures could produce a loss of particle separation efficiency, thus hardly meeting practical process demands [12,13]. Therefore, it is necessary to diagnose the spring failures of a mining vibrating screen for routing maintenance, which can help to ensure safety and reliability.

Figure 1. The SLK3661W double-deck linear mining vibrating screen, unloading side view.

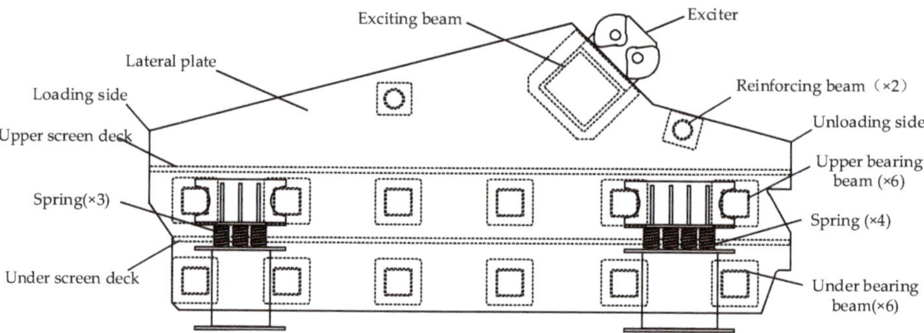

Figure 2. Structures of the SLK3661W double-deck linear mining vibrating screen.

Proper dynamic models provide a basis for diagnosing spring failures. In recent years, even though many studies have reported vibrating screen dynamic models on optimization [14–16], separation [17,18], and particle motion [19], there has been very little research reported on spring failure diagnosis. Aimed at spring failure diagnosis, Rodriguez et al. developed a two-dimensional, three-degree-of-freedom nonlinear model that considered one angular motion and damping, which allowed for the prediction of the behavior of a vibrating screen when there was a reduction in spring stiffness, and they used this model to determine a limit on spring failures before separation efficiency was affected [20]. Peng et al. presented a three-degree-of-freedom rigid plate structure to describe the isolation system, and they also proposed the method of stiffness identification by stiffness matrix disassembly; the numerical simulation results demonstrated the feasibility of the developed method [21]. However, each elastic support of a mining vibrating screen could have spring failures with spring stiffness decreases. The mining vibrating screen operating mode becomes spatial motion with very complicated dynamic characteristics, including multiple degrees of freedom.

The purpose of the present study is to explore the mining vibrating screen dynamic characteristics with spring failures, providing a theoretical basis for spring failure diagnosis. In this paper, a theoretical rigid body of a mining vibrating screen is proposed, the dynamic equation is established, and the steady-state solutions are obtained. Numerical simulations were carried out, and the results showed that the proposed model is feasible. In addition, spring failure simulations were also carried out, and the results indicated that the x, y, and z direction amplitude change rules for all the elastic supports were strongly related to spring failures. Hence, the key for spring failure diagnosis lies in obtaining the amplitude change rules, which can reveal the certain spring failures.

2. Theoretical Rigid Body Model

2.1. The Model

As shown in Figure 3, a six-degree-of-freedom theoretical rigid body model of spatial motion considering three rotations (Roll, Pitch, and Yaw) is proposed for exploring the mining vibrating screen dynamic characteristics with spring failures. The list of symbols is shown in abbreviations section.

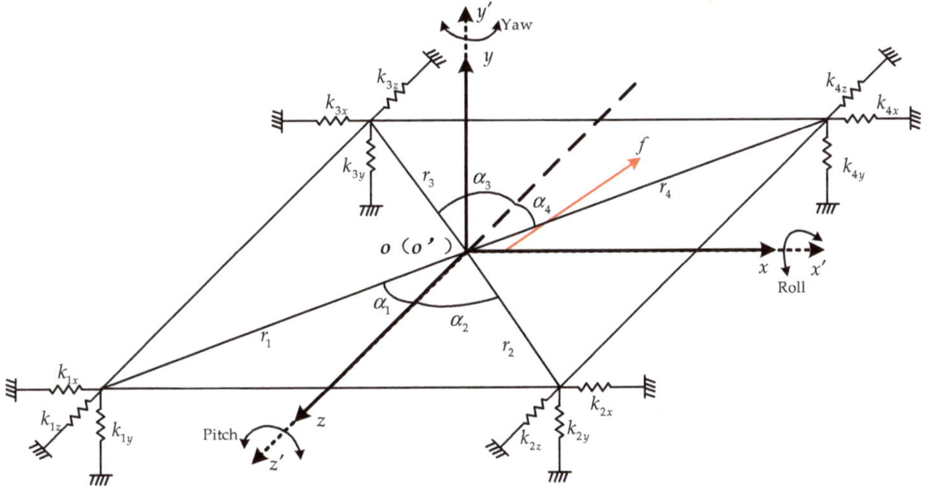

Figure 3. Spatial motion dynamic model of the mining vibrating screen, including three translational degrees of freedom and three rotational degrees of freedom.

The screen box is simplified to a rigid body, and the four elastic support points are individually simplified as three mutually perpendicular springs k_{ix}, k_{iy}, k_{iz}, (where i is the elastic support point number, i = 1, 2, 3, 4). The movement of a rigid body is expressed by the position of a body frame $o'x'y'z'$ relative to the inertial frame $oxyz$. The ox-axis and oz-axis are mutually perpendicular and located in the horizontal plane, and the oy-axis is perpendicular to the horizontal plane. The origin o' of the body frame is located at the mass center of the rigid body, at all times. The $o'x'$-axis and $o'z'$-axis are mutually perpendicular and located in the rigid body plane, and the $o'y'$-axis is perpendicular to the rigid body plane. Initially, the origin o of the inertial frame and the origin o' of the body frame are coincident. The distances between the mass center of the rigid body and the four elastic support points of spring are r_1, r_2, r_3, r_4 and, furthermore, the angles between them and oz-axis are $\alpha_1, \alpha_2, \alpha_3, \alpha_4$. Suppose that the rigid body's mass is m, and the moments of inertia are J_x, J_y, J_z. Define x, y, z as the translation displacements of the rigid body and γ, φ, θ as the angular displacements in the inertial frame. The exciting force f is exerted on the rigid body as an alternating load, with included angles $\beta_x, \beta_y, \beta_z$ between exciting force and the $o'x'$-axis, $o'y'$-axis, and $o'z'$-axis, respectively.

The dynamic equation is established by adopting the Lagrange method, and the processes are as follows. The three rotation angles are small, define $\cos \gamma = \cos \varphi = \cos \theta = 1$, $\sin \gamma = \gamma$, $\sin \varphi = \varphi$, and $\sin \theta = \theta$.

2.2. System Potential Energy

The dynamic system includes three translation motions and three rotation motions. According to the Tait–Bryan angles in the literature [22], the rotation matrix between the body frame and the inertial frame was derived using the rotation system shown in Figure 4.

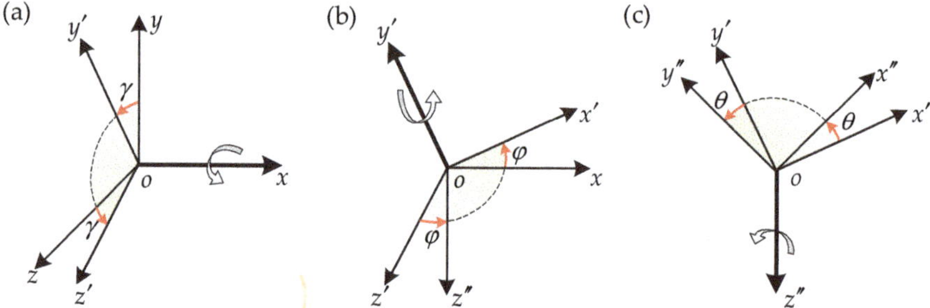

Figure 4. (**a**) Rotation of the inertial frame *oxyz* around the *ox*-axis by angle γ; (**b**) Rotation of the instantaneous system around the *oy'*-axis by angle φ; (**c**) Rotation of the instantaneous system around the *oz''*-axis by angle θ.

As the three rotation angles of the rigid body are small, they can be simplified as rotations around the *oxyz* axis. When rotating the rigid body around the *ox*-axis by the new angle of roll γ, the moment of inertia is J_x, and the rotation matrix is written as:

$$T_x = \begin{bmatrix} 1 & 0 & 0 \\ 0 & \cos\gamma & \sin\gamma \\ 0 & -\sin\gamma & \cos\gamma \end{bmatrix}. \tag{1}$$

When rotating the rigid body around the *oy'*-axis by the new angle of yaw φ, the moment of inertia is J_y, and the rotation matrix is written as:

$$T_y = \begin{bmatrix} \cos\varphi & 0 & -\sin\varphi \\ 0 & 1 & 0 \\ \sin\varphi & 0 & \cos\varphi \end{bmatrix}. \tag{2}$$

When rotating the rigid body around the *oz''*-axis by the new angle of pitch θ, the moment of inertia is J_z, and the rotation matrix is written as:

$$T_z = \begin{bmatrix} \cos\theta & \sin\theta & 0 \\ -\sin\theta & \cos\theta & 0 \\ 0 & 0 & 1 \end{bmatrix}. \tag{3}$$

When rotating the rigid body in the sequence *oz-oy-ox*, the rotation matrix between the body frame and the inertial frame is obtained as:

$$R = T_z T_y T_x = \begin{bmatrix} \cos\theta\cos\varphi & \sin\theta\cos\gamma + \cos\theta\sin\varphi\sin\gamma & \sin\theta\sin\gamma - \cos\theta\sin\varphi\cos\gamma \\ -\sin\theta\cos\varphi & \cos\theta\cos\gamma - \sin\theta\sin\varphi\sin\gamma & \cos\theta\sin\gamma + \sin\theta\sin\varphi\cos\gamma \\ \sin\varphi & -\cos\varphi\sin\gamma & \cos\varphi\cos\gamma \end{bmatrix}. \tag{4}$$

Supposing that the coordinate of the mass center is (x, y, z) in an inertial frame, and any point of the rigid body is (x', y', z') in the body frame, the coordinate of any point of the rigid body in an inertial frame is written as:

$$\begin{bmatrix} x_d \\ y_d \\ z_d \end{bmatrix} = \begin{bmatrix} x \\ y \\ z \end{bmatrix} + R \cdot \begin{bmatrix} x'_d \\ y'_d \\ z'_d \end{bmatrix}. \tag{5}$$

Moreover, the coordinates of the four spring support points in the body frame can be written as:

$$\begin{bmatrix} x'_1 \\ y'_1 \\ z'_1 \end{bmatrix} = \begin{bmatrix} -s_1 \\ 0 \\ c_1 \end{bmatrix}, \begin{bmatrix} x'_2 \\ y'_2 \\ z'_2 \end{bmatrix} = \begin{bmatrix} s_2 \\ 0 \\ c_2 \end{bmatrix}, \begin{bmatrix} x'_3 \\ y'_3 \\ z'_3 \end{bmatrix} = \begin{bmatrix} -s_3 \\ 0 \\ -c_3 \end{bmatrix}, \begin{bmatrix} x'_4 \\ y'_4 \\ z'_4 \end{bmatrix} = \begin{bmatrix} s_4 \\ 0 \\ -c_4 \end{bmatrix}. \tag{6}$$

In (6), $s_1 = r_1 \sin \alpha_1$, $c_1 = r_1 \cos \alpha_1$, $s_2 = r_2 \sin \alpha_2$, $c_2 = r_2 \cos \alpha_2$, $s_3 = r_3 \sin \alpha_3$, $c_3 = r_3 \cos \alpha_3$, $s_4 = r_4 \sin \alpha_4$, and $c_4 = r_4 \cos \alpha_4$.

In the initial state, the coordinate of the mass center is $(0, 0, 0)$ in the inertial frame, and the rotation matrix is $R_0 = [1, 0, 0; 0, 1, 0; 0, 0, 1]^{\mathrm{T}}$. Thus, the coordinates of the four spring support points in the inertial frame can be written as:

$$\begin{bmatrix} x_1 \\ y_1 \\ z_1 \end{bmatrix} = R_0 \begin{bmatrix} -s_1 \\ 0 \\ c_1 \end{bmatrix}, \begin{bmatrix} x_2 \\ y_2 \\ z_2 \end{bmatrix} = R_0 \begin{bmatrix} s_2 \\ 0 \\ c_2 \end{bmatrix}, \begin{bmatrix} x_3 \\ y_3 \\ z_3 \end{bmatrix} = R_0 \begin{bmatrix} -s_3 \\ 0 \\ -c_3 \end{bmatrix}, \begin{bmatrix} x_4 \\ y_4 \\ z_4 \end{bmatrix} = R_0 \begin{bmatrix} s_4 \\ 0 \\ -c_4 \end{bmatrix}. \tag{7}$$

In a motion state, the coordinate of the mass center is (x, y, z) in the inertial frame. Thus, the coordinates of the four spring support points in the inertial frame can be written as:

$$\begin{bmatrix} \Delta x_1 \\ \Delta y_1 \\ \Delta z_1 \end{bmatrix} = \begin{bmatrix} x \\ y \\ z \end{bmatrix} + (R - R_0) \begin{bmatrix} -s_1 \\ 0 \\ c_1 \end{bmatrix}, \begin{bmatrix} \Delta x_2 \\ \Delta y_2 \\ \Delta z_2 \end{bmatrix} = \begin{bmatrix} x \\ y \\ z \end{bmatrix} + (R - R_0) \begin{bmatrix} s_2 \\ 0 \\ c_2 \end{bmatrix},$$

$$\begin{bmatrix} \Delta x_3 \\ \Delta y_3 \\ \Delta z_3 \end{bmatrix} = \begin{bmatrix} x \\ y \\ z \end{bmatrix} + (R - R_0) \begin{bmatrix} -s_3 \\ 0 \\ -c_3 \end{bmatrix}, \begin{bmatrix} \Delta x_4 \\ \Delta y_4 \\ \Delta z_4 \end{bmatrix} = \begin{bmatrix} x \\ y \\ z \end{bmatrix} + (R - R_0) \begin{bmatrix} s_4 \\ 0 \\ -c_4 \end{bmatrix}. \tag{8}$$

The results in (8) are equivalent to the spring compression and, therefore, the system potential energy is obtained as:

$$\begin{aligned} U = &\tfrac{1}{2} k_{1x} [x - c_1(\varphi - \gamma\theta)]^2 + \tfrac{1}{2} k_{2x} [x - c_2(\varphi - \gamma\theta)]^2 + \tfrac{1}{2} k_{3x} [x + c_3(\varphi - \gamma\theta)]^2 \\ &+ \tfrac{1}{2} k_{4x} [x + c_4(\varphi - \gamma\theta)]^2 + \tfrac{1}{2} k_{1y} [y + c_1(\gamma + \varphi\theta) + s_1\theta]^2 \\ &+ \tfrac{1}{2} k_{2y} [y + c_2(\gamma + \varphi\theta) - s_2\theta]^2 + \tfrac{1}{2} k_{3y} [y - c_3(\gamma + \varphi\theta) + s_3\theta]^2 \\ &+ \tfrac{1}{2} k_{4y} [-y + c_4(\gamma + \varphi\theta) + s_4\theta]^2 + \tfrac{1}{2} k_{1z} (z - s_1\varphi)^2 + \tfrac{1}{2} k_{2z} (z + s_2\varphi)^2 \\ &+ \tfrac{1}{2} k_{3z} (z - s_3\varphi)^2 + \tfrac{1}{2} k_{4z} (z + s_4\varphi)^2 \end{aligned} \tag{9}$$

2.3. System Kinetic Energy

According to the literature [23,24], there is a relation expressing a rigid body's spatial motion, which is written as:

$$\begin{cases} \omega_x = \dot{\gamma} - \dot{\theta} \cos\varphi \tan\varphi \\ \omega_y = \dot{\varphi} \cos\gamma + \dot{\theta} \cos\varphi \sin\gamma \\ \omega_z = -\dot{\varphi} \sin\gamma + \dot{\theta} \cos\varphi \cos\gamma \end{cases} \tag{10}$$

Therefore, the system's kinetic energy is obtained as:

$$
\begin{aligned}
E &= \tfrac{1}{2}m\dot{x}^2 + \tfrac{1}{2}m\dot{y}^2 + \tfrac{1}{2}m\dot{z}^2 + \tfrac{1}{2}J_x\omega_x{}^2 + \tfrac{1}{2}J_y\omega_y{}^2 + \tfrac{1}{2}J_z\omega_z{}^2 \\
&= \tfrac{1}{2}m\dot{x}^2 + \tfrac{1}{2}m\dot{y}^2 + \tfrac{1}{2}m\dot{z}^2 + \tfrac{1}{2}J_x\left(\dot{\gamma} - \dot{\theta}\varphi\right)^2 + \tfrac{1}{2}J_y\left(\dot{\varphi} + \dot{\theta}\gamma\right)^2 + \tfrac{1}{2}J_z\left(-\dot{\varphi}\gamma + \dot{\theta}\right)^2 .
\end{aligned}
\tag{11}
$$

2.4. System Force Vector

The mining vibrating screen in this study is equipped with two groups of counter-rotating vibrators, with each group having two pairs of eccentric blocks. Due to manufacturing errors and installation errors, the resultant force f typically does not pass through the center of mass of the screen box in practice. The resultant force f can be equivalent to a force vector. In the body frame, the force vector is written as:

$$
\begin{bmatrix} f'_x \\ f'_y \\ f'_z \end{bmatrix} = \begin{bmatrix} f \cos\beta_x \sin\omega t \\ f \cos\beta_y \sin\omega t \\ f \cos\beta_z \sin\omega t \end{bmatrix} .
\tag{12}
$$

On account of the force vector changing with the rigid body motion, the force vector in a body frame is written as:

$$
\begin{bmatrix} f_x \\ f_y \\ f_z \end{bmatrix} = R \cdot \begin{bmatrix} f'_x \\ f'_y \\ f'_z \end{bmatrix} .
\tag{13}
$$

Meanwhile, supposing that the coordinate of the point exerting force is (x'_f, y'_f, z'_f) in a body frame, the coordinate of the point exerting force in an inertial frame can be written as:

$$
\begin{bmatrix} x_f \\ y_f \\ z_f \end{bmatrix} = \begin{bmatrix} x \\ y \\ z \end{bmatrix} + R \cdot \begin{bmatrix} x'_f \\ y'_f \\ z'_f \end{bmatrix} .
\tag{14}
$$

According to the literature [23,24], there is a relation expressing a rigid body's spatial motion, which is written as:

$$
\begin{bmatrix} M_x \\ M_y \\ M_z \end{bmatrix} = \begin{bmatrix} 0 & -z_f & y_f \\ z_f & 0 & -x_f \\ -y_f & x_f & 0 \end{bmatrix} \cdot \begin{bmatrix} f_x \\ f_y \\ f_z \end{bmatrix} .
\tag{15}
$$

In an inertial frame, the system's force vector is obtained as:

$$
\mathbf{F} = \left[f_x f_y f_z M_x M_y M_z \right]^{\mathrm{T}} .
\tag{16}
$$

2.5. Dynamic Equation

After linearizing, the dynamic equation can be written as:

$$
\mathbf{M\ddot{x}} + \mathbf{Kx} = \mathbf{F} .
\tag{17}
$$

In (17), $\ddot{\mathbf{x}}$ is the acceleration column vector:

$$
\ddot{\mathbf{x}} = \begin{bmatrix} \ddot{x} & \ddot{y} & \ddot{z} & \ddot{\gamma} & \ddot{\varphi} & \ddot{\theta} \end{bmatrix}^{\mathrm{T}} .
\tag{18}
$$

$\mathbf{x}$ is the displacement column vector:

$$
\mathbf{x} = \begin{bmatrix} x & y & z & \gamma & \varphi & \theta \end{bmatrix}^{\mathrm{T}} .
\tag{19}
$$

M is the mass matrix:

$$\mathbf{M} = \begin{bmatrix} m & 0 & 0 & 0 & 0 & 0 \\ 0 & m & 0 & 0 & 0 & 0 \\ 0 & 0 & m & 0 & 0 & 0 \\ 0 & 0 & 0 & J_x & 0 & 0 \\ 0 & 0 & 0 & 0 & J_y & 0 \\ 0 & 0 & 0 & 0 & 0 & J_z \end{bmatrix},$$ (20)

and **K** is the stiffness matrix:

$$\mathbf{K} = \begin{bmatrix} K_{11} & 0 & 0 & 0 & K_{15} & 0 \\ 0 & K_{22} & 0 & K_{24} & 0 & K_{26} \\ 0 & 0 & K_{33} & 0 & K_{35} & 0 \\ 0 & K_{42} & 0 & K_{44} & 0 & K_{46} \\ K_{51} & 0 & K_{53} & 0 & K_{55} & 0 \\ 0 & K_{62} & 0 & K_{64} & 0 & K_{66} \end{bmatrix}.$$ (21)

In (21), $K_{11} = k_{1x} + k_{2x} + k_{3x} + k_{4x}$,
$K_{15} = -c_1 k_{1x} - c_2 k_{2x} + c_3 k_{3x} + c_4 k_{4x}$,
$K_{22} = k_{1y} + k_{2y} + k_{3y} - k_{4y}$,
$K_{24} = c_1 k_{1y} + c_2 k_{2y} - c_3 k_{3y} + c_4 k_{4y}$,
$K_{26} = s_1 k_{1y} - s_2 k_{2y} + s_3 k_{3y} + s_4 k_{4y}$,
$K_{33} = k_{1z} + k_{2z} + k_{3z} + k_{4z}$,
$K_{35} = -s_1 k_{1z} + s_2 k_{2z} - s_3 k_{3z} + s_4 k_{4z}$,
$K_{42} = c_1 k_{1y} + c_2 k_{2y} - c_3 k_{3y} - c_4 k_{4y}$,
$K_{44} = c_1^2 k_{1y} + c_2^2 k_{2y} + c_3^2 k_{3y} + c_4^2 k_{4y}$,
$K_{46} = c_1 s_1 k_{1y} - c_2 s_2 k_{2y} - c_3 s_3 k_{3y} + c_4 s_4 k_{4y}$,
$K_{51} = -c_1 k_{1x} - c_2 k_{2x} + c_3 k_{3x} + c_4 k_{4x}$,
$K_{53} = -s_1 k_{1z} + s_2 k_{2z} - s_3 k_{3z} + s_4 k_{4z}$,
$K_{55} = c_1^2 k_{1x} + c_2^2 k_{2x} + c_3^2 k_{3x} + c_4^2 k_{4x} + s_1^2 k_{1z} + s_2^2 k_{2z} + s_3^2 k_{3z} + s_4^2 k_{4z}$,
$K_{62} = s_1 k_{1y} - s_2 k_{2y} + s_3 k_{3y} - s_4 k_{4y}$,
$K_{64} = c_1 s_1 k_{1y} - c_2 s_2 k_{2y} - c_3 s_3 k_{3y} + c_4 s_4 k_{4y}$,
$K_{66} = s_1^2 k_{1y} + s_2^2 k_{2y} + s_3^2 k_{3y} + s_4^2 k_{4y}$.
Additionally, **F** is the force column vector:

$$\mathbf{F} = \begin{bmatrix} f_x\, f_y\, f_z\, M_x\, M_y\, M_z \end{bmatrix}^{\mathrm{T}}.$$ (22)

According to the dynamic theory [20], the steady-state solutions of a forced vibration system can be written as:

$$\begin{cases} x = X_0 \sin \omega t \\ y = Y_0 \sin \omega t \\ z = Z_0 \sin \omega t \\ \gamma = \Gamma_0 \sin \omega t \\ \varphi = \Phi_0 \sin \omega t \\ \theta = \Theta_0 \sin \omega t \end{cases}.$$ (23)

In (23), ω is the angular speed. It should be noted that the relations are valid for a constant rotational velocity of counter-rotating vibrators, for which there exists a resultant force acting along a straight line towards the body of vibrating screen. Taking the derivative of both sides of (23), the acceleration can be written as:

$$\begin{cases} \ddot{x} = -\omega^2 X_0 \sin \omega t \\ \ddot{y} = -\omega^2 Y_0 \sin \omega t \\ \ddot{z} = -\omega^2 Z_0 \sin \omega t \\ \ddot{\gamma} = -\omega^2 \Gamma_0 \sin \omega t \\ \ddot{\varphi} = -\omega^2 \Phi_0 \sin \omega t \\ \ddot{\theta} = -\omega^2 \Theta_0 \sin \omega t \end{cases} \tag{24}$$

Bringing Equations (23) and (24) into Equation (22), the steady-state solutions can be obtained as:

$$\begin{bmatrix} X_0 \\ Y_0 \\ Z_0 \\ \Gamma_0 \\ \Phi_0 \\ \Theta_0 \end{bmatrix} = \begin{bmatrix} K_{11} - \omega^2 m & 0 & 0 & 0 & K_{15} & 0 \\ 0 & K_{22} - \omega^2 m & 0 & K_{24} & 0 & K_{26} \\ 0 & 0 & K_{33} - \omega^2 m & 0 & K_{35} & 0 \\ 0 & K_{42} & 0 & K_{44} - \omega^2 J_x & 0 & K_{46} \\ K_{51} & 0 & K_{53} & 0 & K_{55} - \omega^2 J_y & 0 \\ 0 & K_{62} & 0 & K_{64} & 0 & K_{66} - \omega^2 J_z \end{bmatrix}^{-1} \begin{bmatrix} f_x \\ f_y \\ f_z \\ M_x \\ M_y \\ M_z \end{bmatrix}. \tag{25}$$

The dynamic equation of a mining vibrating screen in spatial motion, shown above, gives the dynamic characteristics. In the following section, numerical simulations are carried out to verify the proposed model.

3. Simulations

3.1. Numerical Simulations Results

In this paper, numerical simulations are carried out using Matlab, and the programs are available in supplementary materials online. In order to ensure the physical significance of the dynamic equation, the damping is significant. Generally, the damping matrix can be regarded as a linear combination of the mass matrix (Equation (20)) and the stiffness matrix (Equation (21)) in a mechanical system dynamics equation, and can be written as:

$$\mathbf{C} = 0.02\mathbf{M} + 0.02\mathbf{K}. \tag{26}$$

After inserting the damping matrix (Equation (26)) into the dynamic Equation (17), the system dynamic equation can be written as:

$$\mathbf{M}\ddot{\mathbf{x}} + \mathbf{C}\dot{\mathbf{x}} + \mathbf{K}\mathbf{x} = \mathbf{F}. \tag{27}$$

In (27), $\dot{\mathbf{x}}$ is the velocity vector:

$$\dot{\mathbf{x}} = \begin{bmatrix} \dot{x} & \dot{y} & \dot{z} & \dot{\gamma} & \dot{\varphi} & \dot{\theta} \end{bmatrix}^{\mathrm{T}}. \tag{28}$$

This paper intends to use the SLK3661W double-deck linear mining vibrating screen as an exploration object, which has certain parameters, such as screen box mass (18,944 kg), spring stiffness of each unloading side (1,242,400 N/m), spring stiffness of each loading side (931,800 N/m), screen deck dimension (3.6 × 6.1 m), processing capacity (350–400 t/h), motor speed (1480 r/min), and electric power (55 kW). Numerical simulations were carried out, based on the Newmark-β algorithm, and the main parameters used in the simulation are shown in Table 1. Additionally, the coordinate of force action point was (−0.2, 0, 0) in the body frame. The total time of the simulation was t_m, while the time step was dt.

Table 1. Simulation parameters table.

Parameters	m/kg	$J_x/(\text{kg·m}^2)$	$J_y/(\text{kg·m}^2)$	$J_z/(\text{kg·m}^2)$
value	18,900	35,200	30,650	32,600
parameters	$k_{1x}/(\text{N/m})$	$k_{2x}/(\text{N/m})$	$k_{3x}/(\text{N/m})$	$k_{4x}/(\text{N/m})$
value	353,010	470,680	353,010	470,680
parameters	$k_{1y}/(\text{N/m})$	$k_{2y}/(\text{N/m})$	$k_{3y}/(\text{N/m})$	$k_{4y}/(\text{N/m})$
value	931,800	1,242,400	931,800	1,242,400
parameters	$k_{1z}/(\text{N/m})$	$k_{2z}/(\text{N/m})$	$k_{3z}/(\text{N/m})$	$k_{4z}/(\text{N/m})$
value	353,010	470,680	353,010	470,680
parameters	$k_{1z}/(\text{N/m})$	$k_{2z}/(\text{N/m})$	$k_{3z}/(\text{N/m})$	$k_{4z}/(\text{N/m})$
value	353,010	470,680	353,010	470,680
parameters	$r_1/(\text{m})$	$r_2/(\text{m})$	$r_3/(\text{m})$	$r_4/(\text{m})$
value	3	3	3	3
parameters	$\alpha_1/(°)$	$\alpha_2/(°)$	$\alpha_3/(°)$	$\alpha_4/(°)$
value	53.13	53.13	53.13	53.13
parameters	$\beta_x/(°)$	$\beta_y/(°)$	$\beta_z/(°)$	$\omega/(\text{rad/s})$
value	45	45	90	93.12
parameters	$f/(\text{N})$	$t_m/(\text{s})$	$dt/(\text{s})$	-
value	1,800,000	20	1/10,240	-

Under normal conditions, the system vibrations included x, y, and θ, while $z = \varphi = \gamma = 0$. The displacement curves of the mass center are shown in Figure 5.

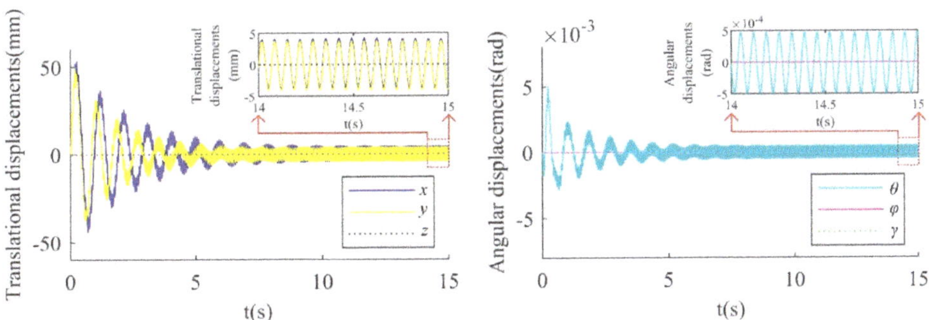

Figure 5. Displacement curves of the mass center under normal conditions, including two translational displacements (x, y) and one angular displacement (θ). Additionally, $z = \varphi = \gamma = 0$.

As shown in Figure 5, the displacements are large initially, then gradually decrease to a stable range. The stable state amplitudes (peak to peak values of displacement) are as follows:

$$x = 7.82 \text{ mm}, \; y = 7.80 \text{ mm}, \; \theta = 4.79 \times 10^{-4} \text{ rad}.$$

Under spring failure conditions, the value of k_{1y} was decreased to 652,260 N/m and the simulation was run again. The system vibrations include x, y, and z, as well as γ, φ, and θ. The displacement curves of the mass center are shown in Figure 6.

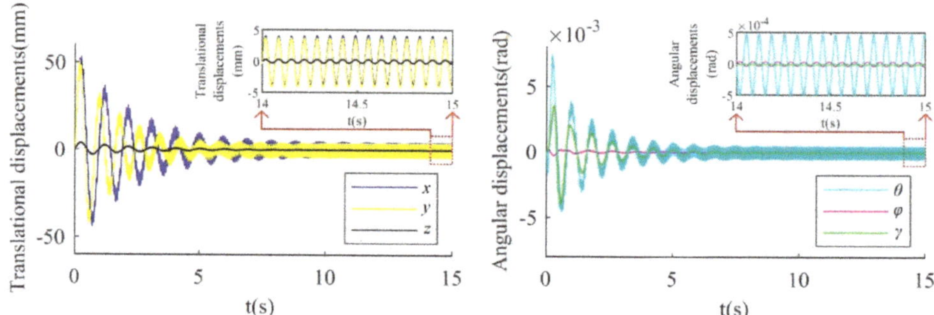

Figure 6. Displacement curves of the rigid body mass center under spring failure conditions, including three translational displacements (x, y, z) and three angular displacements (γ, φ, θ).

As shown in Figure 6, the displacements are large initially, then gradually decrease to a stable range. The stable state amplitudes (peak to peak values of displacement) are as follows:

$x = 7.82$ mm, $y = 7.79$ mm, $z = 0.96$ mm, $\gamma = 0.2 \times 10^{-4}$ rad, $\varphi = 0.75 \times 10^{-4}$ rad, $\theta = 9.40 \times 10^{-4}$ rad.

According to the analysis of the simulations above, the results showed that the four elastic supports of the whole system were symmetric on the x-y plane under normal conditions. The system vibrations included two translations and one rotation; namely, the rigid body only moved in the x–y plane. In addition, the system vibrations changed into a very complicated spatial motion with spring stiffness decrease, which included three translations and three rotations. Meanwhile, the amplitudes changed at the same time.

Therefore, the proposed six-degree-of-freedom model is feasible for exploring the mining vibrating screen dynamic characteristics with spring stiffness decrease caused by spring failures, and vice versa.

3.2. Spring Failure Simulations Results

Under normal conditions, the four elastic support points were symmetrical (point 1 = point 3, point 2 = point 4) in the proposed model. However, this symmetry broke under spring failure conditions, and hence six types of failure were selected for the simulation analysis, as shown in Table 2. Aimed at obtaining the influence rule of the spring failures, only the spring stiffness in the y direction was changed in the simulations.

Table 2. Types of spring failure.

Failures Type	k_1	k_2	k_3	k_4
Single spring failure	$\times$ [1]	$\surd$ [2]	$\surd$	$\surd$
	$\surd$	$\times$	$\surd$	$\surd$
Double spring failure	$\times$	$\times$	$\surd$	$\surd$
	$\times$	$\surd$	$\times$	$\surd$
	$\times$	$\surd$	$\surd$	$\times$
	$\surd$	$\times$	$\surd$	$\times$

Notes: [1] failure; [2] normal.

Due to the difference of each spring's stiffness and stiffness change, the stiffness variation coefficient (SVC) for normalization was defined as:

$$\Delta k_i = \frac{k_{ij0} - k_{ij}}{k_{ij0}} \times 100\%, \ (i = 1,2,3,4 \ ; \ j = 1,2,\ldots,n). \tag{29}$$

In (29), i is the elastic support point sequence number, j is the stiffness sequence number, k_{ij0} is the normal spring stiffness in the y direction (as shown in Table 2), and k_{ij} is the various spring stiffness in the y direction.

Setting the value of β_z as $89°$ in Table 2, the amplitudes of the four elastic support points in all directions were selected to be normal amplitudes. Due to the difference of each amplitude and amplitude change, the amplitude variation coefficient (AVC) for normalization was defined as:

$$\Delta\lambda_{id} = \frac{\lambda_{id0} - \lambda_{id}}{\lambda_{id0}} \times 100\%, \ (i = 1, 2, 3, 4; d = x, y, z). \tag{30}$$

In (30), i is the elastic support point sequence number, d is one of the three directions, λ_{id0} is the normal amplitude of one elastic support point, and λ_{id} is the various amplitudes of the same elastic support point.

3.2.1. Single Spring Failure Simulations Results

In the case of k_1 failures, the spring stiffness variation coefficient (Δk_1) was changed from 0 to 30%, and hence the amplitude variation coefficients of the four elastic support points in all directions changed together.

As shown in Figure 7, if the spring stiffness variation coefficient (Δk_1) increased, the amplitude variation coefficients of all elastic support points in the x direction decreased, while all amplitude variation coefficients in the z direction increased. In the y direction, the amplitude variation coefficients of points 2 and 4 increased, while the amplitude variation coefficients of points 1 and 3 decreased.

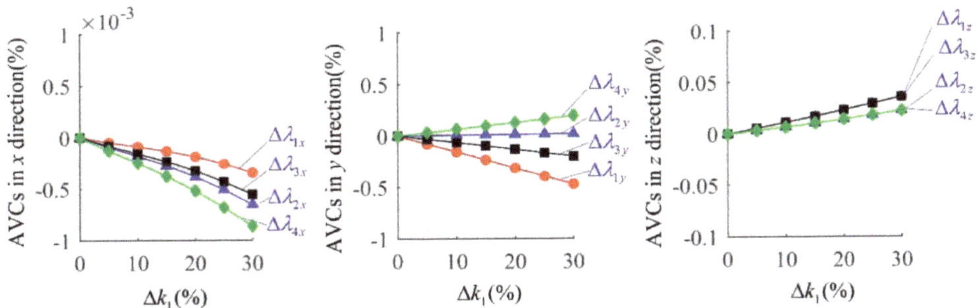

Figure 7. The amplitude variation coefficient curves of four elastic support points, including the amplitude variation coefficients in the x, y, and z directions.

3.2.2. Double Spring Failures Simulations Results

In the case of k_1 and k_2 failures, the spring stiffness variation coefficient (Δk_1 and Δk_2) was changed from 0% to 30%, and hence the amplitude variation coefficients of four elastic support points in all directions changed together.

As shown in Figure 8, if the spring stiffness variation coefficient increased, the amplitude variation coefficients of the four elastic support points in the x direction decreased, increased, or stayed the same (i.e., indeterminate) under the coupling action of k_1 and k_2 failures. The amplitude variation coefficients of all elastic support points in the x direction decreased, increased, or stayed the same (i.e., indeterminate) under the coupling action of k_1 and k_2 failures as well. Meanwhile, the amplitudes of variation coefficients in the z direction always increased, as well as $\Delta\lambda_{1z} = \Delta\lambda_{3z}$ and $\Delta\lambda_{2z} = \Delta\lambda_{4z}$.

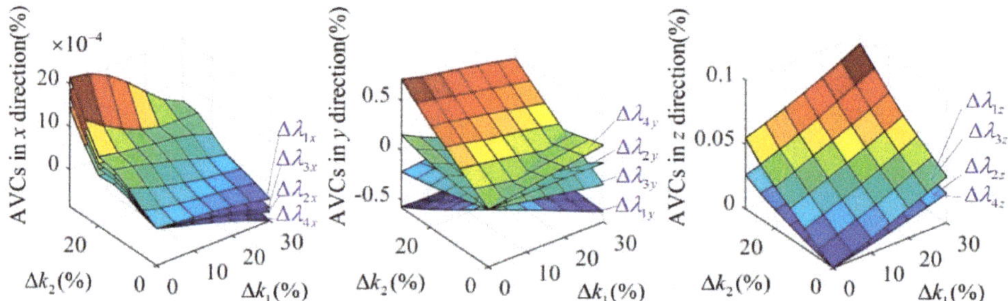

Figure 8. The amplitude variation coefficient surfaces of four elastic support points, including the amplitude variation coefficients in the x, y, and z directions.

3.3. Discussion

Many simulations were carried out with differing failure types, as shown in Table 2, and the change rules between the spring stiffness coefficient and the amplitude variation coefficient were obtained, as shown in Table 3.

Table 3. The change rules between the spring stiffness coefficient and the amplitude variation coefficient.

Stiffness Variation Coefficient	Amplitude Variation Coefficient											
	$\Delta\lambda_{1x}$	$\Delta\lambda_{2x}$	$\Delta\lambda_{3x}$	$\Delta\lambda_{4x}$	$\Delta\lambda_{1y}$	$\Delta\lambda_{2y}$	$\Delta\lambda_{3y}$	$\Delta\lambda_{4y}$	$\Delta\lambda_{1z}$	$\Delta\lambda_{2z}$	$\Delta\lambda_{3z}$	$\Delta\lambda_{4z}$
Δk_1	− [1]	−	−	−	−	+ [2]	−	+	+	+	+	+
Δk_2	+	+	+	+	+	−	+	−	+	+	+	+
$\Delta k_1, \Delta k_2$	± [3]	±	±	±	±	±	±	±	+	+	+	+
$\Delta k_1, \Delta k_3$	−	−	−	−	±	±	+	+	±	±	±	±
$\Delta k_1, \Delta k_4$	±	±	±	±	±	±	±	±	±	±	±	±
$\Delta k_2, \Delta k_4$	+	+	+	+	+	−	+	−	±	±	±	±

[1] decrease; [2] increase; [3] indeterminate.

As shown in Table 3, the amplitude variation coefficient probably increased, decreased or was indeterminate under different spring failures. The change rules of the amplitude variation coefficient are the same as the change rules of the amplitudes, according to Formula (30). Hence, the change rules of the amplitudes of four elastic support points in the x, y, and z directions can be summarized, as follows:

- In the case of spring k_1 failure, the amplitudes of all points in the x direction will decrease, while all amplitudes increase in the z direction. In the y direction, the amplitudes of points 2 and 4 increase, while the amplitudes of points 1 and 3 decrease.

- In the case of spring k_2 failure, all amplitudes will increase in both the x and z directions. In the y direction, the amplitudes of points 1 and 3 increase, while the amplitudes of points 2 and 4 decrease.

- In the case of spring k_1 and spring k_2 failure, all amplitudes will increase in the z direction. In the other directions, the change rules of all amplitudes are indeterminate.

- In the case of spring k_1 and spring k_3 failure, all amplitudes will decrease in the x direction. In the y direction, the amplitudes of points 3 and 4 increase, while the change rules of the amplitudes of points 1 and 2 are indeterminate. In the z direction, the change rules of all amplitudes are indeterminate.

- In the case of spring k_1 and spring k_4 failure, the change rules of all amplitudes are indeterminate in all directions.

- In the case of spring k_2 and spring k_4 failure, all amplitudes will increase in the x direction. In the y direction, the amplitudes of points 1 and 3 increase, while the amplitudes of points 2 and 4 decrease. In the z direction, the change rules of all amplitudes are indeterminate.

The change rules for amplitudes, obtained above, indicated that the spring failures would lead to amplitude change of the four elastic support points in the x, y, and z directions, and the amplitude change rules can reveal certain spring failures. Hence, the amplitude change rules can provide useful information for spring failure diagnosis.

4. Conclusions

The proposed theoretical rigid body model can reveal the dynamic characteristics of a mining vibrating screen, with or without spring failures. From the numerical simulation results, using the Newmark-β method, there are certain relationships between the system amplitudes and different spring failures, which can be used for spring failure diagnosis. This information is useful for operations and maintenance staff, to determine whether it is necessary to change one or more springs. However, further study and experiments need to be done to verify the accuracy of this approach.

Supplementary Materials: The following are available online at http://www.mdpi.com/2227-7390/7/3/246/s1, Matlab Program for dynamic equation establishments and simulations with spring failures.m, Matlab Program for dynamic equation establishments and simulations without spring failures.m.

Author Contributions: Conceptualization, Y.L. and G.M.; Methodology. S.S., L.B., and D.S.; Validation, Y.L. and J.S.; Writing—Original Draft Preparation, Y.L.; Writing—Review & Editing, G.M.; and Visualization, J.S.

Funding: This research was funded by the National key research and development program of China, grant number 2016YFC0600900; the National Natural Science Foundation of China, grant number U13611151; and the Yue Qi Distinguished Scholar Project, China University of Mining & Technology, Beijing.

Conflicts of Interest: The authors declare no conflict of interest.

Abbreviations

The list of symbols is as follows:

k_1	Spring of the elastic support point 1 in y direction
k_2	Spring of the elastic support point 2 in y direction
k_3	Spring of the elastic support point 3 in y direction
k_4	Spring of the elastic support point 4 in y direction
k_{1x}	Spring stiffness of the elastic support point 1 in x direction
k_{1y}	Spring stiffness of the elastic support point 1 in y direction
k_{1z}	Spring stiffness of the elastic support point 1 in z direction
k_{2x}	Spring stiffness of the elastic support point 2 in x direction
k_{2y}	Spring stiffness of the elastic support point 2 in y direction
k_{2z}	Spring stiffness of the elastic support point 2 in z direction
k_{3x}	Spring stiffness of the elastic support point 3 in x direction
k_{3y}	Spring stiffness of the elastic support point 3 in y direction
k_{3z}	Spring stiffness of the elastic support point 3 in z direction
k_{4x}	Spring stiffness of the elastic support point 4 in x direction
k_{4y}	Spring stiffness of the elastic support point 4 in y direction
k_{4z}	Spring stiffness of the elastic support point 4 in z direction
i	Elastic support points number
o	Origin of the inertial frame
x	x-axis
x	Translation displacement of the rigid body in x direction
x	Coordinate of the mass center in the inertial frame
x_d	Coordinate of one point of the rigid body in the inertial frame
$\mathbf{x}$	Displacement column vector
$\dot{\mathbf{x}}$	Velocity column vector
$\ddot{\mathbf{x}}$	Acceleration column vector
y	y-axis
y	*Translation displacement of the rigid body in y direction*

y	Coordinate of the mass center in the inertial frame
y_d	Coordinate of one point of the rigid body in the inertial frame
z	z-axis
z	*Translation displacement of the rigid body in z* direction
z	Coordinate of the mass center in the inertial frame
z_d	Coordinate of one point of the rigid body in the inertial frame
o'	Origin of the body frame
x'	x'-axis
x'	Coordinate of the mass center in the body frame
y'	y'-axis
y'	Coordinate of the mass center in the body frame
z'	z'-axis
z'	Coordinate of the mass center in the body frame
r_1	Distance between the mass center of the rigid body and the elastic support point 1
r_2	Distance between the mass center of the rigid body and the elastic support point 2
r_3	Distance between the mass center of the rigid body and the elastic support point 3
r_4	Distance between the mass center of the rigid body and the elastic support point 4
α_1	Angle between r_1 and z-axis
α_2	Angle between r_2 and z-axis
α_3	Angle between r_3 and z-axis
α_4	Angle between r_4 and z-axis
m	Mass of rigid body
J_x	Moment of inertia of the rigid body rotation around ox-axis
J_y	Moment of inertia of the rigid body rotation around oy-axis
J_z	Moment of inertia of the rigid body rotation around oz-axis
γ	Angular *of the rigid body* rotation around ox-axis
φ	Angular *of the rigid body* rotation around oy-axis
θ	Angular *of the rigid body* rotation around oz-axis
f	Exciting force
F	System force vector
β_x	Angle between exciting force vector and $o'x'$-axis
β_y	Angle between exciting force vector and $o'y'$-axis
β_z	Angle between exciting force vector and $o'z'$-axis
T_x	Rotation matrix of *the rigid body* rotation around ox-axis
T_y	Rotation matrix of *the rigid body* rotation around oy-axis
T_z	Rotation matrix of *the rigid body* rotation around oz-axis
R	Rotation matrix of the rigid body in the sequence of oz-oy-ox
s_1	$= r_1 \sin \alpha_1$
s_2	$= r_2 \sin \alpha_2$
s_3	$= r_3 \sin \alpha_3$
s_4	$= r_4 \sin \alpha_4$
c_1	$= r_1 \cos \alpha_1$
c_2	$= r_2 \cos \alpha_2$
c_3	$= r_3 \cos \alpha_3$
c_4	$= r_4 \cos \alpha_4$
U	System potential energy
E	System kinetic energy
$\mathbf{M}$	Mass matrix
$\mathbf{K}$	Stiffness matrix
C	Damping matrix
ω	Circular frequency of exciting force
ω_x	Angular velocity *of the rigid body* rotation around ox-axis
ω_y	Angular velocity *of the rigid body* rotation around oy-axis
ω_z	Angular velocity *of the rigid body* rotation around oz-axis
X_0	Steady state solution of the forced vibration system

Y_0	Steady state solution of the forced vibration system
Z_0	Steady state solution of the forced vibration system
Γ_0	Steady state solution of the forced vibration system
Φ_0	Steady state solution of the forced vibration system
Θ_0	Steady state solution of the forced vibration system
t_m	Total time of simulation
dt	Time step of simulation
Δk_1	Spring stiffness variation coefficient of spring k_1
Δk_1	Spring stiffness variation coefficient of spring k_2
Δk_1	Spring stiffness variation coefficient of spring k_3
Δk_1	Spring stiffness variation coefficient of spring k_4
Δk_1	Spring stiffness variation coefficient of spring k_1
Δk_1	Spring stiffness variation coefficient of spring k_2
Δk_1	Spring stiffness variation coefficient of spring k_3
Δk_1	Spring stiffness variation coefficient of spring k_4
$\Delta \lambda_{1x}$	Amplitude variation coefficient of elastic support point 1 in x direction
$\Delta \lambda_{1y}$	Amplitude variation coefficient of elastic support point 1 in y direction
$\Delta \lambda_{1z}$	Amplitude variation coefficient of elastic support point 1 in z direction
$\Delta \lambda_{2x}$	Amplitude variation coefficient of elastic support point 2 in x direction
$\Delta \lambda_{2y}$	Amplitude variation coefficient of elastic support point 2 in y direction
$\Delta \lambda_{2z}$	Amplitude variation coefficient of elastic support point 2 in z direction
$\Delta \lambda_{3x}$	Amplitude variation coefficient of elastic support point 3 in x direction
$\Delta \lambda_{3y}$	Amplitude variation coefficient of elastic support point 3 in y direction
$\Delta \lambda_{3z}$	Amplitude variation coefficient of elastic support point 3 in z direction
$\Delta \lambda_{4x}$	Amplitude variation coefficient of elastic support point 4 in x direction
$\Delta \lambda_{4y}$	Amplitude variation coefficient of elastic support point 4 in y direction
$\Delta \lambda_{4z}$	Amplitude variation coefficient of elastic support point 4 in z direction

References

1. Liu, C.S.; Zhang, S.M.; Zhou, H.P.; Li, J.; Xia, Y.F.; Peng, L.P.; Wang, H. Dynamic analysis and simulation of four-axis forced synchronizing banana vibrating screen of variable linear trajectory. *J. Cent. South Univ.* **2012**, *19*, 1530–1536. [CrossRef]
2. Wang, H.; Liu, C.S.; Peng, L.P.; Jiang, X.W.; Ji, L.Q. Dynamic analysis of elastic screen surface with multiple attached substructures and experimental validation. *J. Cent. South Univ.* **2012**, *19*, 2910–2917. [CrossRef]
3. Song, B.C.; Liu, C.S.; Peng, L.P.; Li, J. Dynamic analysis of new type elastic screen surface with multi degree of freedom and experimental validation. *J. Cent. South Univ.* **2015**, *22*, 1334–1341. [CrossRef]
4. Zhang, X.L.; Zhao, C.Y.; Wen, B.C. Theoretical and experimental study on synchronization of the two homodromy exciters in a non-resonant vibrating system. *Shock Vib.* **2013**, *20*, 327–340. [CrossRef]
5. Zhang, Z.R.; Wang, Y.Y.; Fan, Z.M. Similarity Analysis between Scale Model and Prototype of Large Vibrating Screen. *Shock Vib.* **2015**. [CrossRef]
6. Dong, K.J.; Wang, B.; Yu, A.B. Modeling of Particle Flow and Sieving Behavior on a Vibrating Screen: From Discrete Particle Simulation to Process Performance Prediction. *Ind. Eng. Chem. Res.* **2013**, *52*, 11333–11343. [CrossRef]
7. Makinde, O.A.; Ramatsetse, B.I.; Mpofu, K. Review of vibrating screen development trends: Linking the past and the future in mining machinery industries. *Int. J. Miner. Process.* **2015**, *145*, 17–22. [CrossRef]
8. Peng, L.; Liu, C.; Wang, H. Health identification for damping springs of large vibrating screen based on stiffness identification. *J. China Coal Soc.* **2016**, *41*, 1568–1574.
9. Jiang, H.S.; Zhao, Y.M.; Duan, C.L.; Yang, X.L.; Liu, C.S.; Wu, J.D.; Qiao, J.P.; Diao, H.R. Kinematics of variable-amplitude screen and analysis of particle behavior during the process of coal screening. *Powder Technol.* **2017**, *306*, 88–95. [CrossRef]
10. Peng, L.P.; Liu, C.S.; Song, B.C.; Wu, J.D.; Wang, S. Improvement for design of beam structures in large vibrating screen considering bending and random vibration. *J. Cent. South Univ.* **2015**, *22*, 3380–3388. [CrossRef]

11. Zhang, Z.R. Strain modal analysis and fatigue residual life prediction of vibrating screen beam. *J. Meas. Eng.* **2016**, *4*, 217–223. [CrossRef]

12. Li, Z.; Tong, X. Modeling and parameter optimization for vibrating screens based on AFSA-SimpleMKL. *Chin. J. Eng. Des.* **2016**, *23*, 181–187.

13. Xiao, J.; Tong, X. Characteristics and efficiency of a new vibrating screen with a swing trace. *Particuology* **2013**, *11*, 601–606. [CrossRef]

14. Baragetti, S. Innovative structural solution for heavy loaded vibrating screens. *Miner. Eng.* **2015**, *84*, 15–26. [CrossRef]

15. Baragetti, S.; Villa, F. A dynamic optimization theoretical method for heavy loaded vibrating screens. *Nonlinear Dyn.* **2014**, *78*, 609–627. [CrossRef]

16. Peng, L.P.; Fang, R.X.; Fen, H.H.; Zhang, L.; Ma, W.D.; He, X.D. A more accurate dynamic model for dual-side excitation large vibrating screens. *J. Vibroeng.* **2018**, *20*, 858–871.

17. Du, C.L.; Gao, K.D.; Li, J.P.; Jiang, H. Dynamics Behavior Research on Variable Linear Vibration Screen with Flexible Screen Face. *Adv. Mech. Eng.* **2014**, *12*. [CrossRef]

18. Trumic, M.; Magdalinovic, N. New model of screening kinetics. *Miner. Eng.* **2011**, *24*, 42–49. [CrossRef]

19. Dong, K.J.; Yu, A.B.; Brake, I. DEM simulation of particle flow on a multi-deck banana screen. *Miner. Eng.* **2009**, *22*, 910–920. [CrossRef]

20. Rodriguez, C.G.; Moncada, M.A.; Dufeu, E.E.; Razeto, M.I. Nonlinear Model of Vibrating Screen to Determine Permissible Spring Deterioration for Proper Separation. *Shock Vib.* **2016**. [CrossRef]

21. Peng, L.P.; Liu, C.S.; Wu, J.D.; Wang, S. Stiffness identification of four-point-elastic-support rigid plate. *J. Cent. South Univ.* **2015**, *22*, 159–167. [CrossRef]

22. Baranowski, L. Equations of motion of a spin-stabilized projectile for flight stability testing. *J. Theor. Appl. Mech.* **2013**, *51*, 235–246.

23. Bai, L.; Dong, Z.F.; Ge, X.S. The closed-loop kinematics modeling and numerical calculation of the parallel hexapod robot in space. *Adv. Mech. Eng.* **2017**, *9*, 15. [CrossRef]

24. Bai, L.; Ma, L.H.; Dong, Z.F.; Ge, X.S. Kinematics, Dynamics, and Optimal Control of Pneumatic Hexapod Robot. *Math. Probl. Eng.* **2017**. [CrossRef]

Article

Kinematics in the Information Age

Brendon Smeresky [1], Alexa Rizzo [1] and Timothy Sands [2],*

[1] Department of Mechanical and Aerospace Engineering, Naval Postgraduate School, Monterey, CA 93943, USA; bpsmeres@nps.edu (B.S.); akrizzo@nps.edu (A.R.)
[2] Department of Mechanical Engineering, Stanford University, Stanford, CA 94305, USA
*   Correspondence: dr.timsands@stanford.edu; Tel.: +1-831-656-3954

Received: 27 June 2018; Accepted: 21 August 2018; Published: 27 August 2018

Abstract: Modern kinematics derives directly from developments in the 1700s, and in their current instantiation, have been adopted as standard realizations . . . or templates that seem unquestionable. For example, so-called *aerospace sequences* of rotations are ubiquitously accepted as the norm for aerospace applications, owing from a recent heritage in the space age of the late twentieth century. With the waning of the space-age as a driver for technology development, the information age has risen with the advent of digital computers, and this begs for re-evaluation of assumptions made in the former era. The new context of the digital computer defines the use of the term "information age" in the manuscript title and further highlights the novelty and originality of the research. The effects of selecting different Direction Cosine Matrices (DCM)-to-Euler Angle rotations on accuracy, step size, and computational time in modern digital computers will be simulated and analyzed. The experimental setup will include all twelve DCM rotations and also includes critical analysis of necessary computational step size. The results show that the rotations are classified into symmetric and non-symmetric rotations and that no one DCM rotation outperforms the others in all metrics used, yielding the potential for trade space analysis to select the best DCM for a specific instance. Novel illustrations include the fact that one of the ubiquitous sequences (the "313 sequence") has degraded relative accuracy measured by mean and standard deviations of errors, but may be calculated faster than the other ubiquitous sequence (the "321 sequence"), while a lesser known "231 sequence" has comparable accuracy and calculation-time. Evaluation of the 231 sequence also illustrates the originality of the research. These novelties are applied to spacecraft attitude control in this manuscript, but equally apply to robotics, aircraft, and surface and subsurface vehicles.

Keywords: Phoronomics; mechanics; kinetics; kinematics; direction cosines; Euler angles; space dynamics; digital computation; control systems; control engineering

---

1. Introduction

The discipline of kinematics in its current form has a lengthy history that hark back to at least 1775 with Euler's formulations [1], with almost immediate expansion throughout the nineteenth [2–4] and twentieth centuries [5–29]. There was a particular renaissance in the late twentieth century accompanying the race between the then-Soviet Union and the United States to spaceflight, and its accompanying application toward nuclear deterrence, where considerable lessons from that period (both technical and non-technical) have been expressed in subsequent literature [30–62]. From this distinguished lineage, terminology has converged to refer to sequential rotation sequences (e.g., xyz or 123); which are called *aerospace sequences* about non-repeating axes (also referred to as "Tait-Bryan angles"), while the *orbit sequences* have an axis repeated in the rotation sequence (e.g., xyx or 121, also referred to as "proper Euler angles"), [63]. One non-repeating sequence in particular (commonly called either a 321 or 123 sequence) has become the ubiquitous aerospace sequence. These cited manuscripts substantiate specific technical applications of the orbital and aerospace sequences, and those technical applications are the focus of this research in hopes of improved performance. With the rise of digital computation in the Information Age, this research critically evaluates the options (seeking diverging truths for the modern times) by addressing such questions as: Is the ubiquitous aerospace sequence (123 or alternatively 321) the best rotation sequence? Evaluation will be driven by two figures of merit: (1) mean and standard deviations of errors indicating how well each rotation sequence represents true roll, pitch, and yaw angles, and (2) computation time to reveals relative numerical superiority in the context of digital computers of the current state of the art. Analysis and results demonstrate the fact that 321 and 123 rotation sequences result in disparate errors and computation time, with the former being relatively superior. Furthermore, the 123 rotation was significantly slower than all the other rotations. Secondly, the symmetric rotations were on average slower than the non-symmetric rotations, despite the same mathematical process and number of steps to solve for the Euler Angles. Lastly, the fastest non-symmetric rotation was the 321 and the fastest symmetric was the 232, slightly faster than the 121 rotation. Taking all Direction Cosine Matrices (DCM) rotations into account, the 232 rotation was the fastest.

The significance of this research cannot be overstated. The current state of the art uses rotational sequences borne from a different era under a different paradigm, but the success of spaceflight has solidified those older results into the current psyche. This manuscript illustrates that improved errors and computational speed are both possible; and in keeping with the acceptance of the older paradigm by evolution of spaceflight, the context of this research is rotational mechanics [62] applied to spacecraft attitude control systems. These advancements complement advanced algorithms [37–45] for nonlinear adaptive system identification [55–59] and control [46–54] permitting improved performance of space missions [35,36,60] in a time when the United States has a pre-occupation with low-end conflicts in the middle east amidst an increasing belligerent world of threats [30–34]. This realization culminated in the recent edict to create a new military service in the US. purely dedicated to space [61].

2. Materials and Methods

The goal of a spacecraft's Attitude Control System (ACS) is to have a functional system that can move to and hold a specific orientation in three dimensional space, relative to an inertial frame. With regard to classical and rigid body mechanics, the ACS takes into account the Kinetics, Kinematics, Orbital Frame, and Disturbances to control this motion. Figure 1 depicts this process and details the computational steps from desired angle inputs to Euler Angle outputs in the sequence of inputs (from the white blocks in Figure 1) through light grey calculations to dark grey outputs: φ, θ, and ψ. Section 2 will explain the theory behind this control system, Section 3 will detail the experimental setup, and Section 4 will show the results and analysis.

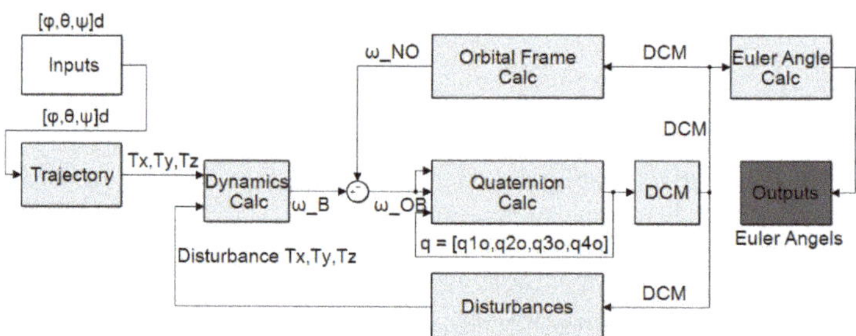

Figure 1. Overall technical roadmap of the overall process: Euler Angle for Euler's Moment Equation driven by a trajectory-fed feedforward controller.

2.1. Theory of Dynamics

Dynamics is synonymous with Mechanics. Newton called dynamics the science of machines which may be divided into two parts: statics (later called kinematics) and kinetics [2]. Chasle's Theorem articulates how a complete description of motion may be described as a screw displacement comprised of translation in accordance with Newton's Law and rotations in accordance with Euler's Moment Equations $T = J\dot{\omega} + \omega \times J\omega$ [6], where $[J]$ is a matrix of mass moments of inertia explained by Kane [23]. Investigation of motion without consideration of the nature of the body that is moved or how the motion is produced is called *Phoronomics*, or "the laws of going", or more commonly but less properly kinematics [13], to be elaborated in Section 2.1.2. The rotation maneuver from one position to another is measured from the inertial reference frame or $[X_I, Y_I, Z_I]$ to the final position, the body reference frame or $[X_B, Y_B, Z_B]$. For this simulation, a model was created to rotate from orientation A, $[X_A, Y_A, Z_A]$ to orientation B, $[X_B, Y_B, Z_B]$. Since the dot product of two unit vectors is the cosine of the angle between them [25], it is referred to in older works as a direction cosine, which may also be used to describe a satellite in an inclined earth orbit [17] or to express the orientation of the perifocal reference frame with respect to the geocentric-equatorial reference frame [26]. Direction angles are the angles between each coordinate axis and the individual components of the vector. The direction cosines are simply the cosines of these angles [28]. The nature of direction cosines matrices is merely to assemble the direction cosines which completely specify the relative orientation of two coordinate systems [18], thus their appeal as universally applicable tools of kinematics.

2.1.1. Kinetics

Kinetics, or Dynamics, is the process of describing the motion of objects with focus on the forces involved. In the inertial frame, Newton's $F = ma$ is applied but becomes Euler's $T = J\dot{\omega}$ when rotation is added, where $T = J\dot{\omega}$ is expressed in the inertial reference frame's coordinates, while $T = J\dot{\omega} + \omega \times J\omega$ from above is still measured in the inertial frame, but expressed in body coordinates.

Combining the Euler and Newton equations, we can account for all six degrees of freedom. In application, when an input angle $[\varphi_d, \theta_d, \psi_d]$ is commanded, the feedforward control uses (1) as the ideal controller with (2) as the sinusoidal trajectory to calculate the required torque $[T_x, T_y, T_z]$ necessary to achieve the desired input angle. The Dynamics calculator then uses (3) to convert the torques into ω_B values, where ω_B is defined as the angular velocity of the body. In order to calculate this, the non-diagonal terms in (4) are neglected, removing coupled motion and leaving only the principle moments of inertia. Then, the inertia matrix J is removed from $J\dot{\omega}$, and the remaining $\dot{\omega}$ is integrated into $[\omega_x, \omega_y, \omega_z]$, which is fed into the Kinematics block of the model to finally determine the outputted Euler Angles.

$$T_d = J\dot{\omega}_d + \omega_d \times J\omega_d \tag{1}$$

$$\theta = \frac{1}{2}\left(A + Asin\left(\omega_f t + \varphi\right)\right) \tag{2}$$

$$T = \dot{H}_i = J\dot{\omega}_i + \omega_i \times J\omega_i \tag{3}$$

2.1.2. Kinematics, Phoronomics, or "The Laws of Going"

Formulation of spacecraft attitude dynamics and control problems involves considerations of kinematics, especially as it pertains to the orientation of a rigid body that is in rotational motion. The subject of kinematics is mathematical in nature, because it does not involve any forces associated with motion. The kinematic representation of the orientation of one reference frame relative to another reference can also be expressed by introducing the time-dependence of Euler Angles. The so-called body-axis rotations involve successive rotations three times about the axes of the rotated body-fixed reference frame resulting in twelve possible sets of Euler angles. The so-called space-axis rotations instead involve three successive rotations using axes fixed in the inertial frame of reference, again producing twelve possible sets of Euler angles. Because the body-axis and space-axis rotations are intimately related, only twelve Euler angle possibilities need be investigated; and the twelve sets from the body-axis sequence are typically used [26]. Consider a rigid body fixed at a stationary point whose inertia ellipsoid at the origin is an ellipsoid of revolution whose center of gravity lies on the axis of symmetry. Rotation around the axis of symmetry does not change the Lagrangian function, so there must-exist a first integral which is a projection of an angular momentum vector onto the axis of symmetry. Three coordinates in the configuration space special orthogonal group (3) may be used to form a local coordinate system, and these coordinates are called the *Euler angles*.

Key tools of kinematics from which the Euler angles may be derived include direction cosines which describe orientation of the body set of axes relative to an external set of axes. Euler's angles may be defined by the following set of rotations: "rotation about x axis by angle and θ, rotation about z' axis by an angle ψ, then rotation about the original z-axis by angle φ". Eulerian angles have several "conventions: Goldstein uses [22] the "x-convention": z-rotation followed by x' rotation, followed by z' rotation (essentially a 3-1-3 sequence). Quantum mechanics, nuclear physics, and particle physics the "y-convention" is used: essentially a 3-2-3 rotation). Both of these have drawbacks, that the primed coordinate system is only slightly different than the unprimed system, such that, φ and ψ become indistinguishable, since their respective axes of rotation (z and z') are nearly coincident. The so-called *Tait-Bryan* convention in Figure 2 therefore gets around this problem by making each of the three rotations about different axes: (essentially a 3-2-1 sequence) [22].

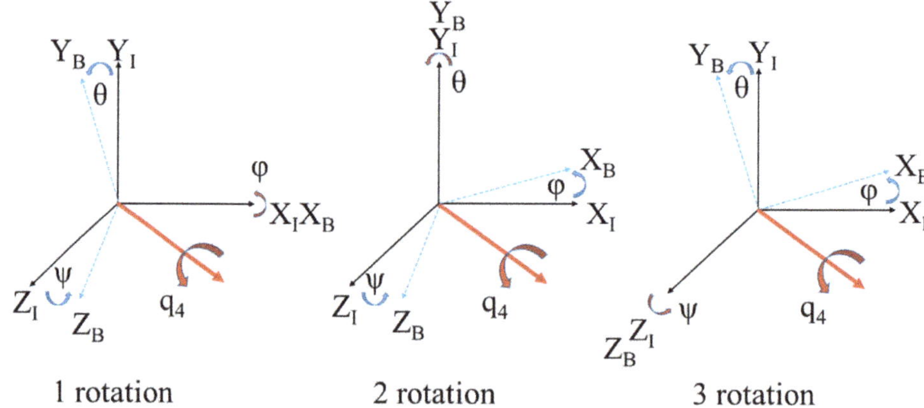

Figure 2. Execution of a 3-2-1 rotation from C^A to C^B (left to right); blue-dotted arrows denote angle rotations. A direct rotation from C^A to C^B can be made about the Euler Axis, q_4 in red. The set of three rotations may be depicted as four rectangular parallelepipeds, where each contains the unit vectors of the corresponding reference frame [29].

Kinematics is the process of describing the motion of objects without focus on the forces involved. The $[\omega_x, \omega_y, \omega_z]$ values from the Dynamics are fed into the Quaternion Calculator where (5) and (6) yield q, the Quaternion vector. The Quaternions define the Euler axis in three dimensional space using $[q_1, q_2, q_3]$. About this axis, a single angle of rotation $[q_4]$ can resolve an object aligned in reference frame A into reference frame B. The Direction Cosine Matrix (DCM) then relates the input ω values to the Euler Angles using one of 12 permutations of possible rotation sequences, where multiple rotations can be made in sequence. Therefore, the rows of the DCM show the axes of Frame A represented in Frame B, the columns show the axes of Frame B represented in Frame A, and φ, θ, and ψ are the angles of rotation that must occur in each axis sequentially to rotate from orientation A to orientation

B, turning C^A to C^B. Figure 2 depicts a 3-2-1 sequence to rotate from C^A to C^B, where the Euler Axis is annotated by the thickest line.

$$\begin{bmatrix} J_{xx}\dot{\omega}_x + J_{xy}\dot{\omega}_y + J_{xz}\dot{\omega}_z - J_{xy}\omega_x\omega_z - J_{yy}\omega_y\omega_z - J_{yz}\omega_z^2 + J_{xz}\omega_x\omega_y + J_{zz}\omega_z\omega_y + J_{yz}\omega_y^2 \\ J_{yx}\dot{\omega}_x + J_{yy}\dot{\omega}_y + J_{yz}\dot{\omega}_z - J_{yz}\omega_x\omega_y - J_{zz}\omega_x\omega_z - J_{xz}\omega_x^2 + J_{xx}\omega_x\omega_z + J_{xy}\omega_z\omega_y + J_{xz}\omega_z^2 \\ J_{zx}\dot{\omega}_x + J_{zy}\dot{\omega}_y + J_{zz}\dot{\omega}_z - J_{xx}\omega_x\omega_y - J_{xz}\omega_y\omega_z - J_{xy}\omega_y^2 + J_{yy}\omega_x\omega_y + J_{yz}\omega_z\omega_x + J_{xy}\omega_x^2 \end{bmatrix} = \begin{bmatrix} T_x \\ T_y \\ T_z \end{bmatrix} \tag{4}$$

$$\begin{bmatrix} \dot{q}_1 \\ \dot{q}_2 \\ \dot{q}_3 \\ \dot{q}_4 \end{bmatrix} = \frac{1}{2}\begin{bmatrix} 0 & \omega_3 & -\omega_2 & \omega_1 \\ -\omega_3 & 0 & \omega_1 & \omega_2 \\ \omega_2 & -\omega_1 & 0 & \omega_3 \\ -\omega_1 & -\omega_2 & -\omega_3 & 0 \end{bmatrix}\begin{bmatrix} q_1 \\ q_2 \\ q_3 \\ q_4 \end{bmatrix} = \frac{1}{2}\begin{bmatrix} q_4 & -q_3 & q_2 & q_1 \\ q_3 & q_4 & -q_1 & q_2 \\ -q_2 & q_1 & q_4 & q_3 \\ -q_1 & -q_2 & -q_3 & q_4 \end{bmatrix}\begin{bmatrix} w_1 \\ w_2 \\ w_3 \\ 0 \end{bmatrix} \tag{5}$$

$$\begin{bmatrix} 1-2(q_2^2+q_3^2) & 2(q_1q_2+q_3q_4) & 2(q_1q_3-q_2q_4) \\ 2(q_2q_1-q_3q_4) & 1-2(q_1^2+q_3^2) & 2(q_2q_3+q_1q_4) \\ 2(q_3q_1+q_2q_4) & 2(q_3q_2-q_1q_4) & 1-2(q_1^2+q_2^2) \end{bmatrix} = \begin{bmatrix} C_2C_3 & C_2S_3 & -S_2 \\ S_1S_2C_3-C_1S_3 & S_1S_2S_3+C_1C_3 & S_1C_2 \\ C_1S_2C_3-S_1S_3 & C_1S_2S_3-S_1C_3 & C_1C_2 \end{bmatrix} \tag{6}$$

2.1.3. The Orbital Frame

In order to more completely represent a maneuvering spacecraft, orbital motion must be included with the Kinematics. This relationship is represented in Figure 1, where the output of the DCM is fed into the Orbital Frame Calculator, and the second column of the DCM is multiplied against the orbital velocity of the spacecraft. The second column of the DCM represents the Y axis of Frame B projected in the X, Y, and Z axes of Frame A. This yields ω^{NO}, the orbital velocity relative to the Inertial Frame. Using (7), this velocity is removed from the velocity of the body relative to the Inertial Frame, leaving only the velocity of the body relative to the Orbital Frame for further calculations.

$$\omega^{OB} = \omega^{NB} - \omega^{NO} \tag{7}$$

2.1.4. Disturbances

Multiple disturbances torques exist that effect the motion of a spacecraft in orbit, two of which are addressed in this paper. The first is the disturbance due to gravity acting upon an object in orbit, where the force due to gravity decreases as the distance between objects increases. The force is applied as a scaling factor to the mass distribution around the Z axis of a spacecraft. This force applied to a mass offset from the center of gravity is calculated through the cross product found in (8) and yields an output torque about the Z axis.

The second disturbance is an aerodynamic torque due to the force of the atmosphere acting upon a spacecraft, which also decreases as the altitude increases. In (9), the force due to air resistance is calculated by scaling the direction of orbital velocity by the atmospheric density, drag coefficient, and magnitude of orbital velocity. This force then acts upon the center of pressure, which is offset from the center of gravity, and yields a torque about the Z axis, due to the cross product in (9).

The disturbances are additive and act upon the dynamics in Figure 1. Because the ideal feedforward controller is the dynamics, an offsetting component equal to the negative anticipated disturbances can be used to negate the disturbance torque. This results in nullifying the disturbances when the two are summed to produce ω^{OB}, the velocity of the body relative to the Inertial Frame.

$$T_g = 3\frac{\mu}{R^3}\hat{z} \times J\hat{z} \tag{8}$$

$$T_a = C_p \times f_a = C_p \times \left[\left(\rho_a V_R^2 A_p\right)\hat{V}_R\right] \tag{9}$$

2.2. Experimental Setup

This experiment implemented and compared the 12 DCM to Euler Angle rotations using a variable step size. An angle of $[\varphi, \theta, \psi] = [30, 0, 0]$ was commanded, with the quaternion and torques initialized as $T = [0, 0, 0]$ and $q = [0, 0, 0, 1]$. The spacecraft had an inertia matrix of $J = [2, 0.1, 0.1; 0.1, 2, 0.1; 0.1, 0.1, 2]$. The orbital altitude was set at 150 km with a drag coefficient of 2.5. Both orbital motion and torque disturbances were turned off.

Each simulation executed over a 5 s quiescent period, 5 s maneuver time, and 5 s post maneuver observation period, totaling 15 s. The sinusoidal trajectory was calculated to have $\omega_f = \pi/2$ and $\varphi = \pi/2$.

The model was built in Matlab and Simulink, where integrations were calculated using the Runge–Kutta solver (ode4) with variable time steps of 0.1, 0.001, and 0.0001 s. Euler Angles were resolved using the 12 unique DCM rotation sequences with the atan2 function.

Three Figures of Merit were used to assess performance. The first two were the mean and standard deviation between the Euler Angles and Body Angles. The third was the calculation time for each rotation as a measure of complexity.

3. Experimental Results and Analysis

3.1. Euler Angle Calculations and Post-Processing

Each of the Euler angles was derived using the DCM and rotation matrices, creating a relationship like (7), but unique to each rotation. φ, θ, and ψ were isolated in this relationship as a method to calculate the Euler Angles. Once calculated, the Euler Angles were implemented in the simulation. However, when a [30, 0, 0] maneuver was commanded, discontinuities due to trigonometric quadrant error manifested. Post-processing removed the error, but yielded output rotation did not match the input command. In order to correct this, the derivations for each Euler Angle were revised to correlate six of the 12 rotations, yielding the results in Figure 3. Therefore, the rotations in Figure 3 are classified into two groups: the upper six non-symmetric rotations and lower six symmetric rotations. An example of symmetric rotations is 121, while a 132 is non-symmetric rotation. The commanded input and output maneuvers were not correlated for the 6 symmetric rotations.

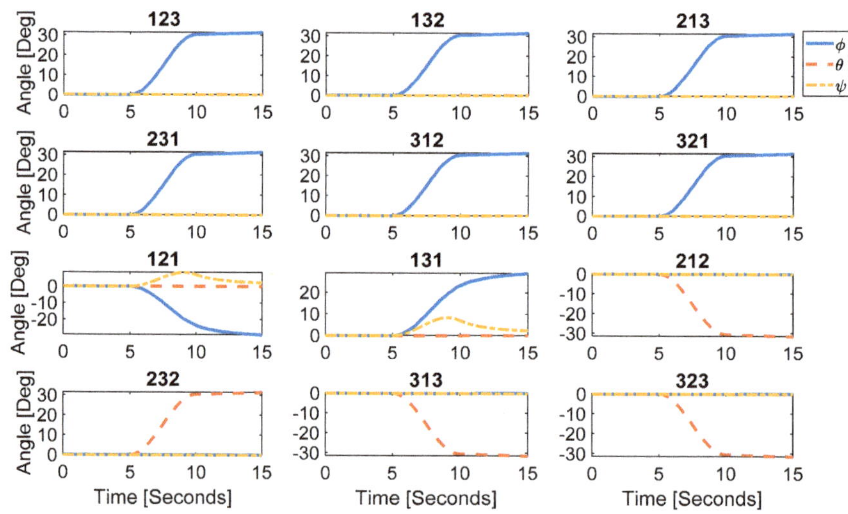

Figure 3. Corrected Euler Angles vs time for all 12 DCM rotations.

3.2. Euler Angle to Body Angle Accuracy

The output Euler Angles are not the same as the commanded Body Angles, but measuring this delta is a method of determining accuracy. Figure 4 depicts the deviation over time and Table 1 provides the associated mean values and standard deviations for each of the rotations.

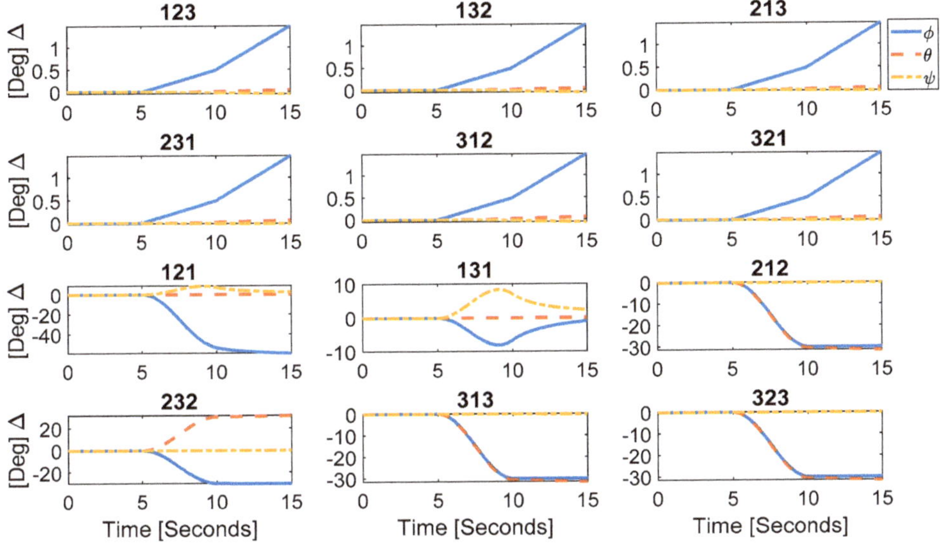

Figure 4. Euler and Body Angle deviation, using a 0.1 step size.

Table 1. Mean and standard deviation for all 12 rotations, using a 0.1 step size.

	Mean			Standard Deviation		
DCM	ϕ	θ	ψ	ϕ	θ	ψ
123	0.413	0.011	0.011	0.462	0.015	0.014
132	0.413	0.010	0.013	0.462	0.013	0.016
213	0.413	0.011	0.005	0.462	0.015	0.006
231	0.413	0.014	0.005	0.462	0.018	0.005
312	0.413	0.016	0.013	0.462	0.021	0.016
321	0.413	0.014	0.005	0.462	0.018	0.005
121	27.544	0.015	2.869	25.804	0.019	2.823
131	2.456	0.015	2.869	2.680	0.019	2.823
212	14.977	15.413	0.010	13.726	14.150	0.010
232	15.010	15.413	0.010	13.757	14.150	0.010
313	14.980	15.413	0.028	13.728	14.150	0.034
323	14.977	15.413	0.010	13.725	14.150	0.010

The six non-symmetric rotations show consistent error in φ, and only begin to deviate beyond the fifth decimal place in both mean error and standard deviation. While φ is commanded to change to 30°, θ and ψ are expected to remain at zero, but show non-zero values due to error incurred by step size.

The six symmetric rotations are substantially harder to draw conclusions from because of the uncorrelated rotations. The mean error and standard deviation values are drastically different from each other in Table 1 and visibly deviate in Figure 4. Therefore, further correlation is required to analyze accuracy. Table 1 values were calculated over the 15 s simulation time, noting that some sequences had not reached steady-state values making their error values even larger compared to others in Table 1 if the simulations has been run until steady state was reached.

3.3. Step Size Versus Accuracy

The simulation step size was altered to determine the effects on accuracy for the 12 rotations. If variable step-size were used, there would be no way to assure a certain level of accuracy, thus fixed step size was utilized and iterated smaller-and-smaller until no discernable accuracy improvement is noted. Figure 5 shows the results of reducing the step size from 0.1 to 0.001 s. Figures 4 and 5 remain comparable, with the 10^2 order of magnitude decrease in step size yielded a comparable 10^2 order of magnitude increase in accuracy. When a step size of 0.0001 s was used, the accuracy increased by another order of magnitude, denoting the trend. Comparing against the accuracy of the rotations, they maintained their relative accuracy; the *132 and 312 remained the most accurate rotations when the step size decreased*, and therefore has limited to no effect.

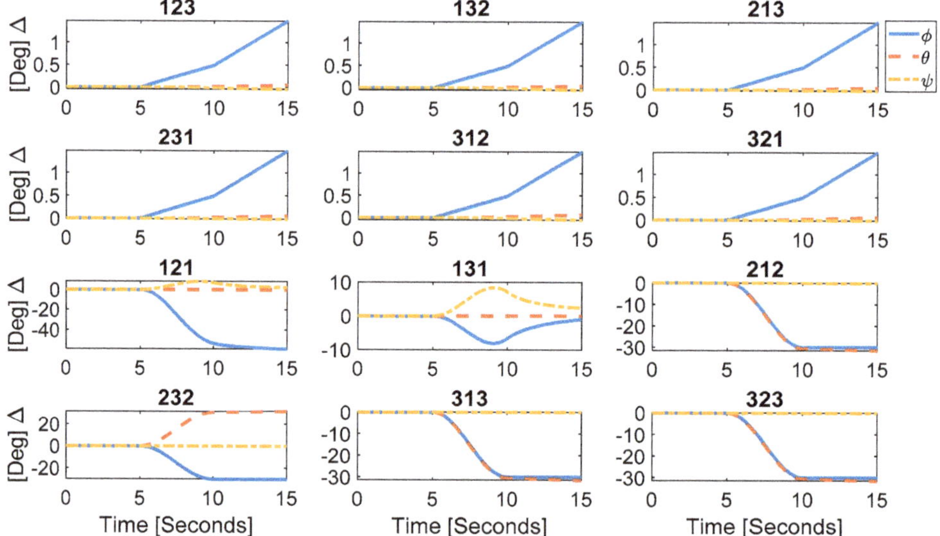

Figure 5. Euler and Body Angle deviation, using a 0.001 step size.

3.4. DCM to Euler Angle Timing

The execution time of each maneuver was standardized at 15 s across all scenarios. Therefore, runtime deviations for each of the 12 rotations are attributable to the complexity of the calculations. Table 2 shows the results for three different step sizes and the runtimes for each rotation. Because the step size affected the simulation timing, comparisons were only valid between rotations of a similar step size; however, relative comparisons between step sizes were valid.

Analyzing the results yields several observations. Firstly, the 123 rotation was significantly slower than all the other rotations. Secondly, the symmetric rotations were on average slower than the non-symmetric rotations, despite the same mathematical process and number of steps to solve for the Euler Angles. Lastly, the fastest non-symmetric rotation was the 321 and the fastest symmetric was the 232, slightly faster than the 121 rotation. Taking all DCM rotations into account, the 232 rotation was the fastest.

Table 2. Simulation run times for all 12 Direction Cosine Matrices (DCM) rotations for a 30° roll maneuver, using 0.1, 0.001, and 0.0001 step sizes.

DCM	Simulation Execution Time [S]		
	0.1 Step size	0.001 Step size	0.0001 Step size
123	8.408	11.836	28.433
132	1.533	6.789	22.187
213	1.419	6.978	22.102
231	1.188	4.436	23.259
312	1.549	4.302	20.971
321	1.018	3.475	21.420
121	0.952	3.715	20.505
131	1.190	4.082	23.331
212	1.015	3.860	21.005
232	0.931	3.710	21.410
313	0.939	3.789	20.908
323	1.091	3.955	22.044

4. Conclusions

This experiment implemented and compared the 12 DCM to Euler Angle rotations using a variable step size. The effects on accuracy, step size, and timing were observed, and the simulation results showed that the DCMs were classified into symmetric and non-symmetric rotations. The non-symmetric rotations were easier to correlate and compare, while the symmetric rotations were not, limiting analysis. Furthermore, no one rotation was the ideal in the analyzed categories. This is beneficial, because trade space analysis can be conducted to determine accuracy, timing, and other high priority design criteria to select the appropriate DCM. The lowest *roll* mean error is obtained by using any of the 123, 132, 213, 231, 312, or 321 rotation sequences, while the lowest *pitch* mean error *cannot be achieved by the ubiquitous 321 sequence, instead the 132 sequence must be used*; while the lowest *yaw* mean error may be achieved with the 213, 231, and 321 sequences. Standard deviations show similar options for selecting different rotation sequences for specific applications. Regarding computational efficiency, the 232 sequence was best, followed by the 313, and then the 121 sequence. The ubiquitously accepted standard 321 sequence was found to be fifth fastest, with four other rotation sequences bearing less computational burden. Novel illustrations include the fact that one of the ubiquitous sequences (the "313 sequence") has degraded relative accuracy measured by mean and standard deviations of errors, but may be calculated faster than the other ubiquitous sequence (the "321 sequence"), while a lesser known "231 sequence" has comparable accuracy and calculation-time. Evaluation of the 231 sequence also illustrates the originality of the research. These novelties are applied to spacecraft attitude control in this manuscript, but can equally be applied to robotics, aircraft, and surface and subsurface vehicles.

Lastly, future research would refine the correlation for symmetric rotations, but furthermore experimental validation will be performed on free-floating spacecraft simulator hardware at the Naval Postgraduate School. The validation will be performed by duplicating one of the specifically cited technical applications (e.g., any of the technical applications in [35–45]) seeking to validate performance improvement.

Author Contributions: B.S. and T.S. conceived and designed the research; B.S. and A.R. conceptualized and developed the methodology, performed the experiments, reviewed the data and validated the results; T.S. performed literature review, wrote the final manuscript, and managed the peer review process. Authorship has been limited to those who have contributed substantially to the work reported.

Funding: This research received no external funding. No sources of funding apply to the study. Grants were not received in support of our research work. No funds were received for covering the costs to publish in open access, instead this was an invited manuscript to the special issue.

Conflicts of Interest: The authors declare no conflict of interest.

References

1. Euler, L. (Euler) Formulae Generales pro Translatione Quacunque Corporum Rigidorum (General Formulas for the Translation of Arbitrary Rigid Bodies), Presented to the St. Petersburg Academy on 9 October 1775. and First Published in *Novi Commentarii Academiae Scientiarum Petropolitanae* 20, 1776, pp. 189–207 (E478) and Was Reprinted in *Theoria motus corporum rigidorum*, ed. nova, 1790, pp. 449–460 (E478a) and Later in His Collected Works Opera Omnia, Series 2, Volume 9, pp. 84–98. Available online: https://math.dartmouth.edu/~euler/docs/originals/E478.pdf (accessed on 22 August 2018).
2. Thompson, W.; Tait, P.G. *Elements of Natural Philosophy*; Cambridge University Press: Cambridge, UK, 1872.
3. Reuleaux, F.; Kennedy Alex, B.W. *The Kinematics of Machinery: Outlines of a Theory of Machines*; Macmillan: London, UK, 1876. Available online: https://archive.org/details/kinematicsofmach00reuluoft (accessed on 22 August 2018).
4. Wright, T.W. *Elements of Mechanics Including Kinematics, Kinetics and Statics*; D. Van Nostrand Company: New York, NY, USA; Harvard University: Cambridge, MA, USA, 1896.
5. Merz, J.T. *A History of European Thought in the Nineteenth Century*; Blackwood: London, UK, 1903; p. 5.
6. Whittaker, E.T. *A Treatise on the Analytical Dynamics of Particles and Rigid Bodies*; Cambridge University Press: Cambridge, UK, 1904.
7. Whittaker, E.T. *A Treatise on the Analytical Dynamics of Particles and Rigid Bodies*; Cambridge University Press: Cambridge, UK, 1917.
8. Whittaker, E.T. *A Treatise on the Analytical Dynamics of Particles and Rigid Bodies*; Cambridge University Press: Cambridge, UK, 1927.
9. Whittaker, E.T. *A Treatise on the Analytical Dynamics of Particles and Rigid Bodies*; Cambridge University Press: Cambridge, UK, 1937.
10. Church, I.P. *Mechanics of Engineering*; Wiley: New York, NY, USA, 1908; p. 111.
11. Wright, T.W. *Elements of Mechanics Including Kinematics, Kinetics, and Statics, with Applications*; Nostrand: New York, NY, USA, 1909.
12. Study, E. ; Delphenich, D.H., Translator; Foundations and goals of analytical kinematics. *Sitzber. d. Berl. Math. Ges.* **1913**, *13*, 36–60. Available online: http://neo-classical-physics.info/uploads/3/4/3/6/34363841/study-analytical_kinematics.pdf (accessed on 14 April 2017).
13. Gray, A. *A Treatise on Gyrostatics and Rotational Motion*; MacMillan: London, UK, 1918; ISBN 978-1-4212-5592-7. (Published 2007).
14. Rose, M.E. *Elementary Theory of Angular Momentum*; John Wiley & Sons: New York, NY, USA, 1957; ISBN 978-0-486-68480-2. (Published 1995).
15. Kane, T.R. *Analytical Elements of Mechanics Volume 1*; Academic Press: New York, NY, USA; London, UK, 1959.
16. Kane, T.R. *Analytical Elements of Mechanics Volume 2 Dynamics*; Academic Press: New York, NY, USA; London, UK, 1961.
17. Thompson, W. *Space Dynamics*; Wiley and Sons: New York, NY, USA, 1961.
18. Greenwood, D. *Principles of Dynamics*; Prentice-Hall: Englewood Cliffs, NJ, USA, 1965; ISBN 9780137089741. (Reprinted in 1988 as 2nd ed.).
19. Fang, A.C.; Zimmerman, B.G. *Digital Simulation of Rotational Kinematics*; NASA Technical Report NASA TN D-5302; NASA: Washington, DC, USA, October 1969. Available online: https://ntrs.nasa.gov/archive/nasa/casi.ntrs.nasa.gov/19690029793.pdf (accessed on 22 August 2018).
20. Henderson, D.M. *Euler Angles, Quaternions, and Transformation Matrices—Working Relationships*; As NASA Technical Report NASA-TM-74839; July 1977. Available online: https://ntrs.nasa.gov/archive/nasa/casi.ntrs.nasa.gov/19770024290.pdf (accessed on 22 August 2018).
21. Henderson, D.M. *Euler Angles, Quaternions, and Transformation Matrices for Space Shuttle Analysis*; Houston Astronautics Division as NASA Design Note 1.4-8-020; 9 June 1977. Available online: https://ntrs.nasa.gov/archive/nasa/casi.ntrs.nasa.gov/19770019231.pdf (accessed on 22 August 2018).
22. Goldstein, H. *Classical Mechanics*, 2nd ed.; Addison-Wesley: Boston, MA, USA, 1981.
23. Kane, T.; David, L. *Dynamics: Theory and Application*; McGraw-Hill: New York, NY, USA, 1985.
24. Huges, P. *Spacecraft Attitude Dynamics*; Wiley and Sons: New York, NY, USA, 1986.
25. Wiesel, W. *Spaceflight Dynamics*, 2nd ed.; Irwin McGraw-Hill: Boston, MA, USA, 1989, 1997.
26. Wie, B. *Space Vehicle Dynamics and Control*; AIAA: Reston, VA, USA, 1998.

27. Slabaugh, G.G. Computing Euler Angles from a Rotation Matrix. January 1999, Volume 6, pp. 39–63. Available online: http://www.close-range.com/docs/Computing_Euler_angles_from_a_rotation_matrix.pdf (accessed on 22 August 2018).

28. Vallado, D. *Fundamentals of Astrodynamics and Applications*, 2nd ed.; Microcosm Press: El Segundo, CA, USA, 2001.

29. Roithmayr, C.M.; Hodges, D.H. *Dynamics: Theory and Application of Kane's Method*; Cambridge University Press: New York, NY, USA, 2016.

30. Sands, T.; Mihalik, R. Outcomes of the 2010 and 2015 nonproliferation treaty review conferences. *World J. Soc. Sci. Hum.* **2016**, *2*, 46–51. Available online: http://pubs.sciepub.com/wjssh/2/2/4/index.html (accessed on 22 August 2018). [CrossRef]

31. Sands, T. Strategies for combating Islamic state. *Soc. Sci.* **2016**, *5*, 39. Available online: www.mdpi.com/2076-0760/5/3/39/pdf (accessed on 22 August 2018). [CrossRef]

32. Mihalik, R.; Camacho, H.; Sands, T. Continuum of learning: Combining education, training and experiences. *Education* **2017**, *8*, 9–13. [CrossRef]

33. Sands, T.; Camacho, H.; Mihalik, R. Education in nuclear deterrence and assurance. *J. Def. Manag.* **2017**, *7*, 166. [CrossRef]

34. Sands, T.; Mihalik, R. Theoretical Context of the Nuclear Posture Review. *J. Soc. Sci.* **2018**, *14*, 124–128. [CrossRef]

35. Sands, T. Satellite electronic attack of enemy air defenses. *Proc. IEEE CDC* **2009**, 434–438. [CrossRef]

36. Sands, T. Space mission analysis and design for electromagnetic suppression of radar. *Int. J. Electromagn. Appl.* **2018**, *8*, 1–25. [CrossRef]

37. Sands, T.; Lu, D.; Chu, J.; Cheng, B. Developments in Angular Momentum Exchange. *Int. J. Aerosp. Sci.* **2018**, *6*, 1–7. [CrossRef]

38. Sands, T.A.; Kim, J.J.; Agrawal, B. 2H Singularity free momentum generation with non-redundant control moment gyroscopes. *Proc. IEEE CDC* **2006**, 1551–1556. [CrossRef]

39. Sands, T. Fine Pointing of Military Spacecraft. Ph.D. Thesis, Naval Postgraduate School, Monterey, CA, USA, 2007.

40. Kim, J.J.; Sands, T.; Agrawal, B.N. Acquisition, tracking, and pointing technology development for bifocal relay mirror spacecraft. *Proc. SPIE* **2007**, *6569*. [CrossRef]

41. Sands, T.A.; Kim, J.J.; Agrawal, B. Control moment gyroscope singularity reduction via decoupled control. *Proc. IEEE SEC* **2009**, 1551–1556. [CrossRef]

42. Sands, T.; Kim, J.J.; Agrawal, B.N. Nonredundant single-gimbaled control moment gyroscopes. *J. Guid. Control Dyn.* **2012**, *35*, 578–587. [CrossRef]

43. Sands, T.; Kim, J.; Agrawal, B. Experiments in Control of Rotational Mechanics. *Int. J. Autom. Control Intell. Syst.* **2016**, *2*, 9–22.

44. Agrawal, B.N.; Kim, J.J.; Sands, T.A. Method and Apparatus for Singularity Avoidance for Control Moment Gyroscope (CMG) Systems without Using Null Motion. U.S. Patent 9567112 B1, 14 February 2017. Available online: https://calhoun.nps.edu/handle/10945/51921 (accessed on 22 August 2018).

45. Sands, T.; Kim, J.J.; Agrawal, B. Singularity Penetration with Unit Delay (SPUD). *Mathematics* **2018**, *6*, 23. Available online: http://www.mdpi.com/2227-7390/6/2/23/pdf (accessed on 22 August 2018). [CrossRef]

46. Sands, T.; Lorenz, R. Physics-Based Automated Control of Spacecraft. In Proceedings of the AIAA Space 2009 Conference and Exposition, Pasadena, CA, USA, 14–17 September 2009.

47. Sands, T. Physics-based control methods. In *Advances in Spacecraft Systems and Orbit Determination*; InTech: London, UK, 2012; Available online: https://www.intechopen.com/books/advances-in-spacecraft-systems-and-orbit-determination/physics-based-control-methods (accessed on 22 August 2018).

48. Sands, T. Improved Magnetic Levitation via Online Disturbance Decoupling. *Phys. J.* **2015**, *1*, 272–280.

49. Nakatani, S. Simulation of spacecraft damage tolerance and adaptive controls. *Proc. IEEE Aerosp.* **2014**, 1–16. [CrossRef]

50. Nakatani, S. Autonomous damage recovery in space. *Int. J. Autom. Control Intell. Syst.* **2016**, *2*, 22–36.

51. Nakatani, S. Battle-damage tolerant automatic controls. *Electr. Electron. Eng.* **2018**, *8*, 10–23. [CrossRef]

52. Heidlauf, P.; Cooper, M. Nonlinear Lyapunov Control Improved by an Extended Least Squares Adaptive Feed forward Controller and Enhanced Luenberger Observer. In Proceedings of the International Conference and Exhibition on Mechanical & Aerospace Engineering, Las Vegas, NV, USA, 2–4 October 2017.

53. Cooper, M.; Heidlauf, P.; Sands, T. Controlling Chaos—Forced van der Pol Equation. *Mathematics* **2017**, *5*, 70. Available online: http://www.mdpi.com/2227-7390/5/4/70/pdf (accessed on 22 August 2018). [CrossRef]

54. Sands, T. Phase Lag Elimination at all Frequencies for Full State Estimation of Spacecraft Attitude. *Phys. J.* **2017**, *3*, 1–12.

55. Sands, T. Nonlinear-adaptive mathematical system identification. *Computation* **2017**, *5*, 47–59. Available online: http://www.mdpi.com/2079-3197/5/4/47/pdf (accessed on 22 August 2018). [CrossRef]

56. Sands, T.; Kenny, T. Experimental piezoelectric system identification. *J. Mech. Eng. Autom.* **2017**, *7*, 179–195. [CrossRef]

57. Sands, T. Space systems identification algorithms. *J. Space Explor.* **2017**, *6*, 138–149.

58. Sands, T. Experimental Sensor Characterization. *J. Space Explor.* **2018**, *7*, 140.

59. Sands, T.; Armani, C. Analysis, correlation, and estimation for control of material properties. *J. Mech. Eng. Autom.* **2018**, *8*, 7–31. Available online: http://www.sapub.org/global/showpaperpdf.aspx?doi=10.5923/j.jmea.20180801.02 (accessed on 22 August 2018). [CrossRef]

60. Sands, T. Satellite Electronic Attack of Enemy Air Defenses. In Proceedings of the IEEE SEC, Atlanta, GA, USA, 5–8 March 2009; pp. 434–438.

61. Remarks by President Trump at a Meeting with the National Space Council and Signing of Space Policy Directive-3. Available online: https://www.whitehouse.gov/briefings-statements/remarks-president-trump-meeting-national-space-council-signing-space-policy-directive-3/ (accessed on 20 June 2018).

62. Sands, T.; Bollino, K.; Kaminer, I.; Healey, A. Autonomous Minimum Safe Distance Maintenance from Submersed Obstacles in Ocean Currents. *J. Mar. Sci. Eng.* **2018**, *6*, 98. Available online: http://www.mdpi.com/2077-1312/6/3/98/pdf (accessed on 22 August 2018). [CrossRef]

63. Kuipers, J.B. Quaternions and Rotation Sequences. In Proceedings of the International Conference on Geometry, Integrability, and Quantization, Varna, Bulgaria, 1–10 September 1999; Coral Press: Sofia, Bulgaria, 2000.

Article

Parameter and State Estimation of One-Dimensional Infiltration Processes: A Simultaneous Approach

Song Bo, Soumya R. Sahoo, Xunyuan Yin, Jinfeng Liu * and Sirish L. Shah

Department of Chemical & Materials Engineering, University of Alberta, Edmonton, AB T6G 1H9, Canada;
sbo@ualberta.ca (S.B.); ssahoo@ualberta.ca (S.R.S.); xunyuan@ualberta.ca (X.Y.); sirish.shah@ualberta.ca (S.L.S.)
* Correspondence: jinfeng@ualberta.ca; Tel.: +1-780-492-1317

Received: 13 December 2019; Accepted:12 January 2020; Published: 16 January 2020

Abstract: The Richards equation plays an important role in the study of agro-hydrological systems. It models the water movement in soil in the vadose zone, which is driven by capillary and gravitational forces. Its states (capillary potential) and parameters (hydraulic conductivity, saturated and residual soil moistures and van Genuchten-Mualem parameters) are essential for the accuracy of mathematical modeling, yet difficult to obtain experimentally. In this work, an estimation approach is developed to estimate the parameters and states of Richards equation simultaneously. In the proposed approach, parameter identifiability and sensitivity analysis are used to determine the most important parameters for estimation purpose. Three common estimation schemes (extended Kalman filter, ensemble Kalman filter and moving horizon estimation) are investigated. The estimation performance is compared and analyzed based on extensive simulations.

Keywords: state estimation; parameter estimation; moving horizon estimation; extended kalman filter; ensemble kalman filter; richards equation; agro-hydrological systems

---

1. Introduction

Water and food scarcities are becoming serious issues worldwide due to population growth and climate change. According to United Nations statistics [1], approximately 70% of all available fresh water is consumed for agricultural activities, with the main consumer being irrigation. Currently, the average water-use efficiency in irrigation worldwide is about 50% as reported in Fischer et al. [2]. Therefore, it is of vital importance to improve irrigation water-use efficiency, in order to address the water crisis. Currently, it is still a common practice to use open-loop irrigation, which leads to excessive consumption of water resources. Closed-loop irrigation is a promising alternative to reduce water consumption and to better maintain the health of crops [3]. In the development of such a closed-loop irrigation system, it is important to have the soil moisture information of the entire field, which is in general very difficult to obtain. One way to overcome this challenge is to estimate the field's soil moisture based on limited sensor measurements. However, this depends on the accuracy of the agro-hydrological model. In this work, we aim to develop a systematic parameter and state estimation scheme that can provide accurate estimates of soil moisture.

Specifically, in this work, we consider simultaneous state and parameter estimation based on agro-hydrological systems modeled using the Richards equation, which describes soil water dynamics. Richards equation is a partial differential equation (PDE) which falls in the family of porous medium equation (PME). The estimation and control problems of this kind of equation were widely studied in chemical engineering [4–7] and meteorology [8,9]. The Richards equation is essentially composed of the continuity equation and Darcy's law, which is incorporated with two algebraic equations of hydraulic conductivity and capillary capacity (derivative of soil-water retention curve) [10]. The parameters of Richards equation are related to soil properties. Different approaches have been developed to estimate soil properties. Soil properties may be estimated in a soil lab by directly fitting the soil-water retention curve and hydraulic conductivity curve using collected field data of soil moisture, hydraulic conductivity and corresponding capillary pressure head [10]. However, soil properties may change over time and it would be expensive to take frequent soil samples for lab analysis especially when a big field is considered. Moreover, the hydraulic conductivity is difficult to measure. As an alternative to direct lab analysis soil parameters can be estimated indirectly based on the Richards equation and some easily-accessible field measurements such as soil moisture or capillary pressure head by minimizing the difference between measured values and model predicted values. This type of indirect approaches are referred to as inverse estimation [11]. Inverse estimation has been widely applied and its applications can be generally classified into two groups: methods based on measurements observed from one-step or multi-step outflow experiments [12–15] or methods based on time-series in-situ measurements [16–20]. These inverse estimation methods can only estimate soil parameters but not soil moisture or capillary pressure head. Moreover, they are mainly applicable to pre-collected datasets and cannot be used for online parameter estimation.

Sequential data assimilation is another widely used approach in estimating soil parameters online, which only requires the current measurement and prior knowledge of the system. In general, it consists of two steps, which are prediction and update steps. In the first step, a dynamical system model is initialized to describe a real process. Due to the limited knowledge about the process, the model may not predict the process accurately. Then, in the update step, an algorithm is designed to determine how to correct the prediction, based on field measurements and the dynamical model. Moreover, sequential data assimilation has the ability to deal with uncertainties in the measurements and the model. Particle filters (PF) [21], extended Kalman filters (EKF) [22] and ensemble Kalman filters (EnKF) [23–27] are common and widely used algorithms in sequential data assimilation for soil parameter estimation. Li and Ren [23] studied parameter estimation by augmenting parameters as states and used EnKF as the estimation algorithm. They also studied the possible factors that affect the performance of EnKF. In Reference [24], dual ensemble Kalman filter (DEnKF) was used to first estimate the states using a standard KF and then to estimate the parameters using an unscented Kalman filter. In Reference [25], two EnKFs were used to estimate the states and parameters, separately, which neglected the complex nonlinear interaction between states and parameters. In Reference [26], the authors compared three ensemble-based simultaneous state and parameter estimation methods, augmented ensemble Kalman filter, DEnKF and simultaneous optimization and data assimilation (SODA) to improve the soil moisture estimation accuracy. It concluded that the augmented EnKF was the most robust method for general conditions and SODA was better at handling complex conditions. However, it was pointed out that SODA required the highest computational resources.

However, one limitation of the above discussed methods is that they cannot handle constraints on the states or parameters and the estimation performance deteriorates when the noise is not Gaussian or the initial guess is not good. Constraints on the states and parameters are important information and may be used to significantly improve estimation performance as will be demonstrated in the simulations of this work. To address the above discussed problem, we can consider the optimization based moving horizon estimation (MHE) method, which is widely used in state estimation of nonlinear systems with explicit constraints taken into account [28–30].

In this work, we first introduce the investigated system and the formulation of the mathematical model in Section 2. The formulations of the estimation methods, MHE, EKF and EnKF for the augmented system are introduced in Section 3. Section 4 includes the methods of identifiability and sensitivity, used to study the significance of parameters. Section 5 shows the synthetic experimental setup, determination of significant parameters and MHE estimation results compared with EKF and EnKF, followed by concluding remarks in Section 6. Some preliminary results of this work were reported in Reference [31]. Compared with Reference [31], this paper provides significantly expanded explanations and significantly extended simulation results.

2. System Description and Problem Formulation

An agro-hydrological system describes the water movements between soil, crop and atmosphere. Figure 1 shows a schematic of an agro-hydrological system, which is a modified version from Reference [32]. The water movements usually involve water transportation within the soil, root water extraction, transpiration and evaporation from the soil and leaves to the air and precipitation including rain and irrigation.

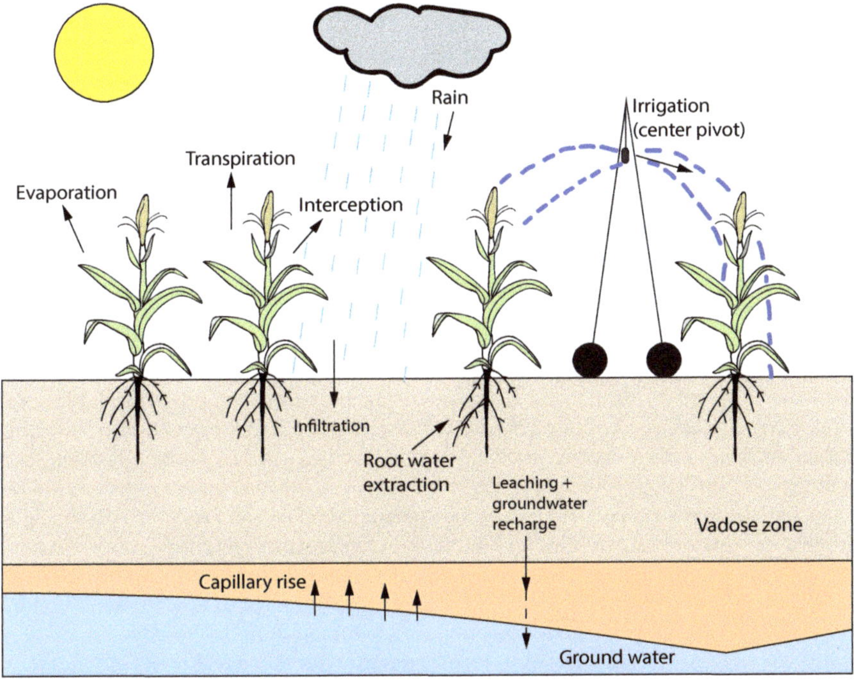

Figure 1. A schematic diagram of an agro-hydrological system.

In this work, we focus on soil that is above the water table (i.e., soil in the vadose zone). Within the vadose zone, the water movement is mainly driven by capillary and gravitational forces and the water dynamics can be modeled using Richards equation under the assumptions: (1) soil properties are spatially homogeneous within the system; (2) irrigation is uniformly applied on the surface of the system; and (3) the horizontal water dynamics are much smaller than the vertical dynamics due to the gravity force and the horizontal water dynamics can be neglected. Then, the 1D Richards equation modeling the vertical water dynamics is shown below [33]:

$$c\left(h\right)\frac{\partial h}{\partial t} = \frac{\partial}{\partial z}\left[K\left(h\right)\left(\frac{\partial h}{\partial z}+1\right)\right].$$

$$(1)$$

In Equation (1), h (m) is the capillary potential in the unsaturated soil, $K(h)$ (m/s) and $c(h)$ (1/m) denote hydraulic conductivity and capillary capacity of the soil, respectively. Note that in Richards equation, the value 1 on the right-hand-side denotes the impact of gravitational force on water in the vertical (z) direction. The upward z-direction is defined as the positive direction.

The van Genuchten-Mualem soil hydraulic model $K(h)$ and $c(h)$, as functions of the capillary potential h, are shown as follows [10]:

$$K\left(h\right) = K_s \left[\left(1+\left(-\alpha h\right)^n\right)^{-\left(1-\frac{1}{n}\right)}\right]^{\frac{1}{2}} \left[1 - \left[1 - \left[\left(1+\left(-\alpha h\right)^n\right)^{-\left(1-\frac{1}{n}\right)}\right]^{\frac{n}{n-1}}\right]^{1-\frac{1}{n}}\right]^2 \tag{2}$$

$$c\left(h\right) = \left(\theta_s - \theta_r\right)\alpha n \left(1 - \frac{1}{n}\right)\left(-\alpha h\right)^{n-1}\left[1+\left(-\alpha h\right)^n\right]^{-\left(2-\frac{1}{n}\right)}, \tag{3}$$

where K_s (m/s), θ_s (m^3/m^3) and θ_r (m^3/m^3) are saturated hydraulic conductivity, saturated soil moisture and residual soil moisture, respectively. The van Genuchten-Mualem parameters α (1/m) and n characterize the properties of the soil, which are proportional to the inverse of the soil air entry pressure and of soil porosity, respectively. These two closed-form expressions are derived by van Genuchten based on his expression of soil-water retention curve and Mualem's open-form expression of hydraulic conductivity. Since Mualem's expression is not studied further, in this work, only van Genuchten's soil-water retention equation is shown below [10]:

$$\theta\left(h\right) = \left(\theta_s - \theta_r\right)\left[\frac{1}{1+\left(-\alpha h\right)^n}\right]^{1-\frac{1}{n}} + \theta_r, \tag{4}$$

where θ (m^3/m^3) denotes volumetric water content in soil.

The five parameters θ_s, θ_r, α, n and K_s determine the properties of a type of soil. With sufficient soil samples, θ_s, θ_r, α and n can be estimated by fitting the soil-water retention curve Equation (4) utilizing soil moisture and capillary potential data sets. Then K_s can be estimated by fitting hydraulic conductivity and capillary potential data sets into Equation (2). By using this approach, we can only get a snapshot of the soil properties at one time instant, however, soil properties do slowly change over time due to agricultural activities [34]. While the experiments can be repeated to get parameter estimates at different times, it is very time consuming and expensive, especially when the investigated field is large and has various soil types over the field. Therefore, online state and parameter estimation based on ease-to-access field measurements provides a favorable approach to estimate soil properties.

In this work, we study the estimation of soil properties based on real-time field measurements: capillary potential h.

Finite Difference Model Development

Richards equation is a nonlinear partial differential equation (PDE) with respect to both the temporal and spatial variables. Because of its complex structure, it is difficult to have a closed-form solution. Therefore a finite difference method is implemented to find a numerical approximation of its solution. Two-point forward difference scheme and two-point central difference scheme are used to approximate the derivatives with respect to the temporal and spatial variables, respectively:

$$\frac{\partial h_i\left(k\right)}{\partial t} = \frac{h_i\left(k+1\right) - h_i\left(k\right)}{\Delta t} \tag{5}$$

$$\frac{\partial}{\partial z}\left[K_i\left(h\left(k\right)\right)\left(\frac{\partial h_i\left(k\right)}{\partial z}+1\right)\right] = \frac{K_{i-\frac{1}{2}}\left(h\left(k\right)\right)\left(\frac{h_{i-1}(k)-h_i(k)}{\frac{1}{2}(\Delta z_{i-1}+\Delta z_i)}+1\right) - K_{i+\frac{1}{2}}\left(h\left(k\right)\right)\left(\frac{h_i(k)-h_{i+1}(k)}{\frac{1}{2}(\Delta z_i+\Delta z_{i+1})}+1\right)}{\Delta z_i}, \tag{6}$$

where $k \in [0, N_t] \subset \mathbb{Z}$ and $i \in [1, N_x] \subset \mathbb{Z}$, representing time and position indices, respectively. N_t and N_x are the total number of time instants and states investigated. $\Delta t = t(k+1) - t(k)$ and Δz_i represents compartment thickness of compartment i. The state i is at the center of the compartment i. The hydraulic conductivity, for example, $K_{i-\frac{1}{2}}$, is linearized explicitly as $K_{i-\frac{1}{2}}(h) = K(\frac{h_{i-1}+h_i}{2})$.

The discrete-time finite difference model at node i and time instant $k+1$ can be obtained by substituting Equations (5) and (6) into Equation (1) as follows:

$$h_i(k+1) = h_i(k) + \Delta t \frac{K_{i-\frac{1}{2}}(h(k))\left(\frac{h_{i-1}(k)-h_i(k)}{\frac{1}{2}(\Delta z_{i-1}+\Delta z_i)}+1\right) - K_{i+\frac{1}{2}}(h(k))\left(\frac{h_i(k)-h_{i+1}(k)}{\frac{1}{2}(\Delta z_i+\Delta z_{i+1})}+1\right)}{\Delta z_i c_i(h(k))}, \qquad (7)$$

where $c_i(h(k))$ is defined as $c(h_i(k))$.

The Neumann boundary condition is utilized to characterize the top and bottom boundaries of the system and are shown below, respectively:

$$\left.\frac{\partial h\,(k)}{\partial z}\right|_T = -1 - \frac{q_T(k)}{K\,(h\,(k))} \qquad (8)$$

$$\left.\frac{\partial\,(h\,(k)+z)}{\partial z}\right|_B = 1, \qquad (9)$$

where the subscripts T and B represent the top and bottom boundary conditions, respectively. The q_T (m/s) is the irrigation rate which is considered as the input of the system and free drainage boundary condition is applied at the bottom.

Before introducing estimation methods, for the sake of simplicity, we obtain the compact form of the model by combining N_x Equation (7) for all spatial nodes and the boundary conditions, Equations (8) and (9). It is shown below:

$$x\,(k+1) = F\,(x\,(k)\,, u\,(k)\,, p\,(k)) + \omega_x\,(k) \qquad (10)$$

where $x(k) \in \mathbb{X} \subset \mathbb{R}^{N_x}$ represents the state vector containing N_x capillary pressure values for corresponding spatial nodes, at the defined time instant k. $p(k) \in \mathbb{P} \subset \mathbb{R}^{N_p}$, represents the parameter vector containing the parameters to be estimated. $u(k) \in \mathbb{U} \subset \mathbb{R}^{N_u}$, $\omega_x(k) \in \mathbb{W}_x \subset \mathbb{R}^{N_{\omega x}}$ denote the input and the model disturbances, respectively.

The general output function, with the measurement noise taken into account, is shown below:

$$y\,(k) = G\,(x\,(k)\,, p\,(k)) + v\,(k)\,, \qquad (11)$$

where $y(k) \in \mathbb{Y} \subset \mathbb{R}^{N_y}$ and $v(k) \in \mathbb{V} \subset \mathbb{R}^{N_v}$ denote the measurement vector and measurement noise. If the volumetric soil moisture θ is measured by the soil moisture sensor, Equation (11) is the general form of Equation (4). On the other hand, if tensiometers are used to measure the water potential h in the soil, Equation (11) simply represents a matrix indicating which states are measured by the tensiometers.

Furthermore, in order to estimate the states and parameters simultaneously, the parameter vector is augmented at the end of the state vector and treated as a part of the augmented state vector, $X = [x,\ p]^T$. An estimation of the augmented state vector X brings the benefit to estimate the states and parameters at the same time. The augmented model can be constructed by augmenting Equation (10) with the following equation:

$$p\,(k+1) = p\,(k) + \omega_p\,(k)\,, \qquad (12)$$

where $\omega_p(k) \in \mathbb{W}_p \subset \mathbb{R}^{N_{\omega p}}$. When the parameter vector p is assumed to be constant during the study, ω_p is equal to 0.

At last, the augmented model and output function used for simultaneous parameter and state estimation are shown below:

$$X\left(k+1\right) = F_a\left(X\left(k\right), u\left(k\right)\right) + \omega_a\left(k\right)$$
$$y\left(k\right) = G_a\left(X\left(k\right)\right) + v\left(k\right)$$

(13)

where $X(k) \in \mathbb{X}_a \subset \mathbb{R}^{N_x+N_p}$, $\omega_a(k) \in \mathbb{W}_a \subset \mathbb{R}^{N_w+N_p}$ and the subscript a of $F(\cdot)$ and $G(\cdot)$ denotes the augmentation.

3. Estimation Methods

In this work, three common estimation schemes, MHE, EKF and EnKF are applied to the augmented model to estimate the states and parameters. The design of these methods are detailed next.

3.1. Moving Horizon Estimation

MHE is an online optimization based estimation method [28]. The MHE optimization problem used in this work is formulated as follows:

$$\min_{\hat{X}(k-N),\cdots,\hat{X}(k),\hat{\omega}_a(k-N),\cdots,\hat{\omega}_a(k-1)} \left\|\hat{X}(k-N) - \bar{X}(k-N)\right\|^2_{P^{-1}} + \sum_{j=k-N}^{k-1}\left\|\hat{\omega}_a(j)\right\|^2_{Q^{-1}} + \sum_{j=k-N}^{k}\left\|\hat{v}(j)\right\|^2_{R^{-1}}$$

(14)

$$\text{s.t. } \hat{X}(j+1) = F_a(\hat{X}(j), u(j)) + \hat{\omega}_a(j), \ \ j \in [k-N, k-1] \subset \mathbb{Z}$$

(15)

$$\hat{v}(j) = y(j) - G_a(\hat{X}(j)), \ \ j \in [k-N, k] \subset \mathbb{Z}$$

(16)

$$\bar{X}(k-N) = \hat{X}(k-N|k-N)$$

(17)

$$\hat{X}(j) \in \mathbb{X}_a, \ \hat{v}(j) \in \mathbb{V}, \ \ j \in [k-N, k] \subset \mathbb{Z}$$

(18)

$$\hat{\omega}_a(j) \in \mathbb{W}_a, \ \ j \in [k-N, k-1] \subset \mathbb{Z}$$

(19)

In the MHE optimization, the objective is to minimize the distance between the predicted and observed measurements which is measured by the term $\|\hat{v}\|^2_{R^{-1}}$ as shown in Equation (14), where the term $\hat{v}$ is defined in Equation (16). The caret sign^indicates that the variable is estimated. The model uncertainty or the process disturbance is taken into account and represented by $\|\hat{\omega}_a\|^2_{Q^{-1}}$, where the term $\hat{\omega}_a$ is defined in Equation (15). The arrival cost, $\|\hat{X} - \bar{X}\|^2_{P^{-1}}$ summarizes the information from the initial state of the model up to the beginning of the estimation window of the MHE. N denotes the length of the estimation window. After each optimization, only the last estimated state within the estimation window is used. $\hat{X}$ and $\hat{\omega}_a$ within the moving window are the decision variables of the optimization problem. The term $\bar{X}$ follows the definition of Equation (17). $\hat{X}(k-N|k-N)$ represents the estimated state $\hat{X}$ at time instant $k-N$, which is estimated at time instant $k-N$. Matrices P, Q, R are positive definite matrices and they are the covariance matrices of state uncertainty, process noise ω_a and measurement noise v, respectively. In addition, MHE takes into account constraints on the states, parameters and model uncertainties as expressed in Equations (18) and (19).

3.2. Extended Kalman Filter

EKF is a common method used for state estimation of nonlinear systems based on successively linearizing the nonlinear system. It can be divided into two steps, which are prediction and update steps. The prediction step predicts the state X and the state covariance matrix P. When a new measurement is available, the Kalman gain K is calculated first and then X and P are updated. The detailed steps are shown below:

1. Prediction step

(a)   State prediction:

$$\hat{X}(k|k-1) = F_a(\hat{X}(k-1|k-1), u(k-1))$$

The model disturbance are not propagated as the states and parameters. Instead, it is explicitly included in the state covariance prediction.

(b)   State covariance prediction:

$$P(k|k-1) = A_a(k)P(k-1|k-1)A_a(k)^T + Q$$

where $A_a(k) = \left.\frac{\partial F_a}{\partial X}\right|_{\hat{X}(k-1|k-1)}$ and Q is the covariance matrix of the model disturbance ω_a.

2.   Update step

(a)   Kalman gain calculation:

$$K(k) = P(k|k-1)C_a(k)^T \left[C_a(k)P(k|k-1)C_a(k)^T + R \right]^{-1}$$

where $C_a(k) = \left.\frac{\partial G_a}{\partial X}\right|_{\hat{X}(k|k-1)}$ and R is the covariance matrix of the measurement noise v.

(b)   State update:

$$\hat{X}(k|k) = \hat{X}(k|k-1) + K(k)\left(y(k) - G_a(\hat{X}(k|k-1)) \right)$$

The augmented state and parameter vector X is updated when a new measurements $y(k)$ is available.

(c)   State covariance update:

$$P(k|k) = \left(I - K(k)C_a(k) \right) P(k|k-1)$$

State covariance matrix P is updated. I is the identity matrix with dimension $N_x + N_p$.

3.3. Ensemble Kalman filter

The EnKF is a method developed by Evensen [35] based on Monte Carlo method. An ensemble of trajectories of the system is generated based on the priori probability distribution of the case. A practical implementation scheme which estimated the probability distribution based on the information embedded within ensembles, instead of propagation of the state covariance matrix P, is discussed in Reference [36]. Unlike EKF, it directly utilizes the nonlinear model Equation (13), which does not require frequent model linearization. In addition, the model disturbance and measurement noise are taken into account at the same time as the states and parameters propagate. It starts with generating the ensembles, then follows with the two steps as the same as in EKF.

1.   Initialization step

(a)   Generating ensembles:

$$\hat{X}^m(0|0) \sim \mathcal{N}(X(0), P(0)), \quad m \in [1, M] \subset \mathbb{Z}$$

where an ensemble containing M initial states $\hat{X}^m(0|0)$, $m = 1, \ldots, M$, is generated and m is the index of the ensemble. The ensemble follows the multivariate normal distribution with mean, $X(0)$ and covariance matrix of the initial state, $P(0)$.

2.   Prediction step

(a)   State prediction:

$$\hat{X}^m(k|k-1) = F_a(\hat{X}^m(k-1|k-1), u(k-1)) + \omega_a^m(k-1), \quad m \in [1, M] \subset \mathbb{Z}$$

where $\omega_a^m(k-1) \sim \mathcal{N}(0, Q)$. Just like generating the ensemble of $\hat{X}^m$, a normally distributed set of ω_a^m are generated with the mean 0 and the covariance matrix Q. Overall M trajectories propagate, with model disturbance explicitly considered.

3.   Update step

(a)   Kalman gain calculation:

$$K(k) = P_{xy}(k|k-1)P_{yy}(k|k-1)^{-1}$$

where $P_{xy}(k|k-1) = \frac{1}{M-1}\sum_{m=1}^{M}[(\hat{X}^m(k|k-1) - \bar{X}(k|k-1))(\hat{y}^m(k|k-1) - \bar{y}(k|k-1))]$
$P_{yy}(k|k-1) = \frac{1}{M-1}\sum_{m=1}^{M}[\hat{y}^m(k|k-1) - \bar{y}(k|k-1)]^2$, $\bar{X}(k|k-1) = \frac{1}{M}\sum_{m=1}^{M}\hat{X}^m(k|k-1)$
and $\bar{y}(k|k-1) = \frac{1}{M}\sum_{m=1}^{M}\hat{y}^m(k|k-1)$. P_{xy} is the cross-covariance matrix of the state and measurement vectors and P_{yy} is the auto-covariance matrix of the measurement vector. The mean of the state or measurement vector is calculated based on the corresponding ensembles.

(b)   State update:

$$\hat{X}^m(k|k) = \hat{X}^m(k|k-1) + K(k)\left[y(k) + v^m(k) - G_a(\hat{X}^m(k|k-1))\right], \quad m \in [1, M] \subset \mathbb{Z}$$

where $v^m(k) \sim \mathcal{N}(0, R)$. All M state vectors are updated, when the new measurement $y(k)$ is available. The measurement uncertainty is taken into account by generating a normally distributed ensemble of measurement noises $v^m(k)$, which has mean 0 and covariance matrix R. At last, the estimated state $\hat{X}(k|k)$ is obtained as the mean of the corresponding ensembles $\hat{X}^m(k|k)$, $m = 1, \ldots, m$.

4. Proposed Procedure to Determine Significant Parameters and Number of Sensors

In reality, it is nearly impossible to measure all states and the parameters can not be determined easily. First, according to Reference [37], it states that the original system of Equation (10) is observable using limited number of measurements. That means the states can be recovered. However, for this work the augmented system of Equation (13) is studied. For this case, it is necessary to ensure that the parameters are also identifiable since they are estimated with the states simultaneously. The proposed procedure to check the identifiability of the parameters, to select appropriate parameters for estimation and to determine the minimum number of sensors is shown in Figure 2. The key steps are explained below.

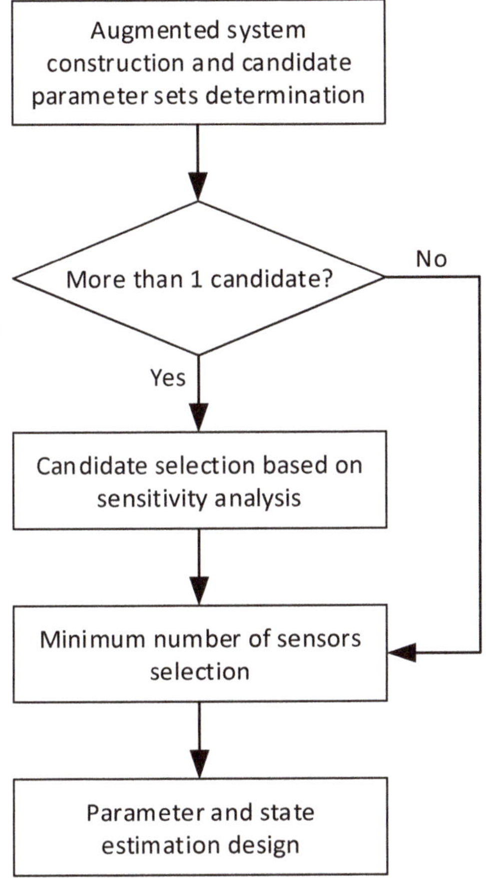

Figure 2. A flowchart of the procedure to determine the significant parameters and number of sensors.

4.1. Determine Candidate Parameter Sets for Estimation

After augmenting the original nonlinear system with the parameters, the entire system may not be observable. In order to determine which parameters can and should be estimated online, we resort to observability analysis [38]. In this step, we assume that all the soil moisture states are measured; that is, $y = x$. This ensures that the observability analysis results depend only on the parameters. If the augmented system is not observable, then the unobservability is caused by the augmentation of the parameters in the state vector.

When checking the observability of the augmented system, we start with the system with all the parameters augmented. If the augmented system is not observable, then one of the parameters is removed from the augmented system. If there are N_p parameters, then there are N_p different ways to remove the one parameter. All these N_p cases are considered. If after removing one parameter and upon finding that the new augmented system is observable, we continue to the next step to determine which parameter set to estimate (described in the next subsection). If we can still not find an observable augmented system after removing one parameter, we continue to remove two parameters from the original augmented system. Again, all the possible cases should be considered. If we can still not find an observable system, we continue to remove three parameters from the original augmented system. This continues until we find at least a system that is observable.

When checking the observability, we propose to use the Popov-Belevitch-Hautus (PBH) observability theory. Other observability tests may also be used. Since the augmented system is a nonlinear system, it should be linearized before PBH can be applied. It is recommended that instead of linearizing the system at one point, it should be linearized at different point along typical operating trajectories as used in Reference [39].

Note that the observability analysis described in this step may generate more than one candidate parameter sets that can be estimated through augmentation of the original agro-hydrological system.

4.2. Sensitivity Analysis

If there is only one candidate parameter set from the previous step, we can continue with the candidate and move to the next subsection to find the minimum number of sensors. However, if there are more than one candidates, we need to determine which parameter set to choose. We propose to use sensitivity analysis to determine the importance of these parameters and pick the set containing the most important parameters for further analysis.

The sensitivity analysis measures how the outputs respond when there is a change in one parameter. The sensitivity matrix $S_y(k)$ shown below contains the information about, at time instant k, how each output is affected by $X(0)$ which is constituted of the initial state $x(0)$ and the parameters p.

$$S_y(k) = \left. \begin{bmatrix} \frac{\partial y_1}{\partial X_1(0)} & \frac{\partial y_1}{\partial X_2(0)} & \cdots & \frac{\partial y_1}{\partial X_{N_x}(0)} & \frac{\partial y_1}{\partial X_{N_x+1}(0)} & \cdots & \frac{\partial y_1}{\partial X_{N_x+N_p}(0)} \\ \frac{\partial y_2}{\partial X_1(0)} & \frac{\partial y_2}{\partial X_2(0)} & \cdots & \frac{\partial y_2}{\partial X_{N_x}(0)} & \frac{\partial y_2}{\partial X_{N_x+1}(0)} & \cdots & \frac{\partial y_2}{\partial X_{N_x+N_p}(0)} \\ \vdots & \vdots & \vdots & \vdots & \vdots & \vdots & \vdots \\ \frac{\partial y_{N_y}}{\partial X_1(0)} & \frac{\partial y_{N_y}}{\partial X_2(0)} & \cdots & \frac{\partial y_{N_y}}{\partial X_{N_x}(0)} & \frac{\partial y_{N_y}}{\partial X_{N_x+1}(0)} & \cdots & \frac{\partial y_{N_y}}{\partial X_{N_x+N_p}(0)} \end{bmatrix} \right|_k$$

The detailed steps to derive the sensitivity matrix is explained below and is inspired by Reference [40]. When performing this sensitivity analysis, we consider the augmented system of Equation (13) without considering the disturbance w_a and v but with $X(0)$ explicitly expressed as shown below:

$$\begin{aligned} X(k+1) &= F_a(X(k), u(k), X(0)) \\ y(k) &= G_a(X(k), X(0)), \end{aligned} \tag{20}$$

where $X(0)$ is considered as an independent variable.

The objective is to check how a change in the initial state x_0 and the parameters p affects the prediction error e, which comes from the difference between the predicted y and the observed measurements y_M. We can represent this as:

$$\frac{\partial e}{\partial X(0)} = \frac{\partial (y - y_M)}{\partial X(0)} = \frac{\partial y}{\partial X(0)} - \frac{\partial y_M}{\partial X(0)}. \tag{21}$$

Because the observed measurement y_M is not affected by the initial state and parameters, the above expression is simplified as below:

$$\frac{\partial e}{\partial X(0)} = \frac{\partial y}{\partial X(0)}. \tag{22}$$

Equation (22) can be derived by taking the partial derivative of Equation (20) with respect to the augmented state vector $X(0)$. And the sensitivity equations with respect to $X(0)$ are shown below:

$$\begin{aligned} \frac{\partial X(k+1)}{\partial X(0)} &= \frac{\partial}{\partial X(0)} F_a(X(k), u(k), X(0)) \\ \frac{\partial y(k)}{\partial X(0)} &= \frac{\partial}{\partial X(0)} G_a(X(k), X(0)). \end{aligned} \tag{23}$$

Because the intermediate variable $X(k)$ depends on the independent variable $X(0)$ as well, the chain rule is applied on the right hand sides of Equation (23) and we can further get that

$$\frac{\partial X(k+1)}{\partial X(0)} = \frac{\partial F_a}{\partial X(k)} \cdot \frac{\partial X(k)}{\partial X(0)} + \frac{\partial F_a}{\partial X(0)}$$
$$\frac{\partial y(k)}{\partial X(0)} = \frac{\partial G_a}{\partial X(k)} \cdot \frac{\partial X(k)}{\partial X(0)} + \frac{\partial G_a}{\partial X(0)}. \tag{24}$$

By defining $S_X(k) = \frac{\partial X(k)}{\partial X(0)}$ and $S_y(k) = \frac{\partial y(k)}{\partial X(0)}$, the above equations can be converted to ordinary differential equations, which are shown below:

$$S_x(k+1) = \frac{\partial F_a}{\partial X(k)} \cdot S_X(k) + \frac{\partial F_a}{\partial X(0)}$$
$$S_y(k) = \frac{\partial G_a}{\partial X(k)} \cdot S_X(k) + \frac{\partial G_a}{\partial X(0)}. \tag{25}$$

Therefore, by giving the initial states of Equations (20) and (25) and solving them at the same time, the sensitivity matrix $S_y(k)$ can be obtained. $S_y(k)$ may be normalized to obtain the normalized sensitivity matrix S_N:

$$S_N(k) = \frac{\partial y(k)}{\partial X(0)} \cdot \frac{X(0)}{y(k)}. \tag{26}$$

Once the sensitivity matrix is obtained, we can use it to determine the relative importance of different parameters. Specifically, we can exam the magnitudes of the elements in the sensitivity matrix. Each parameter corresponds to one column in the sensitivity matrix. We can use, for example, the summation of the absolute values of the elements of each column to compare the relative importance of parameters. A bigger value implies a more important parameter in terms of its impact on the outputs. Among all the candidate parameter sets, we keep the parameter set with the highest sensitivity values.

4.3. Minimum Number of Sensors

After the parameter set to be estimated is determined, the original system is augmented with the parameters, as illustrated in Reference [37], we can use the maximum multiplicity theory [41] to determine the minimum number of sensors required to ensure the observability of the entire system. Then, state estimation techniques can be used to estimate the states and parameters simultaneously.

5. Simulation Results and Discussion

5.1. System Description

In this work, a total length (L) of 67 cm loam soil column is investigated, which is shown in Figure 3. The soil column is equally partitioned into 32 compartments. Correspondingly, Richards equation is spatially discretized into 32 states (N_x) in the z-direction, with each state centered at the corresponding compartment. At the surface of the soil, the irrigation, q_T, is performed at the rate of 2.50 cm/day, from 12:00 PM to 4:00 PM daily. At the bottom, the free drainage boundary condition is used, which means the gradient between the last state and the state at the bottom boundary is 0. The soil column has the homogeneous initial condition ($x(0)$) of -0.514 m capillary pressure head and the parameters of the soil are shown in Table 1 [42].

Table 1. The initial condition and parameters of the investigated loam soil column.

	$x(0)$ (m)	K_s (m/s)	θ_s (m^3/m^3)	θ_r (m^3/m^3)	α (1/m)	n
Loam	-0.514	2.89×10^{-6}	0.430	0.0780	3.60	1.56

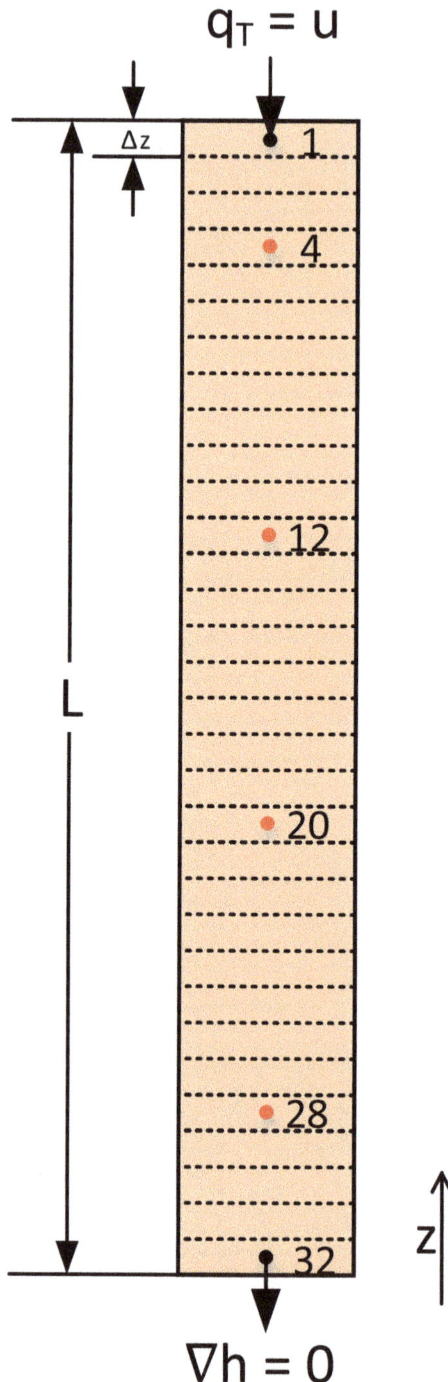

Figure 3. A schematic diagram of the investigated loam soil column.

5.2. Determination of Significant Parameters and Number of Sensors

The augmented system (Equation (13)) is utilized to achieve simultaneous parameter and state estimation. First without knowing the observability of the augmented system, all 5 parameters (K_s, θ_s, θ_r, α and n) are augmented; that is, $N_p = 5$. In addition, all 32 states are assumed to be measured. A 10-day state trajectory, without considering the process and measurement noise, is used in the rest of the subsection for selecting appropriate parameters for estimation and determining the minimum number of sensors. It is assumed that the measurements are available every 1 h.

Following the procedure as discussed in Section 4.1, we apply the PBH observability test on the augmented system to check the identifiability of the parameters. The test is conducted every sampling time, which requires the system to be linearized accordingly. According to the results, the augmented system is not observable. This implies that it is impossible to identify the 5 parameters simultaneously. In order to look for an observable system, parameters are removed from the augmented system. We start with removing only 1 of the parameters and this results in 5 different augmented systems with each one augmented with 4 parameters. Then, the observability of the 5 augmented systems is checked. It was found that 2 of the 5 systems are observable. In these two systems, either θ_s or θ_r is removed. Since observable systems are found, we proceed to the next step to determine the final parameter set.

To determine which parameter set to use, the significance of θ_s and θ_r is compared based on the sensitivity analysis described in Section 4.2. Sensitivity analysis is conducted based on the original augmented system with all the parameters. The initial state of Equation (25) is an identity matrix of size $N_x + N_p$. By comparing the summation of the absolute values of the elements of each column of the normalized sensitivity matrices S_N, it can be found that the summation corresponding to the column $\frac{\partial y_i}{\partial \theta_s}$ (82,674) is much bigger than the one for $\frac{\partial y_i}{\partial \theta_r}$ (14,997). Based on this, θ_s is considered as a more important parameter because it has more impact on the output than θ_r. Therefore, the parameter set containing θ_r is removed and the final parameter set will be used in the remaining analysis is $\{K_s, \theta_s, \alpha, n\}$.

In the above analysis, it was assumed that all the states (soil moisture) are measured for the purpose of determining the parameters for estimation. When the set of parameters is determined, we remove this assumption to determine the minimum number of sensors (measurements) needed to ensure the observability of the augmented system with 4 parameters. Following the method described in Section 4.3, the maximum multiplicity method is conducted, and it can be found that the minimum number of sensors is 4.

5.3. Simultaneous Parameter and State Estimation

According to the minimum number of sensors found above, it is assumed that 4 tensiometers (N_y) are installed. Specifically, we assume that these sensors are installed at 7.30 cm, 24.1 cm, 40.8 cm and 57.6 cm below the surface, which measure the 4th, 12th, 20th and 28th states, respectively. In the simulations, the actual parameter values used are shown in Table 1 and they are assumed to be constant within the investigated temporal domain. Process noise and measurement noise (ω_x and v) are considered in the simulations and they have zero mean and standard deviations 3×10^{-6} m and 8×10^{-3} m, respectively.

In the design of the state and parameter filters (EKF, EnKF) and estimator (MHE), the model augmented with 4 parameters (K_s, θ_s, α and n) is used. The initial guesses of the initial states and parameters in the filters and estimator are listed in Table 2 and compared with those used in the actual system.

For the EKF and EnKF, the weighting matrices Q and R are designed as the auto-covariance matrices of ω_x and v with the standard deviations mentioned before. However, the diagonal elements of Q corresponding to augmented parameters are 0, because the parameters are assumed to be constant. In simulations, 10^{-20} is used to approximate the value 0 and to ensure the positive definiteness of the matrix. The diagonal elements of P corresponding to the states are configured as the square of

3×10^{-3} and those of parameters are configured as the square of 3×10^{-2}. For the designed EnKF, 100 ensembles are used.

Table 2. True values of initial states and parameters of the process and the initial guesses used in filters and estimator.

	$x(0)$ (m)	K_s (m/s)	θ_s (m³/m³)	α (1/m)	n	θ_r (m³/m³)
Loam (true value)	−0.514	2.89×10^{-6}	0.430	3.60	1.56	0.0780
Initial guess	−0.617	3.18×10^{-6}	0.387	3.24	1.72	0.0780

For the design of MHE, the estimation window size is selected to be 8 h. The weighting matrices P, Q, and R retain the same ratio with respect to those used in EKF and EnKF but with a much bigger magnitude to ensure the numerical stability of the associated optimization problem. In addition, the P matrix is constant for all the optimizations. The constraints of the states, parameters and the model uncertainty are listed in Table 3. The upper and lower bounds of the term $\hat{\omega}_p$ are 0 because the parameters are constant.

Table 3. Lower and upper bounds used in moving horizon estimation (MHE).

	$\hat{x}$ (m)	$\hat{K}_s$ (m/s)	$\hat{\theta}_s$ (m³/m³)	$\hat{\alpha}$ (1/m)	$\hat{n}$	$\hat{\omega}_x$	$\hat{\omega}_p$
Lower bounds	−1.00	2.31×10^{-6}	0.344	2.88	1.25	$-\infty$	0.00
Upper bounds	-1.00×10^{-4}	3.47×10^{-6}	0.516	4.32	1.87	∞	0.00

In the following simulations, the root mean square error (RMSE) will be used to evaluate the performance of the MHE, EKF and EnKF. The estimation performance in terms of the states and parameters are evaluated separately. Their equations are shown below:

$$RMSE_x(k) = \sqrt{\frac{\sum_{i=1}^{N_x}(\hat{x}_i(k) - x_i(k))^2}{N_x}} \tag{27}$$

$$RMSE_p(k) = \sqrt{\frac{\sum_{i=1}^{N_p}(\hat{p}_i(k) - p_i(k))^2}{N_p}}. \tag{28}$$

First, we performed simulations assuming that the parameter θ_r (which is not estimated) is known and is the same as the value used in the actual system. Figures 4 and 5 show some representative estimated states and all the parameters using MHE, EKF and EnKF, which are also compared with their true values. Figure 4 shows the state trajectories of the top node and a few middle nodes and one bottom node. From the figure, it can be seen that the top node has more dynamics because it takes time for irrigated water to pass from the upper part and to the lower part. In terms of state estimation performance, from Figure 4, it can be seen that MHE and EnKF give very much more accurate state estimates than the EKF. Note that from Figure 4, it can also be seen that the estimates of the 12th state (h_{12}) converge faster than the other estimates. This is because it is a sensor node.

In terms of parameter estimation, Figure 5 shows the results. From the figure, it can be seen that only MHE is capable of estimating the parameters, whereas those estimated by EKF and EnKF diverge from their true values. This may be because of the constraints used in MHE. These constraints provide more useful information to MHE in addition to the measurements.

The trajectories of the performance indices $RMSE_x$ and $RMSE_p$ associated with the MHE, EnKF and EKF are shown in Figure 6. These trajectories further confirm that the MHE and EnKF have better performance than EKF in estimation of the states and the MHE outperforms both EnKF and EKF in parameter estimation.

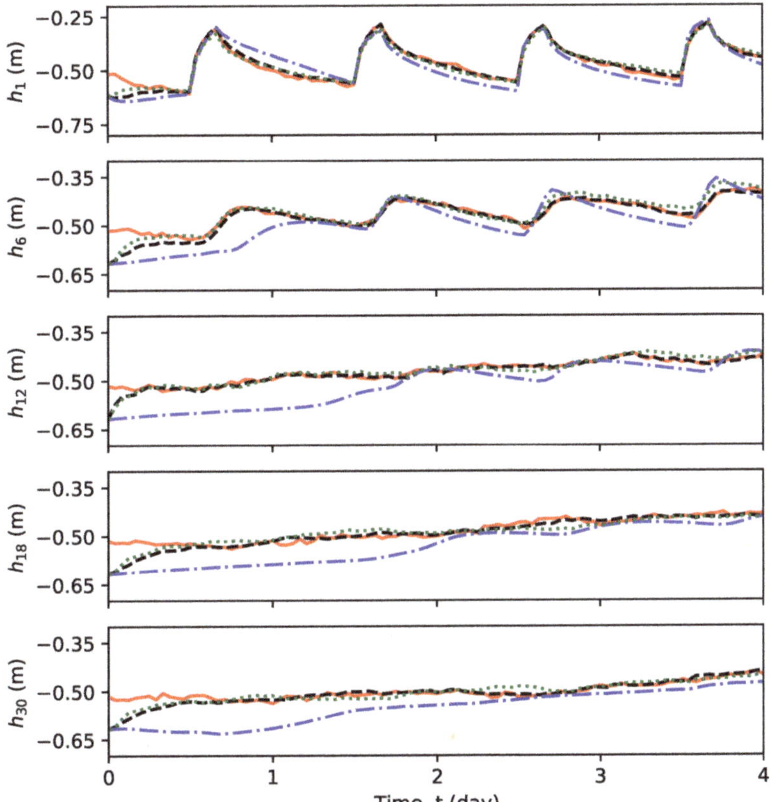

Figure 4. Selected trajectories of the process state and estimated states using MHE, extended Kalman filter (EKF) and ensemble Kalman filter (EnKF).

In the previous set of simulations, the parameter θ_r is assumed to be accurately known and is used in the MHE, EnKF and EKF. However, this assumption may not hold in practice. In this set of simulations, we study how an inaccurate θ_r may affect the state and parameter estimation performance. In this set of simulations, the value of θ_r used in the MHE, EnKF and EKF is assumed to be 10% off from the actual value. The tuning parameters used in the filters and estimator are the same as the ones used in the previous simulations. In this case, the EnKF and EKF cannot give accurate parameter estimates as in the previous case, either. The MHE is still the only estimation method that can give good parameter estimates. Table 4 summarizes the estimated parameters using the MHE in the two sets of simulations. The reported estimated values are the mean estimated values after the estimates have converged. According to the results, a 10% difference of θ_r does not affect the estimation results of other parameters when MHE is used. This verifies that the removal of θ_r has a minor impact on the overall state and parameter estimation performance. This further implies that the proposed method in parameter selection is applicable.

In this work, the spatial heterogeneity in soil properties is not considered. When parameter heterogeneity presents, a 3D Richards equation is needed to describe the water dynamics. The studied MHE algorithm can be extended to handle heterogeneous parameters in a straightforward manner.

It is expected that the weighting matrices should be tuned taking into account the spatial heterogeneity. Also, a system with different soil types may be decomposed into a few subsystems with each subsystem having the same type of soil and distributed or decentralized estimation may be used accordingly. MHE may still be used in the design of the subsystem estimators.

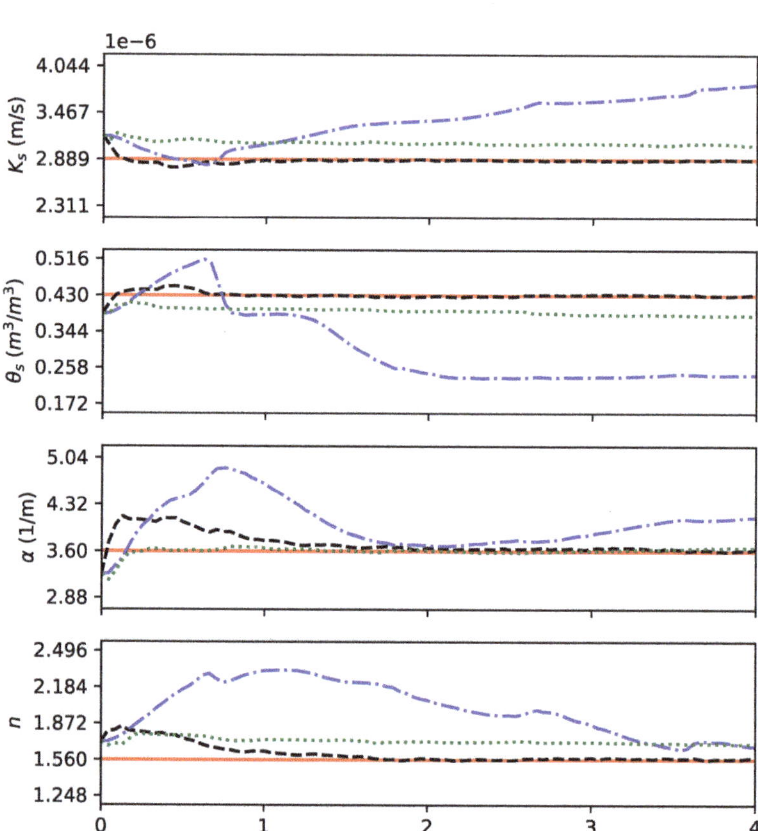

Figure 5. Trajectories of estimated parameters using MHE, EKF and EnKF, compared with their actual values.

Table 4. Comparison of estimated parameters using MHE with their true values, when θ_r is assumed to be accurate and 10% off.

Cases	θ_r (m³/m³)	$\hat{K}_s$ (m/s)	$\hat{\theta}_s$ (m³/m³)	$\hat{\alpha}$ (1/m)	$\hat{n}$
θ_r (true value)	0.0780	2.89×10^{-6}	0.430	3.60	1.56
$\hat{\theta}_r$ ($= \theta_r$)	0.0780	2.89×10^{-6}	0.430	3.60	1.56
$\hat{\theta}_r$ ($= 90\%\theta_r$)	0.0702	2.89×10^{-6}	0.430	3.60	1.56

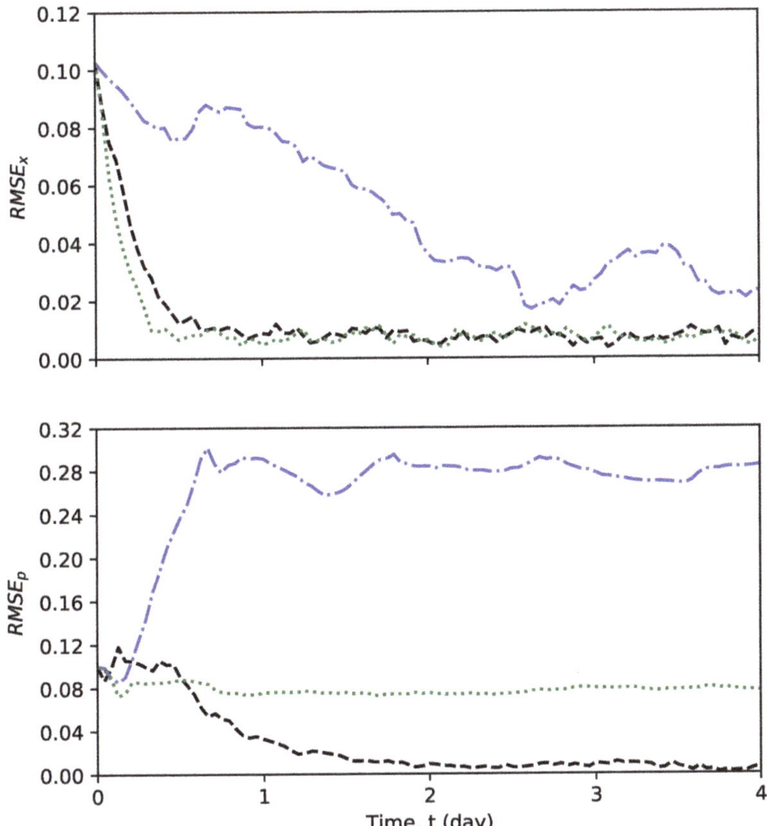

Figure 6. Trajectories of RMSE measuring the estimation performance of MHE, EKF, and EnKF.

5.4. Effects of the Simulation Parameters

In this subsection, we further study the performance of MHE in terms of number of measurements and size of estimation window of MHE.

5.4.1. Effects of Number of Measurements

First, we study the effects of number of measurements on the estimation performance of MHE. In addition to the case with 4 measurements, we also consider cases with 8 and 12 measurements. Figure 7 shows how the two performance indices $RMSE_x$ and $RMSE_p$ evolve over time. From the top plot, it can be seen that the more sensors are used, the faster state estimates converge. This is because the sensors are directly measuring the states. When there are more sensors, it implies that we have more information of the states. For the parameter, there is no obvious difference between the convergence speeds with different number of measurements. Comparing the convergence speed between the state estimates and parameter estimates, the state estimates converge much faster within one day while the parameter estimates take longer time to converge (about 2 days). Overall, from this set of simulations, it can be concluded that 4 sensors are sufficient to estimate all states and parameters accurately.

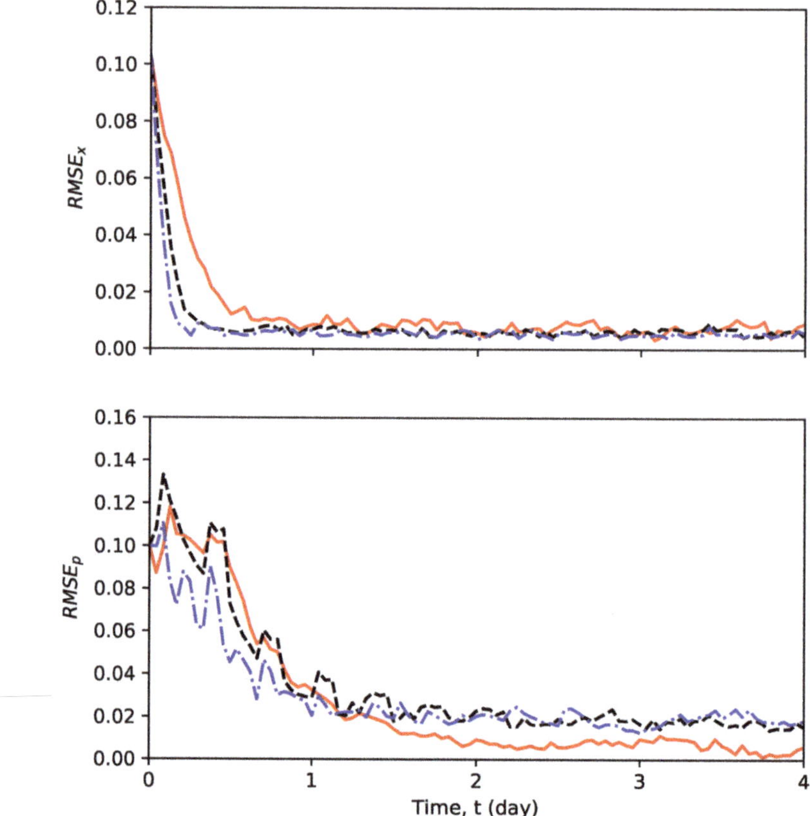

Figure 7. Trajectories of RMSE measuring the error between actual model and estimated states and parameters of MHE using 4, 8 and 12 measurements.

5.4.2. Effects of MHE Estimation Window Size

The effects of the size of the estimation window of MHE on estimation performance are also studied assuming that there are 4 measurements. Figure 8 shows how the two performance indices $RMSE_x$ and $RMSE_p$ evolve over time with different estimation window sizes. From the figure, it can be seen that from both plots that a window size of 8 is sufficient and further increase of the estimation window size does not bring significant performance improvement.

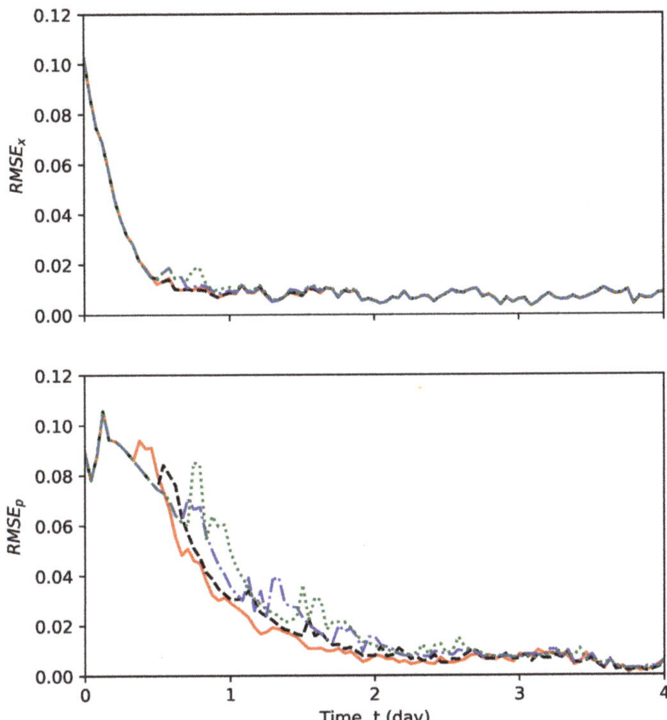

Figure 8. Trajectories of RMSE measuring actual model and estimated states and parameters of MHE with window sizes of 8, 12, 16 and 20.

6. Conclusions

In this work, we have investigated simultaneous state and parameter estimation using moving horizon estimation (MHE), extended Kalman filter (EKF) and ensemble Kalman filter (EnKF) applied to an infiltration process in an agro-hydrological system. First, a procedure was proposed to find the appropriate parameter set for estimation based on the observability of the augmented system and the sensitivity of the outputs to the parameters. It was found that only 4 out of 5 parameters (hydraulic conductivity, saturated soil moisture and van Genuchten-Mualem parameters) can be considered in simultaneous state and parameter estimation. The less important parameter (residual soil moistures) was not considered in parameter estimation. After determining the parameter set for estimation, the minimum number of sensors was also found based on the maximum multiplicity theory. Simulation results showed that the MHE has an overall the best state and parameter estimation performance due to the inclusion of state and parameter constraints in the estimation. It was also found that the uncertainty in the residual soil moisture (which was not estimated) does not affect the overall estimation performance too much. The effects of number of measurements and estimation window size of the MHE were also studied through simulations. It was found that 4 measurements and a window size of 8 for MHE are sufficient to provide accurate state and parameter estimates.

Author Contributions: Conceptualization, methodology, simulation design, S.B., S.R.S. and J.L.; validation and formal analysis, S.B., S.R.S., X.Y. and J.L.; original draft preparation, S.B.; review and editing, S.B., S.R.S., X.Y., J.L. and S.L.S.; supervision, J.L. All authors have read and agreed to the published version of the manuscript.

Funding: This research was funded by Natural Sciences and Engineering Research Council of Canada (NSERC).

Conflicts of Interest: The authors declare no conflict of interest.

References

1. AQUASTAT Main Database. Food and Agriculture Organization of The United Nations (FAO). Available online: http://www.fao.org/nr/water/aquastat/data/query/index.html?lang=en (accessed on 31 October 2019).
2. Fischer, G.; Tubiello, F.N.; van Velthuizen, H.; Wiberg, D.A. Climate change impacts on irrigation water requirements: Effects of mitigation, 1990–2080. *Technol. Forecast. Soc. Chang.* **2007**, *74*, 1083–1107.10.1016/j.techfore.2006.05.021. [CrossRef]
3. Mao, Y.; Liu, S.; Nahar, J.; Liu, J.; Ding, F. Soil moisture regulation of agro-hydrological systems using zone model predictive control. *Comput. Electron. Agric.* **2018**, *154*, 239–247, doi:10.1016/j.compag.2018.09.011. [CrossRef]
4. Narasingam, A.; Siddhamshetty, P.; Kwon, J.S.I. Handling Spatial Heterogeneity in Reservoir Parameters Using Proper Orthogonal Decomposition Based Ensemble Kalman Filter for Model-Based Feedback Control of Hydraulic Fracturing. *Ind. Eng. Chem. Res.* **2018**, *57*, 3977–3989, doi:10.1021/acs.iecr.7b04927. [CrossRef]
5. Aanonsen, S.I. The Ensemble Kalman Filter in Reservoir Engineering—A Review. *SPE J.* **2009**, *14*, 393–412. [CrossRef]
6. Siddhamshetty, P.; Kwon, J.S.I. Model-based feedback control of oil production in oil-rim reservoirs under gas coning conditions. In Proceedings of the 2018 Annual American Control Conference (ACC), Milwaukee, WI, USA, 27–29 June 2018; IEEE: Milwaukee, WI, USA, 2018; pp. 2605–2610, doi:10.23919/ACC.2018.8431419. [CrossRef]
7. Hasan, A.; Foss, B.; Sagatun, S. Flow control of fluids through porous media. *Appl. Math. Comput.* **2012**, *219*, 3323–3335, doi:10.1016/j.amc.2011.07.001. [CrossRef]
8. Bengtsson, L.; Ghil, M.; Källén, E. *Dynamic Meteorology: Data Assimilation Methods*; Springer: New York, NY, USA, 1981.
9. Ghil, M.; Malanotte-Rizzoli, P. Data Assimilation in Meteorology and Oceanography. In *Advances in Geophysics*; Elsevier: Amsterdam, The Netherlands, 1991; Volume 33, pp. 141–266, doi:10.1016/S0065-2687(08)60442-2. [CrossRef]
10. van Genuchten, M.T. A closed-form equation for predicting the hydraulic conductivity of unsaturated soils. *Soil Sci. Soc. Am. J.* **1980**, *44*, 892–898. [CrossRef]
11. Marquardt, D.W. An algorithm for least-squares estimation of nonlinear parameters. *J. Soc. Ind. Appl. Math.* **1963**, *11*, 431–441, doi:10.1137/0111030. [CrossRef]
12. Kool, J.B.; Parker, J.C.; van Genuchten, M.T. Determining soil hydraulic properties from one-step outflow experiments by parameter estimation: I. Theory and numerical studies. *Soil Sci. Soc. Am. J.* **1985**, *49*, 1348–1354, doi:10.2136/sssaj1985.03615995004900060004x. [CrossRef]
13. Toorman, A.F.; Wierenga, P.J.; Hills, R.G. Parameter estimation of hydraulic properties from one-step outflow data. *Water Resour. Res.* **1992**, *28*, 3021–3028. doi:10.1029/92WR01272. [CrossRef]
14. van Dam, J.C.; Stricker, J.N.M.; Droogers, P. Inverse method for determining soil hydraulic functions from one-step outflow experiments. *Soil Sci. Soc. Am. J.* **1992**, *56*, 1042. doi:10.2136/sssaj1992.03615995005600040007x. [CrossRef]
15. Il Hwang, S.; Powers, S.E. Estimating unique soil hydraulic parameters for sandy media from multi-step outflow experiments. *Adv. Water Resour.* **2003**, *26*, 445–456. doi:10.1016/S0309-1708(02)00107-0. [CrossRef]
16. Russo, D.; Bresler, E.; Shani, U.; Parker, J.C. Analyses of infiltration events in relation to determining soil hydraulic properties by inverse problem methodology. *Water Resour. Res.* **1991**, *27*, 1361–1373, doi:10.1029/90WR02776. [CrossRef]
17. Abbaspour, K.C.; van Genuchten, M.T.; Schulin, R.; Schläppi, E. A sequential uncertainty domain inverse procedure for estimating subsurface flow and transport parameters. *Water Resour. Res.* **1997**, *33*, 1879–1892, doi:10.1029/97WR01230. [CrossRef]
18. Ritter, A.; Hupet, F.; MunÄoz-Carpena, R.; Lambot, S.; Vanclooster, M. Using inverse methods for estimating soil hydraulic properties from field data as an alternative to direct methods. *Agric. Water Manag.* **2003**, *59*, 77–96. [CrossRef]

19. Rashid, N.S.A.; Askari, M.; Tanaka, T.; Simunek, J.; van Genuchten, M.T. Inverse estimation of soil hydraulic properties under oil palm trees. *Geoderma* **2015**, *241–242*, 306–312, doi:10.1016/j.geoderma.2014.12.003. [CrossRef]

20. Li, Y.B.; Liu, Y.; Nie, W.B.; Ma, X.Y. Inverse modeling of soil hydraulic parameters based on a hybrid of vector-evaluated genetic algorithm and particle swarm optimization. *Water* **2018**, *10*, 84, doi:10.3390/w10010084. [CrossRef]

21. Montzka, C.; Moradkhani, H.; Weihermüller, L.; Franssen, H.J.H.; Canty, M.; Vereecken, H. Hydraulic parameter estimation by remotely-sensed top soil moisture observations with the particle filter. *J. Hydrol.* **2011**, *399*, 410–421, doi:10.1016/j.jhydrol.2011.01.020. [CrossRef]

22. Lü, H.; Yu, Z.; Zhu, Y.; Drake, S.; Hao, Z.; Sudicky, E.A. Dual state-parameter estimation of root zone soil moisture by optimal parameter estimation and extended Kalman filter data assimilation. *Adv. Water Resour.* **2011**, *34*, 395–406, doi:10.1016/j.advwatres.2010.12.005. [CrossRef]

23. Li, C.; Ren, L. Estimation of unsaturated soil hydraulic parameters using the ensemble Kalman filter. *Vadose Zone J.* **2011**, *10*, 1205–1227, doi:10.2136/vzj2010.0159. [CrossRef]

24. Medina, H.; Romano, N.; Chirico, G.B. Kalman filters for assimilating near-surface observations into the Richards equation - Part 2: A dual filter approach for simultaneous retrieval of states and parameters. *Hydrol. Earth Syst. Sci.* **2014**, *18*, 2521–2541, doi:10.5194/hess-18-2521-2014. [CrossRef]

25. Moradkhani, H.; Sorooshian, S.; Gupta, H.V.; Houser, P.R. Dual state–parameter estimation of hydrological models using ensemble Kalman filter. *Adv. Water Resour.* **2005**, *28*, 135–147, doi:10.1016/j.advwatres.2004.09.002. [CrossRef]

26. Chen, W.; Huang, C.; Shen, H.; Li, X. Comparison of ensemble-based state and parameter estimation methods for soil moisture data assimilation. *Adv. Water Resour.* **2015**, *86*, 425–438, doi:10.1016/j.advwatres.2015.08.003. [CrossRef]

27. Chaudhuri, A.; Franssen, H.J.H.; Sekhar, M. Iterative filter based estimation of fully 3D heterogeneous fields of permeability and Mualem-van Genuchten parameters. *Adv. Water Resour.* **2018**, *122*, 340–354, doi:10.1016/j.advwatres.2018.10.023. [CrossRef]

28. Rao, C.; Rawlings, J.; Mayne, D. Constrained state estimation for nonlinear discrete-time systems: Stability and moving horizon approximations. *IEEE Trans. Autom. Control* **2003**, *48*, 246–258, doi:10.1109/TAC.2002.808470. [CrossRef]

29. Yin, X.; Decardi-Nelson, B.; Liu, J. Subsystem decomposition and distributed moving horizon estimation of wastewater treatment plants. *Chem. Eng. Res. Des.* **2018**, *134*, 405–419, doi:10.1016/j.cherd.2018.04.032. [CrossRef]

30. Yin, X.; Liu, J. Distributed moving horizon state estimation of two-time-scale nonlinear systems. *Automatica* **2017**, *79*, 152–161, doi:10.1016/j.automatica.2017.01.023. [CrossRef]

31. Bo, S.; Sahoo, S.R.; Yin, X.; Liu, J.; Shah, S.L. Simultaneous parameter and state estimation of agro-hydrological systems. In Proceedings of the IFAC 2020 World Congress, Berlin, Germany, 12–17 July 2020.

32. Nahar, J. Closed-loop Irrigation Scheduling and Control. Ph.D. Thesis, University of Alberta, Edmonton, AB, Canada, 2019.

33. Richards, L.A. Capillary conduction of liquids through porous mediums. *Physics* **1931**, *1*, 318–333, doi:10.1063/1.1745010. [CrossRef]

34. Almendro-Candel, M.B.; Lucas, I.G.; Navarro-Pedreño, J.; Zorpas, A.A. Physical Properties of Soils Affected by the Use of Agricultural Waste. In *Agricultural Waste and Residues*; Aladjadjiyan, A., Ed.; IntechOpen: Rijeka, Croatia, 2018; Chapter 2. doi:10.5772/intechopen.77993. [CrossRef]

35. Evensen, G. Sequential data assimilation with a nonlinear quasi-geostrophic model using Monte Carlo methods to forecast error statistics. *J. Geophys. Res.* **1994**, *99*, 10143, doi:10.1029/94JC00572. [CrossRef]

36. Gillijns, S.; Mendoza, O.; Chandrasekar, J.; De Moor, B.; Bernstein, D.; Ridley, A. What is the ensemble Kalman filter and how well does it work? In Proceedings of the 2006 American Control Conference, Minneapolis, MN, USA, 14–16 June 2006; IEEE: Minneapolis, MN, USA, 2006; p. 6.

37. Sahoo, S.R.; Yin, X.; Liu, J. Optimal sensor placement for agro-hydrological systems. *AIChE J.* **2019**, e16795, doi:10.1002/aic.16795. [CrossRef]

38. Villaverde, A.F.; Barreiro, A.; Papachristodoulou, A. Structural identifiability of dynamic systems biology models. *PLoS Comput. Biol.* **2016**, *12*, e1005153, doi:10.1371/journal.pcbi.1005153. [CrossRef]

39. Nahar, J.; Liu, J.; Shah, S.L. Parameter and state estimation of an agro-hydrological system based on system observability analysis. *Comput. Chem. Eng.* **2019**, *121*, 450–464, doi:10.1016/j.compchemeng.2018.11.015. [CrossRef]
40. Eykhoff, P. *System Identification: Parameter and State Estimation*; Wiley: London, UK, 1974.
41. Yuan, Z.; Zhao, C.; Di, Z.; Wang, W.X.; Lai, Y.C. Exact controllability of complex networks. *Nat. Commun.* **2013**, *4*, 2447. [CrossRef] [PubMed]
42. Carsel, R.F.; Parrish, R.S. Developing joint probability distributions of soil water retention characteristics. *Water Resour. Res.* **1988**, *24*, 755–769, doi:10.1029/WR024i005p00755. [CrossRef]

Article

A Non-Newtonian Magnetohydrodynamics (MHD) Nanofluid Flow and Heat Transfer with Nonlinear Slip and Temperature Jump

Jing Zhu *, Yaxin Xu and Xiang Han

School of Mathematics and Physics, University of Science and Technology Beijing, Beijing 100080, China; S20180748@xs.ustb.edu.cn (Y.X.); S20170736@xs.ustb.edu.cn (X.H.)
* Correspondence: zhujing@ustb.edu.cn; Tel.: +86-1368-121-2703

Received: 29 August 2019; Accepted: 28 November 2019; Published: 6 December 2019

Abstract: The velocity and thermal slip impacts on the magnetohydrodynamics (MHD) nanofluid flow and heat transfer through a stretched thin sheet are discussed in the paper. The no slip condition is substituted for a new slip condition consisting of higher-order slip and constitutive equation. Similarity transformation and Lie point symmetry are adopted to convert the derived governed equations to ordinary differential equations. An approximate analytical solution is gained through the homotopy analysis method. The impacts of velocity slip, temperature jump, and other physical parameters on flow and heat transfer are illustrated. Results indicate that the first-order slip and nonlinear slip parameters reduce the velocity boundary layer thickness and Nusselt number, whereas the effect on shear stress is converse. The temperature jump parameter causes a rise in the temperature, but a decline in the Nusselt number. With the increase of the order, we can get that the error reaches 10^{-6} from residual error curve. In addition, the velocity contours and the change of skin friction coefficient are computed through Ansys Fluent.

Keywords: velocity-slip; temperature-jump; homotopy analysis method; nanofluids; power-law fluids

1. Introduction

In a heat transfer mechanism, fluid is a main medium as a heat transfer carrier. Therefore, improving the thermal transfer efficiency of the fluid used is a vital challenge in the industry. Certain experiments have shown that the thermal conductivity of fluids containing metal and oxide particles is higher than that of traditional base liquids such as oil, water, and ethylene glycol [1–3]. For the sake of improving the heat transfer efficiency of the fluid, researchers have added metal and non-metallic nanoparticles into the traditional base liquid to form a new compound "nanofluid". Nanofluids are made up of base fluids and nanoparticles, but not a simple mixture, which are composed of nano-sized solid particle or tubes suspended in the base fluids, are solid–liquid composite materials. Nanoparticles have high surface-activity and tend to aggregate together with time. The idea was first proposed by Choi and Eastman [4]. Nanofluids are important in the fields of energy, chemical, microelectronics, and information. Recently, the flow and conduct heat of nanofluids have been studied by certain scholars. A quick overview is given here. Sheremet et al. [5] discussed natural convection of alumina-water nanofluid in an inclined wavy-walled cavity. Nanofluids flow in microchannels with heat conduction was discussed by Bowers et al. [6]. Hashim et al. [7] discussed the mixed convection and heat conduction of Williamson nanofluids under unsteady condition. Mahdy [8] presented the effects of magnetohydrodynamics (MHD) and variable wall temperature on non-Newtonian Casson nanofluid flow. Asadi et al. [9] presented the latest progress of preparation methods and thermophysical properties of oil-based nanofluids. Pourfattah et al. [10] simulated water/CuO nanofluid fluid flow and heat transfer inside a manifold microchannel. Alarifi et al. [11] investigated the effects of solid concentration of nanoparticles, temperature, and shear rate on the rheological properties of nanofluid. For a traditional base fluid, there are two main types: Newtonian fluids and non-Newtonian fluids. In industry, non-Newtonian fluids play an important role, such as juices, starch solutions, egg whites, and apple pulp. To understand behaviour of non-Newtonian fluids, certain models have been presented. Power law model is relatively simple, widely used among these models. Researchers have further investigated the flow and conduct heat of power law fluids. Javanbakht et al. [12] studied the heat conduction on the surface of a power law fluid. Turan et al. [13] discussed mixed convection of power-law liquids in enclosures. The heat conduction of power law liquid in various section tubes was considered by Zhang et al. [14]. Ahmedet et al. [15] addressed MHD power law liquid flow in a Darcy–Brinkmann porous medium.

In this paper, the base fluid of a nanofluid is power law fluid. When nanoparticles are added into the traditional base liquid, local velocity slip may happen as an effect of high shear force between the fluid and the wall, and the slip condition is no longer negligible in the nanometer or micro scales. The velocity slip is a finite velocity boundary condition between the fluid and the solid [16]. Researchers have done certain studies on the slipping problems of nanofluids. Ramya et al. [17] studied the viscous flow and heat transfer of nanofluid through a stretched sheet with the effect of magnetic field, velocity, and thermal slip. Abbas et al. [18] discussed the stagnation flow of micropolar nanofluids through a cylinder with slip. The effect of heat and velocity slip on the flow of Carson nanofluids through a cylinder was discussed by Usman et al. [19]. Babu et al. [20] investigated the three-dimensional MHD nanofluid flow over a variable thickness slendering stretching sheet with the effect of thermophoresis, Brownian motion, and slip parameter. The above studies all discussed the first-order slip model, whereas higher-order slips should be considered when the velocity and temperature profiles of an average free path are nonlinear. It is now known that the inclusion of higher-order slip yields results closer to those by experiments [21]. Thus, various investigations on higher-order slip flows were published by Uddin et al. [22], Kamran et al. [23], Farooq et al. [24], and Yasin et al. [25]. These all suggest that the power law constitutive equation should be considered on the basis of high order slip for a power law nanofluid.

In the aforementioned literature, there are few papers about the flow and heat transfer of magnetic nanofluids with higher-order slip parameters. Therefore, a new mathematical model is proposed. With the help of similarity transformation variables, governing equations are converted to ordinary differential equations, whose solution is solved using homotopy analysis method. The effects of nanofluid velocity, temperature, concentration, skin friction coefficient and Nusselt number on various physical parameters are simulated. In addition, the fluid flow situation is visualized by the computational fluid dynamics (CFD) software Ansys Fluent.

2. Mathematical Modelling Formulation

2.1. Flow Behavior

Consider a steady, two-dimensional, incompressible MHD fluid flow with copper through a stretching thin plate. All variables mentioned are presented in Tables 1 and 2 [26] gives some physical capabilities of the base liquid and nanoparticles. Meanwhile, a transverse magnetic field is utilized, where the strength is B_x and the presence of surface tension is also considered. Given the above hypotheses, the governing equations composed of continuity equation and momentum equation can be given as

$$\frac{\partial U}{\partial X} + \frac{\partial V}{\partial Y} = 0, \tag{1}$$

$$U\frac{\partial U}{\partial X} + V\frac{\partial U}{\partial Y} = -\frac{1}{\rho_{nf}}\frac{\partial P}{\partial X} + \frac{\partial S_{XX}}{\partial X} + \frac{\partial S_{XY}}{\partial Y} + \frac{\sigma B^2}{\rho_{nf}}(U_e - U), \tag{2}$$

$$U\frac{\partial V}{\partial X} + V\frac{\partial V}{\partial Y} = -\frac{1}{\rho_{nf}}\frac{\partial P}{\partial Y} + \frac{\partial S_{YX}}{\partial X} + \frac{\partial S_{YY}}{\partial Y},\tag{3}$$

$$S_{ij} = 2\mu_{nf}\left(2D_{ml}D_{ml}\right)^{\frac{n-1}{2}}D_{ij},\ D_{ij} = \frac{1}{2}\left(\frac{\partial U_i}{\partial X_j} + \frac{\partial U_j}{\partial X_i}\right).\tag{4}$$

Table 1. Nomenclature.

Symbol	Description	Symbol	Description
B_x	magnetic field strength	c_p	heat capacity
U	field velocity	U_e	free stream speed
T	temperature in the boundary layer	T_∞	temperature far away from the sheet
T_w	unified temperature	C	concentration
C_∞	fluid concentration in the free stream	C_w	unified concentration
S_{ij}	deviatoric part of the stress tensor	δ_{ij}	unit tensor
D_{ij}	rate-of-strain tensor	σ	electrical conductivity
D_T	thermophoresis diffusion coefficient	$\lambda_1, \lambda_2, \lambda_3$	slip parameters of velocity
φ	nanoparticle volume fraction	ρ	density
α	thermal diffusivity	k	thermal conductivity
P	pressure	μ	dynamic viscosity
Nu	Nusselt number	C_f	skin friction coefficients
Pr	Prandtl number	Nt	thermophoresis parameter
Nb	Brownian motion parameter	Sc	Schmidt number
M	Hartmann number	Re	Reynolds number
D_B	Brownian diffusion	Sh	Sherwood number
f	fluid phase	s	solid phase
nf	nanofluid	η	similarity variable
U, V	velocity components	X, Y	Cartesian coordinates

In the above, X and Y are the Cartesian coordinates along and normal to the extension sheet, respectively. **U** is the velocity field. U and V are the x and y components of **U**. P is the pressure, σ the electric conductivity, B_x the magnetic field along the forward direction of Y-axis, U_e the free stream speed, S_{ij} the deviatoric part of the stress tensor $\varsigma_{ij} = -P\delta_{ij} + S_{ij}$, δ_{ij} the unit tensor, and D_{ij} the rate-of-strain tensor. ρ_{nf} the effective density and μ_{nf} the effective dynamic viscosity given by [27]

$$\rho_{nf} = (1-\varphi)\rho_f + \varphi\rho_s,\quad \mu_{nf} = \frac{\mu_f}{(1-\varphi)^{2.5}}.\tag{5}$$

The other parameters of nanofluid $(\rho C_p)_{nf},\ \alpha_{nf},\ k_{nf}$ are given [27]

$$(\rho C_p)_{nf} = (1-\varphi)(\rho C_p)_f + \varphi(\rho C_p)_s,\quad \alpha_{nf} = \frac{k_{nf}}{(\rho C_p)_{nf}},\tag{6}$$

$$\frac{k_{nf}}{k_f} = \frac{k_s + 2k_f - 2\varphi(k_f - k_s)}{k_s + 2k_f + \varphi(k_f - k_s)},\tag{7}$$

where subscripts s, f, and nf represent the solid particle, base liquid, and the thermophysical properties of nanofluid, respectively. φ is the solid volume fraction of nanoparticles, $(\rho C_p)_{nf}$ the effective heat capacity. The thermal conductivity is k_{nf} and the thermal diffusivity is α_{nf}.

For the sake of analyzing the boundary layer in a better way, the following nondimensional variables are introduced,

$$x = \frac{X}{L},\ y = \frac{Y}{\delta},\ u = \frac{U}{U_w},\ v = \frac{LV}{\delta U_w},\ p = \frac{P}{\rho_f U_w^2},\ \tau_{ij} = \frac{S_{ij}}{\rho_f U_w^2},\tag{8}$$

where L and δ represent the characteristic length in the X and Y direction, respectively. U_w denotes the velocity in the X-direction.

Thus, Equations (1)–(4) become

$$\frac{\partial u}{\partial x} + \frac{\partial v}{\partial y} = 0,\tag{9}$$

$$u\frac{\partial u}{\partial x} + v\frac{\partial u}{\partial y} = -\frac{\rho_f}{\rho_{nf}}\frac{\partial p}{\partial x} + \frac{\partial}{\partial y}\left(\frac{\mu_{nf}\rho_f}{\rho_{nf}}\left|\frac{\partial u}{\partial y}\right|^{n-1}\frac{\partial u}{\partial y}\right) + \frac{\rho_f\sigma B^2}{\rho_{nf}}(u_e - u),\tag{10}$$

$$\frac{\partial p}{\partial y} = 0.\tag{11}$$

From Equation (11), it can be concluded that the pressure p is identical with the pressure of mainstream flow.

$$-\frac{\partial p}{\partial x} = u_e\frac{\partial u_e}{\partial x}.\tag{12}$$

For a power law nanofluid, velocity slip effect need be considered. In many investigations, the first-order model is adopted widely. The model is suitable under the assumption that temperature and velocity profiles are linear through a average free path. However, when temperature and velocity profiles are nonlinear through a average free path, higher-order slip would become possible. Mitsuya [28] has obtained a second-order slip model from a physical phenomenon by considering the accommodation coefficient:

$$F = \alpha f_1 m_1\left[\left(\frac{2}{3}\lambda\right)\frac{\partial u}{\partial y} + \frac{1}{2}\left(\frac{2}{3}\lambda\right)^2\frac{\partial^2 u}{\partial y^2} + u_{slip}\right]|_{y=0},\tag{13}$$

where F is the shear stress, α an accommodation coefficient relative to momentum, f_1 the frequency of molecular bombardment, m_1 the molecular mass density, and λ the local molecular average free path.

In this paper, as the base fluid is a power flow fluid, namely, the shear stress $F = \mu_{nf}\left|\frac{\partial u}{\partial y}\right|^{n-1}\frac{\partial u}{\partial y}$, the constitutive equation of a power flow fluid with a higher-order slip is considered. The enhanced slip model is written as

$$u(x,0) = U_w + \left(A_1\frac{\partial u}{\partial y} + A_2\frac{\partial^2 u}{\partial y^2} + A_3\left|\frac{\partial u}{\partial y}\right|^{n-1}\frac{\partial u}{\partial y}\right)|_{y=0},\tag{14}$$

$$v(x,0) = 0, u(x,\infty) = u_e = ax^m,\tag{15}$$

where A_1, A_2, and A_3 denote the velocity slip coefficients; U_w is the the speed of the stretch plate; and $U_w = cx^m$.

For the sake of deriving a simplified model by converting governing equations into ordinary differential equations, a stream function $\psi(x,y)$ is introduced in this paper such that $u = \frac{\partial \psi}{\partial y}, v = -\frac{\partial \psi}{\partial x}$. Then Lie-group transformationsis also introduced to obtain a new set of similar variables.

$$\Gamma : x^* = xe^{\varepsilon\alpha_1}, \quad y^* = ye^{\varepsilon\alpha_2}, \quad \psi^* = \psi e^{\varepsilon\alpha_3}, \quad u^* = ue^{\varepsilon\alpha_4}, \quad v^* = ve^{\varepsilon\alpha_5}, \quad u_e^* = u_e e^{\varepsilon\alpha_6}.\tag{16}$$

Equation (16) can be considered as a point-transformation of coordinates (x,y,ψ,u,v,u_e) into coordinates $(x^*,y^*,\psi^*,u^*,v^*,u_e^*)$. Substituting Equation (16) in Equation (10), we get

$$e^{\varepsilon(\alpha_1+2\alpha_2-2\alpha_3)}\left(\frac{\partial\psi^*}{\partial y^*}\frac{\partial^2\psi^*}{\partial x^*\partial y^*}-\frac{\partial\psi^*}{\partial x^*}\frac{\partial^2\psi^*}{\partial y^{*2}}\right)$$

$$= -e^{\varepsilon(\alpha_1-2\alpha_6)}\frac{\rho_f}{\rho_{nf}}u_e^*\frac{du_e^*}{dx^*}+e^{\varepsilon(3\alpha_2-\alpha_3)}\frac{\mu_{nf}\rho_f}{\rho_{nf}}\frac{\partial^3\psi^*}{\partial y^{*3}}e^{\varepsilon(n-1)(2\alpha_2-\alpha_3)}\left(-\frac{\partial^2\psi^*}{\partial y^{*2}}\right)^{n-1} \tag{17}$$

$$+ \frac{\rho_f\sigma B^2}{\rho_{nf}}\left(e^{-\alpha_6\varepsilon}u_e^*-e^{(\alpha_2-\alpha_3)\varepsilon}\frac{\partial\psi^*}{\partial y^*}\right).$$

The boundary condition Equations (14) and (15) become

$$\frac{\partial\psi^*}{\partial y^*}(x^*,0) = e^{\varepsilon(\alpha_3-\alpha_2-m\alpha_1)}cx^{*m}+A_1e^{\varepsilon\alpha_2}\frac{\partial^2\psi^*}{\partial y^{*2}}+A_2e^{\varepsilon\alpha_2}\frac{\partial^3\psi^*}{\partial y^{*3}}$$

$$+ A_3e^{(n\varepsilon(2\alpha_2-\alpha_3)+\alpha_3-\alpha_2)}\left(-\frac{\partial^2\psi^*}{\partial y^{*2}}\right)^{n-1}\left(-\frac{\partial^2\psi^*}{\partial y^{*2}}\right), \quad at \quad y^*=0; \tag{18}$$

$$\frac{\partial\psi^*}{\partial x^*}(x^*,0) = 0, \quad at \quad y^*=0; \tag{19}$$

$$\frac{\partial\psi^*}{\partial y^*}(x^*,\infty) = e^{\varepsilon(\alpha_3-\alpha_2-m\alpha_1)}ax^{*m}, \quad at \quad y^*\to\infty. \tag{20}$$

The system will remain unaltered under the group of transformations Γ, so the parameters have the following relations, namely,

$$\alpha_2+\alpha_4-\alpha_3 = \alpha_1+\alpha_5-\alpha_3 = \alpha_3-\alpha_1-m\alpha_2 = \alpha_3-\alpha_2-m\alpha_1 = 0, \tag{21}$$

$$2\alpha_2-2\alpha_3+\alpha_1 = n(2\alpha_2-\alpha_3)+\alpha_2 = (n+1)\alpha_2-n\alpha_4 = \alpha_2-\alpha_4-\alpha_5. \tag{22}$$

Thus, Equation (16) becomes

$$\Gamma: \quad x^*=xe^{\varepsilon\alpha_1}, \quad y^*=ye^{\frac{mn-2m+1}{n+1}\alpha_1\varepsilon}, \quad \psi^*=\psi e^{\frac{2mn-m+1}{n+1}\alpha_1\varepsilon},$$

$$u^*=ue^{m\alpha_1\varepsilon}, \quad v^*=ve^{\frac{2mn-m-n}{n+1}\alpha_1\varepsilon}. \tag{23}$$

Based on the above Lie-group transformations, the stream function and similar parameter can be prescribed as follows,

$$\eta = \left(\frac{c^{2-n}}{\mu_f}\right)^{\frac{1}{n+1}}x^{\frac{2m-mn-1}{n+1}}y, \psi = \left(\frac{\mu_f}{c^{1-2n}}\right)^{\frac{1}{n+1}}x^{\frac{2mn+1-m}{n+1}}f(\eta). \tag{24}$$

After further similarity transformations, a nonlinear ordinary differential equation is obtained.

$$nf'''|f''|^{n-1}+m\varphi_1(d^2-f'f')+\varphi_1\varphi_2\frac{2mn-m+1}{n+1}ff''+\varphi_1 M(d-f')=0. \tag{25}$$

The boundary condition Equations (14) and (15) now develop into

$$f(0)=0, f'(\infty)=d, \tag{26}$$

$$f'(0)=1+\lambda_1f''(0)+\lambda_2f'''(0)+\lambda_3\left|f''(0)\right|^{n-1}f''(0), \tag{27}$$

where $d=\frac{a}{c}$, M is the Hartmann number with $M=\frac{\sigma B_0^2}{c}$, λ_1, λ_2, and λ_3 are velocity slip parameters; these parameters and φ_1, φ_2 [27] can now be written as

$$\lambda_1 = A_1\left(\frac{c^{2-n}}{\mu_f}\right)^{\frac{1}{n+1}} x^{\frac{2m-mn-1}{n+1}}, \lambda_2 = A_2\left(\frac{c^{2-n}}{\mu_f}\right)^{\frac{2}{n+1}} x^{\frac{2(2m-mn-1)}{n+1}}, \tag{28}$$

$$\lambda_3 = A_3\left(cx^m\left(\frac{c^{2-n}}{\mu_f}\right)^{\frac{1}{n+1}} x^{\frac{2m-mn-1}{n+1}}\right)^n, \tag{29}$$

$$\varphi_1 = (1-\varphi)^{2.5}, \quad \varphi_2 = 1 - \varphi + \varphi\frac{\rho_s}{\rho_f}, \tag{30}$$

where A_1, A_2, and A_3 are arbitrary positive constants.

2.2. Heat and Mass Transfer Behavior

The heat and mass equations can now be formulated as follows,

$$U\frac{\partial T}{\partial X} + V\frac{\partial T}{\partial Y} = \frac{\partial}{\partial Y}\left(k(T)\frac{\partial T}{\partial Y}\right)$$
$$+ \frac{\tau}{\mu_f}C_f\frac{3}{n+1}(C^3 x^{3m-1})^{\frac{n-1}{n+1}}\left(D_B\frac{\partial C}{\partial Y}\frac{\partial T}{\partial Y} + \frac{D_T}{T_\infty}\left(\frac{\partial T}{\partial Y}\right)^2\right), \tag{31}$$

$$U\frac{\partial C}{\partial X} + V\frac{\partial C}{\partial Y} = \mu_f^{\frac{2}{n+1}}(C^3 x^{3m-1})^{\frac{n-1}{n+1}}\left(D_B\frac{\partial^2 C}{\partial Y^2} + \frac{D_T}{T_\infty}\frac{\partial^2 T}{\partial Y^2}\right), \tag{32}$$

$$k(T) = \frac{k_{nf}}{(\rho C_p)_{nf}}(T_w - T_\infty)^{1-n}U_w^{n-1}\left|\frac{\partial T}{\partial Y}\right|^{n-1}. \tag{33}$$

The boundary conditions are as follows,

$$T(X,0) = T_w + k_{nf}(T_w - T_\infty)^{1-n}\left|\frac{\partial T}{\partial Y}\right|^{n-1}\frac{\partial T}{\partial Y}|_{y=0}, \tag{34}$$

$$C(X,0) = C_w, T(X,\infty) = T_\infty, C(X,\infty) = C_\infty, \tag{35}$$

where T shows temperature in the boundary layer, T_∞ denotes the temperature away from the sheet and is a constant, and T_w indicates the unified temperature of the fluid. C is the concentration of the fluid, C_∞ is the fluid concentration in the free stream, and C_w the unified concentration of the fluid.

For the sake of gaining the similarity solutions of equations, the following similarity variables are introduced,

$$\theta(\eta) = \frac{T - T_\infty}{T_w - T_\infty}, \phi(\eta) = \frac{C - C_\infty}{C_w - C_\infty}. \tag{36}$$

Then, Equations (31)–(33) become

$$n\varphi_4\theta''|\theta'|^{n-1} + \frac{2mn - m + 1}{n+1}Pr\varphi_3 f\theta' + PrNb\varphi_3\phi'\theta' + PrNt\varphi_3\theta'^2 = 0, \tag{37}$$

$$\phi'' + \frac{2mn + 1 - m}{n+1}Scf\phi' + \frac{Nt}{Nb}\theta'' = 0. \tag{38}$$

The boundary conditions Equations (34) and (35) are converted to

$$\theta(0) = 1 + \beta\theta'(0)|\theta'(0)|^{n-1}, \quad \theta(\infty) = 0, \tag{39}$$

$$\phi(0) = 1, \quad \phi(\infty) = 0, \tag{40}$$

where Pr denotes Prandtl number, Nt represents thermophoresis parameter, Sc is Schmidt number, and Nb is Brownian motion parameter. The above parameters, φ_3, φ_4, and β, are defined as

$$Pr = \frac{\mu_f}{\alpha_f}, Nb = \frac{\tau D_B(C_w - C_\infty)}{\mu_f}, Nt = \frac{\tau D_T(T_w - T_\infty)}{\mu_f T_\infty}, \tag{41}$$

$$\varphi_3 = 1 - \varphi + \varphi \frac{(\rho C_p)_s}{(\rho C_p)_f}, \varphi_4 = \frac{k_s + 2k_f - 2\varphi(k_f - k_s)}{k_s + 2k_f + \varphi(k_f - k_s)}, \tag{42}$$

$$\beta = \frac{k_{nf}}{\mu_{nf}^{\frac{n}{n+1}}} (C^{2n-1} X^{2mn-n-m})^{\frac{1}{n+1}}, Sc = \frac{\mu_f}{D_B}. \tag{43}$$

Momentous physical parameters are expressible as follows,

$$C_f = \frac{\mu_{nf} |\frac{\partial u}{\partial y}|^{n-1} \frac{\partial u}{\partial y}|_{y=0}}{\frac{1}{2}\rho_f u_w^2} = \frac{|f''(0)|^{n-1} f''(0)}{(1-\varphi)^{2.5}} Re_x^{-\frac{1}{n+1}}, \tag{44}$$

$$C_f Re_x^{-\frac{1}{n+1}} = \frac{|f''(0)|^{n-1} f''(0)}{(1-\varphi)^{2.5}}, \tag{45}$$

$$Nu_x = -\frac{x k_{nf} \frac{\partial T}{\partial y}|_{y=0}}{k_f(T_w - T_\infty)} = -\frac{k_s + 2k_f - 2\varphi(k_f - k_s)}{k_s + 2k_f + \varphi(k_f - k_s)} Re_x^{\frac{1}{n+1}} \theta'(0), \tag{46}$$

$$Nu_x Re_x^{-\frac{1}{n+1}} = -\frac{k_s + 2k_f - 2\varphi(k_f - k_s)}{k_s + 2k_f + \varphi(k_f - k_s)} \theta'(0), \tag{47}$$

$$Sh_x = -\frac{x D_B \frac{\partial C}{\partial y}|_{y=0}}{D_B(C_w - C_\infty)} = -Re_x^{\frac{1}{n+1}} \phi'(0), \tag{48}$$

$$Sh_x Re_x^{-\frac{1}{n+1}} = -\phi'(0). \tag{49}$$

3. Solution Procedures

In this section, the homotopy analysis method (HAM) [29] is used to solve this problem. The initial guess solutions of velocity, temperature, and concentration, based on boundary conditions, are, respectively,

$$f_0 = B_1 + B_2 e^{-\eta} + B_3 \eta, \theta_0 = B e^{-\eta}, \phi_0 = e^{-\eta}. \tag{50}$$

Three linear operators are selected as

$$L_f = f''' + f'', L_\theta = \theta'' + \theta', L_\phi = \phi'' - \phi. \tag{51}$$

These operators satisfy some properties:

$$L_f(C_1 + C_2 e^{-\eta} + C_3 \eta) = 0, L_\theta(C_4 e^{-\eta} + C_5) = 0, L_\phi(C_6 e^{-\eta} + C_7 e^{\eta}) = 0 \tag{52}$$

where $C_i(i = 1, 2, \cdots, 7)$ are arbitrary constants.

The 0-th order deformation equations and its boundary conditions are derived and the expressions are written as

$$(1-p)L[F(\eta, p) - f_0(\eta)] = p h_f H_f(\eta) N_f[F(\eta, p)], \tag{53}$$

$$(1-p)L[\Theta(\eta, p) - \theta_0(\eta)] = p h_\theta H_\theta(\eta) N_\theta[F(\eta, p), \Theta(\eta, p), \Phi(\eta, p)], \tag{54}$$

$$(1-p)L[\Phi(\eta, p) - \phi_0(\eta)] = p h_\phi H_\varphi(\eta) N_\phi[F(\eta, p), \Theta(\eta, p), \Phi(\eta, p)]; \tag{55}$$

$$F(0,p) = 0, F'(\infty,p) = d, \Theta(\infty,p) = 0, \Phi(0,p) = 1, \Phi(\infty,p) = 0, \tag{56}$$

$$F'(0,p) = 1 + \lambda_1 F''(0,p) + \lambda_2 F'''(0,p) + \lambda_3 |F''(0,p)|^{n-1} F''(0,p), \tag{57}$$

$$\Theta(0,p) = 1 + \beta \Theta_0'(0,p)|\Theta_0'(0,p)|^{n-1}. \tag{58}$$

In the above equations, $p \in [0,1]$ is the embedding parameter; h_f, h_θ, and h_ϕ are auxiliary non-zero parameters; and $H_f(\eta)$, $H_\theta(\eta)$, and $H_\varphi(\eta)$ are nonzero auxiliary functions [30]. Obviously, for $p = 0$ and $p = 1$, we have

$$\begin{aligned}
F(\eta,0) &= f_0(\eta), & F(\eta,1) &= f(\eta), \\
\Theta(\eta,0) &= \theta_0(\eta), & \Theta(\eta,1) &= \theta(\eta), \\
\Phi(\eta,0) &= \phi_0(\eta), & \Phi(\eta,1) &= \phi(\eta).
\end{aligned} \tag{59}$$

As p increases from 0 to 1, $F(\eta,p)$ is from the initial guess $f_0(\eta)$ to the exact solution $f(\eta)$, $\Theta(\eta,p)$ is from the initial guess $\theta_0(\eta)$ to the exact solution $\theta(\eta)$, and $\Phi(\eta,p)$ is from the initial guess $\phi_0(\eta)$ to the exact solution $\phi(\eta)$ [30]. With Taylor's theorem, they can write

$$F(\eta,p) \;=\; F(\eta,0) + \sum_{k=1}^{+\infty} f_k(\eta) p^k, \; f_k(\eta) = \frac{1}{k!} \frac{\partial^k F(\eta,p)}{\partial p^k}\Big|_{p=0}, \tag{60}$$

$$\Theta(\eta,p) \;=\; \Theta(\eta,0) + \sum_{k=1}^{+\infty} \theta_k(\eta) p^k, \quad \theta_k(\eta) = \frac{1}{k!} \frac{\partial^k \Theta(\eta,p)}{\partial p^k}\Big|_{p=0}. \tag{61}$$

$$\Phi(\eta,p) \;=\; \Phi(\eta,0) + \sum_{k=1}^{+\infty} \phi_k(\eta) p^k, \quad \phi_k(\eta) = \frac{1}{k!} \frac{\partial^k \Phi(\eta,p)}{\partial p^k}\Big|_{p=0}. \tag{62}$$

Assuming that the auxiliary parameters h_f, h_θ, and h_ϕ are appropriate chosen, we can obtain convergent solutions in the following form.

$$f(\eta) = f_0(\eta) + \sum_{k=1}^{\infty} f_k(\eta), \theta(\eta) = \theta_0(\eta) + \sum_{k=1}^{\infty} \theta_k(\eta), \phi(\eta) = \phi_0(\eta) + \sum_{k=1}^{\infty} \phi_k(\eta). \tag{63}$$

For the sake of getting the higher order deformation equation, differentiating the 0-th order deformation Equations (53)–(55) k times with regard to p, set $p = 0$ and divide by $k!$, to attain

$$L_f(f_k(\eta) - \chi_k f_{k-1}(\eta)) = h_f H_f(\eta) R_{f,k}(\eta), \tag{64}$$

$$L_\theta(f_\theta(\eta) - \chi_\theta f_{\theta-1}(\eta)) = h_\theta H_\theta(\eta) R_{\theta,k}(\eta), \tag{65}$$

$$L_\phi(f_\phi(\eta) - \chi_\phi f_{\phi-1}(\eta)) = h_\phi H_\phi(\eta) R_{\phi,k}(\eta), \tag{66}$$

where $R_{f,k}(\eta)$, $R_{\theta,k}(\eta)$, and $R_{\phi,k}(\eta)$ are, respectively,

$$\begin{aligned}
R_{f,k}(\eta) \\
= \chi_k \sum_{l=0}^{k-2} f_l''' \sum_{j=2}^{k-l} \sum_{\substack{i_1,i_2,\cdots,i_k=0 \\ i_1+i_2+\cdots i_{k-1}=j-1 \\ i_1+2i_2+\cdots+(k-1)i_{k-1}=k-1-l}}^{k-1} \frac{n(n-1)\cdots(n-j+1)}{i_1!i_2!\cdots i_{k-1}!} |f_0''|^{n-j} \prod_{q=1}^{k-1} |f_q''|^{i_q} \\
+ n f_{k-1}''' |f_0''|^{n-1} - m\varphi_1\varphi_2 \sum_{i=0}^{k-1} f_i' f_{k-1-i}' \\
+ \varphi_1\varphi_2 \frac{2mn-m+1}{n+1} \sum_{i=0}^{k-1} f_i f_{k-1-i}'' - \varphi_1 M f_{k-1}',
\end{aligned} \tag{67}$$

$$
\begin{aligned}
R_{\theta,k}(\eta) &= \chi_k \sum_{l=0}^{k-2} \theta_l'' \sum_{j=2}^{k-l} \sum_{\substack{i_1,i_2,\cdots,i_k=0 \\ i_1+i_2+\cdots i_{k-1}=j-1 \\ i_1+2i_2+\cdots+(k-1)i_{k-1}=k-1-l}}^{k-1} \frac{n(n-1)\cdots(n-j+1)\varphi_4}{i_1!i_2!\cdots i_{k-1}!} |\theta_0'|^{n-j} \prod_{q=1}^{k-1} |\theta_q'|^{i_q} \\
&\quad + n\varphi_4 \theta_{k-1}'' |\theta_0'|^{n-1} + \frac{2mn-m+1}{n+1} \operatorname{Pr}\varphi_3 \sum_{i=0}^{k-1} f_i \theta_{k-1-i}' \\
&\quad + \operatorname{Pr} Nb\varphi_3 \sum_{i=0}^{k-1} \phi_i \theta_{k-1-i}' + \operatorname{Pr} Nt\varphi_3 \sum_{i=0}^{k-1} \theta_i' \theta_{k-1-i}',
\end{aligned}
\tag{68}
$$

$$
R_{\phi,k}(\eta) = \phi_{k-1}'' + \frac{2mn+1-m}{n+1} Sc \sum_{i=0}^{k-1} f_i \phi_{k-1-i}' + \frac{Nt}{Nb} \theta_{k-1}'',
\tag{69}
$$

$$
\chi_k = \begin{cases} 0 & k \leq 1, \\ 1 & k > 1. \end{cases}
\tag{70}
$$

Boundary conditions Equations (56)–(58) become

$$
f_k(0) = 0, \; f_k'(\infty) = 0, \; \theta_k(\infty) = 0, \; \phi_k(0) = 0, \; \phi_k(\infty) = 0,
\tag{71}
$$

$$
\begin{aligned}
f_k'(0) &= \sum_{l=0}^{k-1} f_l''(0) \sum_{j=2}^{k+1-l} \sum_{\substack{i_1,i_2,\cdots,i_k=0 \\ i_1+i_2+\cdots+i_k=j-1 \\ i_1+2i_2+\cdots+ki_k=k-l}}^{k} \frac{\lambda_3(n-1)(n-2)\cdots(n-j+1)}{i_1!i_2!\cdots i_k!} |f_0''(0)|^{n-j} \prod_{q=1}^{k} |f_q''(0)|^{i_q} \\
&\quad + \lambda_3 f_k''(0)|f_0''(0)|^{n-1} + \lambda_1 f_k''(0) + \lambda_2 f_k'''(0),
\end{aligned}
\tag{72}
$$

$$
\begin{aligned}
\theta_k(0) &= \sum_{l=0}^{k-1} \theta_l'(0) \sum_{j=2}^{k+1-l} \sum_{\substack{i_1,i_2,\cdots,i_k=0 \\ i_1+i_2+\cdots+i_k=j-1 \\ i_1+2i_2+\cdots+ki_k=k-l}}^{k} \frac{\beta(n-1)(n-2)\cdots(n-j+1)}{i_1!i_2!\cdots i_k!} |\theta_0'(0)|^{n-j} \prod_{q=1}^{k} |\theta_q'(0)|^{i_q} \\
&\quad + \beta\theta_k'(0)|\theta_0'(0)|^{n-1}.
\end{aligned}
\tag{73}
$$

4. Results and Discussion

In homotopy analysis, the h-curves are ploted to select the effective region of parameter h. For the sake of obtaining the convergent parameters h_f, h_θ, and h_ϕ, Figures 1–3 plot the h-curves of various orders for $f''(0)$, $\theta(0)$ and $\phi(0)$. Ranges of h-curves are $[-0.4, 0]$, $[-0.5, -0.3]$, $[-0.5, 0.3]$, that is, the horizontal segment of the curves, which is called the effective region, so $h_f = h_\theta = h_\phi = h = -0.35$ is selected in the paper.

For the sake of proving the accuracy and effectiveness of homotopy analysis after determining values of h_f, h_θ, and h_ϕ, Figure 4 plots the error curves of various power law index by the "BVPh2.0" procedure software package. As can be seen from Figure 4, the errors have reached 10^{-4} in the second order, meeting the standards of engineering calculation. The larger the order, the smaller the error becomes. Further, surface friction coefficients are compared with the literature [31] for various first-order slip parameter λ_1 in Table 3.

Table 2. The physical capabilities of base fluid and nanoparticles [26].

	Base Fluid (0.0–0.4%)	***Cu***
$C_p/(\text{J} \cdot \text{kg}^{-1} \cdot \text{K}^{-1})$	4179	385
$\rho/(\text{kg} \cdot \text{m}^{-3})$	997.1	8933
$k/(\text{W} \cdot \text{m}^{-1} \cdot \text{K}^{-1})$	0.613	400
$\sigma/(\Omega^{-1} \cdot \text{m}^{-1})$	0.05	5.96×10^7

Table 3. Comparisons of $C_f R_e^{\frac{1}{n+1}}$ for various λ_1 as $n = 1, m = 1, d = 1.5, \lambda_2 = \lambda_3 = 0, \varphi = 0$.

λ_1	$C_f R_e^{\frac{1}{2}}$		
	Ul Haq et al. [31]	**Present Research**	**Percent Difference**
0.5	0.34153	0.341678	0.043%
1	0.34153	0.341215	0.092%

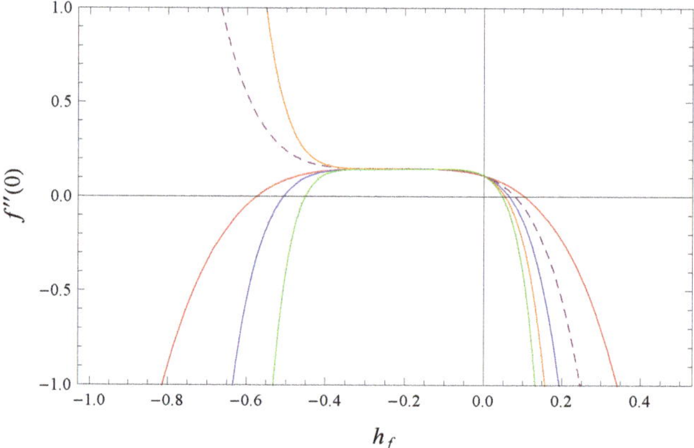

Figure 1. h_f-curves.

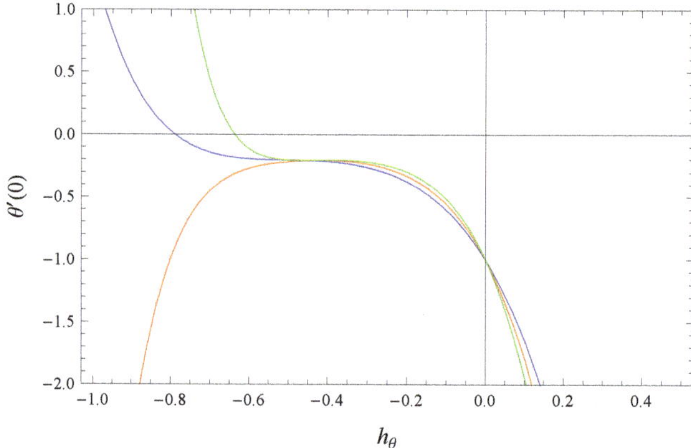

Figure 2. h_θ-curves.

After attesting the accuracy and effectiveness of homotopy analysis, the impacts of various physical parameters are analyzed, such as nondimensional velocity $f'(\eta)$, temperature $\theta(\eta)$, etc.

Meanwhile, the flow of power law nanofluid is numerically simulated by the widely used software Ansys Fluent to further explore the flow properties.

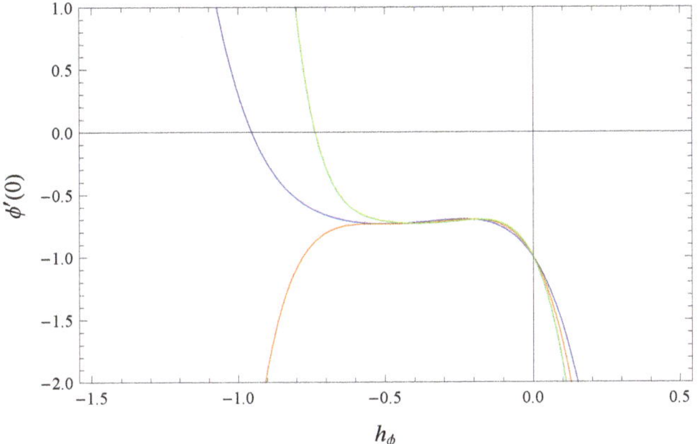

Figure 3. h_ϕ-curves.

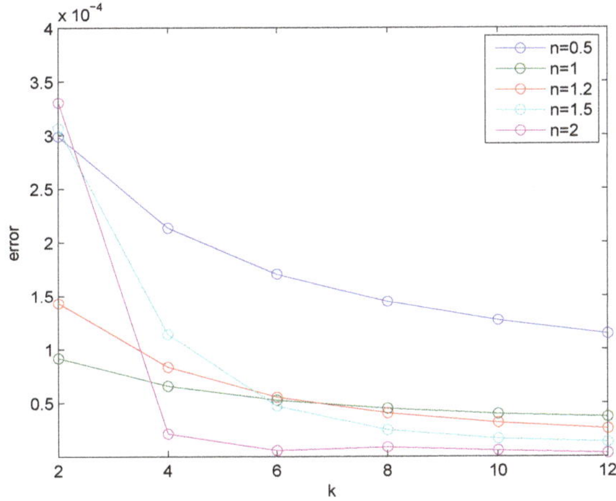

Figure 4. Total error of approximation for various powers n.

4.1. Behavior of Velocity Profiles

Figures 5 and 6 demonstrate effects of power law exponential of the plate m and Hartmann number M on nondimensional velocity $f'(\eta)$. The velocity distribution for various m is showed in Figure 5. By increasing the power exponent of the plate m, the tensile speed of the plate increases. Greater deformation is effected in the fluid, leading to the increase of $f'(\eta)$. As pointed out in [32], the effects of M on $f'(\eta)$ are visible in Figure 5. Recall that Hartmann number M expresses the ratio of electromagnetic force to viscous force. Due to the fact that greater Hartmann number corresponds to larger Lorenz force, the velocity $f'(\eta)$ increases.

When the fluid is pseudoplastic and expansive, impacts of d on $f'(\eta)$ are illustrated in Figure 7. In Figure 7, the velocity of the fluid has upward tendency for various d. Whereas, the velocity of expansive fluid increases slower than that of pseudoplastic fluid due to the increase of the fluid viscosity.

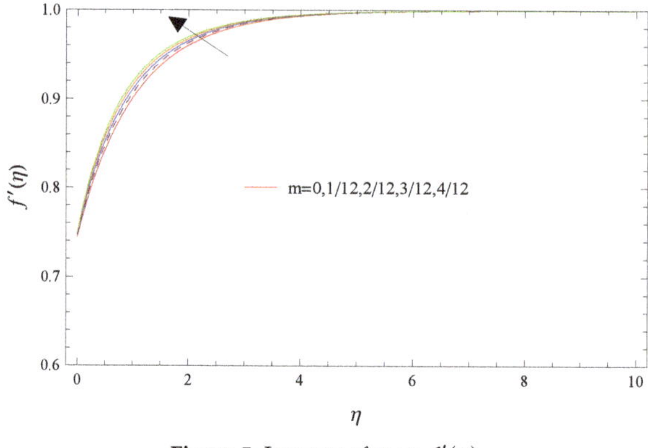

Figure 5. Impacts of m on $f'(\eta)$.

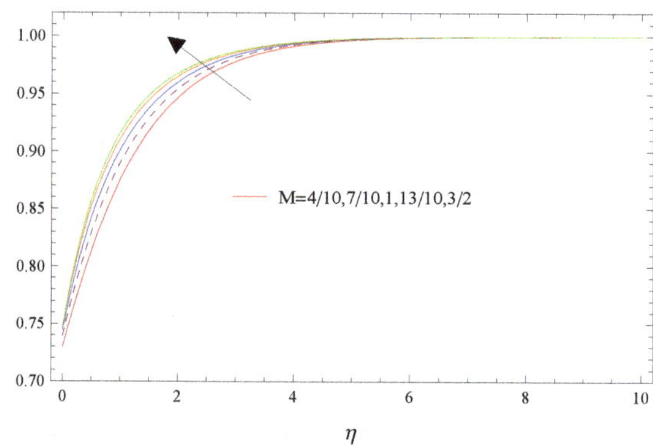

Figure 6. Impacts of M on $f'(\eta)$.

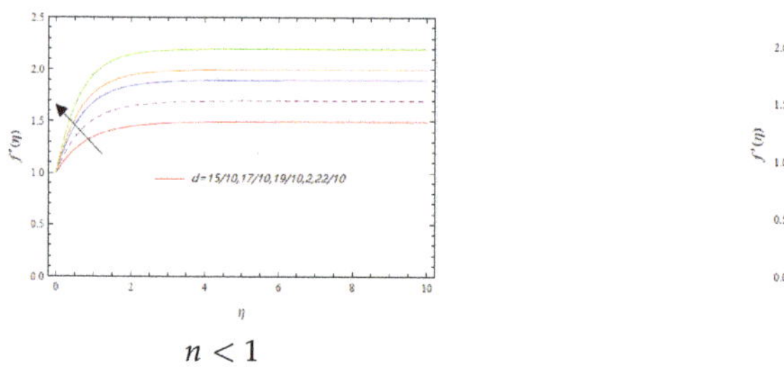

$n < 1$

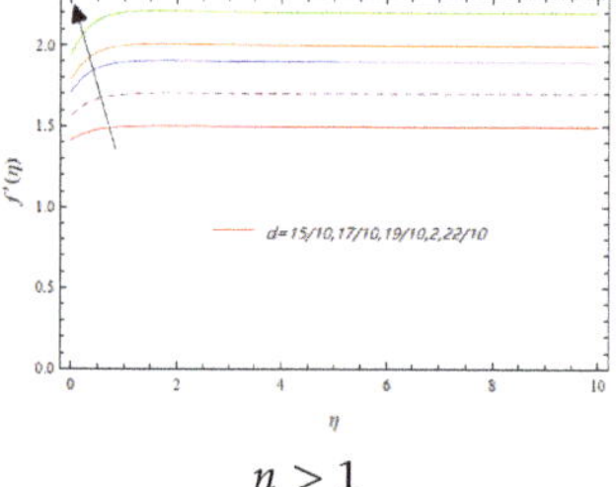

$n > 1$

Figure 7. Impacts of d on $f'(\eta)$ for $n < 1$ and $n > 1$.

Figure 8 clearly presents the impacts of various power law index n on $f'(\eta)$. As seen in Figure 8, the buoyancy becomes larger as the power law index n increases, which causes the increase of velocity.

Influences of different velocity slip parameters λ_1, λ_2, and λ_3 on $f'(\eta)$ are illustrated in Figures 9–11, respectively. Velocity slip mainly affects slip loss and, in a cascade, fluid velocity. With the increases of the second-order slip parameter λ_2, velocity $f'(\eta)$ also increases; however, the results are contradictory when the first-order linear slip parameter λ_1 and nonlinear slip parameter λ_3 increase.

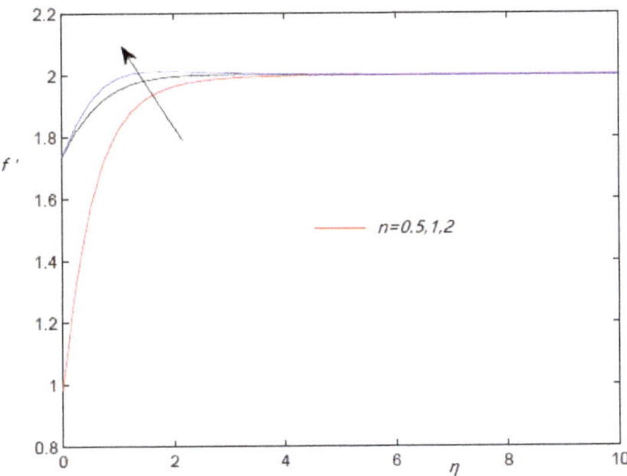

Figure 8. Impacts of n on $f'(\eta)$.

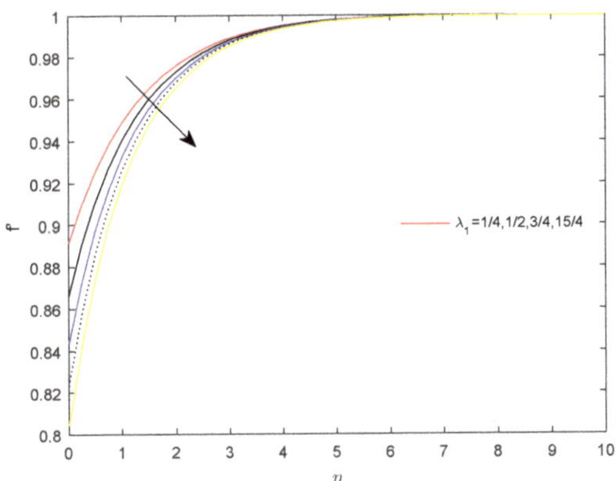

Figure 9. Effects of λ_1 on $f'(\eta)$.

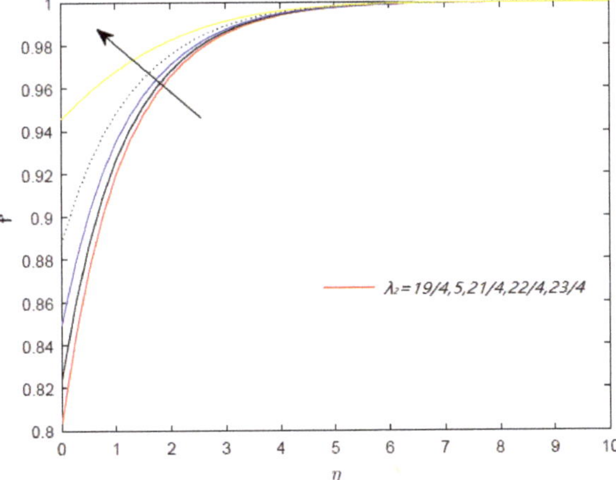

Figure 10. Effects of λ_2 on $f'(\eta)$.

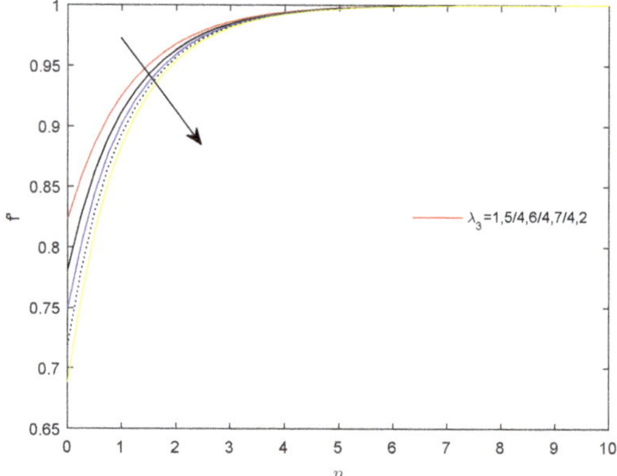

Figure 11. Effects of λ_3 on $f'(\eta)$.

4.2. Behavior of Temperature Profiles

Figures 12 and 13 indicate various temperature behavior for different Nb and Nt. Figure 12 displays the effects of Nb on temperature. Fluid particles generate more heat through random motions when Nb increases, which causes the rise in temperature. Figure 13 clearly shows temperature distribution for various thermophoresis parameter Nt. Thermophoresis indicates that particles move from a high temperature part to a low temperature one in a fluid with temperature gradient. Thus, the temperature increases with the enhancement of the parameter Nt.

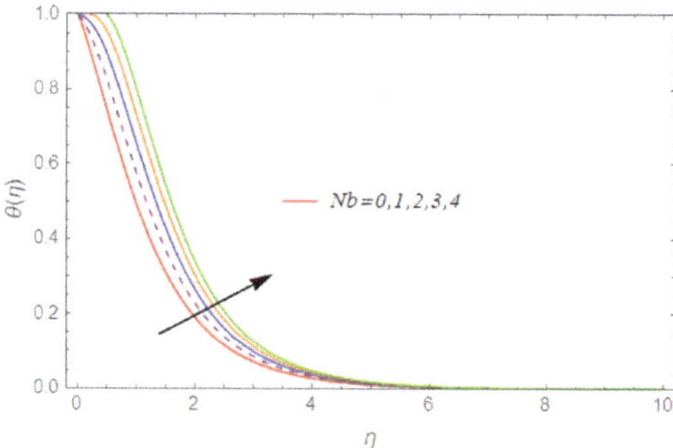

Figure 12. Impacts of Nb on $\theta(\eta)$.

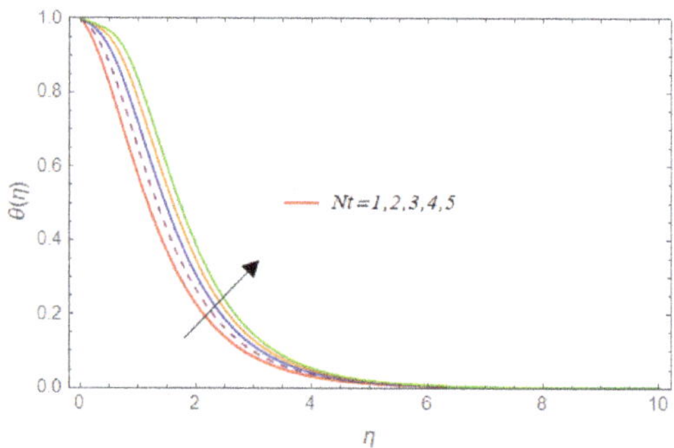

Figure 13. Impacts of Nt on $\theta(\eta)$.

Figures 14 and 15 show temperature distribution for diverse temperature jump parameter β and power law index n. Figure 14 plots the temperature curves for diverse β. Increasing temperature jump parameter β leads to a rise in the thickness of temperature boundary layer. Thus, the temperature has an upward tendency. Figure 15 demonstrates the temperature distribution for various n. The temperature diminish when the power law index rises. In other words, temperature boundary layer becomes thinner with the enhancement of n.

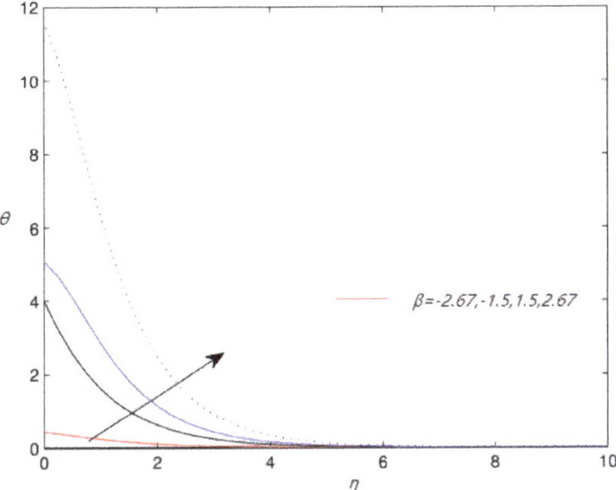

Figure 14. Impacts of β on $\theta(\eta)$.

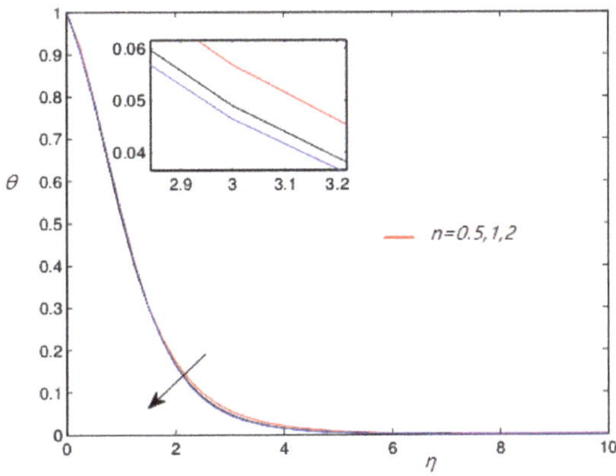

Figure 15. Impacts of n on $\theta(\eta)$.

4.3. Behavior of Concentration Profiles

Figures 16 and 17 show the concentration distribution for diverse values of the Brownian motion parameter Nb and the thermophoresis parameter Nt. From Figure 16, the collision of fluid particles rises with the stronger Brown motion, which leads to the reduction of fluid concentration. Figure 17 indicates the concentration field for various thermophoresis parameter Nt. The magnitude of concentration variation is greater under the influence of thermophoresis parameter.

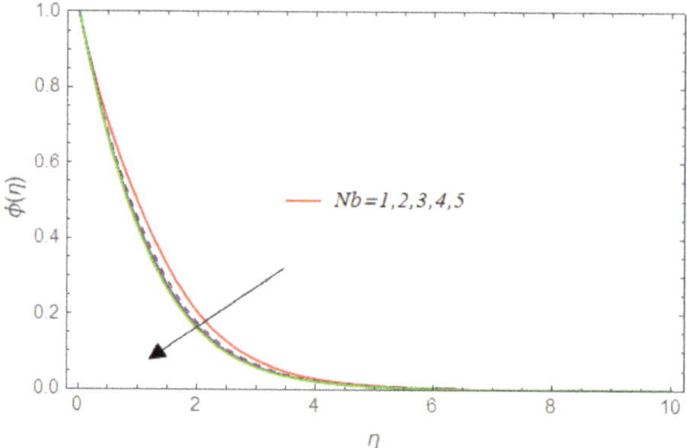

Figure 16. Impacts of Nb on $\phi(\eta)$.

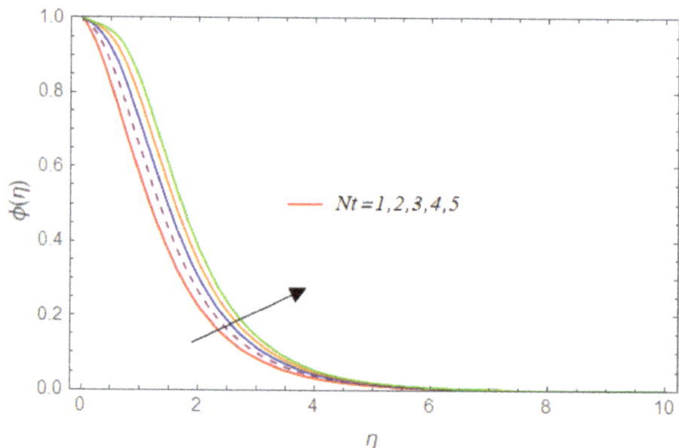

Figure 17. Impacts of Nt on $\phi(\eta)$.

4.4. Analysis of Skin Friction and Nusselt Number

In the study of fluids, vital physical parameters, such as skin friction coefficient and local Nusselt number, are discussed. In this paper, the impacts of various parameters on these two parameters are demonstrated in Table 4. Skin friction coefficients have ascending behavior with the increase of φ, λ_1 and λ_3. On the contrary, the downward trend is seen with the raise of λ_2. For local Nusselt number, when φ and λ_2 rise, the local Nusselt numbers have an upward trend, whereas the local Nusselt numbers diminish with the rise of λ_1, λ_3 and β.

4.5. Simulated Behavior

In this subsection,the laminar model is used to solved governing equations. Ansys Fluent uses the Gauss–Siedel point-by-point iterative method combined with the algebraic multigrid (AMG) method to solve the algebraic equations. The effects of various parameters on the flow of power-law nanofluid over a stretched thin sheet are simulated. The computational results obtained by using CFD solver are compared with the available results of Chen [33] for some limiting conditions. The present results are proved to be in good agreement as shown in Table 5. The effects of various parameters such as power law exponential of the plate m, nanoparticle volume fraction φ, and power law index n on Nusselt number Nu and skin friction coefficient are shown in Figures 18–22.

Table 4. Effects of φ, λ_1, λ_2, λ_3, and β on $C_f Re_x^{\frac{1}{n+1}}$ and $Nu_x Re_x^{-\frac{1}{n+1}}$ for $n = 1/2$, $m = 0$, $M = 1$, $d = 1$, $Pr = 1$, $Nb = 1$, $Nt = 1$, and $Sc = 1$.

φ	λ_1	λ_2	λ_3	β	$C_f Re_x^{\frac{1}{n+1}}$	$Nu_x Re_x^{-\frac{1}{n+1}}$
0	2	1	1	0	0.499647	0.187766
1.5%	2	1	1	0	0.515072	0.194264
3%	2	1	1	0	0.541362	0.200764
0	1/4	5	1	0	0.28862	0.209897
0	3/4	5	1	0	0.366794	0.209846
0	1	5	1	0	0.396051	0.20919
0	1	5	1	0	0.396051	0.20919
0	1	21/4	1	0	0.359076	0.210207
0	1	22/4	1	0	0.297825	0.210503
0	1	5	1	0	0.396051	0.20919
0	1	5	5/4	0	0.452515	0.201034
0	1	5	6/4	0	0.493526	0.188581
0	2	1	1	0	0.499647	0.187766
0	2	1	1	1.5	0.499647	0.137537
0	2	1	1	8/3	0.499647	0.0804748

Table 5. Comparisons of $C_f Re_e^{\frac{1}{n+1}}$ for various n with $m = 0.5$.

n	$C_f Re_e^{\frac{1}{n+1}}$		
	Chen [33]	**Present Research**	**Percent Difference**
0.5	-1.831551	-1.831768	0.012%
1	-1.54073	-1.54079	0.003%
1.5	-1.39441	-1.39578	0.098%

The velocity contours for nonlinear slip are simulated in Figure 18. From these diagrams, the flow produces velocity boundary layer near the entrance. Besides, the velocity boundary layer of the pseudoplastic fluid is thicker than that for a Newton and expansive fluid.

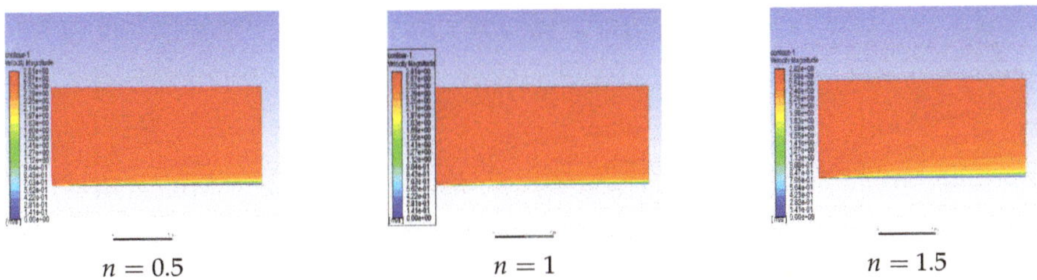

Figure 18. Velocity contours with $n = 0.5$, $n = 1$, $n = 1.5$.

Figures 19 and 20 present the effect of nanoparticle volume fraction on Nusselt number Nu and skin friction coefficient C_f at fixed values of inlet velocity, power law index. From Figure 19, the local Nusselt number increases at any x-location When nanoparticles are added to the base fluid. This is because a lower local temperature difference between the sheet walls and fluid can be achieved. Therefore, the high thermal conductivity of Cu nanoparticles enhances the thermal performance of the fluid. As the viscosity of the liquid can be increased by adding Cu nanoparticles into the base fluid, the C_f along the thin sheet increases when using higher concentrations of nanoparticles, as shown in Figure 20.

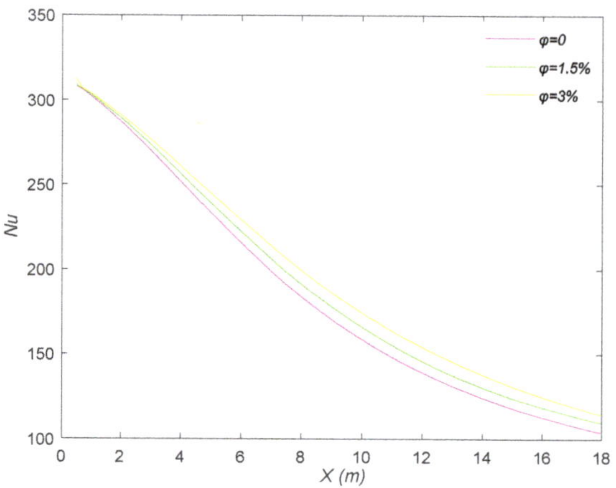

Figure 19. Effect of φ on Nu.

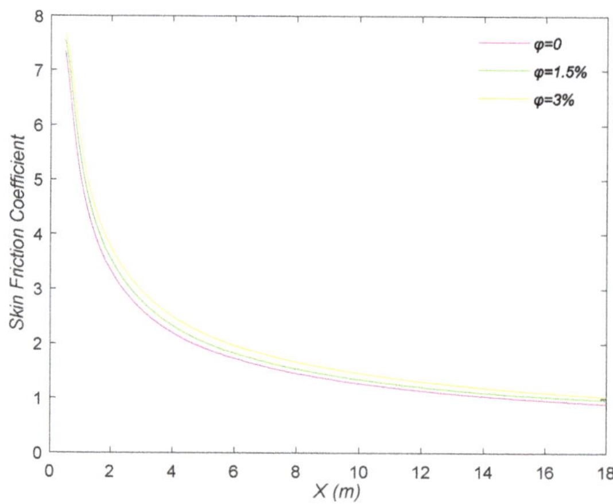

Figure 20. Effect of φ on C_f.

Figure 21 shows the effect of power law index n on skin friction coefficient C_f. The skin friction coefficient decreases with the increase of x-location for a given power law index. However, for a constant x-location, the skin friction coefficient have an upward tendency as the power law index increases.

Figure 22 demonstrates the skin friction coefficient distribution for various φ. The skin friction coefficient increases as the fluid behavior changes from shear-thinning to shear-thickening for a certain φ. As the φ increases the skin friction coefficient increases for a constant power law index.

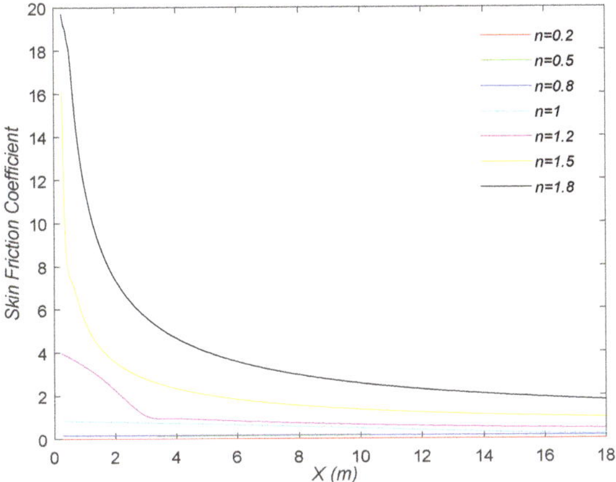

Figure 21. Effect of n on skin friction coefficient.

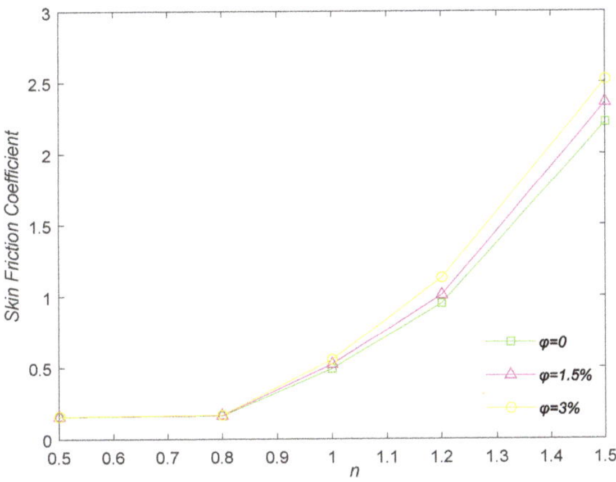

Figure 22. Variation of the skin friction coefficient at different φ.

5. Conclusions

The flow and heat transfer of magnetic nanofluid through a stretched thin sheet with higher-order slip parameters are discussed in the paper. The model contains the influences of Brown motion and thermophoresis impacts. Simplified ODEs are obtained by a series of similarity transformations. The similar solutions are solved through homotopy analysis theory and the stability of the solutions is analyzed. Moreover, the current results are shown to be in good agreement with the literature results, the error of Nusselt number and skin friction coefficient is less than 0.1%. The key conclusions follow.

- Velocity, temperature, and concentration have an upward tendency as the second-order velocity slip parameter, thermophoresis parameter, and temperature jump parameter increase, but a downward trend like the first-order linear slip parameter and nonlinear slip parameters.
- The rise of power law index causes the enhancement of velocity and reduction of temperature.
- Skin friction has increasing behavior due to the enhancement of volume fraction of nanoparticles, the first-order linear slip parameter and nonlinear slip parameter, but decreasing behavior as a result of the second order slip parameter.

- The Nusselt number is found to rise upon the rise of the second order slip parameter, volume fraction, whereas impacts of the first-order linear slip parameter, temperature jump parameter, and nonlinear slip parameter are converse.
- The skin friction coefficient have an upward tendency as the power law index increase at a certain volume fraction of nanoparticles, and also increases as volume fraction of nanoparticles increases at a constant power law index.

Author Contributions: J.Z. conducted the original research, modified the model and contributed analysis tools. X.H. analyzed the data, simulated the modified model and prepared original draft. Y.X. made numerical simulation with ANSYS software. J.Z. and Y.X. revised the manuscript.

Funding: This paper was supported by the National Natural Science Foundation of China (No. 11772046; No. 81870345).

Acknowledgments: The authors would like to express their gratitude to the reviewers for their suggestions.

Conflicts of Interest: The authors declare no conflict of interest.

References

1. Moreira, T.A.; Nascimento, F.J.D.; Ribatski, G. An investigation of the effect of nanoparticle composition and dimension on the heat transfer coefficient during flow boiling of aqueous nanofluids in small diameter channels. *Exp. Therm. Fluid Sci.* **2017**, *89*, 72–89. [CrossRef]
2. Srinivas Rao, S.; Srivastava, A. Whole field measurements to understand the effect of nanoparticle concentration on heat transfer rates in a differentially-heated fluid layer. *Exp. Therm. Fluid Sci.* **2018**, *92*, 326–345. [CrossRef]
3. Ho, M.X.; Pan, C. Experimental investigation of heat transfer performance of molten HITEC salt flow with alumina nanoparticles. *Int. J. Heat Mass Transf.* **2017**, *107*, 1094–1103. [CrossRef]
4. Stephen, U.S.; Choi, J.A.E. Enhancing thermal conductivity of fluids with nanoparticles. *ASME Int. Mech. Eng. Congr. Exp.* **1995**, *66*, 99–105.
5. Sheremet, M.A.; Trimbitas, R.; Grosan, T.; Pop, I. Natural convection of an alumina-water nanofluid inside an inclined wavy-walled cavity with a non-uniform heating using Tiwari and Das' nanofluid model. *Appl. Math. Mech.* **2018**, *39*, 1425–1436. [CrossRef]
6. Bowers, J.; Gao, H.; Qiao, G. Flow and heat transfer behavior of nanofluids in microchannels. *Prog. Nat. Sci.* **2018**, *28*, 225–234. [CrossRef]
7. Hamid, A.; Khan, M. Unsteady mixed convective flow of Williamson nanofluid with heat transfer in the presence of variable thermal conductivity and magnetic field. *J. Mol. Liq.* **2018**, *260*, 436–446.
8. Mahdy, A. Simultaneous impacts of MHD and variable wall temperature on transient mixed Casson nanofluid flow in the stagnation point of rotating sphere. *Appl. Math. Mech.* **2018**, *39*, 1327–1340. [CrossRef]
9. Asadi, A.; Aberoumand, S.; Moradikazerouni, A.; Pourfattah, F.; Zyla, G.; Estelle, P.; Mahian, O.; Wongwises, S.; Nguyen, H.M.; Arabkoohsar, A. Recent advances in preparation methods and thermophysical properties of oil-based nanofluids: A state-of-the-art review. *Powder Technol.* **2019**, *352*, 209–226. [CrossRef]
10. Pourfatta, H.F.; Arani, A.A.A.; Babaie, M.R.; Nguyen, H.M.; Asadi, A. On the thermal characteristics of a manifold microchannel heat sink subjected to nanofluid using two-phase flow simulation. *Int. J. Heat Mass Transf.* **2019**, *143*, 1–13. [CrossRef]
11. Alarifi, I.M.; Alkouh, A.B.; Ali, V.; Nguyen, H.M.; Asadi, A. On the rheological properties of MWCNT-TiO2/oil hybrid nanofluid: An experimental investigation on the effects of shear rate, temperature, and solid concentration of nanoparticles. *Powder Technol.* **2019**, *355*, 157–162. [CrossRef]
12. Javanbakh, T.M.; Moosavi, A. Heat transfer on topographically structured surfaces for power law fluids. *Int. J. Heat Mass Transfer* **2018**, *121*, 857–871. [CrossRef]
13. Turan, O.; Yigit, S.; Liang, R.; Chakraborty, N. Laminar mixed convection of power-law fluids in cylindrical enclosures with heated rotating top wall. *Int. J. Heat Mass Transf.* **2018**, *124*, 885–899. [CrossRef]
14. Zhang, H.; Kang, Y.; Xu, T. Study on Heat Transfer of Non-Newtonian Power Law Fluid in Pipes with Different Cross Sections. *Procedia Eng.* **2017**, *205*, 3381–3388. [CrossRef]
15. Ahmed, F.; Iqbal, M. MHD power law fluid flow and heat transfer analysis through Darcy Brinkman porous media in annular sector. *Int. J. Mech. Sci.* **2017**, *130*, 508–517. [CrossRef]

16. Khan, M.; Hafeez, A. A review on slip-flow and heat transfer performance of nanofluids from a permeable shrinking surface with thermal radiation: Dual solutions. *Chem. Eng. Sci.* **2017**, *173*, 1–11. [CrossRef]

17. Ramya, D.; Raju, R.S.; Rao, J.A. Effects of velocity and thermal wall slip on magnetohydrodynamics (MHD) boundary layer viscous flow and heat transfer of a nanofluid over a non-linearly-stretching sheet: A numerical study. *Propuls. Power Res.* **2018**, *7*, 182–195. [CrossRef]

18. Abbas, N.; Saleem, S.; Nadeem, S. On stagnation point flow of a micro polar nanofluid past a circular cylinder with velocity and thermal slip. *Results Phys.* **2018** *9*, 1224–1232. [CrossRef]

19. Usman, M.; Soomro, F.A.; Ul Haq, R. Thermal and velocity slip effects on Casson nanofluid flow over an inclined permeable stretching cylinder via collocation method. *Int. J. Heat Mass Transf.* **2018**, *122*, 1255–1263. [CrossRef]

20. Jayachandra Badu, M.; Sandeep, N. Three-dimensional MHD slip flow of nanofluids over a slendering stretching sheet with thermophoresis and Brownian motion effects. *Adv. Powder Technol.* **2016**, *27*, 2039–2050. [CrossRef]

21. Beskok, A.; Karniadakis, G.E. Rarefaction and compressibility effects in gas microflows. *J. Fluids Eng.* **1996**, *118*, 448–456. [CrossRef]

22. Uddin, M.J.; Khan, W.A.; Ismail, A.I.M. Melting and second order slip effect on convective flow of nanofluid past a radiating stretching/shrinking sheet. *Propuls. Power Res.* **2018**, *7*, 60–71. [CrossRef]

23. Kamran, M.; Wiwatanaoataphee, B. Chemical reaction and Newtonian heating effects on steady convection flow of a micropolar fluid with second order slip at the boundary. *Eur. J. Mech.-B/Fluids* **2018**, *71*, 138–150. [CrossRef]

24. Farooq, S.; Hayat, T.; AlsaedI; A; Ahmad, B. Numerically framing the features of second-order velocity slip in mixed convective flow of Sisko nanomaterial considering gyrotactic microorganisms. *Int. J. Heat Mass Transf.* **2017**, *112*, 521–532. [CrossRef]

25. Yasin, M.H.M.; Ishak, A.; Pop, I. Boundary layer flow and heat transfer past a permeable shrinking surface embedded in a porous medium with a second-order slip: A stability analysis. *Appl. Therm. Eng.* **2017**, *115*, 1407–1411. [CrossRef]

26. Mustafa, M.; Khan, J.A. Numerical study of partial slip effects on MHD flow of nanofluids near a convectively heated stretchable rotating disk. *J. Mol. Liq.* **2017**, *234*, 287–295. [CrossRef]

27. Hayat, T.; Ijaz, M.; Qayyum, S.; Ayub, M.; Alsaedi, A. Mixed convective stagnation point flow of nanofluid with Darcy-Fochheimer relation and partial slip. *Results Phys.* **2018**, *9*, 771–778. [CrossRef]

28. Mitsuya, Y. Modified Reynolds Equation for Ultra-Thin Film Gas Lubrication Using 1.5-Order Slip-Flow Model and Considering Surface Accommodation Coefficient. *J. Tribol.* **1993**, *115*, 289–294. [CrossRef]

29. Liao, S.J. *Homotopy Analysis Method in Nonlinear Differential Equations*; Shanghai Jiao Tong University: Shanghai, China, 2012.

30. Zhu, J.; Zheng, L.C.; Zhang, X.X. Analytical solution to stagnation-point flow and heat transfer over a stretching sheet based on homotopy analysis. *Appl. Math. Mech.* **2009**, *30*, 463–474. [CrossRef]

31. Ul Haq, R.; Nadeem, S.; Khan, Z.H.; Akbar, N.S. Thermal radiation and slip effects on MHD stagnation point flow of nanofluid over a stretching sheet. *Phys. E Low-Dimens. Syst. Nanostruct.* **2015**, *65*, 17–23. [CrossRef]

32. Lin, Y.H.; Zheng, L.C.; Li, B.T.; Ma, L.X. A new diffusion for laminar boundary layer flow of power law fluids past a flat surface with magnetic effect and suction or injection. *Int. J. Heat Mass Transf.* **2015**, *90*, 1090–1097. [CrossRef]

33. Chen, C.H. Effects of magnetic field and suction/injection on convection heat transfer of non-Newtonian power-law fluids past a power-law stretched sheet with surface heat flux. *Int. J. Therm. Sci.* **2008**, *47*, 954–961. [CrossRef]

Article

An Efficient Micro Grid Optimization Theory

Sooyoung Jung [1], Yong Tae Yoon [1] and Jun-Ho Huh [2,*]

[1] Department of Electrical and Computer Engineering, Seoul National University, Seoul 08826, Korea; sjung7@snu.ac.kr (S.J.); ytyoon@snu.ac.kr (Y.T.Y.)

[2] Department of Data Informatics, Korea Maritime and Ocean University, Busan 49112, Korea

* Correspondence: 72networks@pukyong.ac.kr or 72networks@kmou.ac.kr

Received: 15 January 2020; Accepted: 28 March 2020; Published: 10 April 2020

Abstract: A Micro Grid is an aggregate of many small-scale distributed energy resources (DERs); loads and can be operated independently or together with the existing power grid as a local power grid. The operator of such a grid takes charge of the energy supply and consumption of these resources and loads available in the grid. Meanwhile, the system operator of the grid considers the entire Micro Grid system to be a single load or a generator and assigns the responsibility of its internal management to the operator. The power production from a passive production resource is largely influenced by external environmental factors such as weather conditions, rather than operating conditions. Thus, this study conducted simulations for the cases where four kinds of conditional expressions had not been applied at all or one of them had been applied to compare and evaluate the effectiveness of each expression. As a result, the conditional equations were found to be effective when attempting to optimize the Micro Grids efficiently.

Keywords: Micro Grid; optimization theory; optimization; smart grid; MATLAB simulation

---

1. Introduction

Many small-scale DERs and loads form a Micro Grid, a local power grid that can be operated independently or by tying with the existing power grid [1]. The operator of such grid takes charge of energy supply and consumption for these resources and loads existing within the Micro Grid, whereas the system operator of the grid considers the entire Micro Grid system to be a single load or a generator and assigns the responsibility of its internal management to the operator [2–4].

The energy resources in the Micro Grid are largely divided into passive resources that cannot be controlled by the grid operator and active resources controllable by him/her [5]. These resources are also classified as production, consumption or storage resources in terms of their applications [6]. Typical examples of a passive resource include new and renewable energy resources such as solar rays, wind, etc. The power production using these passive resources can be largely affected by external environmental conditions like weather situations rather than the operator-controlled production process.

Although it would be possible to interrupt power production by disconnecting with the Micro Grid's internal power grid, implementing some kind of control to increase power production arbitrarily is almost impossible as it depends on the external environmental conditions. Even though passive resources have a disadvantage in terms of inability to control, they also have an advantage of being able to achieve eco-friendly power generation without much extra cost after the initial installation cost in majority of the cases. For this reason, these resources are being used as a major power supply resource of a Micro Grid. Their issue of inability to control raises a problem in the balance between power demand and supply, causing sharp fluctuations in it such that an ancillary active resource is often required to solve the problem. One of the most popular passive resources, photovoltaic power generation, has a problem of production variability or excessive production due to its characteristic of insolation-based production. Thus, this study conducted simulations for cases wherein none of the four kinds of conditional equations (i.e., conditional equations for peak control, power use flattening, power demand response and operation of net zero Energy or one of them had been applied to compare them and evaluate the effectiveness of each equation. The result showed that the conditional equations were effective when attempting to optimize the Micro Grids efficiently.

2. Related Research

Typical examples of passive consumption resources of Micro Grid include loads such as household appliances used by the residents on the Micro Grid. Although there are some controllable loads, most of them are used to meet the requirements of residents and beyond the control of Micro Grid operators [7,8].

The active production resources of Micro Grid include fuel cells, combined heat and power generation and most of the other generating facilities that are able to produce power according to the operator's control. Although these resources require some additional costs (fuel expenses) when generating power, they are indispensable in the operation of a Micro Grid wherein power supply and demand or quality issues should be considered as their power output level can be controlled arbitrarily [9–11].

On the other hand, the active consumption resources of Micro Grid include some operator-controllable loads such as lighting fixtures, air-conditioning equipment, etc. Control can be implemented in a way that will not make the residents uncomfortable when attempting to reduce excessive power use in the Micro Grid. Moreover, the active storage resources are useful in keeping the balance between power supply and demand or securing power quality or economy by storing or releasing surplus power. These active storage resources can store or release the desired amount of power in any desired time zone. Likewise, as many of them have rapid responsiveness, it is possible to deal with the variability in power production. Typical examples of these resources include energy storage systems such as batteries, flywheels, combined air energy storage, etc. [12,13].

Meanwhile, Micro Grids are divided into system-connected type or independent type depending on whether they are operating with an external system (power grid) [14]. The grid-connected Micro Grid is operated in a state of establishing a connection with another power system and is able to exchange surplus power to supplement each other. The independent-type Micro Grid (a.k.a. island grid) is operated in a state of separating itself with another power grid and managing the quality (i.e., supply and demand or voltage/current, etc.) by itself. The former can sell the surplus or purchase the power amount lacking by connecting with an external system and some of the typical examples are small-scale building, home and campus Micro Grids. If it is impossible to meet the demand from a consumption resource, these Micro Grids buy the system's surplus power or sell their own surplus to the system when their production is more than enough.

In addition, an external power rates provision system (e.g., KEPCO or Korea Power Exchange) is necessary to settle the power bills incurred from these transactions, further requiring a metering device connected with an outside system along with the external system or market that will be able to pass on the signals in case of blackout or demand response (DR) [15].

The independent Micro Grid manages the power supply and demand balance or power quality within the grid by itself. One of the most typical examples of such grid is one that is being operated in an island, disconnecting itself from a large-scale inland power system and managing its power through its own power production, storage and consumption resources within the Micro Grid to supply power on demand.

There have been a quite number of studies for the optimization of Micro Grids and energy management systems and planning an optimal operating schedule is one of the essential parts of the management: J. Li et al. [16] dealt with the design and implementation of Green Home Service for energy management whereas A. R. Al-Ali et al. [17] focused on the design, implementation and test operation of a smart home using an energy storage system. Y. Zhang et al. [18] presented an optimization algorithm that can be used for a home energy management system in a smart grid whereas

D. I. H Rodriguez et al. [19] and A.C. Luna et al. [20] dealt with the Micro Grid operating system using an optimization algorithm or vice versa and the energy management system for the Micro Grid having its own power generation facility and connected to the existing power grid, respectively. Also, H. Li et al. [21] discussed energy management for the industrial Micro Grid being connected to the existing power grid or operated independently. Further, D. Arcos-Aviles et al. [5] and C. Ju et al. [22] introduced a design of a fuzzy logic-based energy management system for the Micro Grid having new and renewable energy resources and its own energy storage while being connected to the existing power grid and a 2-tier prediction energy management system, respectively.

Meanwhile, for the deduction of schedule [22], H. Kim et al. [8] discussed the minimization of operating costs based on a basic model. F. A. Mohamed et al., [23] and F. A. Mohamed [24] focused on a Micro Grid system model having battery storage and its online management, aiming to minimize the costs while satisfying the demands in a respective system where wind/diesel/ PV generator or fuel cell or battery storage existed. A. Parisio [25] presented a Micro Grid management method based on a model-based predictive control which was to improve calculation results or reduce calculation load by applying mixed-integer linear programming. P. Malysz et al. [26] dealt with the minimization of operating costs of energy storage connected to a grid, predicting future power usage and energy production by using mixed-integer linear programming.

On the other hand, H. Hori et al. [13] and W. Shi et al. [27] proposed a method of using an additional control to deal with the expected errors and the management of distributed energy, respectively. Y. Zhang et al. [15] discussed the model-based predictive control and the operation considering uncertainties.

K. Hoffmann et al. [28] and S. Zhai et al. [29] dealt with the requirements of energy management system information and the flexibility of home appliances used in a household energy management system using smart plugs, respectively. M. M. Eissa et al. [30] discussed the demand-response based on a commercial energy management system. Chee Lim Nge et al. [31] described their real-time energy management system for a PV facility having battery storage. Amin Shokri Gazafroudi et al. [32] introduced their bidding strategy for the automatic housing energy management system. Feras Alasali et al. [33] described their energy management algorithm for the energy storage and crane network. Spyridon Chapaloglou et al. [34] and J. M. G Lopez [35] presented an energy management algorithm for load flattening and peak-reduction and a simulator for the household energy management system loads, respectively.

Yujie Wang et al. [36] presented their rule-based energy management strategy based on the power prediction of a lithium-ion battery and a supercapacitor. Farid Farmani et al. [37] proposed a conceptual model for the residential building energy management system having CCHP. F. Wang, Lidong Zhou et al. [38] introduced their building energy management system considering the unit-price demand-response and other factors such as energy resources, load or storage that change according to time zones.

Meanwhile, Dimitrios et al. [39] proposed an energy management system for the smart building connected to a power system considering the uncertainties of PV generation and the operation schedule of electric vehicles. D. van der Meer et al. [40] described his energy management system that predicts PV generations for charging electric vehicles by detailing the PV generation system. And, H. S. V. S. K Nunna et al. [41] present their energy management strategy when electric vehicles and/or power system are being connected for use. A. Azizvahed et al. [42] dealt with a multi-purpose energy management system that operates a distributed network when there were distributed resources along with energy storage. V. Indragandhi et al. [43] discussed multi-purpose energy management for a new and renewable energy resource-based AC/DC microgrid. V. Pilloni et al. [44] proposed an energy management system for the operation or operating time of home appliances considering the aspect of not only energy cost-saving but also enhancement of user experience quality. I. Ali and S. M Suhail Hussain et al. [45] presented their communication system design for the automated energy management of the Micro Grid involving various types of distributed energy sources. Lastly, W. Ma and J. Wong et al. [46] dealt with the distributed energy management for a networked Micro Grid having uncertainties due to distributed energy resources.

The system operator of a Micro Grid system regards the entire grid as a single load or generator and delegates its internal management responsibility to the grid operator. The power generations based on a passive production resource can be largely affected by the external environmental condition such as weather. Thus, a series of simulations were conducted in this study for evaluation to optimize the efficiency of the Micro Grid. Each one of the four conditional expressions (i.e., peak-zero, power-use flattening, demand-response and net zero operation) was applied to the simulations to compare with the case where any of these expressions were applied. Each performance was evaluated, and the validity of the expressions was determined through MATLAB simulations.

3. Micro Grid Optimization Theory

A Micro Grid consists of new and renewable energy, load and energy storage system. Although there are a number of new and renewable energy resources now available, only photovoltaic power generation was indicated as a representative system for convenience. The power generated by the photovoltaic (PV) system will be consumed by the load or stored in energy storage system (ESS). Their data are saved in the data storage for the estimation of their future values [47–49].

The power grids supply electricity to the ESS or load, whereas the unit cost of power is provided at the power exchange. The event server assumes the role of notifying the situation wherein DR or net zero operation is required. There will be no information about the external operating conditions from the event server in a scenario that does not include any special conditions; otherwise, the constraints and objective function will be changed after receiving external operating condition information.

This section focuses on the operating schedule calculation and prediction functions of ESS in the EMS. The constants used for the calculation of an ESS operating schedule include the PV/load data obtained through prediction, unit price data set by the power exchange, capacity of ESS, maximum/minimum charging/discharging power, etc. The constraints are then set based on these data and charging/discharging schedule, SoC limitations, etc. Lastly, appropriate individual objective functions are set to output an ESS operating (charging/discharging) schedule that minimizes each objective function by using an optimization technique. The resulting schedule presents a method with which the ESS charging/discharging power level in the Micro Grid can be determined in each time zone.

The system flow diagram is shown in Figure 1, where the red arrows show the directions of power to be supplied and the blue arrows represent the movement of each data (information). The system consists of Micro Grid, power grid, power exchange and event server; originally, however, EMS and data storage are also included in such system.

The red arrow in the system diagram (Figure 1) indicates the supply of power whereas the blue arrow is showing the flow of information. The system largely consists of a Micro Grid, power grid, power exchange and event-generating server. The variables used to explain the supply of power in the diagram are as follows: It is assumed that all kinds of powers in each time zone are constant and the time interval is one hour. Although both energy management system (EMS) and data storage belong to the Micro Grid, the Micro Grid, in this case, is one that consists of new and renewable energy sources, loads and energy storage. There are a number of new and renewable energy sources such as photovoltaic (PV) and wind turbine (WT) but only PV was indicated for convenience. The power produced by PV will be stored in the ESS or consumed by the load. Also, PV/Load data are used to predict the future PV/load value after being stored in data storage.

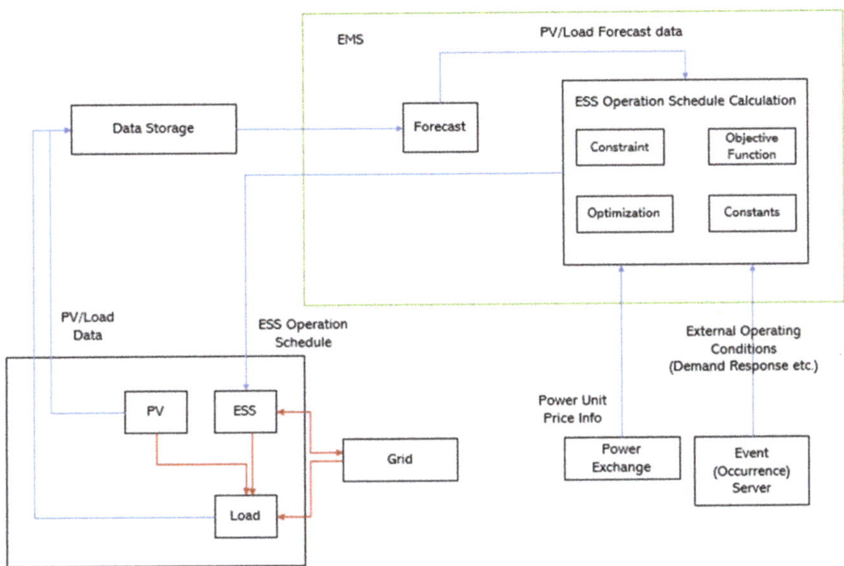

Figure 1. System diagram (red arrow: supply of power; blue arrow: flow of information).

The target Micro Grid consists of new and renewable energy, load and energy storage system. Although there are a number of new and renewable energy resources now available, only photovoltaic power generation was indicated as a representative system for convenience. The power generated by the PV system will be consumed by the load or stored in ESS. Their data are saved in the data storage for the estimation of their future values.

Table 1 below describes the variables to be used to explain (definition) individual cases of supplying power, assuming that all kinds of power in each time zone (one hour) are constant.

Table 1. Definitions of Variables (1).

Variables	Definitions
$P_{PV}[k]$	Power (kW) obtained with new and renewable energy generation in time zone k
$P_{PV,\,ESS}[k]$	Power (kW) transmitted from PV to ESS in time zone k
$P_{PV,\,Load}[k]$	Power (kW) transmitted from PV to load in time zone k
$E_{ESS}[k]$	Amount of power (kWh) stored in time zone k
$P_{ESS}^{dis}[k]$	Power (kW) discharged in time zone k
$P_{ESS,Load}^{dis}[k]$	Power (kW) transmitted from ESS to load in time zone k
$P_{ESS,g}^{dis}[k]$	Power (kW) transmitted from ESS to power grid in time zone k
$P_{ESS}^{chg}[k]$	Power (kW) charged to ESS in time zone k
$P_{g,1}[k]$	Power (kW) received from power grid in time zone k
$P_{g,ESS}[k]$	Power (kW) transmitted from power grid to ESS in time zone k
$P_{g,Load}[k]$	Power (kW) transmitted from power grid to load in time zone k
$P_{g,2}[k]$	Power (kW) transmitted from power grid to in time zone k
$P_{Load}[k]$	Power (kW) consumed by load in time zone k

The relationship between individual variables can be expressed by Equations (1)–(7):

$$P_{PV}[k] = P_{PV,ESS}[k] + P_{PV,Load}[k] \tag{1}$$

Equation (1) indicates that the power generated by PV will be "charged to ESS" or "consumed by load."

$$P_{ESS}^{dis}[k] = P_{ESS,Load}^{dis}[k] + P_{ESS,g}^{dis}[k] \tag{2}$$

Equation (2) indicates that the power discharged from ESS will be "consumed by load" or "sold to the power grid."

$$P_{ESS}^{chg}[k] = P_{g,ESS}[k] + P_{PV,ESS}[k] \tag{3}$$

Equation (3) indicates that the power used to charge ESS had been "received (bought) from the power grid" or "generated by PV."

$$P_{g,1}[k] = P_{g,ESS}[k] + P_{g,Load}[k] \tag{4}$$

Equation (4) indicates that the power received (bought) from the power grid will be "charged to ESS" or "consumed by load."

$$P_{g,2}[k] = P_{ESS,g}^{dis}[k] \tag{5}$$

Equation (5) indicates that the power transmitted (sold) to the power grid had been discharged from ESS.

$$P_{Load}[k] = P_{PV,Load}[k] + P_{ESS,Load}^{dis}[k] + P_{g,Load}[k] \tag{6}$$

Equation (6) shows the balance between demand (right side) and supply (left side) and that the power consumed by load had been supplied from PV, ESS or power grid.

$$E_{ESS}[k+1] = E_{ESS}[k] + P_{ESS}^{chg}[k] \cdot 1h - P_{ESS}^{dis}[k] \cdot 1h \tag{7}$$

Equation (7) explains that the amount of power stored in ESS is determined by adding the charged power to the current power and subtracting the discharged power amount from it.

The variables that will be used to explain (definition) the movement of information in Table 2 are as follows:

Table 2. Definitions of Variables (2).

Variables	Definitions
$d_{PV}[k]$	PV data in time zone k
$d_{Load}[k]$	Load data in time zone k
$D_{PV}[k] = \{d_{PV}[1], d_{PV}[2], \ldots, d_{PV}[k-1]\}$	PV data set in time zone 1–(k-1)
$D_{Load}[k] = \{d_{Load}[1], d_{Load}[2], \ldots, d_{Load}[k-1]\}$	Load data set in time zone 1–(k-1)
$D[k] = D_{PV}[k] \cup D_{Load}[k]$	Data storage in time zone k
$f_{prd,PV}$	Function (or algorithm) to predict PV
$f_{prd,Load}$	Function (or algorithm) to predict load
$PV_{prd}[k]$	PV value calculated based on PV prediction in time zone k
$V_{prd}(k,n){:}D_{PV}[k]$	n PV data predicted based on D_PV [k]
$Load_{prd}[k]$	Load value calculated based on load prediction in time zone k
$Load_{prd}(k,n){:}D_{Load}[k]$	n load data predicted based on $D_{Load}[k]$
C	Set of constants
I_{cost}	Power unit price info
I_{ext}	External operating conditions info
I_{spec}	ESS performance info
I_{ext}	Other constants including SoC range setting in each time zone, etc.
f	Objective function coefficient vector
M	Set of matrices or vectors representing constraints
f_o	Function (or algorithm) for the calculation of optimal solutions
x	ESS operation schedule

The relationship between individual variables can be expressed as:

$$
\begin{aligned}
D_{PV}[k] \cup \{d_{PV}[k]\} &= \{d_{PV}[1], d_{PV}[2], \ldots, d_{PV}[k-1]\} \cup \{d_{PV}[k]\} \\
&= \{d_{PV}[1], d_{PV}[2], \ldots, d_{PV}[k-1], d_{PV}[k]\} \\
&= D_{PV}[k+1]
\end{aligned}
\tag{8}
$$

Equation (8) indicates that the data set in time zone 1–k will be generated by adding Kth PV data to the PV data set in time zone 1–(k-1).

$$
\begin{aligned}
D_{Load}[k] \cup \{d_{Load}[k]\} &= \{d_{Load}[1], d_{Load}[2], \ldots, d_{Load}[k-1]\} \cup \{d_{Load}[k]\} \\
&= \{d_{Load}[1], d_{Load}[2], \ldots, d_{Load}[k-1], d_{Load}[k]\} \\
&= D_{Load}[k+1]
\end{aligned}
\tag{9}
$$

Similar to Equation (8) load data set 1–k can be obtained when Kth load data are added to the PV load data set in time zone 1–(k-1).

$$
\begin{aligned}
D[k] \cup (\{d_{PV}[k]\} \cup \{d_{Load}[k]\}) &= (D_{PV}[k] \cup D_{Load}[k]) \cup (\{d_{PV}[k]\} \cup \{d_{Load}[k]\}) \\
&= (D_{PV}[k] \cup \{d_{PV}[k]\}) \cup (D_{Load}[k] \cup \{d_{Load}[k]\}) \\
&= (D_{PV}[k+1] \cup D_{Load}[k+1]) \\
&= D[k+1]
\end{aligned}
\tag{10}
$$

Equation (10), which can be obtained by using both Equations (8) and (9), indicates that data storage k+1 can be created by adding both PV and load data generated in the same time zone to the data storage in time zone k.

$$
\begin{aligned}
f_{prd,PV}(D_{PV}[k]) &= \{PV_{prd}[k], PV_{prd}[k+1], \ldots, PV_{prd}[k+n-1]\} \\
&= PV_{prd}(k,n)
\end{aligned}
\tag{11}
$$

Equation (11) shows that the PV data collected so far can be used to predict the future PV in n time zone.

$$
\begin{aligned}
f_{prd,Load}(D_{Load}[k]) &= \{Load_{prd}[k], Load_{prd}[k+1], \ldots, Load_{prd}[k+n-1]\} \\
&= Load_{prd}(k,n)
\end{aligned}
\tag{12}
$$

Similar to Equation (11), Equation (12) indicates that the load data collected so far can be used to predict future load in n time zone.

$$
C = PV_{prd}(k,n) \cup Load_{prd}(k,n) \cup I_{cost} \cup I_{ext} \cup I_{spec} \cup I_{etc}
\tag{13}
$$

Equation (13) shows that the union of future PV data, load data, power unit price, external operating conditions, ESS performance information and other constant sets becomes a set of constants.

$$
x = f_o(M(C), f(C))
\tag{14}
$$

Equation (14) shows that an ESS operating schedule can be established by entering the constraints and objective function (coefficient vector) in the optimization function. Nonetheless, it is important to understand that the constraints and objective function are determined by the set of coefficients.

Figure 2 is a diagram that shows how system power usage (blue line) and ESS charging/discharging power (red line) change depending on the external conditions applied. The picture on the upper left is a basic setting and the rest of the pictures in order of bottom left, bottom center, upper left and bottom right show the change when peak control, net zero operation, flattening and demand response have been applied, respectively. A detailed explanation for each diagram will be provided later.

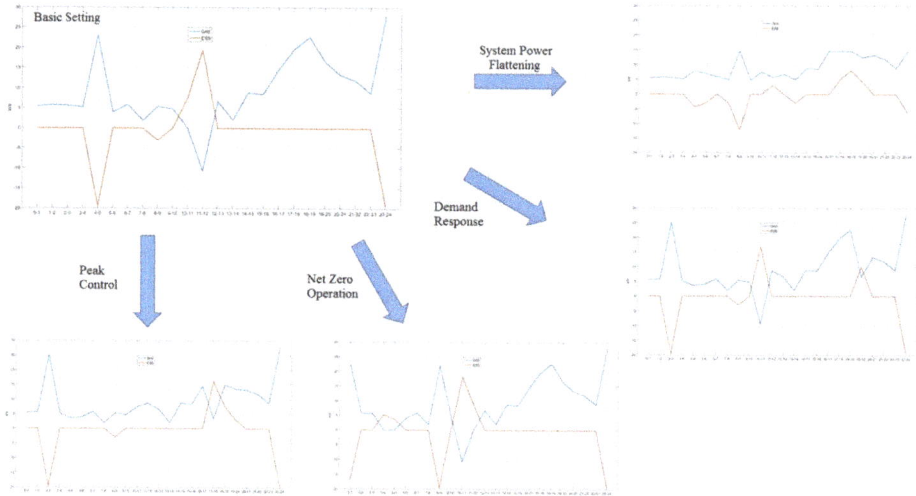

Figure 2. Diagram showing the changes in system power usages, ESS charging/discharging according to the external operating conditions.

Meanwhile, performance evaluations were conducted by comparing a simulation wherein none of the four conditions above had been applied with the simulations to which each of those conditions was applied. For the simulations, MATLAB R2015a was used; the constraints resulting from individual constants and objective functions were used as inputs for the mixed-integer linear programming of MATLAB to show the resultant system power usage, ESS charging/discharging power and total demand with the graphs using a plot function.

The basic setting that does not have any special conditions is as follows: the ranges of ESS charging/discharging and state of charge (SoC) were set at 3–19.5kW and 0.05–0.95, respectively, whereas the ESS capacity was set at 40 kW. The conditions are shown in Table 3.

Table 3. Basic setting for the simulation.

Range of ESS Charging Power	Range of ESS Discharging Power	SoC by Time Zone	ESS Capacity
3–19.5 kW	3–19.5 kW	0.05–0.95	40 kWh

The virtual data in Tables 4–6 is used as load prediction, PV (generation) prediction and power unit cost data. (n)–(n+1) are the time zones for integer (n), ranging from 0 to 23. For example, 1–2 indicates the time zone of 1 o'clock to 2 o'clock. In Table 4, the time zones having low power unit cost and high-power unit cost are denoted in red and blue, respectively. It was also assumed that the power was constant in each time zone (24 time zones/day). In this case, no external conditions were applied.

Table 4. Forecast prices of power demand schedule (virtual data) (kW).

Time Zone	0–1	1–2	2–3	3–4	4–5	5–6	6–7	7–8	8–9	9–10	10–11	11–12
Demand (kW)	5.5	5.8	5.6	5.2	3.6	4	5.9	7.9	11.4	16.8	25.5	26.7

Time Zone	12–13	13–14	14–15	15–16	16–17	17–18	18–19	19–20	20–21	21–22	22–23	23–24
Demand (kW)	24.7	23	23.8	23.5	23.6	24.6	22.7	16.6	13.3	11.9	8.8	8.5

Table 5. Forecast prices of power supply schedule (virtual data) (kW).

Time Zone	0–1	1–2	2–3	3–4	4–5	5–6	6–7	7–8	8–9	9–10	10–11	11–12
Supply (kW)	0	0	0	0	0	0	0	6	9	12	18	18
Time Zone	12–13	13–14	14–15	15–16	16–17	17–18	18–19	19–20	20–21	21–22	22–23	23–24
Supply (kW)	18	21	15	15	9	6	0	0	0	0	0	0

Table 6. Power unit price schedule (virtual data) (Korean won/kWh).

Time Zone	0–1	1–2	2–3	3–4	4–5	5–6	6–7	7–8	8–9	9–10	10–11	11–12
Unit Price for Purchase	66.1	66.1	66.1	66.1	66.1	66.1	66.1	66.1	66.1	96.5	111.3	111.3
Unit Price for Sales	66.1	66.1	66.1	66.1	66.1	66.1	66.1	66.1	66.1	96.5	111.3	111.3
Time Zone	12–13	13–14	14–15	15–16	16–17	17–18	18–19	19–20	20–21	21–22	22–23	23–24
Unit Price for Purchase	96.5	96.5	96.5	96.5	96.5	111.3	111.3	111.3	96.5	96.5	111.3	66.1
Unit Price for Sales	96.5	96.5	96.5	96.5	96.5	111.3	111.3	111.3	96.5	96.5	111.3	66.1

The simulation result from the basic setting is shown in Figure 3, where grid (blue line), ESS (red line) and net demand (yellow line) indicate the system power usage, ESS charging/discharging power and total demand, respectively. Since power demand and supply have to be balanced, the condition net demand-ESS-Grid = 0 must be satisfied. When the grid sign is (+), power is purchased from the system; if the sign is (-), it means that power is sold to the system.

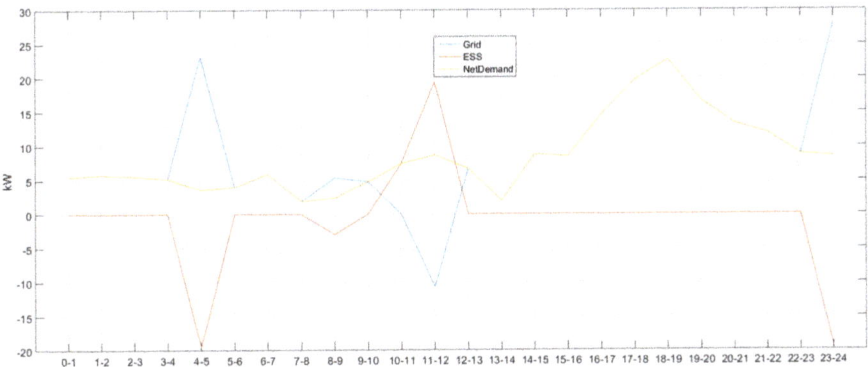

Figure 3. Simulation result when there were no special conditional equations included.

At the same time, the (+) and (−) signs of ESS indicate charging and discharging, respectively. As shown in Figure 4, (n)-(n+1) on the horizontal axis is the time zone for integer n (0–23).

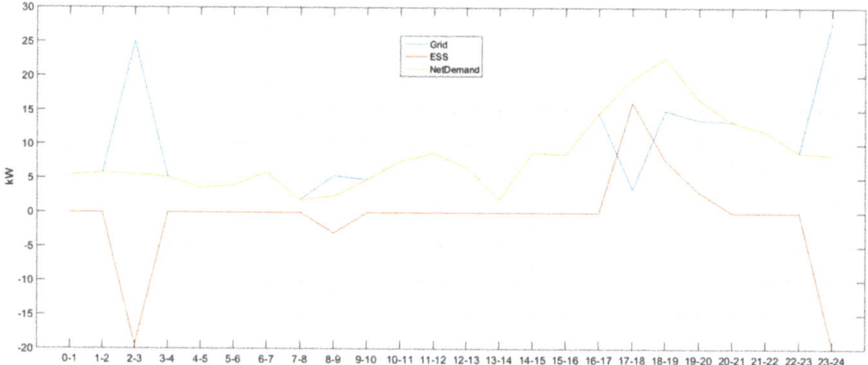

Figure 4. Simulation result when the conditional equation for peak control was included.

Since the value of ESS in time zones 0–1, 1–2, 2–3 and 3–4 was 0, it can be assumed that there was no charging or discharging in these time zones. Moreover, the overlapping yellow and blue lines satisfy the condition net demand = grid, indicating balanced demand and supply. As the sign of the ESS in time zone 4–5 was (−), it can be assumed that charging was required and that power was bought from the system having a (+) grid sign. There was no ESS charging/discharging in time zones 5–6, 6–7 and 7–8 and power demand and supply were balanced out as the condition net demand = grid was satisfied. The same balance was achieved in time zones 9–10 and 10–11 as the individual grid and ESS signs were (+) and (−), respectively, satisfying the conditions net demand = grid and net demand = ESS, respectively. For the latter time zone, it can be understood that power was discharged from ESS as the sign was (+). The ESS sign in time zone 11–12 was also (+), but the grid sign was (−); this means that power was sold to the system. Meanwhile, there was no ESS charging or discharging in time zones 12–13, 13–14, 14–15, 15–16, 16–17, 17–18, 18–19, 19–20, 20–21, 21–22 and 22–23, indicating that power demand and supply have been balanced; thus satisfying the condition net demand = grid. Lastly, the ESS sign in time zone 23–24 was (−), whereas the grid sign was (+), meaning power was purchased from the system.

Figure 3, where no additional conditions have been applied, shows that the ESS was charged in the time zones having the lowest power unit cost of 66.1 (time zones 4–5, 8–9 and 23–24) but discharged in time zones (10–11 and 11–12) having the highest power unit cost of 111.3.

4. Conditional Equation

4.1. Conditional Equation for Peak Control

The following equation should be added to the objective function when the conditional equation is included:

$$\sum_{i \in PC_{g,buy}} [-c_{PC,1}^{g buy}\{1 + (c_{PC,2}^{g buy})^i\}dt \cdot \delta_{PC}^{g buy}(i) + c_{PC,3}^{g buy}dt \cdot p_{PC}^{g buy}(i)] \tag{15}$$

where $c_{PC,1}^{g buy}$, $c_{PC,2}^{g buy}$, $c_{PC,3}^{g buy}$ are the penalty constants for peak control. The possibility of success of peak control will be reflected to the objective function by adding this equation. Likewise, $i \in P_{g,buy}$ means that time zone i will be included in the time zones performing peak control. Since the condition $\delta_{PC}^{g buy}(i)=1$ can be satisfied when peak control is successful, the value of objective function will be decreased $(-c_{PC,1}^{g buy}\{1+(c_{PC,2}^{g buy})^i\}dt \cdot 1)$. In such case, the possibility of success of peak control will be largely reflected when the value of $c_{PC,1}^{g buy}$ is large $(c_{PC,3}^{g buy}dt \cdot p_{PC}^{g buy}(i)=0$ as $0 \leq p_{PC}^{g buy}(i) \leq 0)$. In contrast, the value of the objective function will not be decreased when peak control fails $(-c_{PC,1}^{g buy}\{1+(c_{PC,2}^{g buy})^i\}dt \cdot 0=0$ as

$\delta_{PC}^{gbuy}(i)=0$). This indicates that the value will be increased by $c_{PC,3}^{gbuy} dt \cdot p_{PC}^{gbuy}(i)$ as $0 \leq p_{PC}^{gbuy}(i)$. Moreover, the possibility of success of peak control will be highly reflected when the value of $c_{PC,3}^{gbuy}$ is large.

Figure 4 shows the result of the simulation to which an external operating condition limiting system power usage to 15 kW in time zone 17–20 has been added to the basic setting (Table 7). The conditional equation for peak control seems to be valid as the system power usages in time zone $c_{PC,3}^{gbuy}$ were the same or below 15 kW compared to time zones 17–18, 18–19 and 19–20, where the usages were the same or above 15 kW (note that the blue line indicating system power usage and the yellow line representing total demand are overlapping).

Table 7. Peak control operating condition.

	Time Zone	Peak Setting
Condition	17–18, 18–19, 19–20	15 kW

In Figure 4, there was no charging/discharging in time zones 0–1 and 1–2 as the ESS values were 0. In addition, power demand and supply were balanced as net demand = grid was achieved (i.e., overlapping blue and yellow lines). It can be deduced that ESS has been charged as the sign was (−) in time zone 2–3. In this case, the (+) grid sign means that power was purchased from the system. Power demand and supply were balanced (net demand = grid) in time zones 3–4, 4–5, 5–6, 6–7 and 7–8 where there was no charging/discharging activity. The (+) grid sign and (−) ESS sign in time zone 8–9 indicate that power was purchased from the system to charge ESS. Next, no ESS charging/discharging was performed in time zones 9–10, 10–11, 11–12, 12–13, 13–14, 14–15, 15–16 and 16–17 and power demand and supply were balanced, thus satisfying net demand = grid. The same can be said for time zones 20–21, 21–22 and 22–23, but ESS was charged in time zone 23–24 as its sign was (−), buying power from the system.

In Figure 4, where a condition for peak control was included, it can be deduced that ESS was charged in time zones 2–3 and 8–9 due to their low power unit cost (66.1) and discharged in time zones 17–18, 18–19 and 19–20 where the condition was applied to satisfy it.

4.2. Conditional Equation for Power Usage Flattening

The maximum power demand, as well as power charges, can be reduced by improving the quality of power through power usage flattening, the efficiency of new and renewable energy-based generation and the use of off-peak electricity for the peak time hours during the day.

The following equation should be added to the objective function when the conditional equation is included

$$c_g^{flat}\left(p_g^{max}-p_g^{min}\right)T \tag{16}$$

where c_g^{flat} is a penalty constant for flattening. When this equation is added, the difference between the maximum and minimum system powers $p_g^{max}-p_g^{min}$ will be reflected to the objective function. The value of $p_g^{max}-p_g^{min}$ is reduced to minimize the value of the objective function and system power usage will be flattened. The flattening effect will be largely reflected when the value of c_g^{flat} becomes larger.

The result of a simulation wherein a flattening condition has been added to the present setting is shown in Figure 5. When comparing it with the simulation result that does not include any conditional equation, as the difference between the maximum and minimum system power usages was reduced from 38.68[= 28 − (−10.68)] kW to 9.8(= 14.6–4.8) kW, the conditional equation for power flattening can be considered to be valid. The differences were calculated based on the maximum (minimum) values of 28 (−10.68) and 14.6 obtained from time zones 23–24 (11–12) and 8–9, 16–17, 17–18 and 18–19 (9–10), respectively, in Figures 3 and 5.

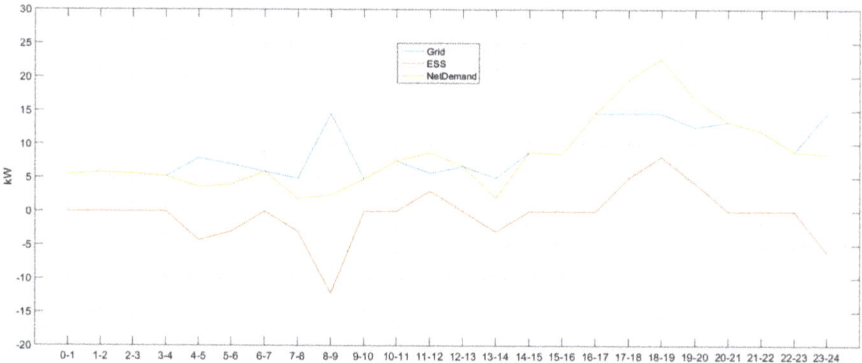

Figure 5. Simulation result when the conditional equation for system power flattening (power use) was included.

In Figure 5, the ESS value in time zones 0–1, 1–2, 2–3 and 3–4 was 0, so there was no ESS charging/discharging at all. Moreover, in these zones, power demand and supply were balanced as the condition net demand = grid was satisfied. The ESS signs in time zones 4–5 and 5–6 were (−), so the ESS was charged. At this time, it can be deduced that power had been purchased from the system as the grid sign was (+). There was no ESS charging/discharging in time zone 6–7, so power demand and supply were well-balanced, thus satisfying the same condition. In addition, the (+) sign of the grid and the (−) sign of the ESS in time zones 7–8 and 8–9 showed that power was purchased from the system to charge the ESS. Power demand and supply were balanced in time zones 9–10 and 10–11 so there was no ESS charging/discharging, but demand and supply were balanced as the condition was satisfied. In time zone 11–12, the sign was (+) for both ESS and grid, so power was purchased from the system for charging. There was no ESS charging/discharging in time zone 12–13 and power demand and supply were balanced as the condition was satisfied. Accordingly, the power transactions for charging/discharging activities in each time zone can be grasped by checking the signs or finding out whether the condition has been satisfied or not.

As such, the flattening operations described in Figure 5 showed that the ESS was charged in time zones 4–5, 5–6, 7–8 and 8–9 when the power unit cost was low (66.1) and discharged in time zones 11–12, 17–18, 18–19 and 19–20 when the cost was high (111.3) to consider cost reduction.

4.3. Conditional Equation for Demand Response Power

Demand response is an activity by the electricity users to control their energy usages by shifting themselves from a passive power-user system to an active one, changing their normal power consumption patterns in response to the incentive(s) obtainable by saving power or the differentiated power rates depending on time zones.

The following equation should be added to the objective function when the conditional equation is included:

$$-c_{DR,1}\{1+(c_{DR,2})^{i}\}\cdot\delta_{DR}+c_{DR,3}\cdot P_{DR} \tag{17}$$

$c_{DR,1}$, $c_{DR,2}$, $c_{DR,3}$ are the penalty constants of demand response. The possibility of success of demand response will be reflected to the objective function by adding this equation. When it becomes successful, the value of the objective function will be decreased based on the calculation $-c_{DR,1}\{1+(c_{DR,2})^{i}\}dt\cdot1$, $\delta_{DR}=1$.

The possibility of success largely depends on the $c_{DR,1}$: $c_{DR,3}\cdot P_{DR}=0$ to be established when $c_{DR,1}$: $c_{DR,3}\cdot P_{DR}=0$. The value will not decrease when demand response fails ($-c_{DR,1}\{1+(c_{DR,2})^{i}\}dt\cdot0=0$, $\delta_{DR}=0$) but will increase instead by $c_{DR,3}\cdot P_{DR}$ ($0\leq P_{DR}$). This suggests that the possibility of failure largely depends on the scale of $c_{DR,3}$.

The result of a simulation wherein the external operating condition listed in Table 8 has been added to time zone 17–20 where 10 kW power is to be discharged is shown in Figure 6. By comparing it with the result obtained from the same time zones in Figure 3 where no conditional equations have been included, this conditional equation for demand response can be considered to be valid as the ESS discharging power in time zone 19–20 was 10 kWh (10 kW*1 h).

Table 8. Demand response operating condition.

-	Time Zone	Demand Response Setting
Condition	17–18, 18–19, 19–20	10 kWh

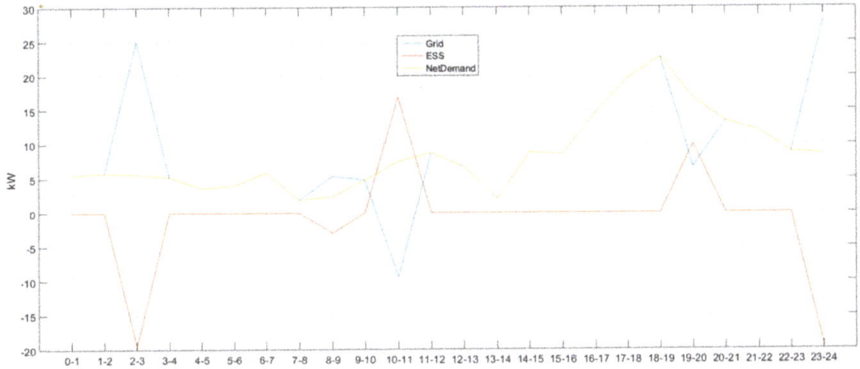

Figure 6. Simulation result when the conditional equation for demand response (power usage) was included.

In Figure 6, there was no ESS charging/discharging in time zones 0–1 and 1–2 as the ESS value was 0. In addition, power demand and supply were balanced as the yellow and blue lines were overlapping (net demand = grid). The (−) ESS sign in time zone 2–3 shows that the ESS had been charged by purchasing power from the system [(+) grid]. Power demand and supply in time zones 3–4, 4–5, 5–6, 6–7 and 7–8 were balanced (net demand = grid) and there was no ESS charging/discharging. Similar to all the other conditions mentioned above, the (+) and (−) signs of either the ESS or the grid explain the power transactions, power balance or ESS charging/discharging events that had taken place.

Added with a demand response condition, Figure 6 also shows that the ESS was charged in time zones 2–3, 8–9 and 23–24 when the power unit cost was low (66.1) and discharged in time zones 10–11 and 19–20 when the cost was high (111.3) to consider cost reduction.

4.4. Conditional Equation for Net Zero Operation

Although achieving net zero energy by utilizing now available new and renewable energy resources or energy-saving equipment is a desirable direction in energy management, establishing a system to realize it can be quite costly.

The following equation should be added to the objective function when the conditional equation is included:

$$\sum_{i\in IO_g}[-c^g_{IO,1}\{1+(c^g_{IO,2})^i\}dt\cdot\delta^g_{IO}(i)+c^g_{IO,3}dt\cdot p^g_{IO}(i)] \tag{18}$$

where $c^g_{IO,1}$, $c^g_{IO,2}$, $c^g_{IO,3}$ are the penalty constants for peak control. The possibility of success of net zero operation will be reflected to the objective function by adding this equation. In addition, $i\in IO_g$ *means* that time zone i will be included in the time zones performing the operation. Since the condition $\delta^g_{IO}(i)=1$ can be satisfied when the operation is successful, the value of the objective function will be decreased $(-c_{IO,1}^{g}\{1+(c_{IO,2}^{g})^{i}\}dt\cdot1)$. In such case, the possibility of success of the operation will be largely reflected when the value of $c_{IO,1}^{g}$ is large. In contrast, the value of the objective function will not be decreased when the operation fails $(-c_{IO,1}^{g}\{1+(c_{IO,2}^{g})^{i}\}dt\cdot0=0,\ \delta_{IO}^{g}(i)=0)$. This indicates that the value will be increased by $c_{IO,3}^{g}dt\cdot p_{IO}^{g}(i)$ as $0\leq p_{IO}^{g}(i)$. Moreover, the possibility of success of the operation will be highly reflected when the value of $c_{IO,3}^{g}$ becomes larger.

The result of a simulation wherein the external operating condition listed in Table 9 has been added to time zone 3–5 for a net zero operation is shown in Figure 7. By comparing it with the result obtained from the same time zones in Figure 3 where system power usage was larger than 0, this conditional equation for the net zero operation can be considered to be valid as the system power usages in time zones 3–4 and 4–5 were 0.

Table 9. Net Zero operating condition.

-	Time Zone
Condition	3–4, 4–5

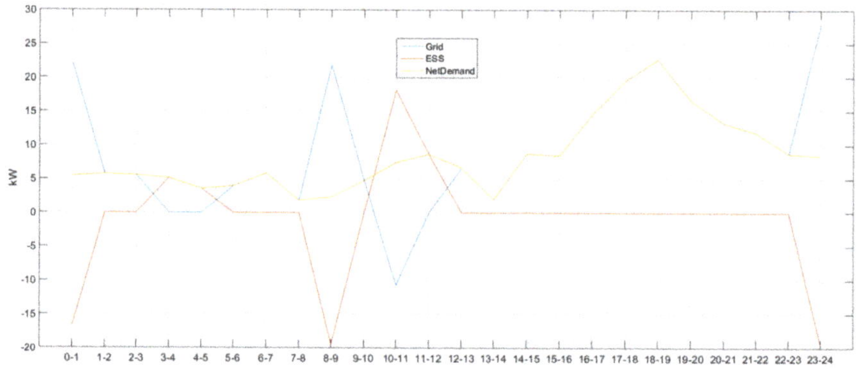

Figure 7. Simulation result when the conditional equation for net zero operation was included.

5. Performance Evaluation

Performance evaluations were conducted by comparing a simulation wherein none of the four conditions above had been applied with the simulations to which each of those conditions was applied. For the simulations, MATLAB R2015a was used; the constraints resulting from individual constants and objective functions were used as inputs for the mixed-integer linear programming of MATLAB to show the resultant system power usage, ESS charging/discharging power and total demand with the graphs using a plot function. Thus, for the optimization of microgrid, four environmental conditions have been put to simulations to find the actual conditions that actually have an impact on improving the optimization performance.

Figure 7 shows all power transactions and operations according to the (+) and (−) signs of both the ESS and grid, whereas the overlapping yellow and blue lines indicate successful establishment of the condition net demand = grid. It is possible to understand what had happened in each time zone.

As such, Figure 7 also shows that the ESS was charged in time zones 0–1, 8–9 and 23–24 when the power unit cost was low (66.1) and discharged in time zone 10–11 when the cost was high (111.3) to achieve net zero operation. The system power usages obtained from a simulation conducted by changing the capacity of ESS from 40 to 120 kW (+10 kW per simulation) are shown in Figures 8–10 and their discharging powers, in Figures 11–13.

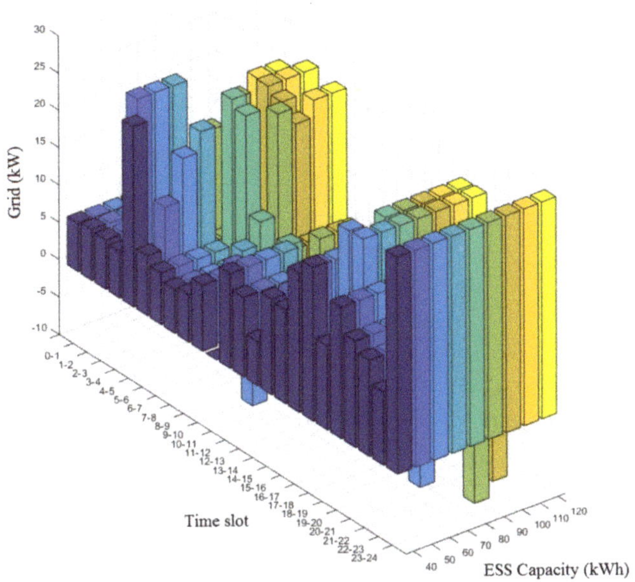

Figure 8. System power usage by ESS capacity (1).

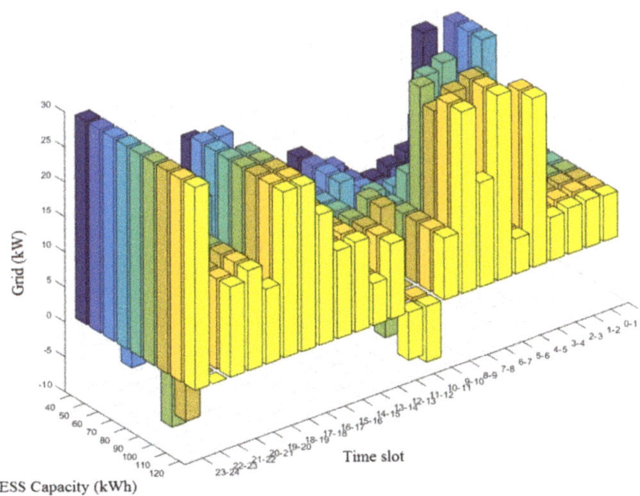

Figure 9. System power usage by ESS capacity (2).

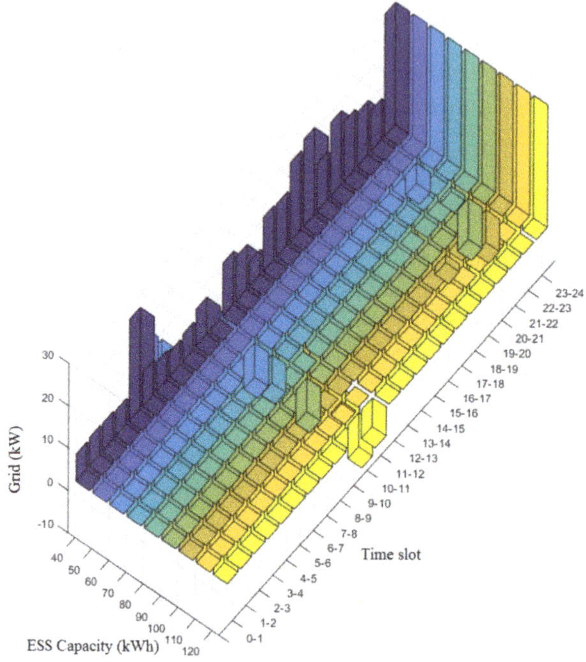

Figure 10. System power usage by ESS capacity (3).

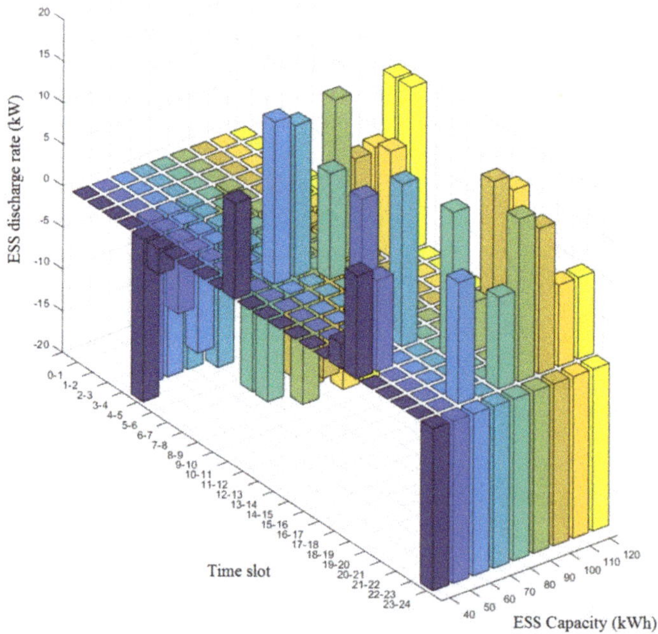

Figure 11. Discharging power by ESS capacity (1).

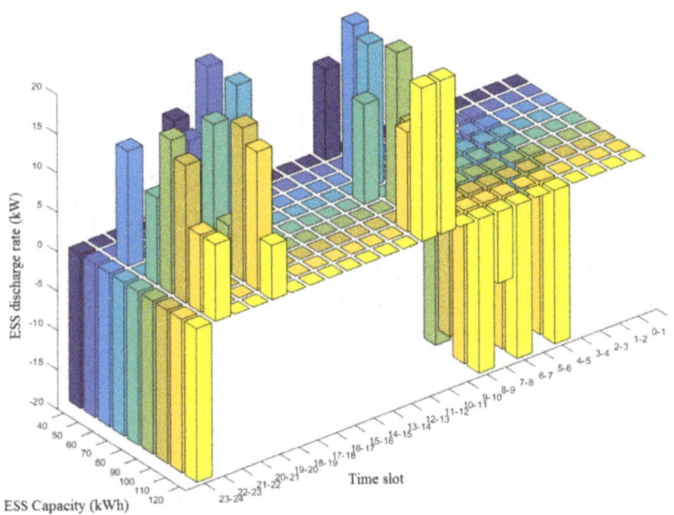

Figure 12. Discharging power by ESS capacity (2).

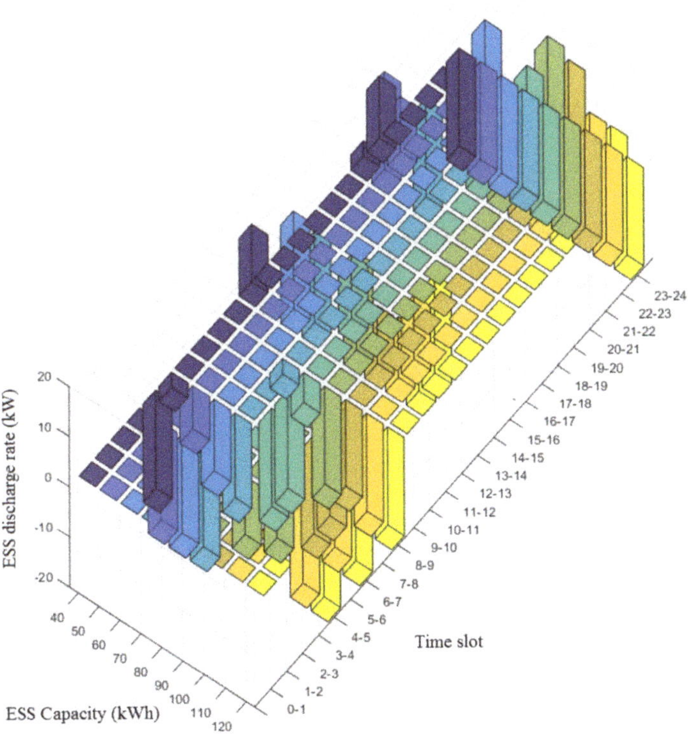

Figure 13. Discharging power by ESS capacity (3).

Figures 9, 10, 12 and 13 are the pictures of Figures 8 and 11, respectively, when viewed from other angles. The vertical axis in Figures 8–10 indicates the system power usages (Grid), whereas the same axis in Figures 11–13 is the ESS charging(-)/discharging(-) power. A long axis, "Time Slot," shows the time zones, with "ESS Capacity" indicating the capacity of the ESS. The bar3 function of MATLAB R2015a was used for graphing.

Observing time zones 4–5 and 8–9 in Figures 8–10, it is possible to recognize the tendency of increasing system power usages in proportion to the ESS capacity. This would mean that the grid is pursuing an economic gain by charging its ESS when the power unit cost is low and selling or reducing its purchase during the time zones when the cost is high.

The time zones where power unit cost is at the lowest or highest level are 9–10 and 12–17. Observing these time zones in Figures 11–13, there were no charging/discharging operations. This also proves that the grid is considering gaining profit by discharging its ESS when the power unit cost is highest and vice versa.

Being a local power grid, a Micro Grid consists of a series of DERs along with loads and is self-sustainable or runs with the existing power grid. In such a power grid, the system operator/administrator managing the entire supply and consumption of these resources assigns some of his/her sublevel tasks to individual operators under him/her. The passive production resources are often affected by the natural environments beyond the control of operators such as weather. A series of simulations were conducted in this study for evaluation to optimize the efficiency of the Micro Grid. Each one of the four conditional expressions (i.e., peak-zero, power-use flattening, demand-response and net zero operation) was applied to the simulations to compare with the case where none of these expressions were applied. Each performance was evaluated, and the validity of the expressions was determined through simulations which proved their effectiveness for the optimization of Micro Grids as a result.

6. Conclusions

This study conducted simulations for cases wherein none of the four kinds of conditional equations (i.e., conditional equations for peak control, power use flattening, power demand response and operation of net zero Energy) or at least one of them had been applied to compare them and evaluate the effectiveness of each equation. The result showed that the conditional equations were found to be effective when attempting to optimize the microgrid's performance efficiently.

The peak-control conditional equation was found to be effective in the simulation as system power usage was decreased below the peak level set at 15 kW. ESS was charged during the daytime when the power cost was low and discharged in the time zones when peak control was implemented.

In the simulation applied with the conditional equation for flattening power use, this equation was found to be effective as the difference between the maximum and minimum system power usages was decreased from 38.68 kW to 9.8 kW. ESS charging was carried out during the daytime when the power cost was low, whereas discharging was performed in the time zones when it was high.

In the simulation applied with the conditional equation for power demand-response, the equation was found to be effective as the power set to be discharged (10 kW) in a fixed time zone was achieved successfully. In this case, ESS was charged during the daytime when the power cost was low.

In the simulation applied with the conditional equation for net zero energy operation, the equation was found to be effective as the system power usage became 0 in a designated time zone, despite the fact that the power cost in that time zone (daytime) was low. As for the rest of the time zones, ESS charging was performed during the daytime, but discharging was carried out in the time zones when the power cost was high.

The results above showed that all the equations were effective in every case and it can be confirmed that all the ESS operating schedules, except net zero energy operation, had been adjusted in such a way that power is charged during the daytime and discharged or sold when the power cost was highest.

Author Contributions: Conceptualization, S.J.; Data curation, S.J.; Formal analysis, S.J.; Funding acquisition, J.-H.H.; Methodology, S.J. and Y.T.Y.; Project administration, S.J., Y.T.Y. and J.-H.H.; Resources, Y.T.Y. and J.-H.H.; Software, S.J.; Supervision, Y.T.Y. and J.-H.H.; Visualization, S.J. and Y.T.Y.; Writing—original draft, S.J. and J.-H.H.; Writing—review & editing, Y.T.Y. and J.-H.H. All authors have read and agreed to the published version of the manuscript.

Funding: This work was supported by the National Research Foundation of Korea (NRF) Grant funded by the Korean Government (MSIT) (No.2017R1C1B5077157). Also, this research was supported by the Energy Cloud R&D Program through the National Research Foundation of Korea (NRF) funded by the Ministry of Science, ICT (NRF-2019M3F2A1073385).

Conflicts of Interest: The authors declare no conflict of interest.

Abbreviations

DERs	distributed energy resources
DR	demand response
EMS	energy management system
PV	photovoltaic
ESS	energy storage system
WT	wind turbine
SoC	state of charge

References

1. Huh, J.-H. *Smart Grid Test Bed Using OPNET and Power Line Communication*; IGI Global: Hershey, PA, USA, 2017; pp. 1–425.
2. Mun, J.H.; Ko, J.S.; Choi, J.S.; Jang, M.G.; Chung, D.H. Efficiency optimization control of SynRM drive using multi-AFLC. In Proceedings of the 2010 International Conference on Electrical Machines and Systems, Incheon, Korea, 11–13 October 2010; IEEE: New York, NY, USA, 2010; pp. 908–913.
3. Choi, J.S.; Ko, J.-S.; Chung, D.-H. Efficiency Analysis of PV tracking system with PSA Algorithm. *J. Korean Inst. Illum. Electr. Install. Eng.* **2009**, *23*, 36–44.
4. Park, D.-M.; Kim, S.-K.; Seo, Y.-S. S-mote: SMART Home Framework for Common Household Appliances in IoT Network. *J. Inf. Process. Syst. KIPS* **2019**, *15*, 449–456.
5. Arcos-Aviles, D.; Pascual, J.; Marroyo, L.; Sanchis, P.; Guinjoan, F. Fuzzy logic-based energy management system design for residential grid-connected microgrids. *IEEE Trans. Smart Grid* **2016**, *9*, 530–543. [CrossRef]
6. Zhang, W.; Ma, Y.; Liu, W.; Ranade, S.J.; Luo, Y. Distributed optimal active power dispatch under constraints for smart grids. *IEEE Trans. Ind. Electron.* **2016**, *64*, 5084–5094. [CrossRef]
7. Gamarra, C.; Guerrero, J.M. Computational optimization techniques applied to microgrids planning: A review. *Renew. Sustain. Energy Rev.* **2015**, *48*, 413–424. [CrossRef]
8. Kim, H.; Kinoshita, T. A new challenge of microgrid operation. *Commun. Comput. Inf. Sci.* **2010**, *78*, 250–260.
9. Mohanpurkar, M.; Ouroua, A.; Hovsapian, R.; Luo, Y.; Singh, M.; Muljadi, E.; Gevorgian, V.; Donalek, P. Real-time co-simulation of adjustable-speed pumped storage hydro for transient stability analysis. *Electr. Power Syst. Res.* **2018**, *154*, 276–286. [CrossRef]
10. Chen, J.; Yang, X.; Zhu, L.; Zhang, M.; Li, Z. Microgrid multi-objective economic dispatch optimization. *Proc. CSEE* **2013**, *33*, 57–66.
11. Saad, W.; Han, Z.; Poor, H.V.; Basar, T. Game-theoretic methods for the smart grid: An overview of microgrid systems, demand-side management, and smart grid communications. *IEEE Signal Process. Mag.* **2012**, *29*, 86–105. [CrossRef]
12. Khan, M.; Basir, R.; Jidin, R.; Pasupuleti, J. Multi-agent based distributed control architecture for microgrid energy management and optimization. *Energy Convers. Manag.* **2016**, *112*, 288–307. [CrossRef]
13. Hori, H.; Ishigaki, Y.; Kimura, Y.; Mai, T.X.; Ozaki, T.; Yokose, T. Energy management system (sEMSA) achieving energy cost minimization. *SEI Tech. Rev.* **2015**, *81*, 56–62. Available online: https://global-sei.com/technology/tr/bn81/pdf/81-11.pdf (accessed on 20 September 2018).
14. Huh, J.-H.; Seo, K. Hybrid AMI design for Smart Grid using the Game Theory model. *Adv. Sci. Technol. Lett.* **2015**, *108*, 86–92.

15. Zhang, Y.; Zhang, T.; Wang, R.; Liu, Y.; Guo, B. Optimal operation of a smart residential microgrid based on model predictive control by considering uncertainties and storage impacts. *Sol. Energy* **2015**, *122*, 1052–1065. [CrossRef]

16. Li, J.; Chung, J.Y.; Xiao, J.; Hong, J.W.-K.; Boutaba, R. On the design and implementation of a home energy management system. In Proceedings of the International Symposium on Wireless and Pervasive Computing, Hong Kong, China, 23–25 February 2011; pp. 1–6.

17. Al-Ali, A.; El-Hag, A.; Bahadiri, M.; Harbaji, M.; El Haj, Y.A. Smart Home Renewable Energy Management System. *Energy Procedia* **2011**, *12*, 120–126. [CrossRef]

18. Zhang, Y.; Zeng, P.; Li, S.; Zang, C.; Li, H. A Novel Multiobjective Optimization Algorithm for Home Energy Management System in Smart Grid. *Math. Probl. Eng.* **2015**, *2015*, 1–19. [CrossRef]

19. Rodriguez, D.I.H.; Braun, M. A comparative study of optimization-and rule-based control for microgrid operation. In Proceedings of the Power and Energy Student Summit (PESS), Dortmund, Germany, 13–14 January 2015.

20. Luna, A.C.; Diaz, N.L.; Graells, M.; Vasquez, J.C.; Guerrero, J.M.; Aldana, N.L.D. Mixed-Integer-Linear-Programming-Based Energy Management System for Hybrid PV-Wind-Battery Microgrids: Modeling, Design, and Experimental Verification. *IEEE Trans. Power Electron.* **2016**, *32*, 2769–2783. [CrossRef]

21. Li, H.; Eseye, A.T.; Zhang, J.; Zheng, D. Optimal energy management for industrial microgrids with high-penetration renewables. *Prot. Control Mod. Power Syst.* **2017**, *2*, 12. [CrossRef]

22. Ju, C.; Wang, P.; Goel, L.; Xu, Y. A Two-Layer Energy Management System for Microgrids with Hybrid Energy Storage Considering Degradation Costs. *IEEE Trans. Smart Grid* **2017**, *9*, 6047–6057. [CrossRef]

23. Mohamed, F.A.; Koivo, H.N. System modelling and online optimal management of microgrid with battery storage. *Renew. Energy Power Qual. J.* **2007**, *1*, 74–78. [CrossRef]

24. Mohamed, F.A.; Koivo, H.N. Online Management of MicroGrid with Battery Storage Using Multiobjective Optimization. In Proceedings of the 2007 International Conference on Power Engineering, Energy and Electrical Drives, Setubal, Portugal, 12–14 April 2007; pp. 231–236.

25. Parisio, A.; Glielmo, L. Energy efficient microgrid management using model predictive control. In Proceedings of the 50th IEEE Conference on Decision and Control and European Control Conference, Orlando, FL, USA, 12–15 December 2011; pp. 5449–5454.

26. Malysz, P.; Sirouspour, S.; Emadi, A. An Optimal Energy Storage Control Strategy for Grid-connected Microgrids. *IEEE Trans. Smart Grid* **2014**, *5*, 1785–1796. [CrossRef]

27. Shi, W.; Xie, X.; Chu, C.-C.; Gadh, R. Distributed Optimal Energy Management in Microgrids. *IEEE Trans. Smart Grid* **2014**, *6*, 1137–1146. [CrossRef]

28. Hoffmann, K.; Menzel, K. A Guideline for the Implementation of an Energy Management System in Facility Management Organisations. *Appl. Mech. Mater.* **2019**, *887*, 247–254. [CrossRef]

29. Zhai, S.; Wang, Z.; Yan, X.; He, G. Appliance Flexibility Analysis Considering User Behavior in Home Energy Management System Using Smart Plugs. *IEEE Trans. Ind. Electron.* **2018**, *66*, 1391–1401. [CrossRef]

30. Eissa, M. Developing incentive demand response with commercial energy management system (CEMS) based on diffusion model, smart meters and new communication protocol. *Appl. Energy* **2019**, *236*, 273–292. [CrossRef]

31. Nge, C.L.; Ranaweera, I.U.; Midtgård, O.-M.; Norum, L. A real-time energy management system for smart grid integrated photovoltaic generation with battery storage. *Renew. Energy* **2019**, *130*, 774–785. [CrossRef]

32. Gazafroudi, A.S.; Soares, J.; Ghazvini, M.A.F.; Pinto, T.; Vale, Z.; Corchado, J.M. Stochastic interval-based optimal offering model for residential energy management systems by household owners. *Int. J. Electr. Power Energy Syst.* **2019**, *105*, 201–219. [CrossRef]

33. Alasali, F.; Haben, S.; Holderbaum, W. Energy management systems for a network of electrified cranes with energy storage. *Int. J. Electr. Power Energy Syst.* **2019**, *106*, 210–222. [CrossRef]

34. Chapaloglou, S.; Nesiadis, A.; Iliadis, P.; Atsonios, K.; Nikolopoulos, N.; Grammelis, P.; Yiakopoulos, C.; Antoniadis, I.; Kakaras, E. Smart energy management algorithm for load smoothing and peak shaving based on load forecasting of an island's power system. *Appl. Energy* **2019**, *238*, 627–642. [CrossRef]

35. Lopez, J.M.G.; Pouresmaeil, E.; Canizares, C.A.; Bhattacharya, K.; Mosaddegh, A.; Solanki, B.V.; Gonzalez, J.M.; Solanki, B. Smart Residential Load Simulator for Energy Management in Smart Grids. *IEEE Trans. Ind. Electron.* **2018**, *66*, 1443–1452. [CrossRef]

36. Wang, Y.; Sun, Z.; Chen, Z. Rule-based energy management strategy of a lithium-ion battery, supercapacitor and PEM fuel cell system. *Energy Procedia* **2019**, *158*, 2555–2560. [CrossRef]
37. Farmani, F.; Parvizimosaed, M.; Monsef, H.; Rahimi-Kian, A. A conceptual model of a smart energy management system for a residential building equipped with CCHP system. *Int. J. Electr. Power Energy Syst.* **2018**, *95*, 523–536. [CrossRef]
38. Wang, F.; Zhou, L.; Ren, H.; Liu, X.; Talari, S.; Shafie-Khah, M.; Catalao, J.P.S. Multi-Objective Optimization Model of Source–Load–Storage Synergetic Dispatch for a Building Energy Management System Based on TOU Price Demand Response. *IEEE Trans. Ind. Appl.* **2018**, *54*, 1017–1028. [CrossRef]
39. Thomas, D.; Deblecker, O.; Ioakimidis, C.S. Optimal operation of an energy management system for a grid-connected smart building considering photovoltaics' uncertainty and stochastic electric vehicles' driving schedule. *Appl. Energy* **2018**, *210*, 1188–1206. [CrossRef]
40. Van Der Meer, D.; Mouli, G.R.C.; Mouli, G.M.-E.; Elizondo, L.R.; Bauer, P.; Morales-Espana, G. Energy Management System with PV Power Forecast to Optimally Charge EVs at the Workplace. *IEEE Trans. Ind. Inform.* **2016**, *14*, 311–320. [CrossRef]
41. Nunna, H.S.V.S.K.; Battula, S.; Doolla, S.; Srinivasan, D. Energy Management in Smart Distribution Systems with Vehicle-to-Grid Integrated Microgrids. *IEEE Trans. Smart Grid* **2016**, *9*, 4004–4016. [CrossRef]
42. Azizivahed, A.; Naderi, E.; Narimani, H.; Fathi, M.; Narimani, M.R. A New Bi-Objective Approach to Energy Management in Distribution Networks with Energy Storage Systems. *IEEE Trans. Sustain. Energy* **2017**, *9*, 56–64. [CrossRef]
43. Indragandhi, V.; Logesh, R.; Subramaniyaswamy, V.; Vijayakumar, V.; Siarry, P.; Uden, L. Multi-objective optimization and energy management in renewable based AC/DC microgrid. *Comput. Electr. Eng.* **2018**, *70*, 179–198.
44. Pilloni, V.; Floris, A.; Meloni, A.; Atzori, L. Smart Home Energy Management Including Renewable Sources: A QoE-driven Approach. *IEEE Trans. Smart Grid* **2016**, *9*, 2006–2018. [CrossRef]
45. Ali, I.; Hussain, S.M.S. Communication Design for Energy Management Automation in Microgrid. *IEEE Trans. Smart Grid* **2016**, *9*, 2055–2064. [CrossRef]
46. Ma, W.-J.; Wang, J.; Gupta, V.; Chen, C. Distributed Energy Management for Networked Microgrids Using Online ADMM With Regret. *IEEE Trans. Smart Grid* **2018**, *9*, 847–856. [CrossRef]
47. Lee, S.Y.; Jin, Y.G.; Yoon, Y.T. Determining the Optimal Reserve Capacity in a Microgrid with Islanded Operation. *IEEE Trans. Power Syst.* **2016**, *31*, 1369–1376. [CrossRef]
48. Eom, S.; Huh, J.-H. The Opening Capability for Security against Privacy Infringements in the Smart Grid Environment. *Mathematics* **2018**, *6*, 202. [CrossRef]
49. Jung, S.; Huh, J.-H. Demand Response Resource Energy Optimization System for Residential Buildings: Smart Grid Approach. In *Advanced Multimedia and Ubiquitous Engineering*; Springer: Berlin/Heidelberg, Germany, 2018; pp. 517–522.

Article

The Opening Capability for Security against Privacy Infringements in the Smart Grid Environment

Sungwook Eom [1] and Jun-Ho Huh [2,*]

[1]   Department of Electrical Engineering, Pohang University of Science and Technology, Pohang 37673, Korea; sweom@postech.ac.kr

[2]   Department of Software, Catholic University of Pusan, Pusan 46252, Korea

*   Correspondence: 72networks@pukyong.ac.kr or 72networks@cup.ac.kr; Tel.: +82-51-510-0662

Received: 10 September 2018; Accepted: 2 October 2018; Published: 14 October 2018

Abstract: It is now known that more information can be leaked into the smart grid environment than into the existing environment. In particular, specific information such as energy consumption data can be exposed via smart devices. Such a phenomenon can incur considerable risks due to the fact that both the amount and the concreteness of information increase when more types of information are combined. As such, this study aimed to develop an anonymous signature technique along with a signature authentication technique to prevent infringements of privacy in the smart grid environment, and they were tested and verified at the testbed used in a previous study. To reinforce the security of the smart grid, a password and anonymous authentication algorithm which can be applied not only to extendable test sites but also to power plants, including nuclear power stations, was developed. The group signature scheme is an anonymous signature schemes where the authenticator verifies the group signature to determine whether the signer is a member of a certain group but he/she would not know which member actually signed in. However, in this scheme, the identity of the signer can be revealed through an "opener" in special circumstances involving accidents, incidents, or disputes. Since the opener can always identify the signer without his/her consent in such cases, the signer would be concerned about letting the opener find out his/her anonymous activities. Thus, an anonymous signature scheme where the signer issues a token when entering his/her signature to allow the opener to confirm his/her identity only from that token is proposed in this study.

Keywords: opening capability; security; smart grid; group signature; anonymous signature

---

1. Introduction

The smart grid is a newly evolving next-generation intelligent power grid, and many technological research works were conducted in different countries over the last decade to increase the efficiency of their power grids. The primary consideration in adopting the smart grid should be protection of the users' privacy [1–3]. In other words, unlike existing security measures, the smart grid system should basically focus on the security of users rather than suppliers. More personal and specific information can be exposed in the smart grid environment by smart devices or hacking attacks when diverse types of information are combined [4–6]. The major issue of personal information protection in the distribution of smart grid technology is that it is possible to infer a user's behavioral pattern based on his/her energy consumption data by collecting and analyzing more detailed personal data, such as the characteristics of the user's energy usage or the frequency of energy production obtained, by applying the latest electric meters and other related equipment and technologies. In addition, the data read by smart meters inevitably require a certain monitoring or surveillance scheme as they are electronically collected and transmitted, rather than manually processed as in the past. The capacity of a meter capable of assessing consumer patterns or types of appliance depends on the frequency with which it collects data and the types of data being collected. Also, the user's behavior at home can make it is easier to infer his/her activity patterns in other places.

The factors associated with privacy intrusion scenarios in a smart grid environment include the following: (1) information concerning the use of a particular medical device or piece of electronic equipment which indicates their activation times and personal patterns, segmented data pertaining to the power consumption of each household appliance and its measurement location, and detailed information on the use of the appliances or equipment in use at a specific location; (2) the possibility of tracking a physical location through newly consumed energy, for instance, the charging of an electric automobile; (3) the activities in a certain house or building can be inferred from the electronic signature or use time pattern upon activation of a device or piece of equipment, which can form the basis for understanding a specific user's activities. Thus, the collection of a consumer's energy use data by a third party should be limited to the information required to serve the third party's purpose and which is authorized by the consumer.

The anonymous signature scheme comprises a function for authenticating signed messages while hiding the actual identity of the signer, which in itself is a common method in current systems that require the input signature to be authenticated. This scheme was developed by Chaum and Heyst in 1991 [7]. As for the group signature scheme, a member of a certain group is able to attach his/her signature in a message to prove that he is actually a member, and the verifier of the message will be able to confirm that person's membership only, without actually identifying the signer. However, it is possible for the opener, who authenticates the input signature, to identify the signer with the information of the signer previously stored in the system. The opener can be an organization or institution that deals with incidents associated with signatures. The group signature scheme is widely used as an anonymous signature scheme because of its reliability [8]. Despite its reliable performance, however, the security of personal information is called into question as many users consider that the opener has sufficient power to identify the signer and obtain the latter's personal information or information on anonymous activities for other purposes. To resolve this problem, Sakai et al. [9] introduced a complementary scheme by adding an "admitter" to the anonymous signature scheme. Thus, Sakai added the admitter and limited the opener's access to the signer's identification only by obtaining the consent of the admitter. In 2013, Ohara et al. [10] resolved the problem raised by Sakai (2012), which was the admitter's limited amount of token issuance.

The group signature scheme is often used when it is necessary for the authenticator to verify that the signer is a member of a particular group without revealing the actual identity. The real identity of the signer can be disclosed to the authenticator only if there are incidents or disputes that need to be solved. Nevertheless, it is quite clear that the signer will feel the burden of revealing his/her identity or anonymous activities to the authenticator without his/her consent and consider that the authority of the authenticator is too great. Thus, to limit the authority of the authenticator while maintaining the effectiveness of the group signature schemes, an anonymous signature scheme which authorizes the authenticator to identify the signer only with the token issued by the signer him/herself when generating a signature is proposed in this study.

2. Related Research

In a conventional power grid where electric power is delivered to the end users via substations (Figure 1), the power generation and distribution processes are centralized by the system, which assumes the role of mapping and visualizing the routine operations while controlling these processes to meet the power supply/demand schedule and its storage.

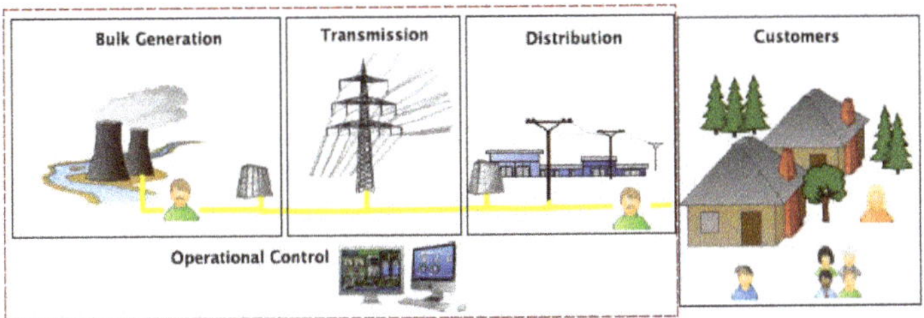

Figure 1. A typical power grid structure.

However, following the rapid development of information technology (IT), such a grid architecture transformed in a way that can provide a more efficient and effective means of power management by integrating with Internet Protocol (IP)-based technologies. The network convergence based on these technologies [11] allows the grid to interwork with an external network(s) by adopting the Transmission Control Protocol (TCP)/IP for a more efficient power management and provision of flexible but efficient service operations.

For the last decade, the development in the hardware, software, and communication technologies led to more advanced and sophisticated information and communications technology (ICT) which were the major factor of widespread mobile smart devices, software applications, or communication architectures [12,13].

The next-generation (21st century) power grid being called the smart grid (Figure 2) enables a smarter, interactive, and dynamic grid management and services based on the ubiquitous computing and advanced ICT technology to respond to the era of the fourth industrial revolution. One of the major advantages of the smart grid is that its bi-directional communication capability can not only improve the power management or operating process but also be utilized for establishing an Internet of things (IoT) system for the users' residences.

The conceptual smart grid model developed by the United States (US) National Institute of Standards and Technology (NIST) defined a smart grid as a complex infrastructure based on a set of seven chief domains [14], namely bulk generation, energy distribution, power transmission, operation and control, market, service providers, and customers and individual domains, composed of heterogeneous elements (e.g., organizations, buildings, individuals, and systems, including system resources and other entities). Also, the backhaul network is essential for achieving smooth but efficient communications between customers and utility companies when advanced power management systems such as advanced metering infrastructure (AMI) are to be embedded into the smart grid [15,16].

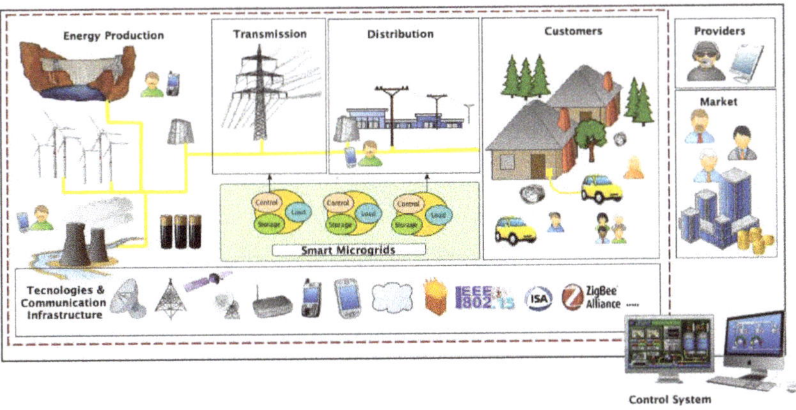

Figure 2. The architecture of a smart grid.

The problem pertaining to breach of privacy is one of the major issues when people are using a service which requires the user to be authenticated. A series of privacy protection schemes were introduced to let users remain anonymous by allowing only encrypted information or minimum user information to be disclosed to the system administrators; however, the security levels and the means of protection provided by those schemes vary and can be inadequate sometimes. The blind signature [17] or the homomorphic encryption [18] scheme is mainly used [19]. To simply describe them, for instance, the former is a scheme where the first party (Party 1) attaches his signature to the message generated by the second party (Party 2) without having any knowledge about the content of the message. Then, the third party (Party 3) can receive the message but the identity of the message sender (Party 2) will remain secure as his/her signature will not be authenticated. Meanwhile, in the latter scheme, a specific mathematical or a computational manipulation is applied to the message or the text to create a ciphertext so that only the authorized party with the right decryption key will be able to decipher the encrypted message. The smart meters usually adopt the latter scheme to encrypt and transmit their requirements to their central control system (utility company administrator) along with a specific encryption function to let the system to decrypt the contents of the requirement with an appropriate decryption key. These schemes were originally developed for the electronic voting systems to conceal the voters' information in the application layer but did not consider the possibility leaking the information from the lower layers (i.e., link layer or network layer) of the protocol stack. It is quite possible that the repeated use of the same IP address overtime may provide access to the identities of the communicating parties or a means for hackers to analyze the traffic [20].

Nonetheless, it is also true that such benefits may be provided at the cost of breaching privacy. That is, a large volume of generated data and its high granularity in which more information is contained would allow any third party with malicious intent to grasp the lifestyles of the customers. Also, there were some claims in some countries that the use of smart meters further endangered the security/privacy of the customers [21]. The balance between achieving an efficient and effective smart metering and guaranteeing the adequate level of personal information protection is always the focus of such a controversy. Using the terminology from Reference [22], the solutions that aim to protect the privacy should guarantee the customers a suitable level of anonymity together with a temporary unlinkability which disconnects them from the metering infrastructure (i.e., disabling power usage reading, etc.). However, the question here is whether the unlinkability can or should be fully achieved even when customers are required to settle their bills at some time or another. The same question can be addressed to unobservability, which refers to the condition where one's power usage cannot be observed by others. Although it is possible to keep the record of the total aggregated amount of one's power usage at the substation level, it still needs to be delivered to the main system for the smart metering system to be fully functional [22,23].

2.1. Anonymous Authentication and Anonymous Signature Schemes

The term "anonymous authentication" refers to a cryptographic technology that allows the person or entity requesting authentication to authenticate him or herself as a legitimate entity while remaining anonymous. Commonly, simple aliases designed to preserve anonymity cannot be used for this type of authentication as the user trail can be traced easily; thus, using them cannot be considered an anonymous authentication scheme. A group signature, anonymous letter of credit, and more were introduced for the purpose of anonymous authentication in a number of research works. The group signature is an electronic signature scheme which the signer can verify him/herself as a member of a particular group without having to reveal his/her identity, thus enabling the authenticator to determine that the person concerned is actually a member without being able to identify him/her. Also, the group signature scheme often involves a credible third-party organization referred to as an "opener", e.g., the police or an internet-related authority. The opener is authorized to identify a signer from the group signature and can track the identity of any user who displays inappropriate behavior (or commits illegal acts) while using anonymous services. The group signature scheme is currently considered the most practical for real-world applications such as web application services as it offers traceable anonymity. In general, the group signature scheme offers anonymity, traceability, and linkability.

Figure 3 shows a schematic representation of the group signature scheme, whose members are normally distinguished as the group manager who sets the parameters, the opener who is authorized to trace a specific signature in a group, the signer, and the authenticator. Each signer in the same group has his/her own private signature key, whereas the authenticator can verify the signatures with an open group key. Also, information that can be used to identify a signer is encrypted in the signature value so that only the opener can trace the identity of a group signer with his open group key.

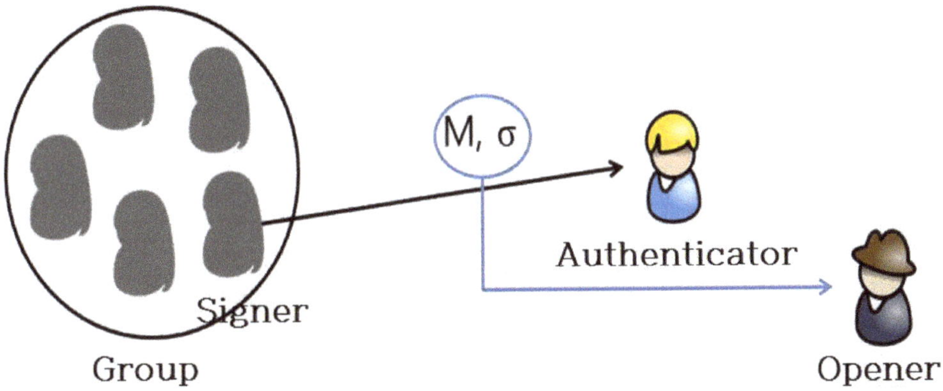

Figure 3. Diagram of the group signature scheme.

Figure 4 represents a group signature scheme that provides linkability, which was studied with a view of applying it to a variety of applications. Linkability is a basis for determining uniformity in a number of signatures so as to determine whether the signatures were written by the same person. Although the linker may detect uniformity in the signatures, he/she is not able to identify the signer.

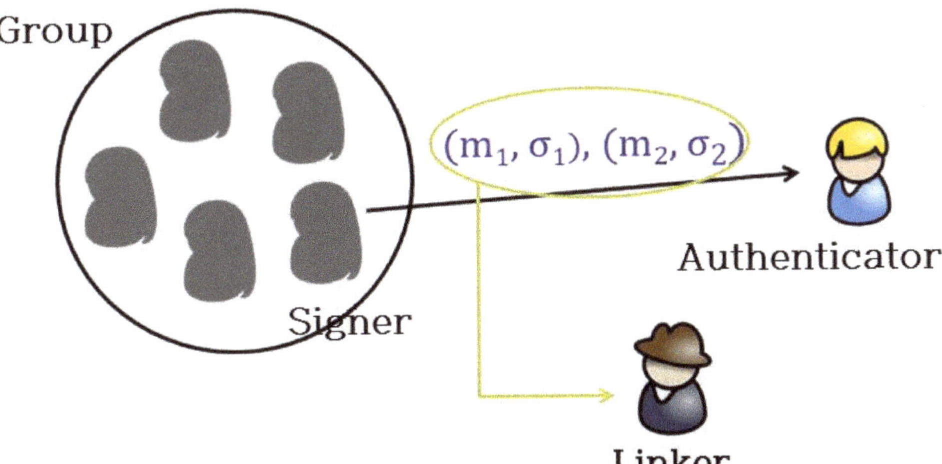

Figure 4. Diagram of group signature scheme with linkability.

In the smart grid environment, service providers can enhance the quality of their services by performing big data analyses of users' data, such as their real-time power usage patterns, etc., and then processing them into meaningful information. Thus, the level of privacy protection can be increased by offering anonymity through group signatures, while the service providers are able to provide flexible services by linking with the data of an anonymous user (signer). Jeong-Yeon Hwang et al. introduced a group signature scheme that provides local linkability [6]; however, in this study, the linker refers to an organization or institution that has a linking key generated by the group manager so that, in general, it becomes the service provider. The linker has the authority to check the link status for all signature values.

Figure 5 shows a group signature scheme offering limited linkability. Unlike existing group signature schemes where the opener is a credible third-party organization, the linker in this scheme is the service provider itself or the organization or institution designated by the service provider, with could result in privacy violations of the service users. For example, let us assume that an anonymous user in the smart grid environment uses a power usage analysis service along with an IoT service. In this case, the service provider will be able to link the power usage information of person A (who just entered his/her signature with the group signature key) with the information about his/her IoT service use. At this time, the service provider does not know the identity of A but it is able to determine whether the user currently using these two services is one and the same, potentially leading to an undesirable breach of privacy. As such, while studies related to existing group signature schemes focused on managing the system for the designated linker so as to be able to test the linkability of all signature values, this study aims to secure a fundamental technology capable of preventing unnecessary information exposures by developing a group signature scheme that allows the designated linker to test the linkability only for those signatures desired by the signer. Thus, in the example shown above, an anonymous signer A can transfer the power usage values to the linker for the linkability test, using a group signature key while transmitting the IoT usage information separately with the same key for the same test. Thus, this scheme can provide a more secure method of preventing privacy breaches by minimizing the level of personal information exposure.

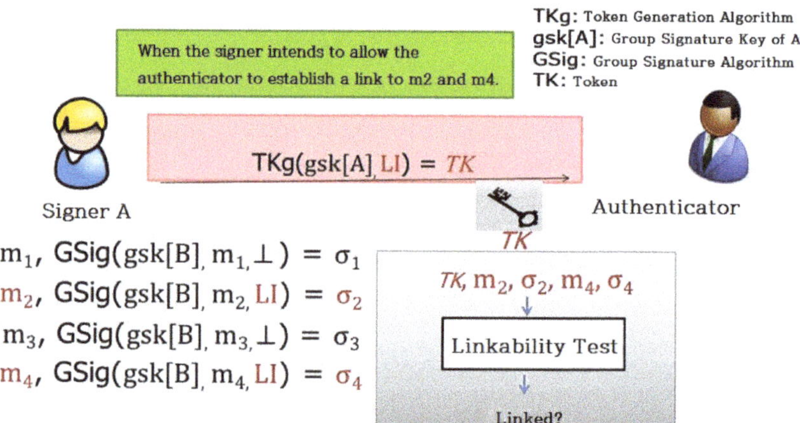

Figure 5. Diagram of group signature scheme with limited linkability.

3. Anonymous Signature with Signer-Controlled Opening Capability

The anonymous signature scheme allows authentication of the signer without revealing his/her identity, whereas the group signature scheme is a method of verifying that the signer is a member of a certain group, also without exposing the signer's identity. Nevertheless, it is possible for an opener to identify an anonymous signer based on the information of the signer, which is neither desirable nor favorable for the signer who wishes his/her signature to be authenticated but does not want to reveal his/her actual identity. Thus, this section discusses a solution whereby the signer obtains a (security) token upon entering his/her signature so that the opener is not able to find the information of the signer without permission.

3.1. Application

The proposed anonymous signature scheme prevents the opener from identifying the signer without his/her permission so that the opener has to obtain a token specifically issued for the signature that the signer wishes to be identified. For example, this scheme can be applied to an anonymous donation system. The identities of the donors are hidden to ensure that the fundraiser cannot know who donated the funds. However, if the donors wish to apply for an income tax deduction, all they have to do is issue a token to the relevant tax administration to prove their donations through signature authentication. Currently, many countries operate an anonymous reporting system against corruption among civil servants, but the problem is that the filed reports and the identity of a whistle-blower or an accuser can be leaked while processing the report, thus endangering that person or making the system useless. The proposed anonymous signature scheme can prevent such an incident by offering a more secure protection mechanism that makes it almost impossible for an intruder or a report handler to find the identity of the person filing the report. If the reporting system requires the accuser to be identified, and if he/she agrees to disclose his/her identity for a final confirmation or compensation, all he/she has to do is issue a token allowing the relevant authority to confirm the true identity.

3.2. Formal Model

The proposed method has the following four algorithms:

GKg$(1^\lambda, 1^n)$: This is the algorithm where the group manager puts the security parameter λ and the number of anonymous signers n to create the signer's signing key gsk_i, the opener's opening key ok, and the public parameters gpk for the system.

GSig(gpk, i, gsk_i, M): This is the algorithm where the anonymous signer uses the group public key gpk, the signer's index i, the signer's signing key gsk_i, and the message M to create the anonymous signature σ, and the token TK_M that permits disclosure.

GVf(gpk, i, gsk_i, M): This is the algorithm where the verifier puts the group public key gpk, the message M, and the anonymous signature σ to verify the signature.

Open$(gpk, ok, M, \sigma, TK_M)$: This is the algorithm where the opener puts the group public key gpk, the opener's opening key ok, the message M, the anonymous signature σ, and the token TK_M to check the signer's identification.

3.3. Security Notion

The four security concepts based on the definition of a general security model [12,13] for the group signature schemes proposed by Mihir Bellare et al. are introduced in the proposed group signature scheme.

- Full anonymity: The identity of a signer should not be accessed unless a token is issued by the signer. Then, the opener, upon receiving the token, is allowed to trace the signer's identity.
- Correctness: A correct signature and a token issued in the proper way should be used for verification when identifying the signer.
- Unforgeability of signature: A valid anonymous signature can only be written by the signer him/herself to attach it to a specific message.
- Unforgeability of token: A valid token can be created and issued to allow the opener to access a specific message or a signature.

3.4. Proposed Scheme

GKg$(1^\lambda, 1^n)$

- Define two hash functions: $H_1: \{0, 1\}^* \to G$, $H_2: \{0, 1\}^* \to Z_p$.
- Select a parameter of the bilinear group (p, G, G_T, e, g).
- Select a random element $u, v, h \in G \backslash \{1\}$, a random integer $\xi_1, \xi_2, \xi_3, \gamma \in Z_p$, and calculate $g_1 = u^{\xi_1} h^{\xi_3}, g_2 = v^{\xi_2} h^{\xi_3}, \omega \leftarrow g^\lambda$.
- Select a random $x_i \in Z_p$ for each signer i $(1 \leq i \leq n)$, then calculate $A_i \leftarrow g^{1/(\gamma+xi)}$.
- Print out the group public key $\quad gpk \leftarrow (p, G, G_T, e, g, g_z, u, v, h, g_1, g_2, \omega, H_1, H_2)$, the opener's opening key $ok \leftarrow (\xi_1, \xi_2, \xi_3, e(A_i, g)_{1 \leq i \leq n})$, and each signer's signing key $gsk_{i(1 \leq i \leq n)} \leftarrow (A_i, x_i)_{1 \leq i \leq n}$.

GSig(gpk, i, gsk_i, M)

- Select a random integer $\alpha, \beta, \rho, \eta, \mu \in Z_p$.
- Calculate $(T_1, T_2, T_3, T_4) \leftarrow (u^\alpha, v^\beta, h^{\alpha+\beta}, g_1{}^\alpha g_2{}^\beta A_i g^\eta)$ and $(T_5, T_6) \leftarrow (g^\rho, e(g^\mu, H_1(M))^\rho)$.
- Select a random integer $r_\alpha, r_\beta, r_\rho, r_\eta, r_x, r_{\alpha x}, r_{\beta x}, r_{\rho x}, r_{\eta x} \in Z_p$.
- Calculate $R_1 \leftarrow u^{r\alpha}; R_2 \leftarrow v^{r\beta}; R_3 \leftarrow h^{r\alpha+r\beta}$ $\quad R_4 \leftarrow e(T_4, g)^{rx} e(g_1, \omega)^{-r\alpha} e(g_1, g)^{-r\alpha x} e(g_2, \omega)^{-r\beta}$ $e(g_2, g)^{-r\beta x} e(g, \omega)^{-r\eta} e(g, g)^{-r\eta x}$ $\quad R_5 \leftarrow g^{r\rho}; R_6 \leftarrow e(g^\mu, H_1(M))^{r\rho} e(g, g)^{-r\eta}$ $\quad R_7 \leftarrow T_1{}^{rx} u^{-r\alpha x}; R_8 \leftarrow T_2{}^{rx} u^{-r\beta x}; R_9 \leftarrow T_5{}^{rx} u^{-r\rho x}$ $\quad R_{10} \leftarrow T_6{}^{rx} e(g^\mu, H_1(M))^{r\rho x} e(g, g)^{-r\eta x}$ $\quad c \leftarrow H_2(M, T_1, \ldots, T_6, R_1, \ldots, R_{10})$ $\quad s_\alpha \leftarrow r_\alpha + c\alpha; s_\beta \leftarrow r_\beta + c\beta; s_\rho \leftarrow r_\rho + c\rho$ $\quad s_\eta \leftarrow r_\eta + c\eta; s_x \leftarrow r_x + cx_i; \leftarrow s_{\alpha x} \leftarrow r_{\alpha x} + c\alpha x_i$ $\quad s_{\beta x} \leftarrow r_{\beta x} + c\beta x_i; s_{\rho x} \leftarrow r_{\rho x} + c\rho x_i; s_{\eta x} \leftarrow r_{\eta x} + c\eta x_i$.
- Print out a signature $\sigma \leftarrow (g^\mu, T_1, \ldots, T_6, c, s_\alpha, s_\beta, s_\rho, s_\eta, s_x, s_{\alpha x}, s_{\beta x}, s_{\rho x}, s_{\eta x})$.
- In addition, calculate and print out the opening-allowed token $TK_M = H_1(M)^\mu$.

GVf(gpk, M, σ)

- Calculate $R_1' \leftarrow u^{s\alpha} T_1{}^{-c}; R_2' \leftarrow v^{s\beta} T_2{}^{-c}; R_3' \leftarrow h^{s\alpha+s\beta} T_3{}^{-c}$ $\quad R_4' \leftarrow e(T_4, g)^{sx} e(g_1, \omega)^{-s\alpha} e(g_1, g)^{-s\alpha x} e(g_2, \omega)^{-s\beta}$ $\cdot\ e(g_2, g)^{-s\beta x} e(g, \omega)^{-s\eta} e(g, g)^{-s\eta x} (e(g, g)/e(T_4, \omega))^{-c}$ $\quad R_5' \leftarrow g^{s\rho} T_5{}^{-c}$; $R_6' \leftarrow e(g^\mu, H_1(M))^{s\rho} e(g, g)^{-s\eta} T_6{}^{-c}$ $\quad R_7' \leftarrow T_1{}^{sx} u^{-s\alpha x}; R_8' \leftarrow T_2{}^{sx} v^{-s\beta x}; R_9' \leftarrow T_5{}^{sx} g^{-s\rho x}$ $\quad R_{10}' \leftarrow T_6{}^{sx} e(g^\mu, H_1(M))^{-s\rho x} e(g, g)^{s\eta x}$.
- Print out "valid" if the equation $c \leftarrow H_2(M, T_1, \ldots, T_6, R_1', \ldots, R_{10}')$ is completed, or "invalid" if the equation is not completed.

Open$(gpk, ok, M, \sigma, TK_M)$

- Verify the signature's validity first using the GVf algorithm. Print out $\perp$ when invalid.
- Verify the token's validity using $e(g^\mu, H_1(M)) = e(g, TK_M)$. Print out $\perp$ when invalid.
- Identify the signer i, who satisfies the equation below when the signature and the token are valid.

$$e(T_4/(T_1{}^{\xi_1} T_2{}^{\xi_2} T_3{}^{\xi_3}), g) \cdot (T_6/e(T_5, TK_M)) = e(A_i, g).$$

- Print out i if there is an i that satisfies the equation. Print out $\perp$ if not.

Based on the assumption that the correctness of the proposed scheme is adequate while the decisional bilinear Diffie–Hellman problem and the decisional linear problem are difficult to solve, full anonymity can be achieved with a random oracle model. Also, the unforgeability of a signature (token) can be dealt with using the same model by assuming that the q-strong (computational) Diffie–Hellman problem is difficult to solve. The details of proof were omitted as they deviate from the research purpose.

In the following section, an anonymous signature scheme is proposed whereby a signer allows the opener to trace his/her identity by accessing his/her information or message to which he/she gave permission by issuing a token. The proposed scheme is expected to raise the level of privacy protection for the signer and can be used in a variety of systems, such as anonymous donation or corruption reporting systems.

4. Group Signature with Signer-Controlled Opening Capability: Separate Token Generator

Group signature schemes are considered a high-security cryptographic signature authentication system for protection of the signer's privacy. The authenticator or the verifier of a signature is provided with a limited amount of information or authority when he/she verifies the signer's affiliation with a certain group without knowing the latter's true identity. Nevertheless, it is still possible for the opener to trace the identity when the situation makes it necessary to deal with malicious accesses. However, concerns about breaches of the signer's privacy through the exposure of his/her personal information still remain. This chapter deals with such a problem by allowing the signer to issue a token with which the opener can access only those messages or items of information, including the signer's identity, whose disclosure is approved.

4.1. Formal Model

The proposed anonymous signature method is composed of the following four algorithms:

KGen(1^λ): This is an algorithm where a trusted third party puts a security parameter λ to create public parameters for the running system gpk, an issuing key for the key issuer ik, and an opening key for the opener ok.

ISS/Join: This is an interactive algorithm between users and issuers that functions as an issuer issues gsk_i to a user in response to a user request.

GSig(gpk, i, gsk_i, M): This is an algorithm where an anonymous signer creates a signature σ using a group public key gpk, an index of the signer i, a signing key of the signer i, gsk_i, and a message M.

TKGen(gpk, i, gsk_i, M): This is an algorithm where an anonymous signer creates an opening-permission token TK_M using a group public key gpk, and an index of a signer i.

GVf(gpk, i, gsk_i, M): This is an algorithm where a signature verifier performs a verification of an anonymous signature using a group public key gpk, a message M, and an anonymous signature σ.

Open(gpk, ok, M, σ, TK_M): This is an algorithm where an opener checks the identity of an anonymous signer from an anonymous signature using an opening key of an opener ok, a message M, an anonymous signature σ, and a token TK_M.

4.2. Security Notion

Mihir Bellare et al. defined the general security model of a group signature method [12,13]. This paper suggests the following four security notions based on Bellare's definition:

Correctness: The proper signature and proper token are always valid when verifying, and the opener with the right signature and the right token can always check the identification from the signature.

Full anonymity: The identity on the anonymous signature must remain inaccessible until the anonymous signer issues a token. When a token is issued, the identity must be inaccessible except by the opener with the token.

Signature unforgeability: Only the proper signer can create a valid anonymous signature for a specific message.

Token unforgeability: Only the proper signer can create a valid token for a specific signature.

4.3. Proposed Scheme

GKg$(1^\lambda, 1^n)$

- Define the two hash functions: H_1: $\{0, 1\}^* \to G$, H_2: $\{0, 1\}^* \to Z_p$.
- Select a parameter of the bilinear group (p, G, G_T, e, g, g_z).
- Select a random element $u, v, h \in G\backslash\{1\}$, a random integer $\xi_1, \xi_2, \xi_3, \gamma \in Z_p$, and calculate $g_1 = u^{\xi_1}h^{\xi_3}$, $g_2 = v^{\xi_2}h^{\xi_3}$, $\omega \leftarrow g^\lambda$.
- Print out the group public key $gpk \leftarrow (p, G, G_T, e, g, g_z, u, v, h, g_1, g_2, \omega, H_1, H_2)$, and issue key λ, the opener's opening key $ok \leftarrow (\xi_1, \xi_2, \xi_3, e(A_i, g)_{1 \leq i \leq n})$.

ISS/Join

- User i selects random $y_i \in Z_p$ and calculates $S_i \leftarrow g_z^{y_i}$.
- User i sends S_i to the issuer.
- Issuer selects random $x_i \in Z_p$ and calculates $\quad A_i \leftarrow (gg_z^{y_i})^{1/(\gamma+xi)}$.
- Issuer sends A_i, x_i to user i.
- User i obtains the signing key $gsk_i = (A_i, x_i, y_i)$.

GSig(gpk, i, gsk_i, M)

- Select a random integer $\alpha, \beta, \rho, \eta \in Z_p$.
- Calculate $(T_1, T_2, T_3, T_4) \leftarrow (u^\alpha, v^\beta, h^{\alpha+\beta}, g_1^\alpha g_2^\beta A_i g^\eta)$ and $(T_5, T_6, T_7) \leftarrow (g^\rho, e(T_5, g)^{y_i}, e(T_5, H_1(M))^{y_i}e(g, g)^{-\eta})$.
- Select a random integer $r_\alpha, r_\beta, r_y, r_\eta, r_x, r_{\alpha x}, r_{\beta x}, r_{\eta x} \in Z_p$.
- Calculate $\quad R_1 \leftarrow u^{r_\alpha}; R_2 \leftarrow v^{r_\beta}; R_3 \leftarrow h^{r_\alpha+r_\beta} \quad R_4 \leftarrow e(T_4, g)^{r_x} e(g_1, \omega)^{-r_\alpha} e(g_1, g)^{-r_{\alpha x}} e(g_2, \omega)^{-r_\beta}$ $\cdot e(g_2, g)^{-r_{\beta x}} e(g, \omega)^{-r_\eta} e(g, g)^{-r_{\eta x}} e(g_z, g)^{-r_y} \quad R_5 \leftarrow e(T_5, g)^{r_y}; R_6 \leftarrow e(T_5, H_1(M))^{r_y}e(g, g)^{-r_\eta} \quad R_7 \leftarrow T_1^{r_x}u^{-r_{\alpha x}}; R_8 \leftarrow T_2^{r_x}u^{-r_{\beta x}}; R_9 \leftarrow T_6^{r_x}e(T_5, g)^{-r_{yx}} \quad R_{10} \leftarrow T_7^{r_x}e(T_5, H_1(M))^{r_{yx}}e(g, g)^{-r_{\eta x}} \quad c \leftarrow H_2(M, T_1, \dots, T_7, R_1, \dots, R_{10}) \quad s_\alpha \leftarrow r_\alpha + c\alpha; s_\beta \leftarrow r_\beta + c\beta; s_y \leftarrow r_y + cy \quad s_\eta \leftarrow r_\eta + c\eta; s_x \leftarrow r_x + cx_i; s_{\alpha x} \leftarrow r_{\alpha x} + c\alpha x_i \quad s_{\beta x} \leftarrow r_{\beta x} + c\beta x_i; s_{\rho x} \leftarrow r_{\rho x} + c\rho x_i; s_{\eta x} \leftarrow r_{\eta x} + c\eta x_i$.
- Print out $\sigma \leftarrow (T_1, \dots, T_7, c, s_\alpha, s_\beta, s_y, s_\eta, s_x, s_{\alpha x}, s_{\beta x}, s_{\rho x}, s_{\eta x})$.

TKGen(gpk, i, gsk_i, M)

- Print out $TK_M = H_1(M)^{y_i}$.

GVf(gpk, M, σ)

- Calculate $\quad R_1' \leftarrow u^{s_\alpha}T_1^{-c}; R_2' \leftarrow v^{s_\beta}T_2^{-c}; R_3' \leftarrow h^{s_\alpha+s_\beta}T_3^{-c} \quad R_4' \leftarrow e(T_4, g)^{s_x} e(g_1, \omega)^{-s_\alpha} e(g_1, g)^{-s_{\alpha x}} e(g_2, \omega)^{-s_\beta} e(g_2, g)^{-s_{\beta x}} \quad \cdot e(g, \omega)^{-s_\eta} e(g, g)^{-s_{\eta x}} e(g_z, g)^{-s_y} (e(g, g)/e(T_4, \omega))^{-c} \quad R_5' \leftarrow e(T_5, g)^{s_y}T_6^{-c}; R_6' \leftarrow e(T_5, H_1(M))^{s_y}e(g, g)^{-s_\eta} T_7^{-c} \quad R_7' \leftarrow T_1^{s_x}u^{-s_{\alpha x}}; R_8' \leftarrow T_2^{s_x}u^{-s_{\beta x}}; R_9' \leftarrow T_6^{s_x}e(T_5, g)^{-s_{yx}} \quad R_{10}' \leftarrow T_7^{s_x}e(T_5, H_1(M))^{s_{yx}}e(g, g)^{-s_{\eta x}}$.
- Print out "valid" if the equation $c \leftarrow H_2(M, T_1, \dots, T_7, R_1', \dots, R_{10}')$ is completed, or "invalid" if the equation is not completed.

Open$(gpk, ok, M, \sigma, TK_M)$

- Verify the signature's validity first using the **GVf** algorithm. Print out $\perp$ if invalid.
- Find out the signer i, who satisfies the equation below, when the signature and the token are valid.

$$e(T_4/(T_1^{\xi_1}T_2^{\xi_2}T_3^{\xi_3}), g)\cdot(T_7/e(T_5, TK_M)) = e(A_i, g).$$

- Print out i if there is an i that satisfies the equation. Print out $\perp$ if not.

Determining correctness in an anonymous signature scheme is not that difficult in the proposed scheme when compared to proving the level of full anonymity. Nevertheless, it can be proven with a random oracle model when an assumption is made that the decisional bilinear Diffie–Hellman problem is not easy to solve. The unforgeability of a signature can be proven with the same model as above when it is assumed that the q-strong Diffie–Hellman problem is not easy to prove. The problem of the unforgeability of a token can be solved in a similar way, but the assumption should be made that the computational Diffie–Hellman problem is not easy to solve. The details of proof were omitted as they deviate from the research purpose.

This chapter provides a solution to signers' concerns about the exposure of their identities in the anonymous signature schemes. The issues pertaining to the excessive authority of the openers were covered by another study in which an admitter was added to the scheme to limit the power of the openers. However, as the possibility of successfully tracing the signer's identity still remained, this study proposed a method by which the signer issues a token him/herself without the intervention of the admitter. It is expected that, if the proposed method is applied to the existing anonymous signature schemes, their level of security will be improved significantly, thus alleviating the users' concerns.

5. Efficiency Comparison

A comparison of theoretical computational costs involved in the algorithms for the generation and verification of the group signatures is shown in Table 1. The group signature scheme in Reference [24] is a sort of a pairing-based group signature scheme which does not offer linkability, and is used for the comparison as a reference scheme. On the other hand, the group signature scheme in Reference [6] offers linkability by allowing the pre-defined linker to check the linkability of all the relevant signatures. Reference [25] introduced a scheme where the signer can control the linkability. When generating the random elements, the respective coefficients (integers) of variables G1, G2, and Zp indicate the individual number of generated random elements (i.e., 2 G1 + 1 G2 + 2 Zp indicates that two random elements were generated for G1, one random element for G2, and two random for Zp). Also, in the calculation formula, P represents the pairing operation; MG1 (or MG2) is the scalar multiplication operation for the group G1 (or G2); EGT is the exponentiation operation in the group GT. As such, the expression 6 P + 9 MG1 + 1 MG2 + 6 EGT implies that six pairings and nine scalar multiplications for G1, one scalar multiplication for G2, and six exponentiations for GT were performed by the algorithm which mainly focuses on the pairing tasks (Table 2), where the pairing operations were performed approximately six times more than the scalar multiplications.

Table 1. The computational costs of the group signatures calculated with the algorithms used by the major group signature schemes.

	Cost	[24]	[6]	[25]	Our Scheme 1	Our Scheme 2
Parameter generation	Elements	2 G1 + 1 G2 + 3 Zp	6 G1 + 1 G2 + 5 Zp	3 G1 + 2 G2 + 3 Zp	3 G1 + 5 Zp	3 G1 + 4 Zp
	Computation	$2\,M_{G1} + 1\,M_{G2}$	$4\,M_{G1} + 3\,M_{G2}$	$2\,M_{G1} + 1\,M_{G2}$	$5\,M_{G1}$	$5\,M_{G1}$
Key generation	Elements	1 Zp	3 Zp	2 Zp	1 Zp	2 Zp
	Computation	$1\,M_{G1}$	$3\,M_{G1}$	$2\,M_{G1}$	$1\,M_{G1}$	$2\,M_{G1}$
Signature generation	Elements	7 Zp	9 Zp	9 Zp	14 Zp	12 Zp
	Computation	$3\,P + 9\,M_{G1} + 3\,E_{GT}$	$7\,P + 16\,M_{G1} + 7\,E_{GT}$	$5\,P + 9\,M_{G1} + 6\,E_{GT} + 2\,M_{G2}$	$11\,P + 22\,M_{G1} + 10\,E_{GT}$	$17\,P + 16\,M_{G1} + 16\,E_{GT}$
Verification	Computation	$5\,P + 8\,M_{G1} + 4\,E_{GT}$	$7\,P + 16\,M_{G1} + 4\,E_{GT}$	$7\,P + 9\,M_{G1} + 7\,E_{GT}$	$13\,P + 16\,M_{G1} + 12\,E_{GT}$	$16\,P + 14\,M_{G1} + 15\,E_{GT}$

Table 2. Performance comparison.

	[24]	[6]	[25]	Our Scheme 1	Our Scheme 2
Parameter generation	0.0306 s	0.0645 s	0.0443 s	0.0151 s	0.0146 s
Key generation	0.0020 s	0.0057 s	0.0038 s	0.0018 s	0.0037 s
Signature generation	0.0498 s	0.1051 s	0.0997 s	0.1579 s	0.2153 s
Verification	0.0653 s	0.0933 s	0.0910 s	0.1653 s	0.1959 s

The operation of each group signature scheme (Table 2) was simulated with the computer (Intel Sandy Bridge i3 2330M 2.2-GHz processor, 4 GB random-access memory (RAM), Ubuntu 12.04), whereas the operations (pairing) were performed using the Python Pairing-Based Cryptography (PYPBC) Library, adopting the d224 curve, specifically. The resulting values are the averages of 100 simulations conducted for the individual schemes. The time required for the proposed scheme to generate and verify the signature was similar to that of References [6,24,25] and the same level of similarity was found in the computational costs. This means that the function "signer-controlled opening capability" being added to the computation process did not actually affect the computational costs much. Meanwhile, the proposed algorithm in this study was developed in a way that it can be adopted in previous research [26–32] pertaining to smart grids.

6. Conclusions

The group signature scheme is an electronic signature scheme with which a signer can prove that he/she is a member of a certain group without revealing his/her own identity, and which allows the authenticator to make a judgment on whether the signature is written by the same person or not, but which does not allow the authenticator to know the identity of the signer. A number of previous studies flexibly applied group signature schemes to various applications.

Meanwhile, the proposed algorithm in this study was developed in a way that it can be adopted in previous research [26–32] pertaining to smart grids.

Thus, two anonymous signature schemes in a smart grid environment were proposed in this study: a scheme where the anonymous signer issues a token to let the opener identify him/her only for the designated signature, and another scheme which requires the signer's consent for identification. In the former, the signer generates the token along with his/her signature using a short-term secret key, whereas, in the latter, the token is generated using a long-term secret key only when the signer agrees to disclose his/her identity after entering the signature. Although there is a possibility of compromising the security a little when the latter scheme is adopted, the burden of the signer having to issue and keep the token all the time can be lightened, improving the convenience of the scheme.

Author Contributions: Conceptualization, S.E. and J.-H.H. Data curation, S.E. and J.-H.H. Formal analysis, S.E. Methodology, J.-H.H. Project administration, J.-H.H. Resources, S.E. and J.-H.H. Visualization, S.E. and J.-H.H. Writing—original draft, S.E. and J.-H.H. Writing—review and editing, J.-H.H.

Funding: This work was supported by the National Research Foundation of Korea (NRF) grant funded by the Korea government (MSIT) (No. 2017R1C1B5077157).

Acknowledgments: The first draft of this paper was presented at The International Conference on Big data, IoT, and Cloud Computing (BIC 2017) [31,33], 22–24 August 2017, Republic of Korea.

Conflicts of Interest: The authors declare no conflict of interest.

References

1.　Huh, J.H. *Smart Grid Test Bed Using OPNET and Power Line Communication*; IGI Global: Hershey, PA, USA, 2017; pp. 64–89.

2.   Freitas, W.; Vieira, J.C.; Morelato, A.; Xu, W. Influence of Excitation System Control Modes on the Allowable Penetration Level of Distributed Synchronous Generators. *IEEE Trans. Energy Convers.* **2005**, *20*, 474–480. [CrossRef]

3.   Dandeno, P.L.; Karas, A.N.; McClymont, K.R.; Watson, W. Effect of High-Speed Rectifier Excitation Systems on Generator Stability Limits. *IEEE Trans. Power Appar. Syst.* **1968**, *PAS-87*, 190–201. [CrossRef]

4.   Wang, D.; Mao, C.; Lu, J. Coordinated Control of EPT and Generator Excitation System for Multidouble-Circuit Transmission-Lines System. *IEEE Trans. Power Deliv.* **2008**, *23*, 371–379. [CrossRef]

5.   Park, S.; Huh, J.H. Effect of Cooperation on Manufacturing IT Project Development and Test Bed for Successful Industry 4.0 Project: Safety Management for Security. *Processes* **2018**, *6*, 88. [CrossRef]

6.   Hwang, J.Y.; Chen, L.; Cho, H.S.; Nyang, D. Short dynamic group signature scheme supporting controllable linkability. *IEEE Trans. Inf. Forensics Secur.* **2015**, *10*, 1109–1124. [CrossRef]

7.   Chaum, D.; van Heyst, E. Group Signatures. In Proceedings of the Workshop on the Theory and Application of Cryptographic Techniques, Brighton, UK, 8–11 April 1991; Springer: Berlin/Heidelberg, Germany, 1991; pp. 257–265.

8.   Hwang, J.Y.; Eom, S.; Chang, K.; Lee, P.J.; Nyang, D. Anonymity-Based Authenticated Key Agreement with Full Binding Property. In Proceedings of the International Workshop on Information Security Applications, Jeju Island, Korea, 16–18 August 2012; Springer: Berlin/Heidelberg, Germany, 2012; pp. 177–191.

9.   Sakai, Y.; Emura, K.; Hanaoka, G.; Kawai, Y.; Matsuda, T.; Omote, K. Group Signatures with Message-Dependent Opening. In Proceedings of the International Conference on Pairing-Based Cryptography, Cologne, Germany, 16–18 May 2012; Springer: Berlin/Heidelberg, Germany, 2012; pp. 270–294.

10.   Ohara, K.; Sakai, Y.; Emura, K.; Hanaoka, G. A group signature scheme with unbounded message-dependent opening. In Proceedings of the 8th ACM SIGSAC Symposium on Information, Computer and Communications Security, Hangzhou, China, 8–10 May 2013; pp. 517–522.

11.   McClanahan, R. SCADA and IP: Is Network Convergence Really Here? *IEEE Ind. Appl. Mag.* **2003**, *9*, 29–36. [CrossRef]

12.   Fan, J.; Borlase, S. The Evolution of Distribution. *IEEE Power Energy Mag.* **2009**, *7*, 63–68.

13.   Farhangi, H. The Path of the Smart Grid. *IEEE Power Energy Mag.* **2010**, *8*, 18–28. [CrossRef]

14.   NIST. *NIST Framework and Roadmap for Smart Grid Interoperability Standards, Release 2.0*; NIST Special Publication: Gaithersburg, MD, USA, 2012.

15.   Mark, J. New Electricity Grids May Be Smart, but Not so Private—The Denver Post. 18 May 2010. Available online: http://www.denverpost.com/business/ci_15106430 (accessed on 9 September 2018).

16.   Siddiqui, F.; Zeadally, S.; Alcaraz, C.; Galvao, S. Smart grid privacy: Issues and solutions. In Proceedings of the 2012 21st International Conference on Computer Communications and Networks (ICCCN), Munich, Germany, 30 July–2 August 2012.

17.   Cheung, J.; Chim, T.; Yiu, S.; Li, V. Credential-based privacypreserving power request scheme for smart grid network. In Proceedings of the IEEE Global Telecommunications Conference, Kathmandu, Nepal, 5–9 December 2011; pp. 1–5.

18.   Marmol, F.; Sorge, C.; Ugus, O.; Perez, G. Do not snoop my habits: Preserving privacy in the smart grid. *IEEE Commun. Mag.* **2012**, *50*, 166–172. [CrossRef]

19.   Zeadally, S.; Pathan, A.; Alcaraz, C.; Badra, M. Towards privacy protection in smart grid. *Wirel. Pers. Commun.* **2013**, *73*, 23–50. [CrossRef]

20.   Badra, M.; Zeadally, S. Design and Performance Analysis of a Virtual Ring Architecture for Smart Grid Privacy. *IEEE Trans. Inf. Forensics Secur.* **2014**, *9*, 321–329. [CrossRef]

21.   Hoenkamp, R.; Huitema, G.B.; de Moor-van Vugt, A.J. The neglected consumer: The case of the smart meter rollout in the Netherlands. *Renew. Energy Law Policy Rev.* **2011**, *2*, 269–282. [CrossRef]

22.   Ptzmann, A.; Hansen, M. A Terminology for Talking about Privacy by Data Minimization: Anonymity, Unlinkability, Undetectability, Unobservability, Pseudonymity, and Identity Management. Available online: http://dud.inf.tu-dresden.de/Anon_Terminology.shtml (accessed on 9 September 2018).

23.   Tudor, V.; Almgren, M.; Papatriantafilou, M. Analysis of the impact of data granularity on privacy for the smart grid. In Proceedings of the 12th ACM Workshop on Workshop on Privacy in the Electronic Society, Berlin, Germany, 4–8 November 2013.

24. Boneh, D.; Boyen, X.; Shacham, H. Short group signature. In Proceedings of the Annual International Cryptology Conference (CRYPTO), Santa Barbara, CA, USA, 15–19 August 2004; Springer: Berlin/Heidelberg, Germany, 2004.

25. Eom, S.; Huh, J.H. Group signature with restrictive linkability: Minimizing privacy exposure in ubiquitous environment. *J. Ambient Intell. Humaniz. Comput.* **2018**, *1*, 1–11. [CrossRef]

26. Yu, C.M.; Chen, C.Y.; Kuo, S.Y.; Chao, H.C. Privacy-Preserving Power Request in Smart Grid Networks. *IEEE Syst. J.* **2014**, *8*, 441–449. [CrossRef]

27. Ciabattoni, L.; Comodi, G.; Ferracuti, F.; Fonti, A.; Giantomassi, A.; Longhi, S. Multi-apartment residential microgrid monitoring system based on kernel canonical variate analysis. *Neurocomputing* **2015**, *170*, 306–317. [CrossRef]

28. Ancillotti, E.; Bruno, R.; Conti, M. The role of communication systems in smart grids: Architectures, technical solutions and research challenges. *Comput. Commun.* **2013**, *36*, 1665–1697. [CrossRef]

29. Kim, S.K.; Huh, J.H. A Study on the Improvement of Smart Grid Security Performance and Blockchain Smart Grid Perspective. *Energies* **2018**, *11*, 1973. [CrossRef]

30. Lee, H.G.; Huh, J.H. A Cost-Effective Redundant Digital Excitation Control System and Test Bed Experiment for Safe Power Supply for Process Industry 4.0. *Processes* **2018**, *6*, 85. [CrossRef]

31. Eom, S.; Huh, J.H. Anonymous Signature with Signer-Controlled Opening Capability. In *Advances in Computer Science and Ubiquitous Computing*; Springer: Singapore, 2017; pp. 878–882.

32. Bellare, M.; Micciancio, D.; Warinschi, B. Foundations of Group Signatures: Formal Definitions, Simplified Requirements, and a Construction Based on General Assumptions. In Proceedings of the International Conference on the Theory and Applications of Cryptographic Techniques (EUROCRYPT), Warsaw, Poland, 4–8 May 2003; Springer: Berlin/Heidelberg, Germany, 2003; pp. 614–629.

33. Eom, S.; Huh, J.H. *Group Signature with Signer-Controlled Opening Capability: Separate Token Generator, In Advances in Computer Science and Ubiquitous Computing*; Springer: Singapore, 2017; pp. 883–887.

Article

Responsive Economic Model Predictive Control for Next-Generation Manufacturing

Helen Durand [†]

Wayne State University, Detroit, MI 48202, USA; helen.durand@wayne.edu; Tel.: +1-313-577-3475
† Current address: 5050 Anthony Wayne Drive, Detroit, MI 48202, USA.

Received: 31 December 2019; Accepted: 11 February 2020; Published: 16 February 2020

Abstract: There is an increasing push to make automated systems capable of carrying out tasks which humans perform, such as driving, speech recognition, and anomaly detection. Automated systems, therefore, are increasingly required to respond to unexpected conditions. Two types of unexpected conditions of relevance in the chemical process industries are anomalous conditions and the responses of operators and engineers to controller behavior. Enhancing responsiveness of an advanced control design known as economic model predictive control (EMPC) (which uses predictions of future process behavior to determine an economically optimal manner in which to operate a process) to unexpected conditions of these types would advance the move toward artificial intelligence properties for this controller beyond those which it has today and would provide new thoughts on interpretability and verification for the controller. This work provides theoretical studies which relate nonlinear systems considerations for EMPC to these higher-level concepts using two ideas for EMPC formulations motivated by specific situations related to self-modification of a control design after human perceptions of the process response are received and to controller handling of anomalies.

Keywords: economic model predictive control; chemical processes; responsive control; artificial intelligence; interpretability; controller verification

---

1. Introduction

The buzz around artificial intelligence (AI), machine learning, and data in recent years has sparked both excitement and skepticism from the process systems engineering community [1,2]. Some of the most prevalent uses of data in the process systems field have included its use in developing models of various processes (e.g., Reference [3]) with potential applications in model-based control [4], in learning control laws [5,6], and in process monitoring [7,8]. Control engineers have debated about whether control itself should be considered to be artificial intelligence, particularly as control laws become more advanced. For example, a particularly intelligent form of control (known as economic model predictive control (EMPC) [9–12]) is an optimization-based control strategy that determines the optimal manner in which to operate a chemical process in the sense that the control actions optimize a profit metric for the process over a prediction horizon, subject to process constraints. The significant potential benefits of this control law for next-generation manufacturing have prompted a wide range of investigations in the context of EMPC, including how it may be used for building temperature regulation [13], wastewater treatment [14], microgrid dispatch [15], and gas pipeline networks [16]. Though chemical processes have traditionally been operated at steady-state, EMPC does not necessarily enforce steady-state operation in its efforts to optimize process economic performance. This has raised key questions for this control design regarding important properties of intelligent systems such as interpretability of its operating strategy and verification that it will work correctly for the real environment that it will need to control and interact with.

Interpretability is a desirable property for artificially intelligent systems. It has been considered in a variety of contexts; for example, the issue of building interpretable data-driven models has been considered to be enhanced by sparse regression, where a model with a small number of available possible terms which could be utilized to build it is derived (with an underlying assumption being that simpler models are more physically realistic and therefore should be more interpretable) [17]. Models identified via sparse regression techniques have been utilized in model predictive control for hydraulic fracturing [18]. Interpretability of other model-building strategies has also been a consideration; for example, for neural networks, where interpretability may be considered to be multidimensional, but to generally constitute whether a human can trace how a neural network obtained its conclusions via how the input information was processed [19], recurrent neural networks with long short-term memory were analyzed for how their cells processed different aspects of character-level language models [20].

It is recognized that interpretability of the control actions computed by an EMPC will be a major determining factor in the adoption of EMPC in the process industries (because, if operators and engineers do not know if the process is in an upset condition, they will likely disable features of the controller that make it difficult to understand due to the need to be sure that safety is maintained at all times). Interpretability for EMPC has not yet received significant focus in the literature. The subset of EMPC formulations which track a steady-state [21] possess a form of interpretability in that the reference behavior is understood by engineers and operators. Reference [22] developed an EMPC formulation in which the desired closed-loop process response specified or restricted by an operator or engineer is tracked by the controller. However, developing the best means for ensuring interpretability for EMPC to appropriately trade off end user understanding with economic optimality remains a largely open question. This work provides new perspectives on this important issue, suggesting that a controller formulation that bridges the human–machine interface by allowing the adjustment of constraints in response to human opinions about the process behavior under the EMPC may provide new avenues of both democratizing advanced control and allowing end users to adjust the response to their liking from an interpretability standpoint.

Another important topic for intelligent control systems is enabling their verification (i.e., certifying that they will perform in practice as intended). Verification can take a significant amount of engineering time and expense, and methods for reducing the time required to validate the controller's performance could reduce the cost of advanced control, could promote operational safety, and could make the controller more straightforward to implement (a lack of ability to verify can prevent an intelligent system from being placed online at all). In the control community, a traditional approach to verification is to design controllers with guaranteed robustness to bounded uncertainty and to use this as a certificate that the controller will be able to maintain closed-loop stability in practice (e.g., References [23–25]). This requires some knowledge of the disturbance characteristics (e.g., upper bounds), which may be difficult to fully determine a priori but is important for EMPC, as the controller could drive the closed-loop state to operate at boundaries of safe operating regions to optimize profits, where the uncertainty in the disturbance characteristics could lead to unsafe conditions. Additional conservatism to account for the uncertainty could lead to over-conservatism that could decrease profits. Other methods for handling disturbances in EMPC have been developed, including methods that account for disturbances probabilistically (making assumptions on their distribution) [26] or adapting models used by the predictive controller online (e.g., References [27–29]). Results on the use of adapting models in EMPC have even included closed-loop stability guarantees when a recurrent neural network that is updated via error triggering is used as the process model [30]. An example of an adaptive control strategy which handles uncertain dynamics in batch processing is that in Reference [31], which uses model predictive control equipped with a probabilistic recursive least squares model parameter update algorithm with a forgetting factor to capture batch process dynamics. In addition, Reference [32] analyzed a learning-based MPC strategy with a terminal constraint for systems with unmodeled dynamics, where performance is enhanced by using a learned model in the MPC but safety goals are met by ensuring that control actions computed via the MPC are stabilizing.

Another direction that has received attention for handling uncertainty is fault tolerance in the sense of controller reconfigurations upon detection of an actuator fault/anomaly (e.g., Reference [33]) or anomaly response cast in a framework of fault-tolerant control handled via fault/anomaly detection followed by updating the model used by a model-based controller [34]. In Reference [35], fault-tolerant control for nonlinear switched systems was analyzed in the context of safe parking for model predictive control with a steady-state tracking objective function for actuator faults. For EMPC, Reference [36] handled faults through error-triggered data-driven model updates in the controller, and the uniting of EMPC with driving the state into safety-based regions in state-space (e.g., References [37,38]) also constitutes a form of fault-handling. Despite these advances in handling anomalies and uncertainty, which are critical for addressing moving toward a verification paradigm for EMPC, verifying the controller today would still be expected to be time-consuming; additional work is needed to explore further ways of considering and establishing verification for the control design.

Another approach in verification of controllers has been online verification via data-driven models complemented by detection algorithms for problematic controller behavior leading to bounds on the time that would elapse before detection of problematic controller behavior [39]. A feature of this direction in verification, therefore, is the combination of data-driven modeling for control (to address model uncertainty) with guarantees that problematic behavior due to model inaccuracies can be flagged within a given time period. In the present work, we take a conceptually similar approach to verification for EMPC using online anomaly handling with a conservative Lyapunov-based EMPC (LEMPC) [24] design approach that facilitates guaranteed detection of significant plant/model mismatch under sufficient conditions and allows upper bounds on the amount of time available until the mismatch would need to be compensated via model updates without compromising closed-loop stability (as well as the characteristics of the resulting control law after model reidentification required to obtain these theoretical results) to be presented. The development of theoretical guarantees on closed-loop stability with data-driven models that can be updated online in LEMPC has some similarities to References [30,40] but is pursued from a different angle that allows the underlying process dynamics to suddenly change and also allows for more general nonlinear data-driven models to be considered (i.e., we do not restrict the modeling methodology to neural networks as in References [30,40]). It also has similarities to the framework for accounting for faults in LEMPC via model updates in Reference [41] but considers a theoretical treatment of anomaly conditions with data-driven LEMPC, which was not explored in that work.

Motivated by the above considerations, this work focuses on advancing both interpretability and verification for EMPC. These are important considerations for human–machine interaction and can be viewed as different aspects of a "responsive" control design in the sense that the controller is made responsive to changing or unexpected conditions like a human would be. We first address the interpretability concept suggested above in an LEMPC framework in which we elucidate conditions under which an LEMPC could be made responsive to potentially inaccurate metrics reflecting the reactions of end users to the LEMPC's behavior without loss of closed-loop stability. We subsequently move in the direction of addressing verification considerations for LEMPC by developing theoretical guarantees which can be made for the controller in the presence of process dynamics anomalies/changes when potentially adapting data-driven models are used in the controller. We evaluate the conditions under which closed-loop stability may be lost in such circumstances, with exploration of bounds on times before which detection and accommodation of the anomaly could be stabilized to avoid potential plant shutdown. Numerical examples utilizing continuous stirred tank reactors (CSTRs) are presented to illustrate major concepts. Throughout, we highlight cases where the proposed methods could interface with other artificial intelligence techniques (e.g., sentiment analysis or image-based sensing) without compromising closed-loop stability, highlighting the range of intelligent techniques which can be used to enhance next-generation control within an appropriate theoretical framework.

This work is organized as follows: in Section 2, preliminaries are presented. These are followed by the main results in Section 3, which consist of controller formulations and implementation strategies, with demonstration via numerical examples, where (1) the controller constraints can be adjusted online in response to potentially inaccurate stimuli without closed-loop stability being lost (Section 3.1) and (2) the control strategy has characterizable properties in the presence of process anomalies resulting in unanticipated changes in the underlying process dynamics (Section 3.2). Section 4 concludes and provides an outlook on the presented results. Proofs for theoretical results associated with the second control strategy noted above are provided in the Appendix. This manuscript is an extended version of Reference [42].

2. Preliminaries

2.1. Notation

The operator $|\cdot|$ denotes the vector Euclidean norm. A function $\alpha : [0, a) \rightarrow [0, \infty)$ is in class $\mathcal{K}$ if it is continuous, if it strictly increases, and if $\alpha(0) = 0$. The notation Ω_ρ defines a level set of a scalar-valued function V (i.e., $\Omega_\rho := \{x \in R^n : V(x) \leq \rho\}$). The operator $'/'$ signifies set subtraction (i.e., $A/B := \{x \in R^n : x \in A, x \notin B\}$). x^T represents the transpose of the vector x. We define a sampling time with the notation $t_k := k\Delta$, $k = 0, 1, \ldots$.

2.2. Class of Systems

This work considers switched nonlinear systems of the following form:

$$\dot{x}_{a,i} = f_i(x_{a,i}(t), u(t), w_i(t)) \tag{1}$$

where $x_{a,i} \in X \subset R^n$ denotes the state vector, $u \in U \subset R^m$ denotes the input vector ($u = [u_1, \ldots, u_m]^T$), and $w_i \in W_i \subset R^z$ denotes the disturbance vector, where $W_i := \{w_i \in R^z : |w_i| \leq \theta_i, \theta_i > 0\}$, for $i = 1, 2, \ldots$. In this notation, the ith model is used for $t \in [t_{s,i}, t_{s,i+1})$, where $x_{a,i}(t_{s,i+1}) = x_{a,i+1}(t_{s,i+1})$ and $t_{s,1} = t_0$. The vector function f_i is assumed to be a locally Lipschitz function of its arguments with $f_1(0, 0, 0) = 0$ and $f_i(x_{a,i,s}, u_{i,s}, 0) = 0$ for $i > 1$ (i.e., the steady-state of the updated models when $w_i = 0$ is at $x_{a,i} = x_{a,i,s}$, $u = u_{i,s}$). The system of Equation (1) with $w_i \equiv 0$ is known as the nominal system. Synchronous measurement sampling is assumed, with measurements available at every $t_k = k\Delta$, $k = 0, 1, \ldots$. It is noted that $t_{s,i}$, $i = 1, 2, \ldots$, is not required to be an integer multiple of t_k. We define $\tilde{x}_{a,i} = x_{a,i} - x_{a,i,s}$ and $\bar{u}_i = u - u_{i,s}$ and define $\tilde{f}_i$ as f_i rewritten to have its origin at $\tilde{x}_{a,i} = 0$, $\bar{u}_i = 0$, $w_i = 0$. Similarly, we define U_i to be the set U in deviation variable form from $u_{i,s}$ and X_i to be the set X in deviation variable form from $x_{a,i,s}$.

We assume that there exists an explicit stabilizing (Lyapunov-based) control law $h_i(\tilde{x}_{a,i}) = [h_{i,1}(\tilde{x}_{a,i}) \ \ldots \ h_{i,m}(\tilde{x}_{a,i})]^T$ that renders the origin of the nominal system of Equation (1) asymptotically stable in the sense that the following inequalities hold:

$$\alpha_{1,i}(|\tilde{x}_{a,i}|) \leq V_i(\tilde{x}_{a,i}) \leq \alpha_{2,i}(|\tilde{x}_{a,i}|) \tag{2}$$

$$\frac{\partial V_i(\tilde{x}_{a,i})}{\partial \tilde{x}_{a,i}} \tilde{f}_i(\tilde{x}_{a,i}, h_i(\tilde{x}_{a,i}), 0) \leq -\alpha_{3,i}(|\tilde{x}_{a,i}|) \tag{3}$$

$$\left| \frac{\partial V_i(\bar{x}_{a,i})}{\partial \bar{x}_{a,i}} \right| \leq \alpha_{4,i}(|\bar{x}_{a,i}|) \tag{4}$$

$$h_i(\bar{x}_{a,i}) \in U_i \tag{5}$$

for all $\bar{x}_{a,i} \in D_i \subseteq R^n$ and $i = 1, 2, \ldots$, where D_i is an open neighborhood of the origin of $\bar{f}_i$, and for a positive definite, sufficiently smooth Lyapunov function V_i. The functions $\alpha_{1,i}$, $\alpha_{2,i}$, $\alpha_{3,i}$, and $\alpha_{4,i}$ are of class $\mathcal{K}$. A level set of V_i denoted by $\Omega_{\rho_i} \subset D_i$ is referred to as the stability region of the system of Equation (1) under the controller $h_i(\bar{x}_{a,i})$. We consider that Ω_{ρ_i} is selected to be contained within X. The Lyapunov-based controller is assumed to be Lipschitz continuous such that the following inequalities hold:

$$|h_{i,j}(x) - h_{i,j}(x')| \leq L_{h,i}|x - x'| \tag{6}$$

for a positive constant $L_{h,i}$ for all $x, x' \in \Omega_{\rho_i}$, and $i = 1, 2, \ldots$, with $j = 1, \ldots, m$.

Lipschitz continuity of f_i and sufficient smoothness of V_i provide the following inequalities, for positive constants M_i, $L_{x,i}$, $L_{w,i}$, $L'_{x,i}$, and $L'_{w,i}$:

$$|\bar{f}_i(x, u, w_i)| \leq M_i \tag{7}$$

$$|\bar{f}_i(x, u, w_i) - \bar{f}_i(x', u, 0)| \leq L_{x,i}|x - x'| + L_{w,i}|w_i| \tag{8}$$

$$\left| \frac{\partial V_i(x)}{\partial x}\bar{f}_i(x, u, w_i) - \frac{\partial V_i(x')}{\partial x}\bar{f}_i(x', u, 0) \right| \leq L'_{x,i}|x - x'| + L'_{w,i}|w_i| \tag{9}$$

for all $x, x' \in \Omega_{\rho_i}$, $u \in U_i$, and $w_i \in W_i$.

As this work considers responses to unexpected conditions, we consider that there may be cases in which the nonlinear model of Equation (1) may not be available, though an empirical model with the following form may be available:

$$\dot{x}_{b,q}(t) = f_{NL,q}(x_{b,q}(t), u(t)) \tag{10}$$

where $f_{NL,q}$ is a locally Lipschitz nonlinear vector function in $x_{b,q} \in \mathbb{R}^n$ and in the input $u \in \mathbb{R}^m$ with $f_{NL,1}(0,0) = 0$ and $f_{NL,q}(x_{b,q,s}, u_{q,s}) = 0$ for $q > 1$ (i.e., the steady-state of the updated models is at $x_{b,q} = x_{b,q,s}$, $u = u_{q,s}$). Here, $q = 1, 2, \ldots$, to allow for the possibility that, as the underlying process dynamics change (i.e., the value of i increases in Equation (1)), it may be desirable to switch the empirical model used to describe the system. However, we utilize the index q instead of i for the empirical model to signify that we do not assume that the empirical model must switch with the same frequency as the process dynamics. When the model of Equation (10) does switch, we assume that the switch occurs at a time $t_{s,NL,q+1}$ in a manner where $x_{b,q}(t_{s,NL,q+1}) = x_{b,q+1}(t_{s,NL,q+1})$. We define $\bar{x}_{b,q} = x_{b,q} - x_{b,q,s}$ and $\bar{u}_q = u - u_{q,s}$ and define $\bar{f}_{NL,q}$ as $f_{NL,q}$, rewritten to have its origin at $\bar{x}_{b,q} = 0$, $\bar{u}_q = 0$, as follows:

$$\dot{\bar{x}}_{b,q}(t) = \bar{f}_{NL,q}(\bar{x}_{b,q}(t), \bar{u}_q(t)) \tag{11}$$

Similarly, we define U_q to be the set U in deviation variable form from $u_{q,s}$ and X_q to be the set X in deviation variable form from $x_{b,q,s}$.

We consider that, for the empirical models in Equation (10), there exists a locally Lipschitz explicit stabilizing controller $h_{NL,q}(\bar{x}_{b,q})$ that can render the origin asymptotically stable in the sense that:

$$\hat{\alpha}_{1,q}(|\bar{x}_{b,q}|) \leq \hat{V}_q(\bar{x}_{b,q}) \leq \hat{\alpha}_{2,q}(|\bar{x}_{b,q}|) \tag{12a}$$

$$\frac{\partial \hat{V}_q(\bar{x}_{b,q})}{\partial \bar{x}_{b,q}} \bar{f}_{NL,q}(\bar{x}_{b,q}, h_{NL,q}(\bar{x}_{b,q})) \leq -\hat{\alpha}_{3,q}(|\bar{x}_{b,q}|) \tag{12b}$$

$$\left| \frac{\partial \hat{V}_q(\bar{x}_{b,q})}{\partial \bar{x}_{b,q}} \right| \leq \hat{\alpha}_{4,q}(|\bar{x}_{b,q}|) \tag{12c}$$

$$h_{NL,q}(\bar{x}_{b,q}) \in U_q \tag{12d}$$

for all $\bar{x}_{b,q} \in D_{NL,q}$ (where $D_{NL,q}$ is a neighborhood of the origin of $\bar{f}_{b,q}$ contained in X), where $\hat{V}_q : \mathbb{R}^n \to \mathbb{R}_+$ is a sufficiently smooth Lyapunov function, $\hat{\alpha}_{i,q}$, $i = 1, 2, 3, 4$, are class $\mathcal{K}$ functions, and $q = 1, 2, \ldots$. We define $\Omega_{\hat{\rho}_q} \subset D_{NL,q}$ as the stability region of the system of Equation (10) under $h_{NL,q}$ and $\Omega_{\hat{\rho}_{safe,q}}$ as a superset of $\Omega_{\hat{\rho}_q}$ contained in $D_{NL,q}$ and X. Lipschitz continuity of $f_{NL,q}$ and sufficient smoothness of $\hat{V}_q$ imply that there exist $M_{L,q} > 0$ and $L_{L,q} > 0$ such that

$$|\bar{f}_{NL,q}(x, u)| \leq M_{L,q} \tag{13a}$$

$$\left| \frac{\partial \hat{V}_q(x_1)}{\partial x} \bar{f}_{NL,q}(x_1, u) - \frac{\partial \hat{V}_q(x_2)}{\partial x} \bar{f}_{NL,q}(x_2, u) \right| \leq L_{L,q}|x_1 - x_2| \tag{13b}$$

$\forall x, x_1, x_2 \in \Omega_{\hat{\rho}_q}, u \in U_q$, and $q = 1, 2, \ldots$.

Furthermore, we define $\bar{x}_{a,i,q} = x_{a,i} - x_{b,q,s}$ as the variable representing the deviation of each $x_{a,i}$ from the steady-state of the qth empirical model of Equation (10) and $\bar{f}_{i,q}$ as the right-hand side of Equation (1) when the model is rewritten in terms of the deviation variables $\bar{x}_{a,i,q}$ and $\bar{u}_q$, as follows:

$$\dot{\bar{x}}_{a,i,q} = \bar{f}_{i,q}(\bar{x}_{a,i,q}(t), \bar{u}_q(t), w_i(t)) \tag{14}$$

We assume that the following holds:

$$|\bar{f}_{i,q}(x, u', w) - \bar{f}_{i,q}(x', u', 0)| \leq L_{x,i,q}|x - x'| + L_{w,i,q}|w| \tag{15}$$

$$\left| \frac{\partial \hat{V}_q(x)}{\partial x} \bar{f}_{i,q}(x, u', w) - \frac{\partial \hat{V}_q(x')}{\partial x} \bar{f}_{i,q}(x', u'', 0) \right| \leq L'_{x,i,q}|x - x'| + L'_{w,i,q}|w| \tag{16}$$

for all x, x', u', u'' and w such that $x + x_{b,q,s} - x_{a,i,s} \in \Omega_{\rho_i}$, $x' + x_{b,q,s} - x_{a,i,s} \in \Omega_{\rho_i}$, $u' + u_q \in U$, $u'' + u_q \in U$, and $w \in W_i$. We define a level set of $\hat{V}_q$ contained in $\Omega_{\hat{\rho}_{safe,q}}$ that is also contained in Ω_{ρ_i} to be $\Omega_{\hat{\rho}_{q,i}}$, and $L_{x,i,q}, L_{w,i,q}, L'_{x,i,q}, L'_{w,i,q} > 0$

2.3. Economic Model Predictive Control

Economic model predictive control (EMPC) [12] is an optimization-based control design formulated as follows:

$$\min_{\bar{u}_i \in S(\Delta)} \int_{t_k}^{t_{k+N}} L_e(\bar{x}_{a,i}(\tau), \bar{u}_i(\tau)) d\tau \tag{17}$$

$$\text{s.t.} \ \dot{\bar{x}}_{a,i}(t) = \bar{f}_i(\bar{x}_{a,i}(t), \bar{u}_i(t), 0) \tag{18}$$

$$\bar{x}_{a,i}(t_k) = x(t_k) \tag{19}$$

$$\bar{u}_i(t) \in U_i, \ \forall t \in [t_k, t_{k+N}) \tag{20}$$

$$\bar{x}_{a,i}(t) \in X_i, \ \forall t \in [t_k, t_{k+N}) \tag{21}$$

where $L_e(\cdot,\cdot)$ represents the stage cost of the EMPC, which can be a general scalar-valued function that is optimized in Equation (17). The notation $u \in S(\Delta)$ signifies that u is a piecewise-constant input trajectory with period Δ. The prediction horizon is denoted by N. Equation (18) represents the nominal process model, with predicted state $\tilde{x}_{a,i}$ for the ith model. Equations (20) and (21) represent the input and state constraints, respectively. We denote the optimal solution of an EMPC at t_k by $u_p^*(t_j|t_k)$, $p = 1,\ldots,m$, $j = k,\ldots,k+N-1$, where each $u_p^*(t_j|t_k)$ holds for $t \in [t_j, t_{j+1})$ within the prediction horizon. $x(t_k)$ in Equation (19) signifies that the state measurement represents the actual system state at t_k placed in deviation variable form with respect to $\tilde{x}_{a,i,s}$. Due to the potential switching of the underlying process dynamics before the model in Equation (18) is updated, the measurement may come from a dynamic system different than the ith model used in Equation (18).

2.4. Lyapunov-Based Economic Model Predictive Control

A variety of variations on the general EMPC formulation in Equations (17)–(21) have been developed. One such variation which will receive focus in this paper is Lyapunov-based EMPC (LEMPC) [24], which is formulated as in Equations (17)–(21) but with the following Lyapunov-based constraints added as well:

$$V_i(\tilde{x}_{a,i}(t)) \leq \rho_{e,i}, \ \forall\, t \in [t_k, t_{k+N}), \ \text{if } t_k \leq t' \text{ and } V_i(x(t_k)) \leq \rho_{e,i} \tag{22}$$

$$\frac{\partial V_i(x(t_k))}{\partial x} \tilde{f}_i(x(t_k), u(t_k), 0) \leq \frac{\partial V_i(x(t_k))}{\partial x} \tilde{f}_i(x(t_k), h_i(x(t_k)), 0), \tag{23}$$
$$\text{if } t_k > t' \text{ or } V_i(x(t_k)) > \rho_{e,i}$$

where $\Omega_{\rho_{e,i}} \subset \Omega_{\rho_i}$ is selected such that the closed-loop state is maintained within Ω_{ρ_i} over time when the process of Equation (1) is operated under the LEMPC of Equations (17)–(23). t' is a time after which the constraint of Equation (23) is always applied, regardless of the value of $V_i(x(t_k))$. The activation conditions of the LEMPC with respect to $V_i(x(t_k))$ ensure that the LEMPC can maintain closed-loop stability within Ω_{ρ_i} as well as recursive feasibility.

2.5. Lyapunov-Based Economic Model Predictive Control with an Empirical Model

Several prior works have developed LEMPC formulations including empirical models [43,44] when the model of Equation (1) is either unknown or undesirable for use (e.g., more computationally intensive than an empirical model). They have the following form:

$$\min_{\bar{u}_q(t) \in S(\Delta)} \int_{t_k}^{t_{k+N}} [L_e(\tilde{x}_{b,q}(\tau), \bar{u}_q(\tau))]d\tau \tag{24a}$$

$$\text{s.t.} \quad \dot{\tilde{x}}_{b,q} = \tilde{f}_{NL,q}(\tilde{x}_{b,q}(t), \bar{u}_q(t)) \tag{24b}$$

$$\tilde{x}_{b,q}(t_k) = x(t_k) \tag{24c}$$

$$\tilde{x}_{b,q}(t) \in X_q, \ \forall\, t \in [t_k, t_{k+N}) \tag{24d}$$

$$\bar{u}_q(t) \in U_q, \ \forall\, t \in [t_k, t_{k+N}) \tag{24e}$$

$$\hat{V}_q(\tilde{x}_{b,q}(t)) \leq \hat{\rho}_{e,q}, \ \forall\, t \in [t_k, t_{k+N}) \text{ if } x(t_k) \in \Omega_{\hat{\rho}_{e,q}} \tag{24f}$$

$$\frac{\partial \hat{V}_q(x(t_k))}{\partial x}(\tilde{f}_{NL,q}(x(t_k), u(t_k))) \leq$$
$$\frac{\partial \hat{V}_q(x(t_k))}{\partial x}(\tilde{f}_{NL,q}(x(t_k), h_{NL,q}(x(t_k)))) \text{ if } x(t_k) \notin \Omega_{\hat{\rho}_{e,q}}$$
$$\text{or } t_k \geq t' \tag{24g}$$

where the notation follows that found in Equations (17)–(23) except that the predictions from the nonlinear empirical model are denoted by $\tilde{x}_{b,q}$ (Equation (24b)) and are initialized from a measurement of the state of the ith system of Equation (1) (i.e., from the state measurement of whichever model describes the process dynamics at t_k). Regardless of which dynamic model describes the underlying process dynamics, the qth empirical model along with the state (Equation (24d)) and Lyapunov-based stability constraints corresponding to that model are used.

3. Responsive Economic Model Predictive Control Design

The next sections present two concepts for moving toward interpretability and verifiability goals for EMPC, cast within a framework of making EMPC more responsive to "unexpected" behavior.

3.1. Automated Control Law Redesign

In this section, we focus on a case in which the process model used does not change over time (i.e., the $i = 1$ process model in Equation (1) is used for all time) and consider the problem that, despite the pushes toward next-generation manufacturing, many companies that may benefit from automation can have difficulty implementing the appropriate advances if they do not have a knowledgeable control engineer on site due to both a lack of knowledge of advanced control as well as a lack of interpretability of the controller's actions. We present one idea for making an LEMPC easier to work with by giving it a "self-design" capability that allows the controller to update its formulation in a manner that satisfies end-user requirements without requiring understanding of the control laws on the part of the end users. Critically, closed-loop stability and recursive feasibility guarantees are retained. This can be considered to be a case in which the human response to the operating strategy is "unexpected" (in the sense that it is not easily predictable by the control designer), but the controller must have the ability to adjust its control law in response to the human reaction.

The first step toward designing an appropriate controller for this scenario is to recognize that the human response to the process behavior is some function of the pattern observed in the state and input data and that the pattern is dictated by the control formulation. For EMPC, for example, it is dictated by the constraints and objective function (though the process model of Equation (18) also plays a role in determining the response, we consider that the model must represent the process at hand and that therefore it cannot be tuned to impact the state/input behavior). Conceptually, then, the solution to handling the "unexpected" response of the end user of the controller is to learn the mapping between the end user's satisfaction with the response and the constraint/objective function formulation and then to use that mapping to find the constraint/objective function formulation that provides "optimal" satisfaction to the end user.

An open question is how to do this and, in particular, how to do it in a manner that provides theoretical guarantees on feasibility/closed-loop stability. To demonstrate this challenge, consider the LEMPC of Equations (17)–(23). The theoretical results for LEMPC which guarantee closed-loop stability and recursive feasibility under sufficient conditions when no changes occur in the underlying process dynamics rely on the constraints of Equations (22) and (23) being present in the control design [24]. Therefore, ad hoc constraint development in an attempt to optimize end-user "satisfaction" with the process response would not be a means for providing closed-loop stability and recursive feasibility guarantees. Instead, any modification of constraints must take place in a more rigorously defined manner.

One approach would be to develop constraints for EMPC which allow "tuning" of the process response but impact neither closed-loop stability nor feasibility as the tuning parameter in these constraints is adjusted. They thus offer some flexibility to the end user in modifying the response but also ensure that the end user's power to adjust the control law is appropriately restricted for feasibility/stability purposes.

An example of constraints which meet this requirement is the input rate of change constraints added to LEMPC in Reference [45]. In the following section, we will discuss in detail how these constraints may be incorporated within the proposed framework for providing an end user with a restricted flexibility in adjusting the process response without losing theoretical properties of LEMPC.

Remark 1. *The question of how the human response may be accurately sensed is outside the scope of the present manuscript. A process example will be provided below in which the end user is assumed to take time to rank his or her "satisfaction" with the process behavior under a number of different controllers to develop a mapping between satisfaction and the tuning parameter of the control law. However, human responses could also be considered to be obtained through other machine learning/artificial intelligence methods, such as sentiment analysis [46].*

Remark 2. *Potential benefits of an approach that adjusts the controller's behavior based on the end user's response (rather than assuming that some type of standard metric for evaluating control performance (e.g., settling time, rise time, or overshoot of the steady-state) is able to capture the desired response) are that (1) EMPC may operate processes in a potentially time-varying fashion, meaning that the closed-loop state may not be driven to a steady-state and that the behavior of the process under the EMPC may not be easily predictable a priori (e.g., without running closed-loop simulations). Therefore, determining what metrics to use to state whether performance under EMPC is acceptable or not may not be intuitive or easily generalizable, unlike in the case where steady-state operation is desired. (2) Again, unlike the steady-state case, not all end users of a given EMPC formulation may have the same definition of "good" behavior. Ideally, the "best" behavior is the one computed by the EMPC when it optimizes the process economics over the prediction horizon in whatever manner is necessary to ensure that the constraints are met but profit is maximized. However, an end user may not find this to constitute the "best" behavior due to other considerations that are perhaps difficult or costly to include in the control law (for example, the most profitable input trajectories from the perspective of the profit metric being used in Equation (17) may be expected to lead to more actuator wear than is desirable, which will be the subject of the example below). Therefore, it may be difficult to set a general metric on "good" behavior under EMPC, as the additional considerations defining "goodness" that are not directly included in the control law may vary between processes. (3) The concept of designing a controller that is responsive to unexpected evaluations of its behavior could have broader implications, if appropriately developed, than the initial goal of achieving desired process behavior for a given control law. Ideally, developments in this direction would serve as a springboard for reducing a priori control design efforts while increasing flexibility for next-generation manufacturing such that end users are able to achieve many goals during production that they may conceive over time as being important to their operation but without needing to interface extensively with vendors or even needing to update their software to achieve these updated process responses. The vision is one where modifications for manufacturing could become as flexible and safe through new responsive and intelligent controller formulations as modifications to codes are for computer scientists who do not work with physical processes and therefore can readily test and evaluate new protocols to advance the field quickly.*

3.1.1. LEMPC with Self-Designing Input Rate of Change Constraints

In Reference [45], an LEMPC formulation with input rate of change constraints was designed with the form in Equations (17)–(23) but with the following rate of change constraints added on the inputs:

$$|u_p(t_k) - h_{1,p}(x(t_k))| \le \epsilon_r, \ p = 1, \ldots, m \tag{25}$$

$$|u_p(t_j) - h_{1,p}(\tilde{\tilde{x}}_{a,i}(t_j))| \le \epsilon_r, \ p = 1, \ldots, m, \ j = k+1, \ldots, k+N-1 \tag{26}$$

where $\epsilon_r \geq 0$. This formulation is demonstrated in Reference [45] to maintain closed-loop stability and recursive feasibility under sufficient conditions and to cause the following constraints to be met:

$$|u_p^*(t_k|t_k) - u_p^*(t_{k-1}|t_{k-1})| \leq \epsilon_{\text{desired}}, \ \forall \ p = 1, \ldots, m \tag{27}$$

$$|u_p^*(t_j|t_k) - u_p^*(t_{j-1}|t_k)| \leq \epsilon_{\text{desired}}, \ \forall \ p = 1, \ldots, m, \ j = k+1, \ldots, k+N-1 \tag{28}$$

where $\epsilon_{\text{desired}} > 0$. The goal of this formulation of LEMPC is to utilize input rate of change constraints to attempt to reduce variations in the inputs between sampling periods that have the potential to cause actuator wear.

However, as noted in Reference [47], despite the intent of the method to prevent actuator wear, there is no explicit relationship between $\epsilon_{\text{desired}}$ or ϵ_r and the amount of actuator wear. Therefore, a control engineer seeking to prevent actuator wear for a given process under the LEMPC of Equations (17)–(23), (25), and (26) might design the value of ϵ_r by performing closed-loop simulations of the process under various values of ϵ_r and then by selecting the one that gives the response that the engineer judges to present a sufficient tradeoff between optimizing economic performance and reducing actuator wear. A company with little control expertise on hand, however, may have difficulties with tuning ϵ_r without vendor assistance. The fact that controllers today cannot readily "fix" their response if engineers who do not have control expertise would like the response to have different characteristics presents a hurdle to the adoption of even simple control laws, let alone the more complex designs which we would like to move into widespread use as part of the next-generation manufacturing paradigm.

These potential negative responses to a lack of on-site control expertise might be prevented by allowing the controller itself to be responsive to end-user preferences. For example, the value of ϵ_r might be designed by allowing a short period of operation under the control law of Equations (17)–(23), (25), and (26) with different values of ϵ_r. The engineers at the plant could then look at time periods in the plant data during which each of the values of ϵ_r were used and could evaluate the performance of the plant through some metric that can be recorded. Then, the value of ϵ_r that is predicted to provide the highest rate of satisfaction (based on some relationship between the value of ϵ_r and the evaluation metrics which can be derived through techniques for fitting appropriate models to the kind of data generated, such as regression or other techniques of machine learning) could be selected for use (and further updated over time through a similar mechanism as necessary).

Remark 3. *One could argue that the algorithm by which a control engineer judges whether a given value of ϵ_r is preferable could be represented mathematically (e.g., as an optimization problem with an objective function representing a tradeoff between penalties on input variation and loss of profit). However, for the reasons noted in Remark 2 above and also with the goal of developing an algorithm which may facilitate interpretability of LEMPC by allowing its control law to be self-adjusted based on how end users feel about the response of the process under the controller, we handle this within the general case of "unexpected" scenarios to which we would like to make EMPC responsive.*

LEMPC with Self-Designing Input Rate of Change Constraints: Theoretical Guarantees

The methodology proposed above incorporates human judgments on the process response for different values of ϵ_r for setting ϵ_r in Equations (17)–(23), (25), and (26). Despite the fact that human judgment is imprecise, the LEMPC formulations of Equations (17)–(23), (25), and (26), by design, maintains closed-loop stability and recursive feasibility under sufficient conditions (proven in Reference [45]) that are unrelated to the value of ϵ_r, demonstrating that the combination of control theory and data-driven models for "unexpected" behavior or human intuition may be possible to achieve with theoretical guarantees.

When the proposed strategy for evaluating ϵ_r online via human responses to different values of the parameter ϵ_r is used, closed-loop stability and feasibility still hold; however, it may not be guaranteed that Equations (27) and (28) hold. Since $\epsilon_{\text{desired}}$ is arbitrary in many respects since it is indirectly tied to actuator wear (primarily though human evaluation), the satisfaction of Equations (27) and (28) may not be significant during the time period that an operator or engineer is evaluating ϵ_r.

There is no guarantee that the proposed method will produce a value of ϵ_r that gives "optimal satisfaction" to the end user. However, this is not considered a limitation of the method, as the end user's satisfaction is subjective and various methods for modeling the relationship between ϵ_r and the end user's satisfaction could be examined if one is found to produce an inadequate result. The value of ϵ_r can also be adjusted further over time if the response after an initial value of ϵ_r is chosen is determined not to be preferable. Reference [45] does guarantee however that, throughout all of the time of operation (both when various values of ϵ_r are tested and when a single value of ϵ_r is selected), closed-loop stability and recursive feasibility can be guaranteed. This is because the value of ϵ_r only impacts whether Equations (27) and (28) are satisfied under the LEMPC of Equations (17)–(23), (25), and (26), and Equations (27) and (28) are only of potential concern for actuator wear and not closed-loop stability or feasibility. Furthermore, because Reference [45] demonstrates that $h_i(\tilde{x}_{a,i}(t_q)), \forall t \in [t_q, t_{q+1}), q = k, \ldots, k + N - 1$ is a feasible solution to Equations (17)–(23), (25), and (26) at every sampling time regardless of the value of ϵ_r because Equations (25) and (26) can be satisfied by $h_i(\tilde{x}_{a,i}(t_q)), t \in [t_q, t_{q+1}), q = k, \ldots, k + N - 1$ for any $\epsilon_r \geq 0$, the value of ϵ_r can change between two sampling periods as ϵ_r is being evaluated and recursive feasibility (and therefore closed-loop stability, since closed-loop stability depends on Equations (22) and (23) and not on Equations (25) and (26)) will be maintained. Finally, though when ϵ_r is being evaluated, the process profit or actuator wear level may not be the same as they would be after the value of ϵ_r is selected, this is not expected to pose significant problems for many processes if it is performed over a short period of time. Furthermore, if there are hard process constraints defined by X_i that must be met in order to ensure that the product produced during the time when ϵ_r is evaluated can be sold, these can be met even as various values of ϵ_r are tried because $\tilde{x}_{a,i}(t) \in \Omega_{\rho_i} \subseteq X_i$ according to Reference [45] for any value of ϵ_r. Furthermore, Reference [45] also guarantees that, even as the values of ϵ_r are adjusted, the closed-loop state can be driven to a neighborhood of a steady-state to avoid production volume losses as ϵ_r is adjusted if necessary.

Remark 4. *The fact that the above stability analysis holds regardless of the value of ϵ_r indicates that the accuracy of the method used in obtaining ϵ_r does not impact closed-loop stability. This is particularly important if the method used in obtaining ϵ_r involves, for example, performing sentiment analysis of human speech data to determine how well humans like a given value of that parameter. We overcome the limitation of interfacing humans with machines by ensuring that the only parameter of the control law design which is modified in response to the algorithm that carries uncertainty is one which, deterministically, does not impact closed-loop stability.*

Remark 5. *Though this section on automated control law redesign has explored only input rate of change constraints, other online redesigns may also be possible in control. For example, in the LEMPC formulation of Equations (17)–(23), the value $\rho_{e,i}$ could be modified over time if an appropriate implementation strategy was developed. Specifically, there exist bounds on $\rho_{e,i}$ given in Reference [24] which are required for closed-loop stability to be maintained for the process of Equation (1) operated under the LEMPC of Equations (17)–(23). Given this, a similar strategy to that presented for the selection of ϵ_r could be utilized to adjust the value of $\rho_{e,i}$ within its bounds online without impacting closed-loop stability. This holds because a value of $\rho_{e,i}$ between the minimum and maximum at a given time would always be utilized. According to Reference [24], the consequence of this is that, at the next sampling time, $\tilde{x}_{a,i}(t_k) \in \Omega_{\rho_i}$. If $\tilde{x}_{a,i}(t) \in \Omega_{\rho_i}$ at the end of every sampling period for any $\rho_{e,i}$ between its minimum and maximum, $\tilde{x}_{a,i}(t) \in \Omega_{\rho_i}$ at all times. If both ϵ_r and $\rho_{e,i}$ were to be simultaneously varied, for example, closed-loop* *stability would again hold, as the value of ϵ_r does not impact closed-loop stability for the reasons noted above and the value of $\rho_{e,i}$ can vary between its minimum and maximum value as just described without impacting closed-loop stability. Recursive feasibility would also not be impacted. This suggests that it may be possible to design more complex control laws with multiple self-tuning parameters that are simultaneously optimized based on human response to develop control laws that behave in a desirable manner online without posing a safety concern due to loss of closed-loop stability.*

EMPC with Self-Designing Input Rate of Change Constraints: Application to a Chemical Process Example

In this section, we employ a process example that demonstrates the concept of self-designing input rate of change constraints. For simplicity, in this example, we do not employ the Lyapunov-based stability constraints of Equations (22) and (23); therefore, no theoretical stability guarantees can be made for this example. However, this does not present problems for illustrating the core concepts of the method of integrating human responses to operating conditions with EMPC.

The process under consideration is an ethylene oxidation process in a continuous stirred tank reactor (CSTR) from Reference [48] with reaction rates from Reference [49]. The following three reactions are considered to occur in the CSTR:

$$C_2H_4 + \tfrac{1}{2}O_2 \rightarrow C_2H_4O \tag{29}$$

$$C_2H_4 + 3O_2 \rightarrow 2CO_2 + 2H_2O \tag{30}$$

$$C_2H_4O + \tfrac{5}{2}O_2 \rightarrow 2CO_2 + 2H_2O \tag{31}$$

Mass and energy balances for the reactor, in dimensionless form, are as follows:

$$\dot{\bar{x}}_1 = \bar{u}_1(1 - \bar{x}_1\bar{x}_4) \tag{32}$$

$$\dot{\bar{x}}_2 = \bar{u}_1(\bar{u}_2 - \bar{x}_2\bar{x}_4) - A_1 e^{\gamma_1/\bar{x}_4}(\bar{x}_2\bar{x}_4)^{0.5} - A_2 e^{\gamma_2/\bar{x}_4}(\bar{x}_2\bar{x}_4)^{0.25} \tag{33}$$

$$\dot{\bar{x}}_3 = -\bar{u}_1\bar{x}_3\bar{x}_4 + A_1 e^{\gamma_1/\bar{x}_4}(\bar{x}_2\bar{x}_4)^{0.5} - A_3 e^{\gamma_3/\bar{x}_4}(\bar{x}_3\bar{x}_4)^{0.5} \tag{34}$$

$$\dot{\bar{x}}_4 = \frac{\bar{u}_1}{\bar{x}_1}(1 - \bar{x}_4) + \frac{B_1}{\bar{x}_1}e^{\gamma_1/\bar{x}_4}(\bar{x}_2\bar{x}_4)^{0.5} + \frac{B_2}{\bar{x}_1}e^{\gamma_2/\bar{x}_4}(\bar{x}_2\bar{x}_4)^{0.25} + \frac{B_3}{\bar{x}_1}e^{\gamma_3/\bar{x}_4}(\bar{x}_3\bar{x}_4)^{0.5} - \frac{B_4}{\bar{x}_1}(\bar{x}_4 - T_c) \tag{35}$$

where the process model parameters are listed in Table 1; the state vector components $\bar{x}_1$, $\bar{x}_2$, $\bar{x}_3$, and $\bar{x}_4$ (i.e., $\bar{x} = [\bar{x}_1\ \bar{x}_2\ \bar{x}_3\ \bar{x}_4]^T$) are dimensionless quantities corresponding to the gas density, ethylene concentration, ethylene oxide concentration, and temperature in the CSTR, respectively; and the input vector components $\bar{u}_1$ and $\bar{u}_2$ are dimensionless quantities corresponding to the feed volumetric flow rate and the feed ethylene concentration. The process of Equations (32)–(35) has a steady-state at $\bar{x}_1 = 0.998$, $\bar{x}_2 = 0.424$, $\bar{x}_3 = 0.032$, $\bar{x}_4 = 1.002$, $\bar{u}_1 = 0.35$, and $\bar{u}_2 = 0.5$.

Table 1. Parameters for the continuous stirred tank reactor (CSTR) of Equations (32)–(35).

Parameter	Value
γ_1	-8.13
γ_2	-7.12
γ_3	-11.07
A_1	92.80
A_2	12.66
A_3	2412.71
B_1	7.32
B_2	10.39
B_3	2170.57
B_4	7.02
T_C	1.0

An EMPC is designed to control this process by maximizing the yield of ethylene oxide, which is defined by the following equation over a time interval from the initial time ($t_0 = 0$) to the final time of operation t_f:

$$Y(t_f) = \frac{\int_0^{t_f} \tilde{u}_1(\tau)\tilde{x}_3(\tau)\tilde{x}_4(\tau)d\tau}{\int_0^{t_f} \tilde{u}_1(\tau)\tilde{u}_2(\tau)d\tau} \tag{36}$$

However, it is assumed that, in addition to the following bounds on the inputs,

$$0.0704 \leq \tilde{u}_1 \leq 0.7042 \tag{37}$$

$$0.2465 \leq \tilde{u}_2 \leq 2.4648 \tag{38}$$

there is also a constraint on the total amount of material which can be fed to the CSTR over time:

$$\int_0^{t_f} \tilde{u}_1(\tau)\tilde{u}_2(\tau)d\tau = 0.175 t_f \tag{39}$$

As Equation (39) fixes the denominator of Equation (36), the stage cost to be minimized using the EMPC is as follows:

$$L_e(x, u) = -\tilde{u}_1(t)\tilde{x}_3(t)\tilde{x}_4(t) \tag{40}$$

To attempt to avoid actuator wear, input rate of change constraints will also be considered. The general form of the EMPC for this example is therefore as follows:

$$\min_{\tilde{u}_1,\tilde{u}_2 \in S(\Delta)} \int_{t_k}^{t_{k+N_k}} -\tilde{u}_1(\tau)\tilde{x}_3(\tau)\tilde{x}_4(\tau)d\tau \tag{41}$$

$$\text{s.t. Equations (32)–(35)} \tag{42}$$

$$\tilde{x}(t_k) = \hat{x}(t_k) \tag{43}$$

$$0.0704 \leq \tilde{u}_1(t) \leq 0.7042, \ \forall \, t \in [t_k, t_{k+N_k}) \tag{44}$$

$$0.2465 \leq \tilde{u}_2(t) \leq 2.4648, \ \forall \, t \in [t_k, t_{k+N_k}) \tag{45}$$

$$\frac{1}{t_v}\int_{rt_v}^{t_k} \tilde{u}_1^*(\tau)\tilde{u}_2^*(\tau)d\tau + \frac{1}{t_v}\int_{t_k}^{t_{k+N_k}} \tilde{u}_1(\tau)\tilde{u}_2(\tau)d\tau = 0.175 \tag{46}$$

$$|\tilde{u}_p(t_j) - \tilde{u}_p(t_{j-1})| \leq \epsilon, \ j = k,\ldots,k+N_k - 1, \ p = 1,2 \tag{47}$$

In this formulation, no Lyapunov-based stability constraints are employed and no closed-loop stability issues arose in the simulations (i.e., the closed-loop state always remained within a bounded region of state-space). Furthermore, due to the lack of Lyapunov-based stability constraints, the input rate of change constraints of Equations (27) and (28) are enforced directly on input differences (i.e., they have the form of Equations Equations (27) and (28) rather than the form of Equations (25) and (26)). $\tilde{x}$ represents the predicted value of the process state according to the model of Equation (42). $\tilde{u}_1^*$ and $\tilde{u}_2^*$ represent the optimal values of $\tilde{u}_1$ and $\tilde{u}_2$ that have been applied in past sampling periods (i.e., $\tilde{u}_1^* = \tilde{u}_1(t_{k-1})$, and $\tilde{u}_2^* = \tilde{u}_2(t_{k-1})$). The values of $\tilde{u}_1(t_{k-1})$ and $\tilde{u}_2(t_{k-1})$ for $k = 0$ are assumed to be the steady-state values of these inputs. N_k is a shrinking prediction horizon in the sense that, at the beginning of every operating period of length $t_v = 46.8$, the value of N_k is reset to 5 but is then reduced by 1 at each subsequent sampling time of the operating period. This shrinking horizon allows the constraint of Equation (39) to be enforced within every operating period to ensure that, by the end of the time of operation, Equation (39) is met. In Equation (46), r signifies the operating periods completed since the beginning of the time of operation (e.g., in the first t_v time units, $r = 0$ because no operating periods have been completed yet).

We assume that the engineers and operators do not know the value of ϵ that they would like to impose in the EMPC of Equations (41)–(46) but plan to determine an appropriate value by assessing the process behavior from the same initial condition under EMPC's with different values of ϵ and by selecting a value that they expect will give the optimal tradeoff between economic performance and actuator wear reduction. To represent the process behavior as ϵ is varied in these experiments, we performed eight closed-loop simulations of the process of Equations (32)–(35) under the EMPC of Equations (41)–(46) from the same initial condition $\tilde{x}_I = [\tilde{x}_{1I}\ \tilde{x}_{2I}\ \tilde{x}_{3I}\ \tilde{x}_{4I}]^T = [0.997\ 1.264\ 0.209\ 1.004]^T$ using eight different input-rate-of-change constraint formulations (the simulations were performed both with no input rate of change constraints and with ϵ values of 0.01, 0.05, 0.1, 0.3, 0.5, 1, and 3). The simulations lasted for 10 operating periods and used a sampling period of $\Delta = 9.36$, an integration step for the model of Equation (42) (i.e., the model used by the controller) of 10^{-4} and an integration step for the model of Equations (32)–(35) (i.e., the model of the plant) of 10^{-5}. The open-source interior point solver Ipopt [50] was used to solve all optimization problems. Figures 1 and 2 show the state and input trajectories for each of the values of ϵ chosen. Table 2 shows how the yield varies with the choice of ϵ. To express the engineer's or operator's judgment of the relative "goodness" of the response that they see when both profit and input variations are considered, the engineers and operators are considered to have ranked the response for a given ϵ on a scale of 1 to 10 as shown in Table 2, with 1 being the worst and 10 being the best.

Figure 3 shows the rankings as a function of ϵ as solid blue circles. From this figure, we postulate that a model that may fit this data has the following form:

$$\text{Ranking} = c_1 e^{(-c_2\epsilon)} \epsilon^{c_3} + c_4 \tag{48}$$

Using the MATLAB function lsqcurvefit, the data from Table 2 for the various values of ϵ reported was fit to the function in Equation (48), resulting in $c_1 = 68.8901$, $c_2 = 3.8356$, $c_3 = 0.8480$, and $c_4 = 0.7933$. The plot of the function fit to the data is shown as the red curve on Figure 3. A more rigorous method could have been utilized to fit the model and the data (involving, for example, more samples and an evaluation of the deviation of the model from the data), but the present method is sufficient for demonstrating the concepts developed in this work.

The utility of the function in Equation (48) is that it provides a mathematical representation of the model that an engineer or operator is using within his or her mind to determine the best value of ϵ to utilize when this engineer or operator is not aware of the model himself or herself. This makes the advanced control design more tractable for the operator or engineer to utilize without advanced control knowledge by fitting the "mind of the human" to a function that can then be utilized in optimizing the control design automatically. To demonstrate this, we determine the "optimal" value of ϵ based on the model of Equation (48) by differentiating the equation with respect to ϵ and by setting it to 0. This gives an "optimal" value of ϵ of c_3/c_2 or 0.22. Simulations were performed for 10 operating periods of the process of Equations (32)–(35) under the EMPC of Equations (41)–(46) with this value of ϵ and initialized from $\tilde{x}_I$, and the resulting state and input trajectories are shown in Figures 4 and 5. The yield is 8.33%.

Table 2. Yield variation with ϵ.

ϵ	Yield (%)	Ranking
0.01	7.17	2
0.05	7.93	5
0.1	8.23	8
0.3	8.37	8
0.5	8.44	7
1	9.03	2
3	9.61	1
No input rate of change constraint	9.61	Not ranked

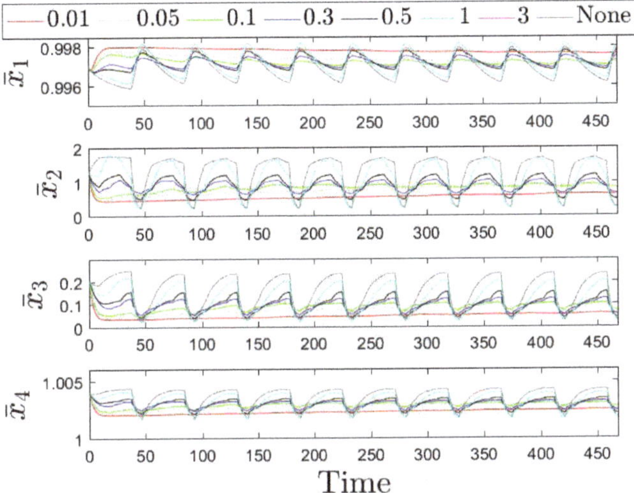

Figure 1. $\bar{x}_1$, $\bar{x}_2$, $\bar{x}_3$, and $\bar{x}_4$ trajectories under economic model predictive controllers (EMPCs) with different values of ϵ specified in the legend (the gray trajectory labeled "None" corresponds to no input rate of change constraint applied).

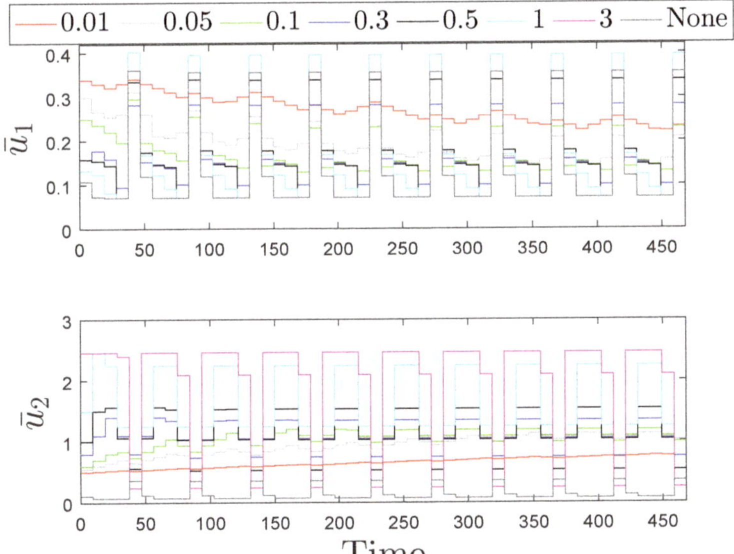

Figure 2. $\bar{u}_1$ and $\bar{u}_2$ trajectories under EMPCs with different values of ϵ specified in the legend (the gray trajectory labeled "None" corresponds to no input rate of change constraint applied).

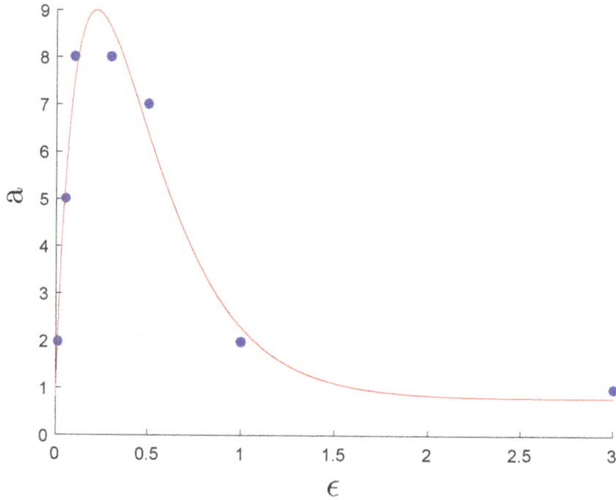

Figure 3. Scatter plot reflecting rankings in Table 2 (solid blue circles) and the curve fit using lsqcurvefit (solid red line).

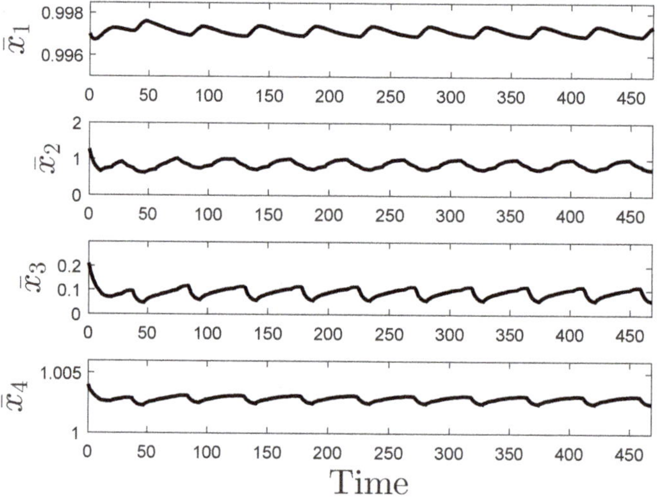

Figure 4. State trajectories under EMPC with $\epsilon = 0.22$.

Remark 6. *The rankings in Table 2 are fabricated to demonstrate the concept that a human judgment could be translated to a modification of an EMPC formulation parameter. They were contrived to display a form to which a reasonable model could be readily fit using lsqcurvefit and, furthermore, are highly simplified (e.g., only a single ranking is provided for each value of ϵ rather than an average ranking with additional information such as standard deviation that might be expected if more than one individual was to rank the response). For an actual process,* *the transformation of human opinion on the response to a function of ϵ would therefore be expected to be more complex and to potentially involve statistics-based techniques or other methods for obtaining models from process data; however, an investigation of such methods is outside of the scope of this paper, and therefore, a simplified ranking model was used to demonstrate the concept that a control law parameter might be decided upon by evaluating characteristics of a response where there is a tradeoff between competing operating objectives where at least one of them (in this case, the actuator wear) is more difficult to quantify with a simple model such that the incorporation of human judgment can make the control law design potentially simpler (than if, for example, a detailed actuator wear model was to be developed to allow the controller to more accurately predict the wear itself to then prevent it through a constraint on wear rather than input rate of change).*

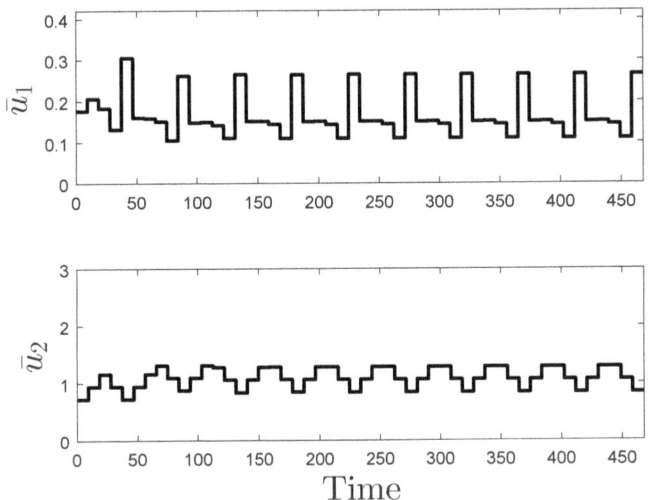

Figure 5. Input trajectories under EMPC with $\epsilon = 0.22$.

3.2. EMPC Response to Unexpected Scenarios via Model Updates

A second case for which we will explore EMPC designs which are responsive to unexpected events considers these "unexpected" events to be defined by a change in the underlying process dynamics (i.e., the value of i increases in Equation (1)). This class of problems covers anomaly responses for EMPC, for which we will adopt the common anomaly-handling strategy (as described in the Introduction section) of updating the process model. Mathematically, we assume that the process model was known with reasonable accuracy before the anomaly (i.e., there is an upper bound on the error between the model used in the LEMPC and the model of Equation (1) with $i = 1$).

We make several points with respect to model updates in this section. First, if the underlying dynamics change, it is possible that the structure of the underlying dynamic model has fundamentally changed. When identifying a new model, it may therefore be preferable to identify the parameters of one with a revised structure; this is a case of seeking to identify a more physics-based model from process data [51]. In keeping with the prior section where the potential was shown for integrating machine learning algorithms known to not be guaranteed to provide accurate data with control, we here highlight that, if machine learning-based sensors (e.g., image-based sensors) are utilized with the process, they may aid in suggesting how to update a process model's structure over time to attempt to keep the structure physically relevant. Because such sensing techniques may not provide correct suggestions, however, a model with a structure suggested by such an algorithm does not need to be automatically implemented in model-based control; instead, engineers could consider multiple models after a machine learning-based algorithm suggests that an anomaly/change in the underlying process model has occurred, where one model to be evaluated is that used until this point and the second is a model that includes any updates implied by the sensing techniques. Subsequently, the prediction accuracy of the two models could be compared, and whichever is most accurate can be considered for use in the LEMPC [52]. Like the methodology in Section 3.1.1, this method limits the ability of any attempts to integrate machine learning (in the sensors) and control from impacting closed-loop stability by using it to complement a rigorous control design approach rather than to dictate it.

Second, at a chemical plant, anomalies may be considered to be either those which pose an immediate hazard to humans and the environment and are considered to require plant shutdown upon detection or those which do not. When the anomaly detected requires plant shutdown, generally the safety system is used to take extreme actions like cutting feeds to shut down the plant as quickly as possible; these generally have a prespecified nature (e.g., closing the feed valve). Anomalies that do not present immediate hazards to humans may either result in sufficiently small plant/model mismatch that the controller is robust against or the plant/model mismatch could cause subsequent control actions to drive the closed-loop state out of the expected region of process operation (at which point, the anomaly may be a hazard). We consider that characterizing conditions under which closed-loop stability is not lost in the second case may constitute steps in moving toward verification of EMPC for the process industries with adaptive model updates in the presence of changing process dynamics.

3.2.1. Automated Response to Anomalies: Formulation and Implementation Strategy

In the next section, we will present theoretical results regarding conditions under which an LEMPC could be conservatively designed to handle anomalies of different types in the sense that closed-loop stability would not be lost upon the occurrence of an anomaly or that impending loss of closed-loop stability could be detected by defining a region $\Omega_{\hat{\rho}_{samp,q}}$ (a superset of $\Omega_{\hat{\rho}_q}$) which the closed-loop state should not leave unless the anomaly has been significant and the model used by the LEMPC should be attempted to be reidentified to try to maintain closed-loop stability. If the closed-loop state leaves $\Omega_{\hat{\rho}_{samp,q}}$, however, it has also left $\Omega_{\hat{\rho}_q}$, so that the LEMPC of Equation (24) may not be feasible. For this reason, the implementation strategy below suggests that, if the closed-loop state leaves $\Omega_{\hat{\rho}_{samp,q}}$, $h_{NL,q}$ should be applied to the process so that a control law with no feasibility issues is used.

The implementation strategy proposed below relies on the existence of two controllers $h_{NL,q}$ and $h_{NL,q+1}$, where $h_{NL,q}$ can stabilize the origin of the nominal closed-loop system of Equation (10) and $h_{NL,q+1}$ can stabilize the origin of the nominal closed-loop system of Equation (10) with respect to the $q+1$th model. Specifically, before the change in the underlying process dynamics that occurs at $t_{s,i+1}$ is detected at $t_{d,q}$, the process is operated under the LEMPC with the qth empirical model. After the change is detected (in a worst case via the closed-loop state leaving $\Omega_{\hat{\rho}_q}$), a worst-case bound $t_{h,q}$ is placed on the time available until the model must be updated at time $t_{ID,q}$ to the $q+1$th empirical model to prevent the closed-loop state from leaving a characterizable operating region.

We consider the following implementation strategy for carrying out the above methodology:

1. At t_0, the $i=1$ first-principles model (Equation (1)) describes the dynamics of the process. The $q=1$ empirical model (Equation (10)) is used to design the LEMPC of Equation (24). An index i_{hx} is set to 0. An index ζ is set to 0. Go to step 2.

2. At $t_{s,i+1}$, the underlying dynamic model of Equation (1) changes to the $i+1$th model. The LEMPC is not yet alerted that the anomaly has occurred; the model used in the LEMPC is not changed despite the change in the underlying process dynamics. Go to step 3.

3. While $t_{s,i+1} < t_k < t_{s,i+2}$, apply a detection method to determine if an anomaly has occurred. If an anomaly is detected, set $\zeta = 1$ and $t_{d,k} = t_k$. Else, $\zeta = 0$. If $x(t_k) \notin \Omega_{\hat{\rho}_q}$ but $\zeta = 0$, set $\zeta = 1$ and $t_{d,k} = t_k$. Go to step 4.

4. If $i_{hx} = 1$, go to step 4a. Else, if $\zeta = 1$, go to step 4b, or if $\zeta = 0$, go to step 4c. If $t_k > t_{s,i+2}$, go to step 5.

    (a) If $x(t_k) \in \Omega_{\hat{\rho}_{q+1}}$, operate the process under the LEMPC of Equation (24) with $q \leftarrow q+1$ and set $i_{hx} = 0$. Else, apply $h_{NL,q+1}(x(t_k))$ to the process. Return to step 3. $t_k \leftarrow t_{k+1}$.

    (b) If $(t_{k+1} - t_{d,q}) < t_{h,q}$, gather online data to develop an improved process model as well as updated functions $\hat{V}_{q+1}$ and $h_{NL,q+1}(x)$ and an updated stability region $\Omega_{\hat{\rho}_{q+1}}$ around the steady-state of the new empirical model but do not yet update the LEMPC and control the process using the prior LEMPC. Else, if $(t_{k+1} - t_{d,q}) \geq t_{h,q}$, set $i_{hx} = 1$ and apply $h_{NL,q+1}(x(t_k))$. Return to step 3. $t_k \leftarrow t_{k+1}$.

    (c) Operate the process under the LEMPC of Equation (24) that was used at the prior sampling time. Return to step 3. $t_k \leftarrow t_{k+1}$.

5. If $t_k > t_{s,i+2}$, a process dynamics change occurred at $t_{s,i+2}$. Set $t_{s,i+1} \leftarrow t_{s,i+2}$ and $t_k \leftarrow t_{k+1}$. Return to step 2 with $\zeta = 0$ and $i_{hx} = 0$. Else, if $t_k < t_{s,i+2}$, $t_k \leftarrow t_{k+1}$; return to step 3.

We note that we do not specify the detection method to be used in step 3, but the use of a sufficiently conservative $\Omega_{\hat{\rho}_q}$ (in a sense to be clarified in the following section) allows a worst-case detection mechanism to be that the closed-loop state exits $\Omega_{\hat{\rho}_q}$ in step 3. We consider that each $t_{s,i+1}$ and $t_{s,i+2}$ are separated by a sufficient period of time such that no second change in the underlying process dynamics occurs before the first change has resulted in an update in the dynamic model and the closed-loop state is within $\Omega_{\hat{\rho}_{q+1}}$.

Remark 7. *A significant difference between the proposed procedure and that in References [53,54], which also involves switched systems under LEMPC, is that Reference [53] assumes that the time at which the model is to be switched is known a priori. In handling of anomalies, this cannot be known; therefore, the proposed approach corresponds to LEMPC for switched systems with unknown switching times. We place bounds in the next section on a number of properties of the LEMPC of Equation (24) for this case to demonstrate the manner in which closed-loop stability guarantees depend on, for example, how large the possible changes in the process model could be when they occur. The goal is to provide a perspective on the timeframes available for detecting various anomalies without loss of closed-loop stability, which could aid in verification and self-design studies for EMPC.*

3.2.2. Automated Response to Anomalies: Stability and Feasibility Analysis

According to the implementation strategy above, when an anomaly occurs that changes the underlying process dynamics, one of two things will happen: (1) the model used in Equation (24b) remains the same or (2) the change in the underlying process dynamics is detected and the model used in Equation (24b) is changed within a required timeframe to a new model (i.e., q is incremented by one in Equation (10)). In this section, we present the conditions under which closed-loop stability can be maintained in either case. For readability, proofs of theorems presented in this section are available in the Appendix.

We first present several propositions. The first defines the maximum difference between the process model of Equation (1) and that of Equation (10) over time when the two models are initialized from the same state, as long as the states of both systems are kept within a level set of $\hat{V}_q$ which is also contained within the stability region around the steady-state for the model of Equation (1) and as long as there is no change in the underlying dynamics. The second sets an upper bound on the difference between the value of $\hat{V}_q$ at any two points in $\Omega_{\hat{\rho}_q}$. The third provides the closed-loop stability properties of the closed-loop system of Equation (10) under the controller $h_{NL,q}$.

Proposition 1 ([51]). *Consider the systems*

$$\dot{\tilde{x}}_{a,i,q} = \bar{f}_{i,q}(\tilde{x}_{a,i,q}(t), \bar{u}_q(t), w_i(t)) \tag{49a}$$

$$\dot{\tilde{x}}_{b,q} = \bar{f}_{NL,q}(\tilde{x}_{b,q}(t), \bar{u}_q(t)) \tag{49b}$$

with initial states $\tilde{x}_{a,i,q}(t_0) = \tilde{x}_{b,q}(t_0) = \tilde{x}(t_0)$ *contained within* $\Omega_{\hat{\rho}_{q,i}}$, *with* $t_0 = 0$, $\bar{u}_q \in U_q$, *and* $w_i \in W_i$. *If* $\tilde{x}_{a,i,q}(t)$ *and* $\tilde{x}_{b,q}(t)$ *remain within* $\Omega_{\hat{\rho}_{q,i}}$ *for* $t \in [0, T]$, *then there exists a function* $f_{W,i,q}(\cdot)$ *such that:*

$$|\tilde{x}_{a,i,q}(t) - \tilde{x}_{b,q}(t)| \leq f_{W,i,q}(t) \tag{50}$$

with:

$$f_{W,i,q}(t) := \frac{L_{w,i,q}\theta_i + M_{err,i,q}}{L_{x,i,q}}(e^{L_{x,i,q}t} - 1) \tag{51}$$

where $M_{err,i,q} > 0$ *is defined by:*

$$|\bar{f}_{i,q}(x, u, 0) - \bar{f}_{NL,q}(x, u)| \leq M_{err,i,q} \tag{52}$$

for all x contained in $\Omega_{\hat{\rho}_{q,i}}$ *and* $u \in U_q$.

Proposition 2 ([24,55]). *Consider the Lyapunov function* $\hat{V}_q(\cdot)$ *of the nominal system of Equation (10) under the controller* $h_{NL,q}(\cdot)$ *that meets Equation (12). There exists a quadratic function* $f_{V,q}(\cdot)$ *such that:*

$$\hat{V}_q(x) \leq \hat{V}_q(x') + f_{V,q}(|x - x'|) \tag{53}$$

for all $x, \tilde{x}' \in \Omega_{\hat{\rho}_{safe,q}}$ *with*

$$f_{V,q}(s) := \hat{\alpha}_{4,q}(\hat{\alpha}_{1,q}^{-1}(\hat{\rho}_q))s + M_{v,q}s^2 \tag{54}$$

where $M_{v,q}$ *is a positive constant.*

Proposition 3 ([51]). *Consider the closed-loop system of Equation (10) under* $h_{NL,q}(\tilde{x}_{b,q})$ *that satisfies the inequalities of Equation (12) in sample-and-hold. Let* $\Delta > 0$, $\hat{\epsilon}_{W,q} > 0$, *and* $\hat{\rho}_{safe,q} > \hat{\rho}_q > \hat{\rho}_{e,q} > \hat{\rho}_{min_q} > \hat{\rho}_{s,q} > 0$ *satisfy the following:*

$$-\hat{\alpha}_{3,q}(\hat{\alpha}_{2,q}^{-1}(\hat{\rho}_{s,q})) + L_{L,q}M_{L,q}\Delta \leq -\hat{\epsilon}_{W,q}/\Delta \tag{55}$$

$$\hat{\rho}_{min_q} := \max\{\hat{V}_q(\tilde{x}_{b,q}(t+\Delta)) \; : \; \hat{V}_q(\tilde{x}_{b,q}(t)) \leq \hat{\rho}_{s,q}\}. \tag{56}$$

If $\tilde{x}_{b,q}(0) \in \Omega_{\hat{\rho}_{safe,q}}$, *then,*

$$\hat{V}_q(\tilde{x}_{b,q}(t_{k+1})) - \hat{V}_q(\tilde{x}_{b,q}(t_k)) \leq -\hat{\epsilon}_{W,q} \tag{57}$$

for $\tilde{x}_{b,q}(t_k) \in \Omega_{\hat{\rho}_{safe,q}}/\Omega_{\hat{\rho}_{s,q}}$ *and the state trajectory* $\tilde{x}_{b,q}(t)$ *of the closed-loop system is always bounded in* $\Omega_{\hat{\rho}_{safe,q}}$ *for* $t \geq 0$ *and is ultimately bounded in* $\Omega_{\hat{\rho}_{min_q}}$.

The next proposition bounds the error between the actual process state and a prediction of the process state using an empirical model initialized from the same value of the process state over a period of time in which the underlying process dynamics change, but the empirical model is not updated. This requires overlap in stability regions for the ith and $i + 1$th models of Equation (1) and for the qth model of Equation (10) within $\Omega_{\hat{\rho}_{q,i}}$ while the qth model is used. The proof of this proposition is available in Appendix A.

Proposition 4. *Consider the following systems:*

$$\dot{\tilde{x}}_{a,i,q} = \tilde{f}_{i,q}(\tilde{x}_{a,i,q}(t), \bar{u}_q(t), w_i(t)) \tag{58}$$

$$\dot{\tilde{x}}_{b,q} = \tilde{f}_{NL,q}(\tilde{x}_{b,q}(t), \bar{u}_q(t)) \tag{59}$$

$$\dot{\tilde{x}}_{a,i+1,q} = \tilde{f}_{i+1,q}(\tilde{x}_{a,i+1,q}(t), \bar{u}_q(t), w_{i+1}(t)) \tag{60}$$

with initial states $\tilde{x}_{a,i,q}(t_0) = \tilde{x}_{b,q}(t_0) \in \Omega_{\hat{\rho}_{q,i}}$ *with* $t_0 = 0$, $\bar{u}_q \in U_q$, $w_i \in W_i$, *and* $w_{i+1} \in W_{i+1}$. *Also,* $\tilde{x}_{a,i,q}(t_{s,i+1}) = \tilde{x}_{a,i+1,q}(t_{s,i+1})$. *If* $\tilde{x}_{a,i,q}(t)$, $\tilde{x}_{b,q}(t)$, $\tilde{x}_{a,i+1,q}(t) \in \Omega_{\hat{\rho}_{q,i}}$ *for* $t \in [0, t_1]$ *and*

$$|\tilde{f}_{i+1,q}(\tilde{x}_{a,i+1,q}(s), \bar{u}_q(s), w_{i+1}(s)) - \tilde{f}_{i,q}(\tilde{x}_{a,i,q}(s), \bar{u}_q(s), w_i(s))| \leq M_{change,i,q} \tag{61}$$

for all $\tilde{x}_{a,i,q}, \tilde{x}_{a,i+1,q} \in \Omega_{\hat{\rho}_{q,i}}$, $\bar{u}_q \in U_q$, $w_i \in W_i$, *and* $w_{i+1} \in W_{i+1}$, *then*

$$|\tilde{x}_{a,i,q}(t) - \tilde{x}_{b,q}(t)| \leq f_{W,i,q}(t) \tag{62}$$

where $f_{W,i,q}(t)$ *is defined in Equation* (51) *for* $t \in [0, t_{s,i+1}]$ *and*

$$|\tilde{x}_{a,i+1,q}(t) - \tilde{x}_{b,q}(t)| \leq f_{W,i,q}(t_{s,i+1} - t_0) + (M_{change,i,q})(t - t_{s,i+1}) + \frac{L_{w,i,q}\theta_i + M_{err,i,q}}{L_{x,i,q}}(e^{L_{x,i,q}t} - e^{L_{x,i,q}t_{s,i+1}}) \tag{63}$$

for $t \in [t_{s,i+1}, t_1]$.

The following theorem provides the conditions under which, when no change in the underlying dynamic model occurs throughout the time of operation and $x(t_k) \in \Omega_{\hat{\rho}_q}$, the LEMPC of Equation (24) designed based on $h_{NL,q}$ and the qth empirical model of Equation (10) guarantees that the closed-loop state is maintained within $\Omega_{\hat{\rho}_q}$ over time and is ultimately bounded in a neighborhood of the origin of the model of Equation (10).

Theorem 1 ([51]). *Consider the closed-loop system of Equation* (1) *under the LEMPC of Equation* (24) *based on the controller* $h_{NL,q}(x)$ *that satisfies the inequalities in Equation* (12)*. Let* $\epsilon_{W,i,q} > 0$, $\Delta > 0$, $N \geq 1$, *and* $\hat{\rho}_q > \hat{\rho}_{e,q} > \hat{\rho}_{min,i,q} > \hat{\rho}_{s,q} > 0$ *satisfy the following:*

$$-\hat{\alpha}_{3,q}(\hat{\alpha}_{2,q}^{-1}(\hat{\rho}_{e,q})) + \hat{\alpha}_{4,q}(\hat{\alpha}_{1,q}^{-1}(\hat{\rho}_q))M_{err,i,q} + L'_{x,i,q}M_i\Delta + L'_{w,i,q}\theta_i \leq -\epsilon_{W,i,q}/\Delta \tag{64}$$

$$\hat{\rho}_{e,q} \leq \hat{\rho}_q - f_{V,q}(f_{W,i,q}(\Delta)) \tag{65}$$

If $x(0) \in \Omega_{\hat{\rho}_q}$ *and Proposition* 3 *is satisfied, then the state trajectory* $\tilde{x}_{a,i,q}(t)$ *of the closed-loop system is always bounded in* $\Omega_{\hat{\rho}_q}$ *for* $t \geq 0$. *Furthermore, if* $t > t'$ *and*

$$-\hat{\alpha}_{3,q}(\hat{\alpha}_{2,q}^{-1}(\hat{\rho}_{s,q})) + \hat{\alpha}_{4,q}(\hat{\alpha}_{1,q}^{-1}(\hat{\rho}_q))M_{err,i,q} + L'_{x,i,q}M_i\Delta + L'_{w,i,q}\theta_i \leq -\epsilon_{W,i,q}/\Delta \tag{66}$$

then the state trajectory $x_{a,i}(t)$ *of the closed-loop system is ultimately bounded in* $\Omega_{\hat{\rho}_{min,i,q}}$ *and defined as follows:*

$$\hat{\rho}_{min,i,q} := \max\{\hat{V}_q(\tilde{x}_{a,i,q}(t + \Delta)) \mid \hat{V}_q(\tilde{x}_{a,i,q}(t)) \leq \hat{\rho}_{s,q}\} \tag{67}$$

The prior theorem provided conditions under which the closed-loop state is maintained within $\Omega_{\hat{\rho}_q}$ in the absence of changes in the dynamic model. In the following theorem, we provide sufficient conditions under which the closed-loop state is maintained in $\Omega_{\hat{\rho}_q}$ after $t_{s,i}$. The proof of this result is presented in Appendix B.

Theorem 2. *Consider the closed-loop system of Equation* (1) *under the LEMPC of Equation* (24) *with* $h_{NL,q}$ *meeting Equation* (12), *where the conditions of Propositions* 3 *and* 4 *hold and where* $\Omega_{\hat{\rho}_{safe,q}}$ *is contained in both* Ω_{ρ_i} *and* $\Omega_{\rho_{i+1}}$. *If* $t_{s,i+1} \in [t_k, t_{k+1})$, *such that, after* $t_{s,i+1}$, *the system of Equation* (1) *is controlled by the LEMPC of Equation* (24), *where* $x_{a,i}(t_{s,i+1}) = x_{a,i+1}(t_{s,i+1}) \in \Omega_{\hat{\rho}_q}$, *and if the following hold true,*

$$-\hat{\alpha}_{3,q}(\hat{\alpha}_{2,q}^{-1}(\hat{\rho}_{e,q})) + \hat{\alpha}_{4,q}(\hat{\alpha}_{1,q}^{-1}(\hat{\rho}_q))M_{err,p,q} + L'_{x,p,q}M_p\Delta + L'_{w,p,q}\theta_p \leq -\epsilon_{W,p,q}/\Delta \tag{68}$$

$$\hat{\rho}_{e,q} \leq \hat{\rho}_q - f_{V,q}(f_{W,p,q}(\Delta)) \tag{69}$$

for both $p = i$ *and* $p = i+1$, *and*

$$\hat{\rho}_{e,q} + f_{V,q}(f_{W,i,q}\Delta + (M_{change,i})\Delta + \frac{L_{w,i,q}\theta_i + M_{err,i,q}}{L_{x,i,q}}(e^{L_{x,i,q}\Delta} - e^{L_{x,i,q}t_{s,i+1}})) \leq \hat{\rho}_q \tag{70}$$

$$-\hat{\alpha}_{3,q}(\hat{\alpha}_{2,q}^{-1}(\hat{\rho}_{e,q})) + \hat{\alpha}_{4,q}(\hat{\alpha}_{1,q}^{-1}(\hat{\rho}_q))M_{err,i,q} + L'_{x,i,q}M_i\Delta + L'_{w,i,q}\theta_i + \hat{\alpha}_{4,q}(\hat{\alpha}_{1,q}^{-1}(\hat{\rho}_q))M_{change,i,q} + L'_{x,i+1,q}M_{i+1}\Delta$$
$$+L'_{w,i+1,q}\theta_{i+1} \leq -\epsilon'_{W,i,q}/\Delta \tag{71}$$

then the closed-loop state is bounded in $\Omega_{\hat{\rho}_q}$ *for all* $t \geq 0$.

We highlight that these conditions are conservative and not intended to form the least conservative bounds possible. However, they do help to elucidate some of the factors which impact whether a model used in an LEMPC will need to be reidentified to continue to maintain closed-loop stability when the underlying dynamics change, such as the extent to which the dynamics change. The above theorem indicates that, if $\Omega_{\hat{\rho}_q}$ is initially chosen in a sufficiently conservative fashion and the empirical model is sufficiently close to the underlying process dynamics before the model change, closed-loop stability may be maintained even after the underlying dynamics change if the model changes are such that the empirical model remains sufficiently close to the new dynamic model after the change. In general, anomalies may occur that could violate the conditions of Theorem 2. The result of this could be that the closed-loop state may leave $\Omega_{\hat{\rho}_q}$. In this case, it is helpful to characterize conditions under which changes in the underlying dynamics that could be destabilizing could be detected, triggering a model update and controller redesign for the new dynamic model to stabilize the closed-loop system. Therefore, the following theorem characterizes the length of time that the closed-loop state can remain in $\Omega_{\hat{\rho}_{safe,q}}$ after a change in the underlying process dynamics occurs if the conditions of Theorem 2 are not met. This can be used in determining how quickly a model reidentification algorithm would need to successfully provide a new model for the LEMPC of Equation (24) for closed-loop stability to be maintained as a function of factors such as the extent that the new model deviates from the empirical model used in the LEMPC when the underlying dynamics change, the sampling period, and the conservatism in the selection of $\hat{\rho}_q$. The proof of this theorem is presented in Appendix C.

Theorem 3. *Consider the closed-loop system of Equation* (1) *under the LEMPC of Equation* (24) *with* $h_{NL,q}$ *meeting Equation* (12) *and Proposition* 3, *where* $\Omega_{\hat{\rho}_{safe,q}}$ *is contained in both* Ω_{ρ_i} *and* $\Omega_{\rho_{i+1}}$. *If at* $t = t_{s,i+1}$, *where* $t_{s,i+1} \in [t_k, t_{k+1})$, *such that, after* $t_{s,i+1}$, *the system of Equation* (1) *is controlled by the LEMPC of Equation* (24), *where* $x_{a,i}(t_{s,i+1}) = x_{a,i+1}(t_{s,i+1}) \in \Omega_{\hat{\rho}_{safe,q}}$, *then if the following hold true with* $\hat{\rho}_{safe,q} > \hat{\rho}_{samp,q} > \hat{\rho}_q > \hat{\rho}_{q,e}$, $\hat{\rho}_{q,e} > \hat{\rho}_{min,q,i} > \hat{\rho}_{s,q} > 0$, *and* $\hat{\rho}_{q,e} > \hat{\rho}_{min,i+1,q} > \hat{\rho}_{s,q} > 0$:

$$-\hat{\alpha}_{3,q}(\hat{\alpha}_{2,q}^{-1}(\hat{\rho}_{s,q})) + \hat{\alpha}_{4,q}(\hat{\alpha}_{1,q}^{-1}(\hat{\rho}_q))M_{err,i+1,q} + L'_{x,i+1,q}M_{i+1}\Delta + L'_{w,i+1,q}\theta_{i+1} \leq \epsilon_{W,i+1,q}/\Delta \tag{72}$$

$$\hat{\rho}_{e,q} + f_{V,q}(f_{W,i,q}\Delta + (M_{change,i,q})\Delta + \frac{L_{w,i,q}\theta_i + M_{err,i,q}}{L_{x,i,q}}(e^{L_{x,i,q}\Delta} - e^{L_{x,i,q}t_{s,i+1}})) \leq \hat{\rho}_{samp,q} \tag{73}$$

$$\hat{\rho}_q + f_{V,q}(f_{W,i,q}\Delta + (M_{change,i,q})\Delta + \frac{L_{w,i,q}\theta_i + M_{err,i,q}}{L_{x,i,q}}(e^{L_{x,i,q}\Delta} - e^{L_{x,i,q}t_{s,i+1}})) \leq \hat{\rho}_{samp,q} \tag{74}$$

$$\hat{\rho}_{e,q} + f_{V,q}(f_{W,i+1,q}(\Delta)) \leq \hat{\rho}_{samp,q} \tag{75}$$

$$\hat{\rho}_q + \epsilon_{W,i+1,q} \leq \hat{\rho}_{samp,q} \tag{76}$$

as well as Equations (65)–(67), then if $x(t_{s,i+1}) \in \Omega_{\hat{\rho}_q}$ and $\Omega_{\hat{\rho}_{min,i+1,q}} \subset \Omega_{\hat{\rho}_{samp,q}}$ and the change to the model is not detected until a sampling time $t_{d,q}$ with $\bar{x}(t_{d,q}) \in \Omega_{\hat{\rho}_{safe,q}}/\Omega_{\hat{\rho}_q}$ ($\bar{x}(t_{d,q}) \in \Omega_{\hat{\rho}_{samp,q}} \subset \Omega_{\hat{\rho}_{safe,q}}$) after which $h_{NL,q}$ is used to control the system in sample-and-hold, then the number of sampling periods between $t_{ID,q}$ and $t_{d,q}$ within which the model in the LEMPC can be updated to a new model meeting Equation (65) with i replaced by $i+1$ and q replaced by $q+1$ without the closed-loop state exiting $\Omega_{\hat{\rho}_{safe,q}}$ is given by $t_{h,q} = floor(\frac{(\hat{\rho}_{safe,q}-\hat{\rho}_{samp,q})}{\epsilon_{W,i,q}})$, where floor represents the "floor" function that returns the largest integer less than the value of the argument. $\bar{x}(t)$ refers either to $\bar{x}_{a,i+1,q}(t)$ or $\bar{x}_{a,i,q}(t)$, depending on whether $t_{s,i+1}$ is within the sampling period preceding the closed-loop state exiting $\Omega_{\hat{\rho}_q}$.

The following theorem provides the conditions under which the closed-loop state is maintained within $\Omega_{\hat{\rho}_{safe,q+1}}$ for all times after $t_{ID,q}$ and is driven into $\Omega_{\hat{\rho}_{q+1}}$ after the model reidentification. The proof of the result is presented in Appendix D.

Theorem 4. *If $\Omega_{\hat{\rho}_{safe,q}} \subset \Omega_{\hat{\rho}_{safe,q+1}}$ and if both $\Omega_{\hat{\rho}_{safe,q}}$ and $\Omega_{\hat{\rho}_{safe,q+1}}$ are contained in Ω_{ρ_i} and $\Omega_{\rho_{i+1}}$, then if $h_{NL,q+1}$ is used to control the system after $t_{ID,q}$ while $x(t_k) \in \Omega_{\hat{\rho}_{safe,q+1}}/\Omega_{\hat{\rho}_{q+1}}$ with the conditions of Equations (65) and (66) met for the $q+1$th empirical model for the $i+1$th dynamic system and the LEMPC of Equation (24) using the $q+1$th empirical model of Equation (10) is used to control the system for all times after $x(t_k) \in \Omega_{\hat{\rho}_{q+1}}$, then the closed-loop state is then maintained within $\Omega_{\hat{\rho}_{safe,q+1}}$ until it enters $\Omega_{\hat{\rho}_{q+1}}$ and is then maintained in $\Omega_{\hat{\rho}_{q+1}}$ for all subsequent sampling times.*

Remark 8. *From a verification standpoint, the proofs above move toward addressing the question of what may happen if a controller is designed and even tested for certain conditions, but the process dynamics change. It provides a theoretical characterization of conditions under which action would subsequently need to be taken as well as indications of the time available to take the subsequent action. However, the results above may be difficult to utilize directly in developing an online monitoring scheme, as many of the theoretical conditions rely on knowing properties of the current and updated models that would likely not be characterizable or would not be known until after the anomaly occurred. However, these still may aid in gaining an understanding of different possibilities. For example, a conservative stability region $\Omega_{\hat{\rho}_q}$ suggests that larger anomalies could still be detected and mitigated by a combined detection and reidentification procedure without loss of closed-loop stability. Earlier detection may provide more time for reidentification.*

Remark 9. *If there is an indication from detection methods that are not based on the closed-loop state leaving the stability region that the underlying dynamics may have changed but that the closed-loop state has not yet left $\Omega_{\hat{\rho}_q}$, then until the closed-loop state leaves $\Omega_{\hat{\rho}_q}$, online experiments (e.g., modifying the objective function as in Reference [51]) could be performed if they do not impact the constraint set to attempt to probe whether the dynamics are more consistent with the prior process model or the potential model postulated after the anomaly is suggested. This may be a method for attempting to detect the changes before the closed-loop state leaves $\Omega_{\hat{\rho}_q}$, which could allow larger changes in the process model to be handled practically than could be guaranteed to be handled in the theorems above, as the magnitude of the deviations in the dynamic model allowed above without loss of closed-loop stability depends on the distance between $\Omega_{\hat{\rho}_{safe,q}}$ and $\Omega_{\hat{\rho}_{samp,q}}$. However, it is also highlighted that the above is a conservative result, meaning that, in general, larger changes may be able to be handled without loss of closed-loop stability.*

Remark 10. *The above results can be used to comment on why giving greater flexibility to the process after an anomaly to handle it could introduce additional complexity. Specifically, consider the possibility that some actuators may not typically be used for control but could be considered for use after an anomaly (similar to how safety systems activate for chemical processes, but in this case, they would not act according to a prespecified logic but might be able to be manipulated in either an on-off or continuous manner to give the process additional capabilities for handling the anomaly). It is noted that this would constitute dynamics not previously considered. According to the proofs above, one way to guarantee closed-loop stability in the presence of sufficiently small disturbances is to cause the dynamics after they change to not differ too radically from those assumed before the change and used in the prior dynamic model in the EMPC. If additional flexibility is given to the system, this would be an additional model that would have to match up well.*

Remark 11. *The results above suggest that, if a model identification algorithm could be guaranteed to provide an accurate model with a small amount of data that could be gathered between when the closed-loop state leaves $\Omega_{\hat{\rho}_q}$ but before it leaves $\Omega_{\hat{\rho}_{safe,q}}$ (where the amount of data available in that timeframe could be known a priori by the number of measurements available in a given sampling period), then the model could be reidentified and placed within the LEMPC in a manner that is stabilizing.*

Remark 12. *Instead of changes to the underlying dynamic model, anomalies may present changes in the constraint set (e.g., anomalies may change equipment material limitations (e.g., maximum shear stresses, which can change with temperature) used to place constraints on the state in an LEMPC). Because the above results assume that the stability region is fully contained within the state constraint set, the detection and response procedure above would need to ensure that there is no time at which the stability region is no longer fully included within the state constraint set under the new dynamic model. This may be handled by making $\Omega_{\hat{\rho}_{safe,q}}$ sufficiently conservative such that the closed-loop state never exits a region where the state constraints can be met under different dynamic models.*

3.2.3. Automated Response to Unexpected Hazards: Application to a Chemical Process Example

In this section, we demonstrate concepts described above through a process example. This example considers a nonisothermal reactor in which an $A \rightarrow B$ reaction takes place, but the reactant inlet concentration C_{A0} and the heat rate Q supplied by a jacket are adjusted by an LEMPC. The process model is as follows:

$$\dot{C}_A = \frac{F}{V}(C_{A0} - C_A) - k_0 e^{-\frac{E}{R_g T}} C_A^2 \tag{77}$$

$$\dot{T} = \frac{F}{V}(T_0 - T) - \frac{\Delta H k_0}{\rho_L C_p} e^{-\frac{E}{R_g T}} C_A^2 + \frac{Q}{\rho_L C_p V} \tag{78}$$

where the parameters are listed in Table 3 and include the reactor volume V, inlet reactant temperature T_0, pre-exponential constant k_0, solution heat capacity C_p, solution density ρ_L, feed/outlet volumetric flow rate F, gas constant R_g, activation energy E, and heat of reaction ΔH. The state variables are the reactant concentration C_A and temperature T in the reactor, which can be written in deviation form from the operating steady-state vector $C_{As} = 1.22$ kmol/m^3, $T_s = 438.2$ K, $C_{A0s} = 4$ kmol/m^3, and $Q_s = 0$ kJ/h as $x = [x_1\ x_2]^T = [C_A - C_{As}\ T - T_s]^T$ and $u = [u_1\ u_2]^T = [C_{A0} - C_{A0s}\ Q - Q_s]^T$. The model of Equations (77) and (78) has the following form:

$$\dot{x} = \tilde{f}(x) + g(x)u \tag{79}$$

where $\tilde{f}$ represents a vector function derived from Equations (77) and (78) that is not multiplied by u and where $g(x) = [g_1\ g_2]^T = [\frac{F}{V}\ 0;\ 0\ \frac{1}{\rho_L C_p V}]^T$ represents the vector function which multiplies u in these equations.

Table 3. Parameters for the CSTR model of Equations (77) and (78).

Parameter	Value	Unit
V	1	m^3
T_0	300	K
k_0	8.46×10^6	$\text{m}^3/\text{h·kmol}$
C_p	0.231	kJ/kg·K
ρ_L	1000	kg/m^3
F	5	m^3/h
R_g	8.314	kJ/kmol·K
E	5×10^4	kJ/kmol
ΔH	-1.15×10^4	kJ/kmol

The EMPC utilized to adjust the manipulated inputs C_{A0} and Q utilizes the following stage cost (to maximize the production rate of the desired product) and physical bounds on the inputs:

$$L_e = -k_0 e^{-E/(R_g T(\tau))} C_A(\tau)^2 \tag{80}$$

$$0.5 \leq C_{A0} \leq 7.5 \; \text{kmol/m}^3 \tag{81}$$

$$-5 \times 10^5 \leq Q \leq 5 \times 10^5 \; \text{kJ/h} \tag{82}$$

Lyapunov-based stability constraints are also enforced (where a constraint of the form of Equation (22) is enforced at the end of every sampling time if $x(t_k) \in \Omega_{\hat{\rho}_e}$, and the constraint of the form of Equation (23) is enforced at t_k when $x(t_k) \in \Omega_{\hat{\rho}}/\Omega_{\hat{\rho}_e}$ but then followed by a constraint of the form of Equation (22) at the end of all sampling periods after the first).

We will consider several simulations to demonstrate the developments above. In the first, we explore several aspects of the case in which there is a change in the underlying dynamics while the process is operated under LEMPC that is minor such that the closed-loop state does not leave $\Omega_{\hat{\rho}}$ after the change in the underlying dynamics. For this case, the Lyapunov function selected was $\hat{V}_q = x^T P x$, with P given as follows:

$$P = \begin{bmatrix} 1200 & 5 \\ 5 & 0.1 \end{bmatrix} \tag{83}$$

The Lyapunov-based controller $h_{NL,1}(x)$ was designed such that its first component $h_{NL,1,1}(x) = 0$ kmol/m^3 and its second component $h_{NL,1,2}(x)$ is computed as follows (Sontag's formula [56]):

$$h_{NL,1,2}(x) = \begin{cases} -\dfrac{L_{\tilde{f}}\hat{V}_q + \sqrt{L_{\tilde{f}}\hat{V}_q^2 + L_{\tilde{g}_2}\hat{V}_q^4}}{L_{\tilde{g}_2}\hat{V}_q}, & \text{if } L_{\tilde{g}_2}\hat{V}_q \neq 0 \\ 0, & \text{if } L_{\tilde{g}_2}\hat{V}_q = 0 \end{cases} \tag{84}$$

Then, it is saturated at the input bounds of Equation (82) if they are met. $L_{\tilde{f}}\hat{V}_q$ and $L_{\tilde{g}_2}\hat{V}_q$ are Lie derivatives of $\hat{V}_q$ with respect to the vector functions $\tilde{f}$ and $\tilde{g}_2$, respectively. $\hat{\rho}$ and $\hat{\rho}_e$ were taken from Reference [57] to be 300 and 225, respectively. The process state was initialized at $x_{init} = [-0.4 \; \text{kmol/m}^3 \; 8 \; \text{K}]^T$, with controller parameters $N = 10$ and $\Delta = 0.01$ h. The process model of Equations (77) and (78) was integrated with the explicit Euler numerical integration method using an integration step size of 10^{-4} h within the LEMPC and of 10^{-5} h to simulate the process.

For this first simulation, we assume that a change in the underlying process dynamics occurs at 0.5 h that does not compromise closed-loop stability. Specifically, at 0.5 h, it is assumed that an additional source of heat arises outside the reactor such that the right-hand side of Equation (78) is modified by the addition of another term $Q_{extra} = 300$ K/h. Figures 6 and 7 show the process responses when the LEMPC is not aware of the change in the process dynamic model when it occurs and when it is aware of the change in the process dynamic model after it occurs such that it is fully compensated (i.e., an accurate process model is used in the LEMPC at all times, even after the dynamics change). In both cases, the closed-loop state was maintained within the stability region at all times. These simulations were carried out in MATLAB R2016b using fmincon with the default settings except for the increased iterations/function evaluations allowed, scaling u_2 down by 10^5 and providing the steady-state input values as the initial guess for the optimization problem solution at each sampling time. No attempt was made to check whether the LEMPCs in the simulations located globally optimal solutions to the LEMPC optimization problems. However, the profit was higher than that at the steady-state around which the LEMPC was designed.

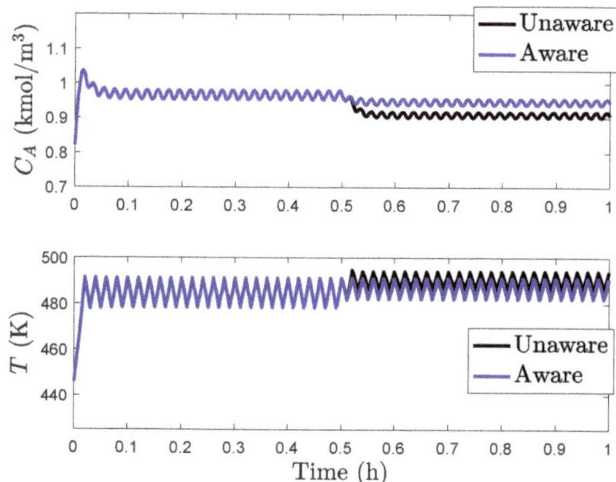

Figure 6. State trajectories under Lyapunov-based EMPC (LEMPC) with $Q_{extra} = 300$ K/h starting at 0.5 h, where the LEMPC has not been made aware ("Unaware") and has been made aware ("Aware") of the change in the energy balance.

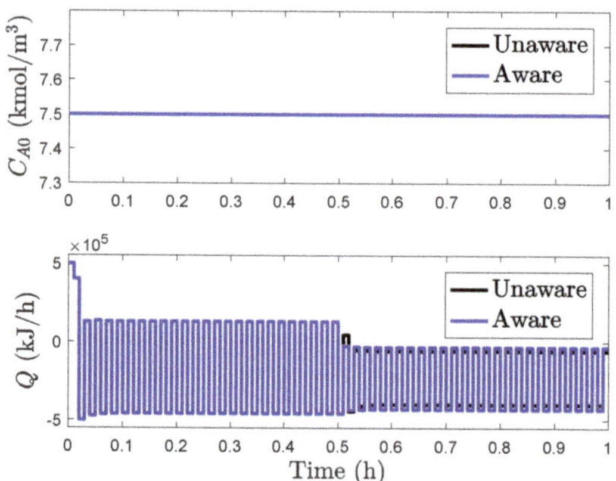

Figure 7. Input trajectories under LEMPC with $Q_{extra} = 300$ K/h starting at 0.5 h, where the LEMPC has not been made aware ("Unaware") and has been made aware ("Aware") of the change in the energy balance.

The oscillatory behavior of the states before 0.5 h is caused by the fact that the profit is maximized for this process at the boundary of $\Omega_{\hat{\rho}_e}$. Without plant-model mismatch, the LEMPC is able to maintain the closed-loop state exactly on the boundary of $\Omega_{\hat{\rho}_e}$ and therefore always operates the process using the constraint of Equation (22); however, when the plant-model mismatch occurs (induced by the use of different integration steps to simulate the process dynamic model within the LEMPC and for the simulation of the process under the computed control actions), the closed-loop state then exits $\Omega_{\hat{\rho}_e}$ when the LEMPC predicts it will stay inside of it under the control actions computed by the controller. The result is that the constraint of Equation (23) is then activated until the closed-loop state reenters $\Omega_{\hat{\rho}_e}$. This process of entering $\Omega_{\hat{\rho}_e}$, attempting to operate at its boundary, and then being kicked out only to be driven back in is the cause of the oscillatory response of the states and inputs in Figures 6 and 7. It is noted, however, that though this behavior may be undesirable from, for example, an actuator wear perspective, it does not reflect a loss of closed-loop stability or a malfunction of the controller. The controller is in fact maintaining the closed-loop state within $\Omega_{\hat{\rho}}$ as it was designed to do; the fact that it does so in perhaps a visually unfamiliar fashion means that we have not specified in the control law that it should not do that, so it is not aware that an end user would find that behavior strange (if the oscillatory behavior is deemed undesirable, one could consider, for example, input rate of change constraints and potentially the benefits of the human response-based input rate of change strategy in the prior section for handling unexpected events).

In the case that the LEMPC is not aware of the change in the process dynamics, the profit is 32.7103, whereas when the LEMPC is aware of the change in the dynamics, the profit is 32.5833. Though these values are very close, an interesting note is that the profit when the LEMPC is not aware of the change in the underlying dynamics is slightly higher than when it is aware. Intuitively, one would expect an LEMPC with a more accurate process model to be able to locate a more economically optimal trajectory for the closed-loop state to follow than an LEMPC that cannot provide as accurate predictions. Part of the reason for the enhanced optimality in the case without knowledge of the change in the underlying dynamics, however, comes from the two-mode nature of LEMPC. In the case that the LEMPC is aware of the change in the underlying dynamics, it drives the closed-loop state to an operating condition that remains closer to the boundary of $\Omega_{\hat{\rho}_e}$ after 0.5 h than when it is not aware of the change in the underlying dynamics due to the plant/model mismatch being different in the different cases. The result is that the process accesses regions of state-space that lead to higher profits when the LEMPC does not know about the change in the dynamics than if the LEMPC knows more about the process dynamics.

The remainder of this example focuses on elucidating the conservativeness of the proposed approach. Specifically, we now consider the Lyapunov function selected as $\hat{V}_q = x^T P x$, with P given as follows:

$$P = \begin{bmatrix} 2000 & -10 \\ -10 & 3 \end{bmatrix} \tag{85}$$

Again, $h_{NL,1}(x)$ is designed such that $h_{NL,1,1}(x) = 0$ kmol/m^3, and $h_{NL,1,2}(x)$ is computed via Sontag's formula but saturated at the input bounds of Equation (82) if they are met. $\hat{\rho}$ and $\hat{\rho}_e$ were taken to be 1300 and 975, respectively, and $\hat{\rho}_{safe}$ was set to 1800. The process state was initialized at $x_{init} = [0 \text{ kmol/m}^3 \; 0 \text{ K}]^T$, with controller parameters $N = 10$ and $\Delta = 0.01$ h. The process model of Equations (77) and (78) was integrated with the explicit Euler numerical integration method using an integration step size of 10^{-4} h within the EMPC and with an integration step size of 10^{-5} h to simulate the process. The constraint of the form of Equation (23) is enforced at t_k when $x(t_k) \in \Omega_{\hat{\rho}}/\Omega_{\hat{\rho}_e}$ but then followed by a constraint of the form of Equation (22) at the end of all sampling periods.

At 0.5 h, it is assumed that an additional source of heat arises outside the reactor such that the right-hand side of Equation (78) is modified by the addition of another heat term $Q_{extra} = 500$ K/h. In this case, with no change in the process model used by the EMPC or even in the control law (i.e., in contrast to the implementation strategy in Section 3.2.1, $h_{NL,1}$ is not employed when the closed-loop state exits $\Omega_{\hat{\rho}}$), the behavior in Figure 8 results. Notably, the closed-loop state does not leave $\Omega_{\hat{\rho}_{safe}}$, and no infeasibility issues occurred. In contrast, if we begin to utilize $h_{NL,1}$ when the closed-loop state leaves $\Omega_{\hat{\rho}}$, the closed-loop state will eventually leave $\Omega_{\hat{\rho}_{safe}}$ (Figure 9). While we can obtain a new empirical model (in this case, we assume that the dynamics become fully known at 0.54 h and are accounted for completely to demonstrate the result) and can use that to update $h_{NL,1}$ to $h_{NL,2}$ (i.e., $h_{NL,1}$ but with modified saturation bounds to reflect design around the new steady-state of the system with $Q_{Added} = 500$ K/h) before the closed-loop state leaves $\Omega_{\hat{\rho}_{safe}}$ as suggested in the implementation strategy in Section 3.2.1 (creating the profile shown in Figure 10 corresponding to 2 h of operation in which the closed-loop state is driven back to the origin under $h_{NL,2}$), the fact that the closed-loop state would not have left the stability region if the controller had not been adjusted illustrates the conservativeness of the approach. We note that Figure 10 does not complete the implementation strategy in Section 3.2.1 (which would involve the use of a new LEMPC after the closed-loop state reenters $\Omega_{\hat{\rho}}$ for this example) because that part of the implementation strategy will be demonstrated in the discussion for a slightly different LEMPC presented below.

Finally, we provide a result where the LEMPC computes a time-varying input policy due to the desire to enforce a constraint on the amount of reactant available in the feed over an hour (i.e., a material/feedstock constraint) as follows:

$$\frac{1}{1\,h} \int_{t=0\,h}^{t=1\,h} u_1(\tau)d\tau = 0\,\text{kmol/m}^3 \tag{86}$$

This constraint is enforced via a soft constraint formulation by introducing slack variables s_1 and s_2 that are penalized in a modified objective function as follows:

$$\int_{t_k}^{t_{k+N}} \left[-k_0 e^{-\frac{E}{R_g T(\tau)}} C_A(\tau)^2 \right] d\tau + 100(s_1^2 + s_2^2) \tag{87}$$

They are used in the following constraints:

$$\sum_{i=0}^{k-1}(u_1^*(t_i|t_i)) + \sum_{i=k}^{k+N_k}(u_1(t_i|t_k)) - 3.5\delta(100 - \frac{t_k}{\Delta} - N) \leq s_1 \tag{88}$$

$$-\sum_{i=0}^{k-1}(u_1^*(t_i|t_i)) - \sum_{i=k}^{k+N_k}(u_1(t_i|t_k)) - 3.5\delta(100 - \frac{t_k}{\Delta} - N) \leq s_2 \tag{89}$$

where $N_k = N$ and $\delta = 1$ when $t_k < 0.9$ h and where $\delta = 0$ and N_k is the number of sampling periods left in a 1 h operating period when $t_k \geq 0.9$ h. These constraints are developed based on Reference [12]. $u_1^*(t_i|t_i)$ signifies the value of u_1 applied to the process at a prior sampling time, and $u_1(t_i|t_k)$ reflects the value of u_1 predicted at the current sampling time t_k to be applied for $t \in [t_i, t_{i+1})$, $i = k, \ldots, k + N_k$. The upper and lower bounds on s_1 and s_2 were set to 2×10^{19} and -2×10^{19}, respectively, to allow them to be effectively unbounded. The initial guesses of the slack variables were set to 0 at each sampling time.

When the LEMPC with the above modifications is applied to the process with $Q_{Added} = 500$ K/h starting at 0.5 h, the closed-loop state again exits $\Omega_{\hat{\rho}}$ for some time after 0.5 h but reenters it and also does not exit $\Omega_{\hat{\rho}_{safe}}$, once again reflecting the conservatism from a closed-loop stability standpoint of a strategy that updates the process model whenever the closed-loop state leaves $\Omega_{\hat{\rho}}$. Furthermore, if $h_{NL,1}$ is utilized after it is detected that the closed-loop state leaves $\Omega_{\hat{\rho}}$ (the first sampling time at which this occurs is 0.51 h), then it exits $\Omega_{\hat{\rho}_{safe}}$ by 0.52 h, showing that the length of the sampling period or the size of $\Omega_{\hat{\rho}}$ with respect to $\Omega_{\hat{\rho}_{safe}}$ is not sufficiently small enough to impose model updates before closed-loop stability is jeopardized because measurements are only available every sampling time. If instead, however, $\hat{\rho}$ is updated to be 1200 and $\hat{\rho}_e$ is set to 900, then the closed-loop state remains in $\Omega_{\hat{\rho}}$ between 0.51 and 0.52 h.

If at 0.52 h, we assume that the new dynamics (i.e., with $Q_{Added} = 500$ K/h) become available and are used in designing $h_{NL,2}$ (used from 0.52 h until the first sampling time at which $x(t_k) \in \Omega_{\hat{\rho}}$ again) and that a second LEMPC designed based on the updated model is used after the closed-loop state has reentered $\Omega_{\hat{\rho}}$, the state-space trajectory in Figure 11 results.

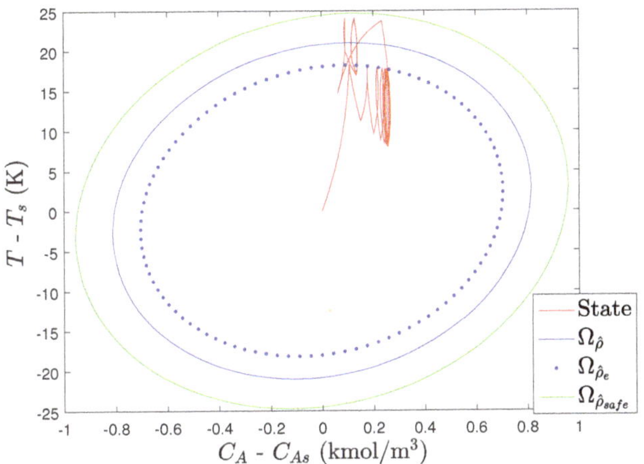

Figure 8. State-space plot under LEMPC with $Q_{extra} = 500$ K/h starting at 0.5 h and no change in the control law or model in response.

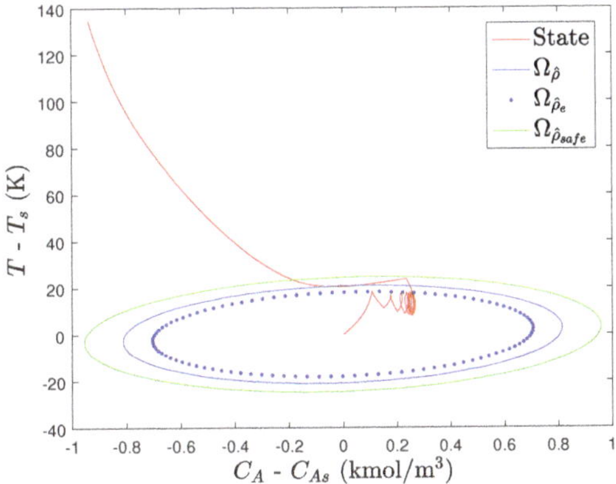

Figure 9. State-space plot under LEMPC with $Q_{extra} = 500$ K/h starting at 0.5 h and the control law switched to $h_{NL,1}$ in response to the closed-loop state leaving $\Omega_{\hat{\rho}}$.

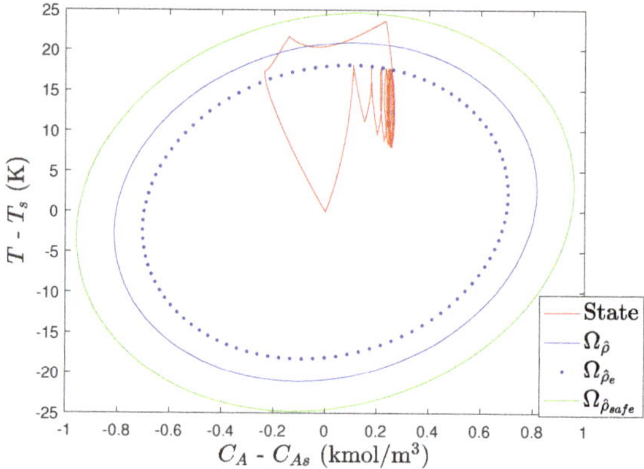

Figure 10. State-space plot under LEMPC with $Q_{extra} = 500$ K/h starting at 0.5 h and the control law switched to $h_{NL,1}$ in response to the closed-loop state leaving $\Omega_{\hat{\rho}}$ and then switched to $h_{NL,2}$ at 0.54 h.

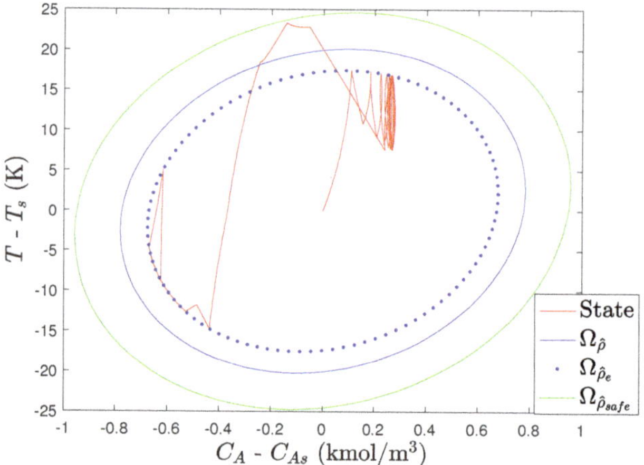

Figure 11. State-space plot under LEMPC with $Q_{extra} = 500$ K/h starting at 0.5 h and the control law switched to $h_{NL,1}$ in response to the closed-loop state leaving $\Omega_{\hat{\rho}}$, then switched to $h_{NL,2}$ at 0.52 h, and finally switched back to an LEMPC incorporating an updated process model after the closed-loop state reenters $\Omega_{\hat{\rho}}$.

4. Conclusions

This work developed a Lyapunov-based EMPC framework for handling unexpected considerations of different types. One of the types of considerations handled was end-user response to how a control law operates a process, providing a controller self-update capability through input rate of change constraints that allows even uncertain or imprecise information about the end-user response to be used in optimizing the controller formulation without loss of closed-loop stability or feasibility. The second type of consideration was the occurrence of anomalies, where conditions which would guarantee that the closed-loop state can be stabilized in the presence of an anomaly that changes the underlying process dynamics as long as a detection method identifies a new process model sufficiently quickly, were presented that uses the LEMPC stability properties in developing an anomaly detection mechanism. Chemical process examples were presented for both cases to demonstrate the proposed approach.

The work above provides insights into interpretability and verification considerations for EMPC from a theoretical perspective. However, these remain significant challenges for this control design. For example, there is no guarantee that adjusting a given constraint (e.g., adjusting the upper bound on an input rate of change constraint) will cause process behavior to appear interpretable to an end user before it approaches steady-state behavior, which may reduce the benefits of using EMPC. Furthermore, the results related to anomaly handling were demonstrated via process examples to be highly conservative. No methods were presented for practically ascertaining time (online) until an anomaly would result in the closed-loop state leaving a known region of state-space after detection to facilitate appropriate actions to be taken. Further work on these issues needs to be undertaken to develop practical EMPC designs with appropriate safety and interpretability properties with low time required to verify the designs before putting them into the field for different processes.

Funding: Financial support from the National Science Foundation CBET-1839675, from the Air Force Office of Scientific Research award number FA9550-19-1-0059, and from Wayne State University is gratefully acknowledged.

Conflicts of Interest: The author declares no conflict of interest.

Appendix A. Proof of Proposition 4

Proof. The result in Equation (62) is stated in Proposition 1; therefore, it remains to prove that Equation (63) holds. To derive the result of Equation (63), Equations (59) and (60) are integrated as follows:

$$\tilde{x}_{a,i+1,q}(t) = \tilde{x}_{a,i,q}(t_{s,i+1}) + \int_{t_{s,i+1}}^{t} \bar{f}_{i+1,q}(\tilde{x}_{a,i+1,q}(s), \bar{u}_q(s), w_{i+1}(s))ds \tag{A1}$$

$$\tilde{x}_{b,q}(t) = \tilde{x}_{b,q}(t_{s,i+1}) + \int_{t_{s,i+1}}^{t} \bar{f}_{NL,q}(\tilde{x}_{b,q}(s), \bar{u}_q(s))ds \tag{A2}$$

for $t \in [t_{s,i+1}, t_1]$. Subtracting Equation (A2) from Equation (A1) and taking norms of both sides of the resulting equation gives the following:

$$
\begin{aligned}
|\tilde{x}_{a,i+1,q}(t) - \tilde{x}_{b,q}(t)| \quad &= |\tilde{x}_{a,i,q}(t_{s,i+1}) - \tilde{x}_{b,q}(t_{s,i+1}) + \int_{t_{s,i+1}}^{t} [\bar{f}_{i+1,q}(\tilde{x}_{a,i+1,q}(s), \bar{u}_q(s), w_{i+1}(s)) - \bar{f}_{NL,q}(\tilde{x}_{b,q}(s), \bar{u}_q(s))]ds| \\
&\leq |\tilde{x}_{a,i,q}(t_{s,i+1}) - \tilde{x}_{b,q}(t_{s,i+1})| + \int_{t_{s,i+1}}^{t} |\bar{f}_{i+1,q}(\tilde{x}_{a,i+1,q}(s), \bar{u}_q(s), w_{i+1}(s)) - \bar{f}_{NL,q}(\tilde{x}_{b,q}(s), \bar{u}_q(s))|ds \\
&\leq f_{W,i,q}(t_{s,i+1} - t_0) + \int_{t_{s,i+1}}^{t} |\bar{f}_{i+1,q}(\tilde{x}_{a,i+1,q}(s), \bar{u}_q(s), w_{i+1}(s)) - \bar{f}_{i,q}(\tilde{x}_{a,i,q}(s), \bar{u}_q(s), w_i(s)) \\
&\qquad + \bar{f}_{i,q}(\tilde{x}_{a,i,q}(s), \bar{u}_q(s), w_i(s)) - \bar{f}_{NL,q}(\tilde{x}_{b,q}(s), \bar{u}_q(s))|ds \\
&\leq f_{W,i,q}(t_{s,i+1} - t_0) + \int_{t_{s,i+1}}^{t} |\bar{f}_{i+1,q}(\tilde{x}_{a,i+1,q}(s), \bar{u}_q(s), w_{i+1}(s)) - \bar{f}_{i,q}(\tilde{x}_{a,i,q}(s), \bar{u}_q(s), w_i(s))|ds \\
&\qquad + \int_{t_{s,i+1}}^{t} |\bar{f}_{i,q}(\tilde{x}_{a,i,q}(s), \bar{u}_q(s), w_i(s)) - \bar{f}_{NL,q}(\tilde{x}_{b,q}(s), \bar{u}_q(s))|ds
\end{aligned}
\tag{A3}
$$

From Equations (15), (52), and (61), we have the following:

$$
\begin{aligned}
|\tilde{x}_{a,i+1,q}(t) - \tilde{x}_{b,q}(t)| \quad &\leq f_{W,i,q}(t_{s,i+1} - t_0) + \int_{t_{s,i+1}}^{t} M_{change,i,q}ds + \int_{t_{s,i+1}}^{t} |\bar{f}_{i,q}(\tilde{x}_{a,i,q}(s), \bar{u}_q(s), w_i(s)) - \bar{f}_{i,q}(\tilde{x}_{b,q}(s), \bar{u}_q(s), 0) \\
&\qquad + \bar{f}_{i,q}(\tilde{x}_{b,q}(s), \bar{u}_q(s), 0) - \bar{f}_{NL,q}(\tilde{x}_{b,q}(s), \bar{u}_q(s))|ds \\
&\leq f_{W,i,q}(t_{s,i+1} - t_0) + M_{change}(t - t_{s,i+1}) \\
&\qquad + \int_{t_{s,i+1}}^{t} |\bar{f}_{i,q}(\tilde{x}_{a,i,q}(s), \bar{u}_q(s), w_i(s)) - \bar{f}_{i,q}(\tilde{x}_{b,q}(s), \bar{u}_q(s), 0)|ds \\
&\qquad + \int_{t_{s,i+1}}^{t} |\bar{f}_{i,q}(\tilde{x}_{b,q}(s), \bar{u}_q(s), 0) - \bar{f}_{NL,q}(\tilde{x}_{b,q}(s), \bar{u}_q(s))|ds \\
&\leq f_{W,i,q}(t_{s,i+1} - t_0) + M_{change}(t - t_{s,i+1}) + \int_{t_{s,i+1}}^{t} (L_{x,i,q}|\tilde{x}_{a,i,q}(s) - \tilde{x}_{b,q}(s)| + L_{w,i,q}|w_i(s)|)ds \\
&\qquad + \int_{t_{s,i+1}}^{t} M_{err,i,q}ds
\end{aligned}
\tag{A4}
$$

Using Equation (50) we get the following,

$$
\begin{aligned}
|\bar{x}_{a,i+1,q}(t) - \bar{x}_{b,q}(t)| \quad &\le f_{W,i,q}(t_{s,i+1} - t_0) + M_{change,i,q}(t - t_{s,i+1}) + (L_{w,i,q}\theta_i + M_{err,i,q})\int_{t_{s,i+1}}^{t}(e^{L_{x,i,q}s} - 1)ds \\
&\quad + \int_{t_{s,i+1}}^{t}(L_{w,i,q}\theta_i + M_{err,i,q})ds \\
&\le f_{W,i,q}(t_{s,i+1} - t_0) + M_{change,i,q}(t - t_{s,i+1}) + (L_{w,i,q}\theta_i + M_{err,i,q})\int_{t_{s,i+1}}^{t}(e^{L_{x,i,q}s} - 1)ds \\
&\quad + (L_{w,i,q}\theta_i + M_{err,i,q})(t - t_{s,i+1}) \\
&\le f_{W,i,q}(t_{s,i+1} - t_0) + M_{change,i,q}(t - t_{s,i+1}) + \frac{(L_{w,i,q}\theta_i + M_{err,i,q})}{L_{x,i,q}}(e^{L_{x,i,q}t} - e^{L_{x,i,q}t_{s,i+1}})
\end{aligned}
\tag{A5}
$$

$\square$

Appendix B. Proof of Theorem 2

Proof. To guarantee the results, recursive feasibility of the LEMPC must hold. Feasibility of the LEMPC of Equation (24) follows from Theorem 1 when $x(t_k) \in \Omega_{\hat{\rho}_q}$. Subsequently, closed-loop stability must be proven both when $t_{s,i+1} = t_k$ and when $t_{s,i+1} \in (t_k, t_{k+1})$.

Consider first the case that $t_{s,i+1} = t_k$. In this case, if Equation (68) holds with $p = i+1$ and $x(t_k) \in \Omega_{\hat{\rho}_q}$, then $x(t) \in \Omega_{\hat{\rho}_q}$ from Theorem 1 for $t \ge 0$. Consider second the case that $t_{s,i+1} \in (t_k, t_{k+1})$. In this case, until $t_{s,i+1}$, if Equations (68) and (69) hold for $p = i$, the closed-loop state is maintained within $\Omega_{\hat{\rho}_q}$ from Theorem 1. To guarantee that the closed-loop state is maintained in $\Omega_{\hat{\rho}_q}$ after $t_{s,i+1}$ until t_{k+1}, it is first noted that, if $x(t_k) \in \Omega_{\hat{\rho}_{e,q}}$ and $t_{s,i+1} \in (t_k, t_{k+1})$, then from Proposition 2, we have the following:

$$
\hat{V}_q(\bar{x}_{a,i+1,q}(t)) \le \hat{V}_q(\bar{x}_{b,q}(t_{k+1})) + f_{V,q}(|\bar{x}_{a,i+1,q}(t) - \bar{x}_{b,q}(t_{k+1})|)
\tag{A6}
$$

if $\bar{x}_{a,i+1,q}(t), \bar{x}_{b,q}(t) \in \Omega_{\hat{\rho}_q}$ for $t \in [t_k, t_{k+1}]$. If Proposition 4 holds, then from Equation (24f), we have the following:

$$
\hat{V}_q(\bar{x}_{a,i+1,q}(t)) \le \hat{\rho}_{e,q} + f_{V,q}\left(f_{W,i,q}(t_{s,i+1} - t_k) + (M_{change,i,q})(t - t_{s,i+1}) + \frac{L_{w,i,q}\theta_i + M_{err,i,q}}{L_{x,i,q}}(e^{L_{x,i,q}t} - e^{L_{x,i,q}t_{s,i+1}})\right)
\tag{A7}
$$

If Equation (70) holds, then $\hat{V}_q(\bar{x}_{a,i+1,q}(t)) \le \hat{\rho}_q$ for $t \in [t_{s,i+1}, t_{k+1}]$.

If instead $x(t_k) \in \Omega_{\hat{\rho}_q}/\Omega_{\hat{\rho}_{e,q}}$ and if Equations (68) and (69) hold, the closed-loop state is maintained within $\Omega_{\hat{\rho}_q}$ from Theorem 1 until $t_{s,i+1}$. To guarantee that the closed-loop state is maintained in $\Omega_{\hat{\rho}_q}$ after $t_{s,i+1}$ until t_{k+1}, it is first noted that the following is true:

$$
\begin{aligned}
&\frac{\partial \hat{V}_q(x(t_k))}{\partial x}(\bar{f}_{NL,q}(x(t_k), \bar{u}_q(t_k))) \\
&\le \frac{\partial \hat{V}_q(x(t_k))}{\partial x}(\bar{f}_{NL,q}(x(t_k), h_{NL,q}(x(t_k)))) \le -\hat{\alpha}_{3,q}(|x(t_k)|)
\end{aligned}
\tag{A8}
$$

from Equation (12b) and Equation (24g). When $t_k \le t < t_{s,i+1}$, then from Reference [51], if Equation (68) and the conditions of Theorem 2 hold with $p = i$, the following is true:

$$
\begin{aligned}
&\frac{\partial \hat{V}_q(\bar{x}_{a,i,q}(\tau))}{\partial x}(\bar{f}_{i,q}(\bar{x}_{a,i,q}(\tau), \bar{u}_q(t_k), w_i(\tau))) \\
&\le -\hat{\alpha}_{3,q}(\hat{\alpha}_{2,q}^{-1}(\hat{\rho}_{e,q})) + \hat{\alpha}_{4,q}(\hat{\alpha}_{1,q}^{-1}(\hat{\rho}_q))M_{err,i,q} + L'_{x,i,q}M_i\Delta + L'_{w,i,q}\theta_i
\end{aligned}
\tag{A9}
$$

for $\tau \in [t_k, t_{s,i+1})$, and

$$
\hat{V}_q(\bar{x}_{a,i,q}(t_{s,i+1})) \le \hat{V}_q(x(t_k))
\tag{A10}
$$

Given that $\bar{x}_{a,i,q}(t_{s,i+1}) = \bar{x}_{a,i+1,q}(t_{s,i+1})$, the following holds:

$$
\begin{aligned}
&\frac{\partial \hat{V}_q(\bar{x}_{a,i+1,q}(t_{s,i+1}))}{\partial x}\,(\bar{f}_{i+1,q}(\bar{x}_{a,i+1,q}(t_{s,i+1}),\bar{u}_q(t_k),0)) \\
&= \frac{\partial \hat{V}_q(\bar{x}_{a,i+1,q}(t_{s,i+1}))}{\partial x}\,(\bar{f}_{i+1,q}(\bar{x}_{a,i+1,q}(t_{s,i+1}),\bar{u}_q(t_k),0)) + \frac{\partial \hat{V}_q(\bar{x}_{a,i,q}(t_{s,i+1}))}{\partial x}\,(\bar{f}_{i,q}(\bar{x}_{a,i,q}(t_{s,i+1}),\bar{u}_q(t_k),0)) \\
&\quad - \frac{\partial \hat{V}_q(\bar{x}_{a,i,q}(t_{s,i+1}))}{\partial x}\,(\bar{f}_{i,q}(\bar{x}_{a,i,q}(t_{s,i+1}),\bar{u}_q(t_k),0)) \\
&\leq -\hat{\alpha}_{3,q}(\hat{\alpha}_{2,q}^{-1}(\hat{\rho}_{e,q})) + \hat{\alpha}_{4,q}(\hat{\alpha}_{1,q}^{-1}(\hat{\rho}_q))M_{err,i,q} + L'_{x,i,q}M_i\Delta + L'_{w,i,q}\theta_i + \left| \frac{\partial \hat{V}_q(\bar{x}_{a,i+1,q}(t_{s,i+1}))}{\partial x}\,(\bar{f}_{i+1,q}(\bar{x}_{a,i+1,q}(t_{s,i+1}),\bar{u}_q(t_k),0)) \right. \\
&\quad \left. - \frac{\partial \hat{V}_q(\bar{x}_{a,i,q}(t_{s,i+1}))}{\partial x}\,(\bar{f}_{i,q}(\bar{x}_{a,i,q}(t_{s,i+1}),\bar{u}_q(t_k),0)) \right| \\
&\leq -\hat{\alpha}_{3,q}(\hat{\alpha}_{2,q}^{-1}(\hat{\rho}_{e,q})) + \hat{\alpha}_{4,q}(\hat{\alpha}_{1,q}^{-1}(\hat{\rho}_q))M_{err,i,q} + L'_{x,i,q}M_i\Delta + L'_{w,i,q}\theta_i \\
&\quad + \left| \frac{\partial \hat{V}_q(\bar{x}_{a,i,q}(t_{s,i+1}))}{\partial x} \right| \left| \bar{f}_{i+1,q}(\bar{x}_{a,i+1,q}(t_{s,i+1}),\bar{u}_q(t_k),0) - \bar{f}_{i,q}(\bar{x}_{a,i,q}(t_{s,i+1}),\bar{u}_q(t_k),0) \right| \\
&\leq -\hat{\alpha}_{3,q}(\hat{\alpha}_{2,q}^{-1}(\hat{\rho}_{e,q})) + \hat{\alpha}_{4,q}(\hat{\alpha}_{1,q}^{-1}(\hat{\rho}_q))M_{err,i,q} + L'_{x,i,q}M_i\Delta + L'_{w,i,q}\theta_i + \hat{\alpha}_{4,q}(|\bar{x}_{a,i,q}(t_{s,i+1})|)M_{change,i,q} \\
&\leq -\hat{\alpha}_{3,q}(\hat{\alpha}_{2,q}^{-1}(\hat{\rho}_{e,q})) + \hat{\alpha}_{4,q}(\hat{\alpha}_{1,q}^{-1}(\hat{\rho}_q))M_{err,i,q} + L'_{x,i,q}M_i\Delta + L'_{w,i,q}\theta_i + \hat{\alpha}_{4,q}(\hat{\alpha}_{1,q}^{-1}(\hat{\rho}_q))M_{change,i,q}
\end{aligned}
\tag{A11}
$$

where the last inequality follows from the fact that $\bar{x}_{a,i,q}(t_{s,i+1}) \in \Omega_{\hat{\rho}_q}$ if $x(t_k) \in \Omega_{\hat{\rho}_q}$ when Equations (68) and (69) hold according to Theorem 1.

Finally, for $\tau \in [t_{s,i+1}, t_{k+1})$,

$$
\begin{aligned}
&\frac{\partial \hat{V}_q(\bar{x}_{a,i+1,q}(\tau))}{\partial x}\,(\bar{f}_{i+1,q}(\bar{x}_{a,i+1,q}(\tau),\bar{u}_q(t_k),w_{i+1}(\tau))) \\
&= \frac{\partial \hat{V}_q(\bar{x}_{a,i+1,q}(\tau))}{\partial x}\,(\bar{f}_{i+1,q}(\bar{x}_{a,i+1,q}(\tau),\bar{u}_q(t_k),w_{i+1}(\tau))) + \frac{\partial \hat{V}_q(\bar{x}_{a,i+1,q}(t_{s,i+1}))}{\partial x}\,(\bar{f}_{i+1,q}(\bar{x}_{a,i+1,q}(t_{s,i+1}),\bar{u}_q(t_k),0)) \\
&\quad - \frac{\partial \hat{V}_q(\bar{x}_{a,i+1,q}(t_{s,i+1}))}{\partial x}\,(\bar{f}_{i+1,q}(\bar{x}_{a,i+1,q}(t_{s,i+1}),\bar{u}_q(t_k),0)) \\
&\leq -\hat{\alpha}_{3,q}(\hat{\alpha}_{2,q}^{-1}(\hat{\rho}_{e,q})) + \hat{\alpha}_{4,q}(\hat{\alpha}_{1,q}^{-1}(\hat{\rho}_q))M_{err,i,q} + L'_{x,i,q}M_i\Delta + L'_{w,i,q}\theta_i + \hat{\alpha}_{4,q}(\hat{\alpha}_{1,q}^{-1}(\hat{\rho}_q))M_{change,i,q} \\
&\quad + \left| \frac{\partial \hat{V}_q(\bar{x}_{a,i+1,q}(\tau))}{\partial x}\,(\bar{f}_{i+1,q}(\bar{x}_{a,i+1,q}(\tau),\bar{u}_q(t_k),w_{i+1}(\tau))) - \frac{\partial \hat{V}_q(\bar{x}_{a,i+1,q}(t_{s,i+1}))}{\partial x}\,(\bar{f}_{i+1,q}(\bar{x}_{a,i+1,q}(t_{s,i+1}),\bar{u}_q(t_k),0)) \right| \\
&\leq -\hat{\alpha}_{3,q}(\hat{\alpha}_{2,q}^{-1}(\hat{\rho}_{e,q})) + \hat{\alpha}_{4,q}(\hat{\alpha}_{1,q}^{-1}(\hat{\rho}_q))M_{err,i,q} + L'_{x,i,q}M_i\Delta + L'_{w,i,q}\theta_i + \hat{\alpha}_{4,q}(\hat{\alpha}_{1,q}^{-1}(\hat{\rho}_q))M_{change,i,q} \\
&\quad + L'_{x,i+1,q}|\bar{x}_{a,i+1,q}(\tau) - \bar{x}_{a,i+1,q}(t_{s,i+1})| + L'_{w,i+1,q}\theta_{i+1} \\
&\leq -\hat{\alpha}_{3,q}(\hat{\alpha}_{2,q}^{-1}(\hat{\rho}_{e,q})) + \hat{\alpha}_{4,q}(\hat{\alpha}_{1,q}^{-1}(\hat{\rho}_q))M_{err,i,q} + L'_{x,i,q}M_i\Delta + L'_{w,i,q}\theta_i + \hat{\alpha}_{4,q}(\hat{\alpha}_{1,q}^{-1}(\hat{\rho}_q))M_{change,i,q} \\
&\quad + L'_{x,i+1,q}M_{i+1}\Delta + L'_{w,i+1,q}\theta_{i+1}
\end{aligned}
\tag{A12}
$$

If Equation (71) holds, then integrating Equation (A12) gives that $\hat{V}_q(\bar{x}_{a,i+1,q}(t)) \leq \hat{V}_q(\bar{x}_{a,i,q}(t_{s,i+1}))$, for all $t \in [t_{s,i+1}, t_{k+1}]$. Since $\bar{x}_{a,i+1,q}(t_{s,i+1}) \in \Omega_{\hat{\rho}_q}$, this guarantees that the closed-loop state remains in $\Omega_{\hat{\rho}_q}$ even after the switch in the process model occurs, regardless of whether it occurs at a sampling time or throughout a sampling period, when the conditions of the theorem hold. $\square$

Appendix C. Proof of Theorem 3

Proof. This proof consists of several parts. First, recursive feasibility of the LEMPC of Equation (24) until $t_{d,q}$ is presented. Second, it is demonstrated that, after $t_{s,i+1}$ and before $t_{d,q}$, the closed-loop state is maintained in $\Omega_{\hat{\rho}_{samp,q}}$ under the conditions of the theorem. Third, it is demonstrated that, after $t_{d,q}$, the closed-loop state will be maintained in $\Omega_{\hat{\rho}_q}$ for a number of sampling periods given by $t_{h,q}$.

Part 1. Until $t_{d,q}$, each state measurement provided to the LEMPC of Equation (24) is within $\Omega_{\hat{\rho}_q}$. From Reference [51], under the conditions of Equations (65) and (66), this guarantees feasibility of the LEMPC of Equation (24). After $t_{d,q}$, when the closed-loop state exits $\Omega_{\hat{\rho}_q}$, feasibility is no longer guaranteed for the LEMPC of Equation (24) but $h_{NL,q}$ is then used instead according to the statement of the theorem so that a characterizable control law is always used.

Part 2. Until $t_{s,i+1}$, closed-loop stability within $\Omega_{\hat{\rho}_q}$ is guaranteed under the LEMPC of Equation (24) under the conditions in Equations (65) and (66) from Reference [51]. Subsequently, until $t_{d,q}$, it must be demonstrated that, if the state measurement is contained within $\Omega_{\hat{\rho}_q}$ at t_k, then $x(t) \in \Omega_{\hat{\rho}_{samp,q}} \subset \Omega_{\hat{\rho}_{safe,q}}$,

$t \in [t_k, t_{k+1}]$. Here, one of two cases holds: either $x(t_k) \in \Omega_{\hat{\rho}_{e,q}}$ or $x(t_k) \in \Omega_{\hat{\rho}_q}/\Omega_{\hat{\rho}_{e,q}}$. The state of the underlying model before $t_{s,i+1}$ is denoted by $\bar{x}_{a,i,q}$ and, after, is $\bar{x}_{a,i,q+1}$.

If $x(t_k) \in \Omega_{\hat{\rho}_{e,q}}$ and if $t_{s,i+1} \in [t_k, t_{k+1})$, from Propositions 1 and 2 and Equation (24f), we have the following:

$$
\begin{aligned}
\hat{V}_q(\bar{x}_{a,i,q}(t)) &\leq \hat{V}_q(\bar{x}_{b,q}(t)) + f_{V,q}(|\bar{x}_{a,i,q}(t) - \bar{x}_{b,q}(t)|) \\
&\leq \hat{\rho}_{e,q} + f_{V,q}(f_{W,i,q}(\Delta)) \leq \hat{\rho}_q
\end{aligned}
\tag{A13}
$$

for $t \in [t_k, t_{s,i+1})$ when Equation (65) holds, and

$$
\begin{aligned}
\hat{V}_q(\bar{x}_{a,i+1,q}(t)) &\leq \hat{V}_q(\bar{x}_{b,q}(t)) + f_{V,q}(|\bar{x}_{a,i+1}(t) - \bar{x}_{b,q}(t)|) \\
&\leq \hat{\rho}_{e,q} + f_{V,q}(f_{W,i,q}(t_{s,i+1} - t_k) + (M_{change,i,q})(t - t_{s,i+1}) + \frac{L_{w,i,q}\theta_i + M_{err,i,q}}{L_{x,i,q}}(e^{L_{x,i,q}t} - e^{L_{x,i,q}t_{s,i+1}}))
\end{aligned}
\tag{A14}
$$

for $t \in [t_{s,i+1}, t_{k+1})$ from Proposition 4. From the conditions in Equation (73), this gives that $\hat{V}_q(x(t))$ is maintained within $\Omega_{\hat{\rho}_{samp,q}}$ for all $t \in [t_k, t_{k+1})$.

If instead $t_{s,i+1}$ occurs before or at t_k, then $\bar{x}_{b,q}(t_k) = \bar{x}_{a,i+1,q}(t_k)$ and Propositions 1 and 2 and Equation (24f) give the following:

$$
\begin{aligned}
\hat{V}_q(\bar{x}_{a,i+1,q}(t)) &\leq \hat{V}_q(\bar{x}_{b,q}(t)) + f_{V,q}(f_{W,i+1,q}(\Delta)) \\
&\leq \hat{\rho}_{e,q} + f_{V,q}(f_{W,i+1,q}(\Delta))
\end{aligned}
\tag{A15}
$$

for all $t \in [t_k, t_{k+1})$. From the conditions in Equation (75), this gives that $\hat{V}_q(x(t))$ is maintained within $\Omega_{\hat{\rho}_{samp,q}}$ for all $t \in [t_k, t_{k+1})$.

If $x(t_k) \in \Omega_{\hat{\rho}_q}/\Omega_{\hat{\rho}_{e,q}}$, then the constraint of Equation (24g) is used. In this case, we consider the cases where $t_{s,i+1} \in [t_k, t_{k+1})$ and the case where $t_{s,i+1}$ occurs before t_k, separately.

When $t_{s,i+1} \in [t_k, t_{k+1})$, then before $t_{s,i+1}$, Equation (24g) holds. From Reference [51], Equation (66) with Equation (67) cause $\bar{x}_{a,i,q}(t) \in \Omega_{\hat{\rho}_q}$ for $t \in [t_k, t_{s,i+1})$. Subsequently, this result no longer holds because the underlying dynamic model changed so that Equation (24g) no longer provides an indication of the conditions which the closed-loop state meets, and a worst-case scenario in which the closed-loop state could subsequently move out of $\Omega_{\hat{\rho}_q}$ is considered. Specifically, the first inequality in Equation (A14) continues to hold. Equation (24f) does not necessarily hold but instead it is guaranteed [51] that $\bar{x}_{b,q}(t) \in \Omega_{\hat{\rho}_q}$ under Equations (66) and (67), so that $\hat{V}_q(\bar{x}_{b,q}) \leq \hat{\rho}_q$. Then, if Equation (74) holds, extending the first inequality in Equation (A14) guarantees that $\hat{V}_q(\bar{x}_{a,i+1,q}(t)) \leq \hat{\rho}_{samp,q}$, for $t \in [t_{s,i+1}, t_{k+1})$. Therefore, throughout a sampling period containing $t_{s,i+1}$, the closed-loop state does not leave $\Omega_{\hat{\rho}_{samp,q}}$. If instead $t_{s,i+1}$ is before t_k, then Equation (24g) is activated at t_k and when $\bar{x}_{a,i+1,q}(t_k) \in \Omega_{\hat{\rho}_q}/\Omega_{\hat{\rho}_{s,q}}$ [51]:

$$
\begin{aligned}
\frac{\partial \hat{V}_q(\bar{x}_{a,i+1,q}(\tau))}{\partial x} & (\bar{f}_{i+1,q}(\bar{x}_{a,i+1,q}(\tau), \bar{u}_q(t_k), w_{i+1}(\tau)) \\
&\leq -\hat{\alpha}_{3,q}(\hat{\alpha}_{2,q}^{-1}(\hat{\rho}_{s,q})) + \hat{\alpha}_{4,q}(\hat{\alpha}_{1,q}^{-1}(\hat{\rho}_q))M_{err,i+1,q} + L'_{x,i+1,q}M_{i+1}\Delta + L'_{w,i+1,q}\theta_{i+1}
\end{aligned}
\tag{A16}
$$

When Equation (72) is satisfied,

$$
\frac{\partial \hat{V}_q(\bar{x}_{a,i+1,q}(\tau))}{\partial x} (\bar{f}_{i+1,q}(\bar{x}_{a,i+1,q}(\tau), \bar{u}_q(t_k), w_{i+1}(\tau)) \leq \epsilon_{W,i+1,q}/\Delta
\tag{A17}
$$

or

$$
\hat{V}_q(\bar{x}_{a,i+1,q}(t)) \leq \hat{V}_q(\bar{x}_{a,i+1,q}(t_k)) + \frac{\epsilon_{W,i+1,q}}{\Delta}(t - t_k)
\tag{A18}
$$

This indicates that $\hat{V}_q$ is guaranteed to increase at a worst-case rate along the closed-loop state trajectories under the control actions determined by the LEMPC of Equation (24) if the condition of Equation (72) is satisfied after an anomaly occurs. To ensure that, at the end of the sampling period, $\hat{V}_q(\bar{x}_{a,i+1,q}(t)) \leq \hat{\rho}_{samp,q}$, given that $\hat{V}_q(\bar{x}_{a,i+1,q}(t_k)) \leq \hat{\rho}_q$, Equation (76) must hold. If $t_{s,i+1}$ is before t_k but $\bar{x}_{a,i+1,q}(t_k) \in \Omega_{\hat{\rho}_{s,q}}$, then if $\hat{\rho}_{min,i+1,q} \subset \hat{\rho}_{samp,q}$, then $\bar{x}_{a,i+1,q}(t) \in \Omega_{\hat{\rho}_{samp,q}}$ from Equation (67).

Thus, whether $x(t_k) \in \Omega_{\hat{\rho}_{e,q}}$ or $x(t_k) \in \Omega_{\hat{\rho}_q}/\Omega_{\hat{\rho}_{e,q}}$, $x(t_{k+1}) \in \Omega_{\hat{\rho}_{samp,q}}$. Applying this recursively indicates that, from $t_{s,i+1}$ until $t_{d,q}$, the closed-loop state is maintained within $\Omega_{\hat{\rho}_{samp,q}}$. This also indicates that $\hat{V}_q(\bar{x}_{a,i+1,q}(t_{d,q})) \leq \hat{\rho}_{samp,q}$. Because $\Omega_{\hat{\rho}_{samp,q}} \subset \Omega_{\hat{\rho}_{safe,q}}$, $\bar{x}_{a,i+1,q}(t_{d,q}) \in \Omega_{\hat{\rho}_{safe,q}}$ as well.

Part 3. At $t_{d,q}$, $h_{NL,q}$ in sample-and-hold begins to be used to control the process. Again, Equations (A16)–(A18) hold.

The time $t_{out,q}$ at which the closed-loop state reaches $\Omega_{\hat{\rho}_{safe,q}}$ (i.e., when $\hat{V}_q(\bar{x}_{a,i+1,q}(t_{out,q})) = \hat{\rho}_{safe,q}$) when initialized from $\hat{V}_q(\bar{x}_{a,i+1,q}(t_k)) = \hat{\rho}_{samp,q}$, where $\hat{\rho}_{samp,q} \leq \hat{\rho}_{safe,q}$, is at least $\frac{(\hat{\rho}_{safe,q}-\hat{\rho}_{samp,q})\Delta}{\epsilon_{W,i+1,q}} + t_k$. To ensure that the time between t_k and $t_{out,q}$ is no greater than $\frac{(\hat{\rho}_{safe,q}-\hat{\rho}_{samp,q})\Delta}{\epsilon_{W,i+1,q}}$, the number of sampling periods available after $t_{d,q}$ until the model needs to be updated with one which meets the conditions in Equation (66) with i set to $i + 1$ and q set to $q + 1$ is floor$\left(\frac{(\hat{\rho}_{safe,q}-\hat{\rho}_{samp,q})}{\epsilon_{W,i+1,q}}\right)$. $\square$

Appendix D. Proof of Theorem 4

Proof. If $h_{NL,q+1}$ is used to control the system after $t_{ID,q}$ and the conditions of Theorem 4 are met, then $x_{a,i+1,q}(t_{ID,q}) = x_{a,i+1,q+1}(t_{ID,q})$, which lies in both $\Omega_{\hat{\rho}_{safe,q}}$ and in $\Omega_{\hat{\rho}_{safe,q+1}}$ so that the closed-loop state has not left either region. From Reference [51], if Equation (66) is met for the $q + 1/i + 1$ model combination, then $h_{NL,q+1}$ causes $\hat{V}_{q+1}$ to decrease so that it will not leave $\Omega_{\hat{\rho}_{safe,q+1}}$ before the closed-loop state enters $\Omega_{\hat{\rho}_{q+1}}$. Once the closed-loop state enters $\Omega_{\hat{\rho}_{q+1}}$, then the LEMPC of Equation (24) is used with the $q + 1$ model, and if Equations (65) and (66) are met for the $q + 1/i + 1$ model combination, the closed-loop state is maintained in $\Omega_{\hat{\rho}_{q+1}}$ from Reference [51]. $\square$

References

1. Lee, J.H.; Shin, J.; Realff, M.J. Machine learning: Overview of the recent progresses and implications for the process systems engineering field. *Comput. Chem. Eng.* **2018**, *114*, 111–121. [CrossRef]
2. Venkatasubramanian, V. The promise of artificial intelligence in chemical engineering: Is it here, finally? *AIChE J.* **2019**, *65*, 466–478. [CrossRef]
3. Bangi, M.S.F.; Kwon, J.S.I. Deep hybrid modeling of chemical process: Application to hydraulic fracturing. *Comput. Chem. Eng.* **2020**, *134*, 106696. [CrossRef]
4. Wu, Z.; Christofides, P.D. Economic Machine-Learning-Based Predictive Control of Nonlinear Systems. *Mathematics* **2019**, *7*, 494. [CrossRef]
5. Lovelett, R.J.; Dietrich, F.; Lee, S.; Kevrekidis, I.G. Some manifold learning considerations towards explicit model predictive control. *arXiv* **2018**, arXiv:1812.01173.
6. Lucia, S.; Karg, B. A deep learning-based approach to robust nonlinear model predictive control. *IFAC-PapersOnLine* **2018**, *51*, 511–516. [CrossRef]
7. Tong, C.; Palazoglu, A.; Yan, X. Improved ICA for process monitoring based on ensemble learning and Bayesian inference. *Chemom. Intell. Lab. Syst.* **2014**, *135*, 141–149. [CrossRef]
8. Chiang, L.H.; Kotanchek, M.E.; Kordon, A.K. Fault diagnosis based on Fisher discriminant analysis and support vector machines. *Comput. Chem. Eng.* **2004**, *28*, 1389–1401. [CrossRef]
9. Rawlings, J.B.; Angeli, D.; Bates, C.N. Fundamentals of economic model predictive control. In Proceedings of the IEEE Conference on Decision and Control, Maui, HI, USA, 10–13 December 2012; pp. 3851–3861.

10. Grüne, L. Economic receding horizon control without terminal constraints. *Automatica* **2013**, *49*, 725–734. [CrossRef]

11. Huang, R.; Harinath, E.; Biegler, L.T. Lyapunov stability of economically oriented NMPC for cyclic processes. *J. Process Control* **2011**, *21*, 501–509. [CrossRef]

12. Ellis, M.; Durand, H.; Christofides, P.D. A tutorial review of economic model predictive control methods. *J. Process Control* **2014**, *24*, 1156–1178. [CrossRef]

13. Patel, N.R.; Risbeck, M.J.; Rawlings, J.B.; Wenzel, M.J.; Turney, R.D. Distributed economic model predictive control for large-scale building temperature regulation. In Proceedings of the American Control Conference, Boston, MA, USA, 6–8 July 2016; pp. 895–900.

14. Zhang, A.; Yin, X.; Liu, S.; Zeng, J.; Liu, J. Distributed economic model predictive control of wastewater treatment plants. *Chem. Eng. Res. Des.* **2019**, *141*, 144–155. [CrossRef]

15. Zachar, M.; Daoutidis, P. Nonlinear Economic Model Predictive Control for Microgrid Dispatch. *IFAC-PapersOnLine* **2016**, *49*, 778–783. [CrossRef]

16. Gopalakrishnan, A.; Biegler, L.T. Economic nonlinear model predictive control for periodic optimal operation of gas pipeline networks. *Comput. Chem. Eng.* **2013**, *52*, 90–99. [CrossRef]

17. Brunton, S.L.; Proctor, J.L.; Kutz, J.N. Discovering governing equations from data by sparse identification of nonlinear dynamical systems. *Proc. Natl. Acad. Sci. USA* **2016**, *113*, 3932–3937. [CrossRef]

18. Narasingam, A.; Kwon, J.S.I. Data-driven identification of interpretable reduced-order models using sparse regression. *Comput. Chem. Eng.* **2018**, *119*, 101–111. [CrossRef]

19. Chakraborty, S.; Tomsett, R.; Raghavendra, R.; Harborne, D.; Alzantot, M.; Cerutti, F.; Srivastava, M.; Preece, A.; Julier, S.; Rao, R.M.; et al. Interpretability of deep learning models: A survey of results. In Proceedings of the IEEE Smart World Congress, San Francisco, CA, USA, 4–8 August 2017.

20. Karpathy, A.; Johnson, J.; Li, F.-F. Visualizing and understanding recurrent networks. *arXiv* **2015**, arXiv:1506.02078.

21. Qin, S.J.; Badgwell, T.A. A survey of industrial model predictive control technology. *Control Eng. Pract.* **2003**, *11*, 733–764. [CrossRef]

22. Kheradmandi, M.; Mhaskar, P. Prescribing Closed-Loop Behavior Using Nonlinear Model Predictive Control. *Ind. Eng. Chem. Res.* **2017**, *56*, 15083–15093. [CrossRef]

23. Bayer, F.A.; Müller, M.A.; Allgöwer, F. Tube-based robust economic model predictive control. *J. Process Control* **2014**, *24*, 1237–1246. [CrossRef]

24. Heidarinejad, M.; Liu, J.; Christofides, P.D. Economic model predictive control of nonlinear process systems using Lyapunov techniques. *AIChE J.* **2012**, *58*, 855–870. [CrossRef]

25. Diehl, M.; Bjornberg, J. Robust dynamic programming for min-max model predictive control of constrained uncertain systems. *IEEE Trans. Autom. Control* **2004**, *49*, 2253–2257. [CrossRef]

26. Mesbah, A. Stochastic model predictive control: An overview and perspectives for future research. *IEEE Control Syst. Mag.* **2016**, *36*, 30–44.

27. Das, B.; Mhaskar, P. Lyapunov-based offset-free model predictive control of nonlinear process systems. *Can. J. Chem. Eng.* **2015**, *93*, 471–478. [CrossRef]

28. Vaccari, M.; Pannocchia, G. A modifier-adaptation strategy towards offset-free economic MPC. *Processes* **2017**, *5*, 2. [CrossRef]

29. Adetola, V.; DeHaan, D.; Guay, M. Adaptive model predictive control for constrained nonlinear systems. *Syst. Control Lett.* **2009**, *58*, 320–326. [CrossRef]

30. Wu, Z.; Rincon, D.; Christofides, P.D. Real-Time Adaptive Machine-Learning-Based Predictive Control of Nonlinear Processes. *Ind. Eng. Chem. Res.* **2019**, in press. [CrossRef]

31. Aumi, S.; Mhaskar, P. Adaptive data-based model predictive control of batch systems. In Proceedings of the American Control Conference, Montreal, QC, Canada, 27–29 June 2012.

32. Aswani, A.; Gonzalez, H.; Sastry, S.S.; Tomlin, C. Provably safe and robust learning-based model predictive control. *Automatica* **2013**, *49*, 1216–1226. [CrossRef]

33. El-Farra, N.H.; Gani, A.; Christofides, P.D. Fault-tolerant control of process systems using communication networks. *AIChE J.* **2005**, *51*, 1665–1682. [CrossRef]

34. Perk, S.; Shao, Q.M.; Teymour, F.; Cinar, A. An adaptive fault-tolerant control framework with agent-based systems. *Int. J. Robust Nonlinear Control* **2012**, *22*, 43–67. [CrossRef]
35. Du, M.; Mhaskar, P. Uniting safe-parking and reconfiguration-based approaches for fault-tolerant control of switched nonlinear systems. In Proceedings of the 2010 American Control Conference, Baltimore, MD, USA, 30 June–2 July 2010; pp. 2829–2834.
36. Alanqar, A.; Durand, H.; Christofides, P.D. Fault-Tolerant Economic Model Predictive Control Using Error-Triggered Online Model Identification. *Ind. Eng. Chem. Res.* **2017**, *56*, 5652–5667. [CrossRef]
37. Bø, T.I.; Johansen, T.A. Dynamic safety constraints by scenario based economic model predictive control. *IFAC Proc. Vol.* **2014**, *47*, 9412–9418. [CrossRef]
38. Albalawi, F.; Alanqar, A.; Durand, H.; Christofides, P.D. A feedback control framework for safe and economically-optimal operation of nonlinear processes. *AIChE J.* **2016**, *62*, 2391–2409. [CrossRef]
39. Zhang, X.; Clark, M.; Rattan, K.; Muse, J. Controller verification in adaptive learning systems towards trusted autonomy. In Proceedings of the ACM/IEEE Sixth International Conference on Cyber-Physical Systems, Seattle, WA, USA, 14–16 April 2015; pp. 31–40.
40. Wu, Z.; Rincon, D.; Christofides, P.D. Real-time machine learning for operational safety of nonlinear processes via barrier-function based predictive control. *Chem. Eng. Res. Des.* **2020**, *155*, 88–97. [CrossRef]
41. Alanqar, A.; Durand, H.; Christofides, P.D. Error-triggered on-line model identification for model-based feedback control. *AIChE J.* **2017**, *63*, 949–966. [CrossRef]
42. Durand, H.; Messina, D. Enhancing practical tractability of Lyapunov-based economic model predictive control. In Proceedings of the American Control Conference, Denver, CO, USA, 1–3 July 2020.
43. Alanqar, A.; Durand, H.; Christofides, P.D. On identification of well-conditioned nonlinear systems: Application to economic model predictive control of nonlinear processes. *AIChE J.* **2015**, *61*, 3353–3373. [CrossRef]
44. Alanqar, A.; Ellis, M.; Christofides, P.D. Economic model predictive control of nonlinear process systems using empirical models. *AIChE J.* **2015**, *61*, 816–830. [CrossRef]
45. Durand, H.; Ellis, M.; Christofides, P.D. Economic model predictive control designs for input rate-of-change constraint handling and guaranteed economic performance. *Comput. Chem. Eng.* **2016**, *92*, 18–36. [CrossRef]
46. Nasukawa, T.; Yi, J. Sentiment analysis: Capturing favorability using natural language processing. In Proceedings of the Second International Conference on Knowledge Capture, Sanibel Island, FL, USA, 23–25 October 2003; pp. 70–77.
47. Durand, H.; Christofides, P.D. Economic model predictive control: Handling valve actuator dynamics and process equipment considerations. *Found. Trends Syst. Control* **2018**, *5*, 293–350. [CrossRef]
48. Özgülşen, F.; Adomaitis, R.A.; Çinar, A. A numerical method for determining optimal parameter values in forced periodic operation. *Chem. Eng. Sci.* **1992**, *47*, 605–613. [CrossRef]
49. Alfani, F.; Carberry, J.J. An exploratory kinetic study of ethylene oxidation over an unmoderated supported silver catalyst. *Chim. Ind.* **1970**, *52*, 1192–1196.
50. Wächter, A.; Biegler, L.T. On the implementation of an interior-point filter line-search algorithm for large-scale nonlinear programming. *Math. Program.* **2006**, *106*, 25–57. [CrossRef]
51. Giuliani, L.; Durand, H. Data-Based Nonlinear Model Identification in Economic Model Predictive Control. *Smart Sustain. Manuf. Syst.* **2018**, *2*, 61–109. [CrossRef]
52. Kheradmandi, M.; Mhaskar, P. Model predictive control with closed-loop re-identification. *Comput. Chem. Eng.* **2018**, *109*, 249–260. [CrossRef]
53. Heidarinejad, M.; Liu, J.; Christofides, P.D. Economic model predictive control of switched nonlinear systems. *Syst. Control. Lett.* **2013**, *62*, 77–84. [CrossRef]
54. Heidarinejad, M.; Liu, J.; Christofides, P.D. Distributed model predictive control of switched nonlinear systems with scheduled mode transitions. *AIChE J.* **2013**, *59*, 860–871. [CrossRef]
55. Mhaskar, P.; Liu, J.; Christofides, P.D. *Fault-Tolerant Process Control: Methods and Applications*; Springer: London, UK, 2013.

56. Lin, Y.; Sontag, E.D. A universal formula for stabilization with bounded controls. *Syst. Control. Lett.* **1991**, *16*, 393–397. [CrossRef]
57. Durand, H. On accounting for equipment-control interactions in economic model predictive control via process state constraints. *Chem. Eng. Res. Des.* **2019**, *144*, 63–78. [CrossRef]

Article

Inframarginal Model Analysis of the Evolution of Agricultural Division of Labor

Xueping Jiang [1], Jen-Mei Chang [2] and Hui Sun [2,*]

[1] College of Mathematics and Informatics, South China Agricultural University, Guangzhou 510642, China; jxuep@scau.edu.cn
[2] Department of Mathematics and Statistics, California State University, Long Beach, CA 90840, USA; jen-mei.chang@csulb.edu
* Correspondence: hui.sun@csulb.edu; Tel.: +1-(562)-985-5609

Received: 24 October 2019; Accepted: 25 November 2019; Published: 1 December 2019

Abstract: Division of labor plays a critical role in many parts of agriculture. For example, a specialized division of labor can lead to the improvement of labor productivity, the reduction of production costs, and the innovation of production technology and organization. At the heart of agricultural management is how the comparative advantages of farmers impact their production decision-making behavior, and, consequently, influence the division of labor structure. In this paper, we apply an infra-marginal model to interpret the selection logic of heterogeneous farmers' specialized production with exogenous comparative technical advantages and transaction costs. Solving the nonlinear programming problem of the utility function within each respective labor structure leads to a corner equilibrium. Under reasonable assumptions of the model, we reduced the number of possible production–consumption decision modes from the maximum of 64 to an optimal of 3. Through this analysis, we discovered the ranges for transaction efficiency coefficients and preference parameter under which each structure can achieve general equilibrium. Our theoretical model thereby explains the structural evolution of agricultural division of labor.

Keywords: comparative advantage; transaction cost; specialized production; infra-marginal model; agricultural division of labor

---

1. Introduction

In agriculture, each farmer behaves as a limited and rational production decision maker; farmers allocate resources rationally similar to entrepreneurs. In the traditional agriculture where profit maximization is the farmers' ultimate behavioral goal, it is relatively rare to see an inefficient allocation of production factors [1]. As long as farmers prefer the principle of manufacturer in management, they may allocate resources to the most efficient production field, thus bringing specialization and division of labor. Specialized division of labor has a direct impact on economics that leads to the improvement of labor productivity and the reduction of production costs, and an indirect impact that leads to the innovation of production technology and organization. Together, these impacts lead to the saving of factor resources and the improvement of labor efficiency [2]. Under the appropriate external economic conditions, the development of the division of labor within the household will naturally devote labor and capital on a few business activities, or even one. As a result, farmers generally increase the amount of capital, technology, or land input in the original factor combination, forming an intensive management based on a certain factor. Therefore, the rational production decision of farmers is to pursue the division of labor economy formed by the comparative advantage.

China's current policies focus on promoting moderate scale and specialized agricultural operation to improve the scale economy and division of labor economy, and promoting the transformation of agricultural management methods. An underlying aspect of such policies is to encourage farmers to switch from small and full to specialized management. The heterogeneity of the farmers is assumed in the heart of agricultural management. That is, the farmers have their own comparative advantages under the conditions of open management. With this assumption, we study the effect of the comparative advantages on the production decision-making behavior, which then influences the kind of division of labor structure we present. This forms the basis of our paper.

Our study uses the Ricardian model [3–5] to include the comparative advantages of farmers and market transaction costs. David Ricardo's theory of comparative advantage is considered the cornerstone of modern trade theory. However, due to the presence of corner solutions, traditional marginal analysis cannot be applied to the Ricardo model [6]. For this reason, the model has not received its due attention [4]. If we used the absolute separation between pure consumers and enterprises, we would generate multiple general equilibria based on multiple corner and interior point solution structures. However, under the Walras system, companies do not care which structure they choose, and pure consumers cannot choose the production structure. Hence, partial equilibrium may be a general equilibrium in each structure. This multiplicity of the general equilibrium makes comparatively static analysis of general equilibrium impossible [7]. Now, if the Smith framework is used for analysis, each individual can be a producer–consumer, and can choose its level of specialization. That is, the general equilibrium is one of the multiple corner equilibria. The general equilibrium is an effective compromise between the division of labor economies generated by exogenous comparative technological advantages and transaction costs [6].

In the literature, there are exogenous and endogenous comparative advantages, as well as comprehensive comparative advantages [8–10]. Based on the Ricardo model, we construct a mathematical model on farmers' participation in the division of labor with exogenous comparative technical advantages and transaction costs by taking into consideration the simplification of the model and the simplicity of the structure. In our work, we pioneer the use of the infra-marginal model to study the evolution of agricultural division of labor, which is about farmers' specialization and the change of their agricultural economic organization.

The infra-marginal model provides a powerful tool to study the division of labor and professionalization of the economy. The concept was initiated in the 1950s and 1960s [11–13] and further developed by Yang [14,15]. In such a model, it is assumed that business decisions can be categorized into two classes: marginal and infra-marginal. Marginal decisions are concerned with the extent to which resources are allocated to a pre-determined set of activities, while infra-marginal decisions are about what activities to engage in (or whether or not to engage in an activity). In the context of social division of labor, the infra-marginal decisions of individuals allow the formation of a network division of labor of various sizes. The infra-marginal analysis is concerned with the optimal infra-marginal network decisions and the outcome of these decisions. The optimal infra-marginal network decisions rely on the total cost–benefit analysis across different network patterns of specialization and trade connections as well as the marginal analysis of resource allocation for a given network pattern. Mathematically speaking, infra-marginal analysis transcends into non-classical mathematical programming problems (e.g., linear and nonlinear programming, mixed integer programming, dynamic programming, and control theory) that allow corner solutions [16].

The infra-marginal model finds a variety of applications. For example, it can be coupled with the Ricardian model to study the mechanisms for economic development as well as the evolution of trade policy regimes [3–5,17]. Infra-marginal analysis was applied to the Dixit–Krugman model to explain the evolution of trade pattern determined by the interplay between endogenous and exogenous comparative advantages [18]. It was also used in the Dixit–Stiglitz model to predict the tests of scale effects [19]. The aforementioned applications are all on international trade. Moreover, dynamic infra-marginal analysis was applied in the Yang and Borland (Y–B) model to obtain the dynamic general equilibrium based on corner solutions. It can also be seen in the areas of economic growth and development theory [8,20]. More applications of the infra-marginal model can be found in the studies of the firm, contract and property rights, insurance, e-business, money, capital and business cycle [21–27], and urbanization and industrialization such as the relationships among the division of labor, agglomeration, and land rentals [28,29]. Despite the rich applications of the infra-marginal model, its application to the special topic of agricultural division of labor is generally lacking. A major contribution of our work is the development of a framework that helps to explain the selection logic of farmers' specialized production and the structural evolution of agricultural division of labor through the construction of an infra-marginal model.

The rest of the paper is organized as follows. In Section 2, we construct an infra-marginal model by considering agricultural comparative advantages. The model consists of four possible division of labor structures. In Section 3, we set up and develop the corner equilibrium solutions to the nonlinear utility optimization problems that are associated with the four structures. In Section 4, we analyze the conditions in the parameter space that lead to various general equilibria as well as explain the division selection logic and decision mechanism of farmers with comparative advantage. We conclude our work with a summary and discussion in Section 5.

2. Materials and Methods—An Infra-Marginal Model

Based on the Ricardo model with exogenous comparative technical advantages and transaction cost, we construct an infra-marginal model of farmers' comparative advantage and specialization choice, which reveals the selection logic of farmers' specialized production and the structural evolution rule of agricultural division of labor.

2.1. Model Definition

Our mathematical model inherits a set of reasonable assumptions. The economy is composed of two producer–consumer integrated farmers and each farmer has a comparative advantage due to their heterogeneity. Two different farmers, Farmers 1 and 2, both consume two agricultural products x and y (x and y may also be labor services in agricultural production links) and determine their own patterns of production and trading activities.

With these assumptions, the farmer production system (as a production–consumer integration in the model) can be constructed. In general, we have at our disposal many utility functions (e.g., linear, Leontief, constant elasticity substitution, Cobb–Douglas (C-D), etc.). Each comes with a set of restrictions. In our agricultural model, the two labor services or products are both necessary and indispensable to the final product. This assumption is enforced with a zero utility if one of the necessary services or products has a value of zero. Since C-D utility function is the only one among those described above that satisfies this requirement, it is used in our model.

The utility function of farmer i ($i = 1, 2$) is:

$$U_i = (x_i + kx_i^d)^\beta (y_i + ky_i^d)^{1-\beta}, \tag{1}$$

where x_i and y_i are the respective self-sufficiency quantities of agricultural products (or production link), x_i^d and y_i^d are the respective demand quantities of farmers, k is the transaction efficiency coefficient, and β is the preference parameter of farmers.

The production functions of farmer i ($i = 1, 2$) are:

$$x_i^p = x_i + x_i^s = a_{ix}l_{ix}, \quad \text{and} \quad y_i^p = y_i + y_i^s = a_{iy}l_{iy}. \tag{2}$$

Here, x_i^p and y_i^p are, respectively, the output level of two kinds of agricultural products produced by farmers (or labor services engaged in two production links); and x_i^s and y_i^s are, respectively, the supply quantity of farmers' products or labor service. Moreover, l_{ij} ($i = 1, 2; j = x, y$) is the amount of labor used by farmer i to produce agricultural product (or labor services) j, which is called the level of specialization of farmer i when producing agricultural product (or labor services) j. In addition, coefficient a_{ij} is the labor productivity of farmer i when producing agricultural product (or labor services) j.

Under these definitions, the case where Farmer 1 has a comparative advantage in the production of agricultural product (or labor services) x can be represented mathematically by $a_{1x}/a_{1y} > a_{2x}/a_{2y}$. It means that, compared to Farmer 2, Farmer 1 has a higher relative productivity on x over y; therefore, Farmer 1's opportunity cost for product x is smaller.

Moreover, we can use the labor endowment constraint to measure farmers' level of specialization. In particular, the labor endowment constraint of farmer i is given by $l_{ix} + l_{iy} = 1$. For example, $l_{ix} = 0$ means that farmer i devotes all of their labor to produce product y, making them a specialized producer of y.

Farmers' consumption, production, and trading decisions involve six non-negative variables $x_i, x_i^s, x_i^d, y_i, y_i^s$, and y_i^d, resulting in a total of $2^6 = 64$ combinations.

With market clearing (supply equals demand), the budget constraint reads $p_x x_i^s + p_y y_i^s = p_x x_i^d + p_y y_i^d$. To avoid needless trade cost, it is prohibited to buy and sell the same product (or service). As a result, x_i^s and x_i^d cannot be positive at the same time, i.e., $x_i^s x_i^d = 0$. The same holds true for product y to arrive at $y_i^s y_i^d = 0$. Overall, the budget constraint simplifies to $(y_i^s = x_i^d = 0, p_x x_i^s = p_y y_i^d)$ or $(x_i^s = y_i^d = 0, p_x x_i^d = p_y y_i^s)$.

2.2. Optimal Decision Mode and Division of Labor Structure

Based on the budget constraint and the other constraints that we present next, most of the optimal decisions from the 64 possible combinations can be excluded. Consequently, only a few division of labor structures are deduced. In all cases, any combination of the six variables should meet the budget constraints and the condition of positive utility.

For the convenience of the analysis, we write the variables into a 6-tuple $Z_i = (x_i, x_i^s, x_i^d, y_i, y_i^s, y_i^d)$. We use the notations 0 or $+$ inside the 6-tuple to denote zero or positive values. For example, $(\ ,0,+,\ ,\ ,\ )$ denotes the case $x_i^s = 0$ and $x_i^d > 0$. The cases that violate budget constraint are: $(\ ,0,\ ,\ ,\ ,+)$, $(\ ,+,\ ,\ ,\ ,0)$, $(\ ,\ ,0,\ ,+,\ )$, and $(\ ,\ ,+,\ ,0,\ )$. There are a total of $2^4 + 2^3 + 2^4 + 2^3 = 48$ such combinations. Moreover, there are four cases with $(\ ,+,+,\ ,+,+)$ that involve selling and buying the same product, which are inefficient cases, because they introduce unnecessary transaction costs. In the remaining 12 combinations, there are seven combinations with either the form $(0,\ ,0,\ ,\ ,\ )$ or $(\ ,\ ,\ ,0,\ ,0)$, which do not meet the positive utility constraint $U_i > 0$. The remaining five cases can be summarized into three decision modes: self-sufficient mode $(+,0,0,+,0,0)$, semi-specialized mode $((+,+,0,+,0,+)$ and $(+,0,+,+,+,0))$, and complete-specialization mode $((+,+,0,0,0,+)$ and $(0,0,+,+,+,0))$. These three decision modes are assigned to Farmer 1 or Farmer 2. We call a combination of modes for both farmers a structure. With the comparative advantage assumption of Farmer 1 producing product x, certain structures need to be avoided. For example, the structures with either $Z_1 = (0,0,+,+,+,0)$ (meaning Farmer 1 specializes in production y) or $Z_2 = (+,+,0,0,0,+)$ (meaning Farmer 2 specializes in production x) violate the comparative disadvantage of individual farmers, hence need to be excluded from consideration. Now, we analyze in detail the three modes that make up the various types of structures.

1. Self-sufficiency mode is generally expressed as $(xy)_i$ and defined as $Z_i = (+,0,0,+,0,0)$, for $i = 1, 2$. This indicates that all agricultural products or labor services are self-sufficient. With an economy of two farmers, this kind of social organization structure is called self-sufficiency Structure A.

2. Semi-specialized mode is when farmers produce products or services with comparative advantages, generally expressed as $(xy/y)_1$ and $(xy/x)_2$. The mode $(xy/y)_1$ corresponds to the case where $Z_1 = (+,+,0,+,0,+)$, meaning that Farmer 1 produces certain self-sufficient quantities of products x and y, sells products x, and purchases products y. Consider the example of the labor of plant protection and weeding in agricultural production. Farmer 1 purchases a small portable spraying machine to spray chemicals on his own and others' crops and he takes care of his own weeding partially. In addition, he also purchases weeding labor services

from other farmers. Namely, he outsources the labor services of weeding. The mode $(xy/x)_2$ corresponds similarly to $Z_2 = (+, 0, +, +, +, 0)$.

3. Complete-specialization mode is when farmers produce products or services with comparative advantage, expressed as $(x/y)_1$ and $(y/x)_2$. The mode $(x/y)_1$, or $Z_1 = (+, +, 0, 0, 0, +)$, represents the case where Farmer 1 specializes in producing goods or services x and is self-sufficient in selling x and buying goods or services y. The mode $(y/x)_2$, or $Z_2 = (0, 0, +, +, +, 0)$, represents that Farmer 2 specializes in producing goods or services y and is self-sufficient in selling y and buying goods or services x.

In addition to self-sufficiency Structure A, the decisions of Farmers 1 and 2 to their own production and trading activities also involve two partial division of labor structures: Ba, composed of $(xy/y)_1$ and $(y/x)_2$; Bb, composed of $(x/y)_1$ and $(xy/x)_2$; and a complete division of labor Structure C, which is composed of $(x/y)_1$ and $(y/x)_2$. The above modes and structures are demonstrated in Figure 1.

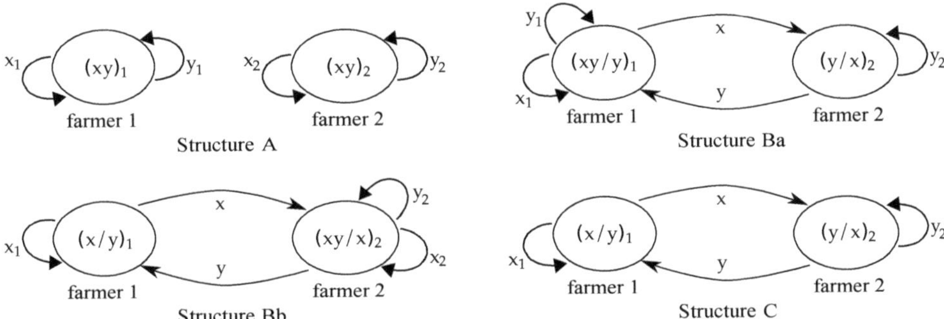

Figure 1. A schematic view of the four possible division of labor structures, under the assumption of Farmer 1 being comparative advantageous in production of x. A self-looping arrow indicates that a farmer consumes the products that he/she makes. A forward arrow from farmer i to farmer j means that farmer i produces certain products (indicated by the symbol above or below the arrow) and sells them to farmer j.

3. Optimization Analysis—Decision and Corner Equilibrium

To analyze the comparative advantages of exogenous technology of farmers and how the transaction costs affect the division of labor, that is, how the social organization structure evolves from self-sufficiency to partial division of labor and then to complete division of labor, it is necessary to analyze the decision-making strategies by first maximizing individual utility based on the infra-margin, to obtain partial or corner equilibriums for each given structure. The general equilibrium is one of the four corner equilibria with the maximum utility. To do this, we first use nonlinear programming to solve the problem of maximization of farmers' individual benefits, then use the market clearing conditions to solve the partial equilibrium of each of the four structures, and finally use the total return-cost analysis method to determine the general equilibrium.

3.1. The Selection of Self-Sufficiency Mode

The selection of self-sufficiency mode $(xy)_1$ can be formulated as:

$$\max_{x_1, y_1, l_{1x}, l_{1y}} U_1 = x_1^{\beta} y_1^{1-\beta} \tag{3}$$

$$\text{s.t.} \quad x_1 = a_{1x} l_{1x}, \quad y_1 = a_{1y} l_{1y}, \quad l_{1x} + l_{1y} = 1 \tag{4}$$

We use the marginal analysis method to solve this problem by first substituting the constraints into the objective function, and then setting the first derivative to zero. This yields the solution $l_{1x} = \beta$,

$l_{1y} = 1 - \beta$, $x_1 = a_{1x}\beta$, $y_1 = a_{1y}(1 - \beta)$, with the utility of self-sufficient Farmer 1 being $U_{A_1} = \beta^\beta(1 - \beta)^{1-\beta}a_{1x}^\beta a_{1y}^{1-\beta}$. Similarly, the utility of self-sufficient Farmer 2 is $U_{A_2} = \beta^\beta(1 - \beta)^{1-\beta}a_{2x}^\beta a_{2y}^{1-\beta}$.

3.2. The Selection of Semi-Specialized Mode

Given the partial division of labor Structure Ba, the utility maximization problem for semi-specialized mode $(xy/y)_1$ is:

$$\max_{x_1,y_1,x_1^s,y_1^d,l_{1x},l_{1y}} U_1 = x_1^\beta(y_1 + ky_1^d)^{1-\beta} \tag{5}$$

$$\text{s.t.} \quad x_1 + x_1^s = a_{1x}l_{1x}, \quad y_1 = a_{1y}l_{1y}, \quad l_{1x} + l_{1y} = 1, \quad y_1^d = px_1^s, \tag{6}$$

where $p \equiv p_x/p_y$ is the relative price of product or service x compared to y. Similarly, to solve this problem, we substitute all the variables in the objective function with l_{1x} and x_1^s using the four constraints. The first-order derivatives are

$$\frac{\partial U_1}{\partial x_1^s} = \left(-\frac{\beta}{a_{1x}l_{1x} - x_1^s} + \frac{kp(1 - \beta)}{a_{1y}(1 - l_{1x}) + kpx_1^s}\right)U_1 \tag{7}$$

$$\frac{\partial U_1}{\partial l_{1x}} = \left(\frac{a_{1x}\beta}{a_{1x}l_{1x} - x_1^s} - \frac{a_{1y}(1 - \beta)}{a_{1y}(1 - l_{1x}) + kpx_1^s}\right)U_1 \tag{8}$$

Setting both derivatives zero then requires $p = a_{1y}/(ka_{1x})$. It then follows naturally that $x_1 = \beta a_{1x}$, $x_1^s = a_{1x}(l_{1x} - \beta)$, $y_1 = a_{1y}(1 - l_{1x})$, $y_1^d = a_{1y}(l_{1x} - \beta)/k$, and $U_1 = \beta^\beta(1 - \beta)^{1-\beta}a_{1x}^\beta a_{1y}^{1-\beta}$. We refer the interested reader to Appendix A.1 for more details. Interestingly, the maximizer variables x_1^s, y_1, and y_1^d are functions of l_{1x}, while the maximal utility U_1 is independent of l_{1x}. The above equilibrium solution relies on a fixed relative market price p. Our analysis in the following remarks shows that this relative market price determines the mode choice of Farmer 1.

Remark 1. *If $p > a_{1y}/(ka_{1x})$, with the optimal value of x_1^s given by $\partial U_1/\partial x_1^s = 0$, we have $\partial U_1/\partial l_{1x} > 0$. This means the utility of Farmer 1 can always be improved by improving l_{1x}. That is, the utility of Farmer 1 will always increase with the increase of labor allocation to x (the specialization level generating x). Therefore, the optimal value of l_{1x} is its upper limit value. Due to the constraint of farmers' endowment of working hours, if the upper limit $l_{1x} = 1$ is taken, the farmer should not produce y, but should be specialized in the production of x. That is, when $p > a_{1y}/(ka_{1x})$, Farmer 1 will choose the mode $(x/y)_1$ instead of the mode $(xy/y)_1$. Similarly, when $p < a_{1y}/(ka_{1x})$, Farmer 1 will choose the mode $(xy)_1$ instead of the mode $(xy/y)_1$. Only when the relative price of market p is $a_{1y}/(ka_{1x})$ will Farmer 1 select mode $(xy/y)_1$. This condition is similar to the zero-profit condition in the standard general equilibrium with the same scale return.*

Remark 2. *It is seen that, if the relative price p of the transaction cost, after any discount, in the market is lower than the marginal conversion rate $a_{1y}/(ka_{1x})$ of Farmer 1 in self-sufficiency, the optimal decision of farmers is to be self-sufficient and produce two products or services x and y at the same time. If $p > a_{1y}/(ka_{1x})$, the marginal utility of the level of specialization of the Farmer 1 always increases with the increase of l_{1x}, so the optimal decision is to specialize in producing x. However, when p is $a_{1y}/(ka_{1x})$, self-sufficiency mode and semi-specialized mode $(xy/y)_1$ produce the same effect. Thus, if the market clearing conditions in the general equilibrium can ensure that demand and supply can be achieved in mode $(xy/y)_1$, the farmer will choose this mode. In this decision-making solution, the optimal value of l_{1x} is uncertain, and its equilibrium value will be determined by the conditions for market clearing.*

3.3. The Selection of Complete-Specialized Mode

The utility maximization problem for Farmer 2 with mode $(y/x)_2$ is:

$$\max_{x_2^d, y_2, y_2^s} U_2 = (kx_2^d)^\beta y_2^{1-\beta} \tag{9}$$

$$\text{s.t.} \quad y_2 + y_2^s = a_{2y}l_{2y}, \quad y_2^s = px_2^d, \quad l_{2y} = 1. \tag{10}$$

In the context of Structure Ba, $(y/x)_2$ is selected jointly with $(xy/y)_1$, and $p = a_{1y}/(ka_{1x})$ is the equilibrium relative price. The system yields the optimal solution $x_2^d = k\beta a_{2y}a_{1x}/a_{1y}$, $y_2 = (1-\beta)a_{2y}$, $y_2^s = \beta a_{2y}$. The market clearing conditions $x_1^s = x_2^d$ lead to $l_{1x} = \beta + k\beta a_{2y}/a_{1y}$. The condition $l_{1x} < 1$ is met if and only if $a_{2y}/a_{1y} < (1-\beta)/(k\beta)$, in which case Structure Ba is selected. At this point, the maximum utility of Farmer 2, that is, the real income per capita, is $U_2 = \beta^\beta(1-\beta)^{1-\beta}(k^2 a_{1x}/a_{1y})^\beta a_{2y}$.

In the context of Structure C, the maximization utility problem for Farmer 1 with mode $(x/y)_1$ is:

$$\max_{x_1, x_1^s, y_1^d} U_1 = x_1^\beta (ky_1^d)^{1-\beta}, \tag{11}$$

$$\text{s.t.} \quad x_1 + x_1^s = a_{1x}l_{1x}, \quad y_1^d = px_1^s, \quad l_{1x} = 1, \tag{12}$$

with solution $x_1 = \beta a_{1x}$, $x_1^s = (1-\beta)a_{1x}$, and $y_1^d = (1-\beta)pa_{1x}$. Similarly, we can establish the maximization problem for Farmer 2 with mode $(y/x)_2$. The market clearing condition $x_1^s = x_2^d$ sets the equilibrium relative price $p = \dfrac{\beta a_{2y}}{(1-\beta)a_{1x}}$. Under this condition, the maximum utility of Farmer 1 in Structure C is $U_1^c = \beta a_{1x}^\beta (ka_{2y})^{1-\beta}$ and the maximum utility of Farmer 2 is $U_2^c = (1-\beta)(ka_{1x})^\beta a_{2y}^{1-\beta}$. For more details about the derivation, see Appendix A.2.

The comparative advantage of farmers and the equilibrium of four corner points in the model of division of labor selection are summarized in Table 1.

Table 1. Four corner equilibria of the model of farmers' comparative advantage and division of labor selection.

Structure	Relative Price p	Relative Parameter Interval	Real Income per Capita (Utility)	
			Farmer 1	Farmer 2
A	N.A.		$U_1(A) = $ $\beta^\beta(1-\beta)^{1-\beta}a_{1x}^\beta a_{1y}^{1-\beta}$	$U_2(A) = $ $\beta^\beta(1-\beta)^{1-\beta}a_{2x}^\beta a_{2y}^{1-\beta}$
Ba	$\dfrac{a_{1y}}{ka_{1x}}$	$k < k_1 < 1$ with $k_1 = \dfrac{(1-\beta)a_{1y}}{\beta a_{2y}}$	$U_1(A)$	$(1-\beta)^{1-\beta}\left(\dfrac{\beta k^2 a_{1x}}{a_{1y}}\right)^\beta a_{2y}$
Bb	$\dfrac{ka_{2y}}{a_{2x}}$	$k < k_2 < 1$ with $k_2 = \dfrac{\beta a_{2x}}{(1-\beta)a_{1x}}$	$\beta^\beta\left(\dfrac{(1-\beta)k^2 a_{2y}}{a_{2x}}\right)^{1-\beta}a_{1x}$	$U_2(A)$
C	$\dfrac{\beta a_{2y}}{(1-\beta)a_{1x}}$		$\beta(ka_{2y})^{1-\beta}a_{1x}^\beta$	$(1-\beta)(ka_{1x})^\beta a_{2y}^{1-\beta}$

4. Selection Logic and Structural Evolution of Division of Labor

If heterogeneous farmers have exogenous comparative technical advantages, under the influence of market transaction cost, the choice of production and consumption will be made in the four division of labor structures listed above. As each of the four division of labor structures leads to a corner equilibrium (cf. Table 1), general equilibrium is among the corner equilibria. Under this corner equilibrium relative price, no farmer has incentive to deviate from the model he/she chooses.

To explore the influence of comparative advantage and transaction cost on the division of labor choice of farmers, we find the conditions for each division of labor structure that lead to the general equilibrium. This can be accomplished using the total cost–benefit analysis method and the definition of general equilibrium. Furthermore, by studying the relationship between comparative advantage and transaction efficiency coefficient, we can deduce the varying relationships in farmers' equilibrium in their division of labor. These analyses help explain the division selection logic and decision mechanism of farmers with comparative advantage.

4.1. General Equilibrium and Comparative Static Analysis

Let us take the partial division of labor Structure Ba as an example. If the following conditions are met, Structure Ba is a general equilibrium.

1. With the corner equilibrium relative price $p = a_{1y}/(ka_{1x})$ for this structure, Farmer 2 prefers $(y/x)_2$, rather than mode $(xy)_2$ or $(x/y)_2$, given that: (1) $U_2(y/x) > U_2(A)$, which is equivalent to $k > k_0 = \sqrt{a_{2x}a_{1y}/a_{1x}a_{2y}}$; and (2) $U_2(y/x) > U_2(x/y)$, which is equivalent to $k > k_3 = \sqrt[2\beta]{a_{2x}a_{1y}/a_{1x}a_{2y}}$.
2. Farmer 1 prefers mode $(xy/y)_1$ than any other mode. This requires: (1) $U_1(xy/x) > U_1(x/y)$, which is true if $a_{1y}/a_{1x} > kp$; and (2) $U_1(xy/x) > U_1(y/x)$, which is true if $k < 1$.
3. Farmers are semi-specialized rather than fully specialized in producing products or services. This requires $l_{1x} < 1$, which is equivalent to $k < k_1 = (1-\beta)a_{1y}/(\beta a_{2y})$.

Notice that $k_3 < k_0$ and $k_0 < k_1$ are true if and only if $(1-\beta)/\beta > \sqrt{a_{2x}a_{1y}/a_{1x}a_{2y}}$. Hence, when $k \in (k_0, k_1)$, the three conditions above are true, and the corner equilibrium in Structure Ba is the general equilibrium. In this case, although Farmer 1 has an exogenous technological comparative advantage in the production of product or service x, he is unwilling to give up production of product or service y, because his relative preference for product or service y is greater than a threshold, which is the square root of the reciprocal of comparative advantage. Meanwhile, farmers are faced with a low market transaction efficiency coefficient, that is, farmers need to pay higher transaction costs to purchase the products or services they need, which sets farmers' preference to produce a part of their own products or services. The comparative static analysis for other structures (A, Bb, and C) can be carried out in a similar way, which is summarized in Table 2 with

$$k_0 = \sqrt{a_{2x}a_{1y}/a_{1x}a_{2y}}, \quad k_1 = (1-\beta)a_{1y}/\beta a_{2y}, \quad \text{and} \quad k_2 = \beta a_{2x}/((1-\beta)a_{1x}). \tag{13}$$

Table 2. General equilibrium and infra-marginal comparative static analysis of farmers' comparative advantage and division of labor.

Parameter Interval	$k < k_0$	$k > k_0$			
		$\dfrac{1-\beta}{\beta} > \left(\dfrac{a_{2x}a_{1y}}{a_{1x}a_{2y}}\right)^{\frac{1}{2}}$		$\dfrac{1-\beta}{\beta} < \left(\dfrac{a_{2x}a_{1y}}{a_{1x}a_{2y}}\right)^{\frac{1}{2}}$	
		$k_0 < k < k_1$	$k_1 < k < 1$	$k_0 < k < k_2$	$k_2 < k < 1$
Equilibrium Structure	A	Ba	C	Bb	C

4.2. The Logic and Decision Mechanism of Farmers Participating in the Division of Labor

Comparing various structures in the corresponding parameter subspace, as shown in Table 2, we can obtain relevant conclusions about the division of labor selection logic and decision mechanism of farmers with comparative advantages. In summary:

1. If $k < k_0$ (k is the transaction efficiency coefficient between farmers), the general equilibrium structure is self-sufficient, and the farmers produce two products or services themselves.
2. If $k > k_0$ and $(1 - \beta)/\beta > \sqrt{a_{2x}a_{1y}/a_{1x}a_{2y}}$, Ba or C is selected. When $k < k_1$, the general equilibrium structure is Ba, in which Farmer 1 produces both products or services, while Farmer 2 specializes in producing y, and the transaction is carried out between Farmer 1 selling x and Farmer 2 selling y. When $k > k_1$, the general equilibrium structure is C, in which Farmers 1 and 2 specialize in the production of products or services x and y, respectively, forming a pair of trading partners through market transactions.
3. If $k > k_0$ and $(1 - \beta)/\beta < \sqrt{a_{2x}a_{1y}/a_{1x}a_{2y}}$, Bb or C is selected. When $k < k_2$, the general equilibrium structure is Bb, in which Farmer 1 specializes in the production of comparative advantage products or services x and Farmer 2 produces both products, and the transaction is carried out between Farmer 1 selling x and Farmer 2 selling y. When $k > k_2$, the general equilibrium structure is C.

According to the table of the general equilibrium of marginal comparative static analysis (the equilibrium structure and the endogenous parameters with the parameter changes and the discontinuous jump between different corner points equilibrium), as the transaction efficiency coefficient between farmers increases from a low value to k_0, and then to k_1 or k_2, the general equilibrium jumps from self-sufficiency to partial division of labor, and then to complete division of labor. As for whether the intermediate transformation structure is Ba or Bb, it depends on the comparison of relative preferences and relative productivity among farmers.

4.3. The Function Logic of Comparative Advantage on the Choice of Farmer Specialization

It is also worth mentioning how the farmers' exogenous technology comparative advantages play a role in choosing division of labor and structure. It has been assumed in the model that $a_{1x}/a_{2x} > a_{1y}/a_{2y}$, that is, Farmer 1 has a comparative advantage in the production of product or service x. The degree of comparative advantage of exogenous technologies is denoted as $r = r_1 r_2$, with $r_1 = a_{1x}/a_{2x}$ and $r_2 = a_{2y}/a_{1y}$.

The three critical values for transaction efficiency coefficient in the parameter subspace in Equation (13) are obtained via partial differentiations. The results are as follows.

1. From the fact that $\partial k_0/\partial r_1 < 0$ and $\partial k_0/\partial r_2 < 0$, we observe that the higher is the degree of farmers' exogenous technology comparative advantage, the smaller is the critical value of transaction efficiency coefficient. That means that, if the "threshold" of crossing the self-sufficient structure is lowered, it urges the division of labor to take place under the condition of low transaction efficiency. In this way, farmers can strive for the benefits of comparative advantage and division of labor to make up for the loss of advantages and benefits in the self-sufficient structure.
2. With $\partial k_1/\partial r_2 < 0$ and $\partial k_2/\partial r_1 < 0$, we know that, in the structural selection of partial and complete division of labor, the higher is the degree of farmers' exogenous technology comparative advantage, the more likely it is to have $k > k_i (i = 1, 2)$ for a given transaction efficiency coefficient, that is, the equilibrium level of division of labor may be higher. That means that farmers are more likely to choose the production modes that allow them to maximize their comparative advantages and specialize in producing superior products or services, to avoid the efficiency loss of resource allocation in the partial division of labor structure and gain more comparative benefits and division of labor economy.

3. The level of specialization in Structure C is higher than that of Structure Ba or Structure Bb. Since the level of division of labor is positively correlated with individual specialization level, the division of labor in Structure C is obviously higher than that in other structures. Therefore, the complete division of labor Structure C becomes a general equilibrium, that is the farmers choose to specialize in the production of products or services with the comparative advantages of exogenous technology, satisfying

$$\sqrt{a_{2x}a_{2y}/(a_{1x}a_{1y})} < (1-\beta)/\beta < a_{2y}/a_{1y}, \quad \text{or} \quad a_{2x}/a_{1x} < (1-\beta)/\beta < \sqrt{a_{2x}a_{2y}/(a_{1x}a_{1y})}.$$

This means that the greater is the degree of balance between farmers' relative preference and relative productivity, the higher is the level of division of labor, and the more inclined farmers are to specialize in division of labor.

Finally, in the equilibrium, the productivity level selected by the farmers will be improved endogenously with the improvement of the trading conditions, so that even if there is no scale economy in the model, there is a division economy with the "one plus one greater than two" effect [30,31]. This means that the overall equilibrium productivity of an economy will improve as the size of the equilibrium labor division network increases.

5. Summary

The division of labor in agriculture is influenced by factors such as the innate characteristics of the crops, variations of the seasons, duration of a product's shelf life, and the interconnectedness of the production process; these are all heavily interlinked, making it difficult to completely separate the factors in searching for farmer's maximum profit. Moreover, marginal analysis in economics cannot be used to model the division of labor mathematically. Our work here is the first attempt to analyze the division of labor using infra-marginal model in agriculture by treating heterogeneous farmers as a single producer–consumer integrated unit.

One of our major contributions in this study is to apply the corner equilibrium analysis in studying farmers' selection logic. When we impose reasonable budget constraints, positive utility, and comparative advantage, the number of possible production–consumption decision modes can be reduced from the maximum of 64 to an optimal of 3. If we assume that at least one of the farmers selects a specialized mode and each farmer prefers a different production–consumption mode, then four division of labor structures can be derived. Solving the nonlinear programming problem of the utility function within each respective labor structure leads to a corner equilibrium. We discovered the ranges for transaction efficiency coefficients, k, and preference parameter, β, under which each structure can achieve general equilibrium. Our work is concluded by showing how farmers' exogenous comparative advantage influence the way labor is divided and labor structures are selected.

The general equilibrium is determined by the relative productivity, relative preferences, and transaction efficiency levels of the two farmers. When other parameters are set, the improvement of transaction efficiency causes the general equilibrium to jump from self-sufficiency to partial division of labor and then to complete division of labor. Given the terms of the transaction and the relative preference for the two products, the greater is the comparative advantage of the farmer, the higher is the level of division of labor. Given the conditions of trade, the more balanced are the relative preferences compared with relative productivity, the higher is the equilibrium division of labor. With the improvement of the level of equilibrium division of labor, the equilibrium aggregate productivity of the economy in which the farmer is located increases. The aforementioned super-marginal comparative static analysis of general equilibrium explains the selection logic and decision path for the participation of superior farmers in the division of labor, and also provides a general equilibrium mechanism for the development of agricultural economy. In this mechanism, exogenous comparative advantage and transaction efficiency are the driving forces of agricultural economic development.

It is worth pointing out that our simplified model only takes into consideration the exogenous comparative technical advantages in understanding farmers' decision-making and selection logic. Further research to investigate the role of endogenous comparative advantages, which are obtained through one's practices and experiences, with the improvement of production and trading environment is much needed. On the other hand, the applicability of our work can be strengthened and validated with numerical studies of actual field data. Data that are currently collected from large-scale agricultural production activities in China will be extremely useful for this purpose.

In reality, there are many critical factors such as the demographic population and the factor endowment of the farmers, the level of expertise in the agricultural production, and the market transaction efficiency that can influence the selection space of farmers and the ultimate division of labor structure. A brand new set of mathematical models and accompanying analysis would most likely be needed to provide a more comprehensive result in this area.

Author Contributions: Conceptualization, X.J.; methodology, X.J.; formal analysis, X.J.; investigation, X.J.; writing—original draft preparation, X.J.; writing—review and editing, X.J. and J.-M.C.; and supervision, H.S.

Funding: X.J. was funded by National Natural Science Foundation of China grant numbers 71742003 and 71703041, Ministry of Education Humanities and Social Sciences Youth Fund Project grant number 15YJC790036, and China Scholarship Council grant number 201808440675. H.S. was funded by Simons Foundation Collaborative Award grant number 522790.

Acknowledgments: The authors gratefully acknowledge the valuable comments and numerous suggestions from the editor and the anonymous reviewers.

Appendix A

Appendix A.1

In this section, we provide the details to the derivation of solution to the utility maximization problem for semi-specialized mode $(xy/y)_1$, as presented in Section 3.2. Note that the case $(xy/y)_1$ corresponds to $Z_i = (x_i, x_i^s, x_i^d, y_i, y_i^s, y_i^d) = (+, 0, +, +, +, 0)$. Hence, the constrained utilities maximization problem can be formulated as

$$\max_{x_1, y_1, x_1^s, y_1^d, l_{1x}, l_{1y}} U_1 = x_1^\beta (y_1 + ky_1^d)^{1-\beta} \tag{A1}$$

$$\text{s.t.} \quad x_1 + x_1^s = a_{1x} l_{1x}, \quad y_1 = a_{1y} l_{1y}, \quad l_{1x} + l_{1y} = 1, \quad y_1^d = p x_1^s, \tag{A2}$$

Note that the four constraints allow us to express the variables x_1, y_1, l_{1y}, and y_1^d in terms of x_1^s and l_{1x}. Then, the constraint optimization problem in Equations (A1) and (A2) can be rewritten as

$$\max_{x_1^s, l_{1x}} U_1 = (a_{1x} l_{1x} - x_1^s)^\beta (a_{1y}(1 - l_{1x}) + kp x_1^s)^{1-\beta} \tag{A3}$$

$$\text{s.t.} \quad 0 \leq x_1^s \leq a_{1x} l_{1x}, \quad \text{and} \quad 0 \leq l_{1x} \leq 1. \tag{A4}$$

First-order derivatives on the utility function U_1 gives us

$$\frac{\partial U_1}{\partial x_1^s} = \left(-\frac{\beta}{a_{1x} l_{1x} - x_1^s} + \frac{kp(1-\beta)}{a_{1y}(1 - l_{1x}) + kp x_1^s} \right) U_1, \tag{A5}$$

$$\frac{\partial U_1}{\partial l_{1x}} = \left(\frac{a_{1x}\beta}{a_{1x} l_{1x} - x_1^s} - \frac{a_{1y}(1-\beta)}{a_{1y}(1 - l_{1x}) + kp x_1^s} \right) U_1. \tag{A6}$$

Setting both derivatives equal to zero yields the condition on relative price $p = a_{1y}/(ka_{1x})$. With this relative price, Equation (A5) then simplifies to $x_1^s = a_{1x}(l_{1x} - \beta)$. Moreover, the constraints in Equation (A4) are equivalent to $\beta \leq l_{1x} \leq 1$. Furthermore, the original constraints in Equation (A2) lead to

$$x_1 = a_{1x}l_{1x} - x_1^s = \beta a_{1x} \tag{A7}$$

$$y_1 = a_{1y}l_{1y} = a_{1y}(1 - l_{1x}) \tag{A8}$$

$$y_1^d = px_1^s = \frac{a_{1y}(l_{1x} - \beta)}{k}. \tag{A9}$$

Substituting the appropriate variables in the utility function leads to

$$U_1 = \beta^\beta (1 - \beta)^{1-\beta} a_{1x}^\beta a_{1y}^{1-\beta} \tag{A10}$$

Appendix A.2

In this section, we explain the solution derivation to the utility maximization problem for fully specialized mode $(x/y)_1$, in the context of Structure C, i.e.,

$$\max_{x_1, x_1^s, y_1^d} U_1 = x_1^\beta (k y_1^d)^{1-\beta}, \tag{A11}$$

$$\text{s.t.} \quad x_1 + x_1^s = a_{1x}l_{1x}, \quad y_1^d = px_1^s, \quad l_{1x} = 1. \tag{A12}$$

Replacing x_1 and y_1^d in terms of x_1^s, the constrained optimization problem in Equations (A11) and (A12) may be reformulated as

$$\max_{0 \leq x_1^s \leq a_{1x}} U_1 = (a_{1x} - x_1^s)^\beta (kpx_1^s)^{1-\beta}. \tag{A13}$$

The first-order derivative of the utility function reads

$$\frac{dU_1}{dx_1^s} = \left(-\frac{\beta}{a_{1x} - x_1^s} + \frac{1-\beta}{x_1^s} \right) U_1. \tag{A14}$$

At the critical point, the first-order derivative vanishes, and this implies

$$x_1^s = (1 - \beta)a_{1x}. \tag{A15}$$

Furthermore, we have

$$x_1 = \beta a_{1x} \quad \text{and} \quad y_1^d = (1 - \beta)pa_{1x}. \tag{A16}$$

The value of the utility function is

$$U_1 = \beta^\beta (1 - \beta)^{1-\beta} (pk)^{1-\beta} a_{1x}. \tag{A17}$$

In the context of Structure C, we have the market clearing condition $x_1^s = x_2^d$, which sets the equilibrium relative price $p = \frac{\beta a_{2y}}{(1-\beta)a_{1x}}$. Hence, the critical utility value for Farmer 1 in Structure C is $U_1^c = \beta a_{1x}^\beta (ka_{2y})^{1-\beta}$. The corresponding utility of Farmer 2 is $U_2^c = (1 - \beta)(ka_{1x})^\beta a_{2y}^{1-\beta}$.

References

1. Schultz, T.W. *Transforming Traditional Agriculture*; Commercial Press: Beijing, China, 1987; pp. 35–78.
2. Sheng, H. *Division of Labor and Transactions*; Shanghai People's Press: Shanghai, China, 2006; pp. 45–82.
3. Cheng, W.; Liu, M.; Yang, X. A Ricardo model with endogenous comparative advantage and endogenous trade policy regime. *Econ. Rec.* **2000**, *76*, 172–182. [CrossRef]
4. Cheng, W.; Sachs, J.; Yang, X. An inframarginal analysis of the Ricardian model. *Rev. Int. Econ.* **2000**, *8*, 208–220. [CrossRef]
5. Zhang, D. A Note on "An Inframarginal Analysis of The Ricardian Model". *Pac. Econ. Rev.* **2006**, *11*, 505–512. [CrossRef]
6. Yang, X.; Zhang, Y. *New Classical Economics and Infra-marginal Analysis*; Social Science Literature Press: Beijing, China, 2003; pp. 240–268.
7. Dixit, A.; Norman, V. *Theory of International Trade: A Dual, General Equilibrium Approach*; Cambridge University Press: Cambridge, UK, 1980.
8. Yang, X.; Borland, J. A Microeconomic Mechanism for Economic Growth. *J. Political Econ.* **1991**, *3*, 460–482. [CrossRef]
9. Jiang, X. Outsourcing of Production Process—Measurement Model Based on Expert Questionnaire. *South. Econ.* **2014**, *12*, 96–104.
10. Pang, C. Integration, Outsourcing and Economic Evolution—Infra-marginal New Classical General Equilibrium Analysis. *Econ. Res.* **2010**, *3*, 114–128.
11. Buchanan, J.; Stubblebine, W. Externality. *Economica* **1962**, *29*, 371–384. [CrossRef]
12. Koopman, T. *Three Essays in the State of Economic Science*; McGraw-Hill: New York, NY, USA, 1957.
13. Arrow, K.; Enthoven, A.; Hurwicz, L. *Studies in Linear and Nonlinear Programming*; Uzawa, H., Ed.; Stanford University Press: Stanford, CA, USA, 1958.
14. Yang, X. *Introduction to Economic Cybernetics*; Hunan People's Press: Changsha, China, 1984.
15. Borland, J.; Yang, X. Specialization and a New Approach to Economic Organization and Growth. *Am. Econ. Rev.* **1992**, *82*, 386–391.
16. Cheng, W.; Yang, X. Inframarginal analysis of division of labor A survey. *J. Econ. Behav. Organ.* **2004**, *55*, 137–174.
17. Yang, X. Driving Force I—Exogenous Comparative Advantage and Trading Efficiency. In *Economic Development and the Division of Labor*; Blackwell Publishers: Malden, MA, USA, 2003; pp. 57–95.
18. Sachs, J.; Yang, X.; Zhang, D. Pattern of trade and economic development in the model of monopolistic competition. *Rev. Dev. Econ.* **2002**, *6*, 1–25. [CrossRef]
19. Yang, X.; Zhang, D. Economic development, international trade, and income distribution. *J. Econ.* **2003**, *78*, 163–190. [CrossRef]
20. Wen, M.; King, S. Push or pull? The relationship between development, trade and primary resource endowment. *J. Econ. Behav. Organ.* **2004**, *53*, 569–591. [CrossRef]
21. Liu, P.; Yang, X. Theory of irrelevance of the size of the firm. *J. Econ. Behav. Organ.* **2000**, *42*, 145–165. [CrossRef]
22. Li, K. A General equilibrium model with impersonal networking decisions and bundling sales. In *The Economics of E-Commerce and Networking Decisions: Applications and Extensions of Inframarginal Analysis*; Ng, Y.-K., Shi, H., Sun, G., Eds.; Macmillan: London, UK, 2003.
23. Sun, G.; Yang, X. Agglomeration economies, division of labor and the urban land-rent escalation. *Austr. Econ. Pap.* **2002**, *41*, 164–184. [CrossRef]
24. Yang, X. *Economic Development and the Division of Labor*; Blackwell: Cambridge, MA, USA, 2003.
25. Yang, X. An equilibrium model of hierarchy. In *The Economics of E-Commerce and Networking Decisions: Applications and Extensions of Inframarginal Analysis*; Ng, Y.-K., Shi, H., Sun, G., Eds.; Macmillan: London, UK, 2003.
26. Yang, X. The division of labor, investment, and capital. *Metroeconomica* **1999**, *50*, 301–324. [CrossRef]
27. Du, J. Endogenous, efficient long-run cyclical unemployment, endogenous long-run growth, and division of labor. *Rev. Dev. Econ.* **2003**, *7*, 266–278. [CrossRef]
28. Yang, X.; Yeh, Y. A general equilibrium model with endogenous principal—Agent relationship. *Austr. Econ. Pap.* **2002**, *41*, 15–36. [CrossRef]

29. Shi, H.; Yang, X. A new theory of industrialization. *J. Comp. Econ.* **1995**, *20*, 171–189. [CrossRef]
30. Rosen, S. Substitution and the division of labor. *Economica* **1978**, *45*, 235–250. [CrossRef]
31. Mussa, M.; Rosen, S. Monopoly and Product Quality. *J. Econ. Theory* **1978**, *18*, 301–317. [CrossRef]

Article

Weighted Block Golub-Kahan-Lanczos Algorithms for Linear Response Eigenvalue Problem

Hongxiu Zhong [1,*], Zhongming Teng [2] and Guoliang Chen [3]

[1] School of Science, Jiangnan University, Wuxi 214122, China
[2] College of Computer and Information Science, Fujian Agriculture and Forestry University, Fuzhou 350002, China; peter979@163.com
[3] School of Mathematical Sciences, Shanghai Key Laboratory of PMMP, East China Normal University, Shanghai 200241, China; glchen@math.ecnu.edu.cn
* Correspondence: zhonghongxiu@126.com

Received: 30 November 2018; Accepted: 29 December 2018; Published: 7 January 2019

Abstract: In order to solve all or some eigenvalues lied in a cluster, we propose a weighted block Golub-Kahan-Lanczos algorithm for the linear response eigenvalue problem. Error bounds of the approximations to an eigenvalue cluster, as well as their corresponding eigenspace, are established and show the advantages. A practical thick-restart strategy is applied to the block algorithm to eliminate the increasing computational and memory costs, and the numerical instability. Numerical examples illustrate the effectiveness of our new algorithms.

Keywords: linear response eigenvalue problem; block methods; weighted Golub-Kahan-Lanczos algorithm; convergence analysis; thick restart

AMS Subject Classification: 65F15; 15A18

1. Introduction

In this paper, we are interested in solving the linear response eigenvalue problem (LREP):

$$\mathbf{H}z := \begin{bmatrix} 0 & M \\ K & 0 \end{bmatrix} \begin{bmatrix} u \\ v \end{bmatrix} = \lambda \begin{bmatrix} u \\ v \end{bmatrix} = \lambda z,$$

where K and M are $N \times N$ real symmetric positive definite matrices. Such a problem arises from studying the excitation energy of many particle systems in computational quantum chemistry and physics [1–3]. It also known as the Bethe-Salpeter (BS) eigenvalue-problem [4] or the random phase approximation (RPA) eigenvalue problem [5]. There has immense past and recent work in developing efficient numerical algorithms and attractive theories for LREP [6–15].

Since all the eigenvalues of $\mathbf{H}$ are real nonzero and appear in pairs $\{\lambda, -\lambda\}$ [6], thus we order the eigenvalues in ascending order, i.e.,

$$-\lambda_1 \leq \cdots \leq -\lambda_N < \lambda_N \leq \cdots \leq \lambda_1.$$

In this paper, we focus on a small portion of the positive eigenvalues for LREP, i.e., λ_i, $i = k, k+1, \cdots, \ell$ with $1 \leq k \leq \ell \leq N$ and $\ell - k + 1 \ll N$, and their corresponding eigenvectors. We only consider the real case, all the results can be easily applied to the complex case.

The weighted Golub-Kahan-Lanczos method (**wGKL**) for LREP was introduced in [16]. It produces recursively a much small projection $\mathbf{B}_j = \begin{bmatrix} 0 & B_j \\ B_j^T & 0 \end{bmatrix}$ of $\mathbf{H}$ at j-th iteration, where $B_j \in \mathbb{R}^{j \times j}$ is upper bidiagonal. Afterwards, the eigenpairs of $\mathbf{H}$ can be constructed by the singular value decomposition of B_j. The convergence analysis performs that running k iterations of **wGKL** is equivalently running $2k$ iterations of a weighted Lanczos algorithm for **H** [16]. Actually, B_j can be also a lower bidiagonal matrix, and the same discussion can be taken place as in the case of B_j is upper bidiagonal. In the following, we only consider the upper bidiagonal case.

It is well known that the single-vector Lanczos method is widely used for searching a small number of extreme eigenvalues, and it may encounter very slow convergence when the wanted eigenvalues stay in a cluster [17]. Instead, a block Lanczos method with a suitable block size is capable of computing a cluster of eigenvalues including multiple eigenvalues very quickly. Motivated by this idea, we are going to develop a block version of **wGKL** in [16] in order to find efficiently all or some positive eigenvalues within a cluster for LREP. Based on the standard block Lanczos convergence theory in [17], the error bounds of approximation to an eigenvalue cluster, as well as their corresponding eigenspace are established to illustrate the advantage of our weighted block Golub-Kahan-Lanczos algorithm (**wbGKL**).

As the increasing size of the Krylov subspace, the storage demands, computational costs, and numerical stability of a simple version of a block Lanczos method may be affected [18]. Several kinds of efficiently restarting strategies to eliminate these effects are developed for the classic Lanczos method, such as, implicitly restart method [19], thick restart method [20]. In order to make our block method more practical, and using the special structure of LREP, we consider the thick restart strategy to our block method.

The rest of this paper is organized as follows. Section 2 gives some necessary preliminaries for our later use. In Section 3, the weighted block Golub-Kahan-Lanczos algorithm (**wbGKL**) for LREP is presented, and its convergence analysis is discussed. Section 4 proposed the thick restart weighted block Golub-Kahan-Lanczos algorithm (**wbGKL-TR**). The numerical examples are tested in Section 5 to illustrate the efficiency of our new algorithms. Finally, some conclusions are given in Section 6.

Throughout this paper, $\mathbb{R}^{m \times n}$ is the set of all $m \times n$ real matrices, $\mathbb{R}^n = \mathbb{R}^{n \times 1}$, and $\mathbb{R} = \mathbb{R}^1$. I_n (or simply I if its dimension is clear from the context) is the $n \times n$ identity matrix, and $0_{m \times n}$ is an $m \times n$ matrix of zero. The superscript $"T"$ denotes transpose. $\| \cdot \|_F$ denotes the Frobenius norm of a matrix, and $\| \cdot \|_2$ denotes the 2-norm of a matrix or a vector. For a matrix $X \in \mathbb{R}^{m \times n}$, $rank(X)$ denotes the rank of X, and $\mathcal{R}(X) = span(X)$ denotes the column space of X; the submatrices $X_{i:j,:}$ and $X_{:,k:\ell}$ of X composed by the intersections of row i to row j and column k to column ℓ, respectively. For matrices or scalars X_i, $diag(X_1, \cdots, X_k)$ denotes the block diagonal matrix with the i-th diagonal block X_i.

2. Preliminaries

For a symmetric positive definite matrix $W \in \mathbb{R}^{N \times N}$, the W-inner product is defined as following

$$\langle x, y \rangle_W := y^T W x, \quad \forall x, y \in \mathbb{R}^N.$$

If $\langle x, y \rangle_W = 0$, then we denote it by $x \perp_W y$, and call it with x and y are W-orthogonal. The projector Π_W is called the W-orthogonal projector onto $\mathcal{Y}$ if for any $y \in \mathbb{R}^N$,

$$\Pi_W y \in \mathcal{Y}, \quad (I - \Pi_W) y \perp_W \mathcal{Y}.$$

For two subspaces $\mathcal{X}, \mathcal{Y} \subseteq \mathbb{R}^N$, and suppose $k = dim(\mathcal{X}) \leq dim(\mathcal{Y}) = \ell$, if $X \in \mathbb{R}^{N \times k}$ and $Y \in \mathbb{R}^{N \times \ell}$ are W-orthonormal basis of $\mathcal{X}$ and $\mathcal{Y}$, respectively, i.e.,

$$X^T W X = I_k, \ \mathcal{X} = \mathcal{R}(X) \quad \text{and} \quad Y^T W Y = I_\ell, \ \mathcal{Y} = \mathcal{R}(Y),$$

and ν_j for $j = 1, \cdots, k$ with $\nu_1 \leq \cdots \leq \nu_k$ are the singular values of $Y^T W X$, then the W-canonical angles $\theta_W^{(j)}(\mathcal{X}, \mathcal{Y})$ from $\mathcal{X}$ to $\mathcal{Y}$ are defined by

$$0 \leq \theta_W^{(j)}(\mathcal{X}, \mathcal{Y}) = \arccos \nu_j \leq \pi/2, \quad \text{for } j = 1, \cdots, k.$$

If $k = \ell$, these angles can be said between $\mathcal{X}$ and $\mathcal{Y}$. Obviously, $\theta_W^{(1)}(\mathcal{X}, \mathcal{Y}) \geq \cdots \geq \theta_W^{(k)}(\mathcal{X}, \mathcal{Y})$. Set

$$\Theta_W(\mathcal{X}, \mathcal{Y}) = diag(\theta_W^{(1)}(\mathcal{X}, \mathcal{Y}), \cdots, \theta_W^{(k)}(\mathcal{X}, \mathcal{Y})).$$

Especially, if $k = 1$, X is a vector, there is only one W-canonical angle from $\mathcal{X}$ to $\mathcal{Y}$. In the following, we may use a matrix in one or both arguments of $\Theta_W(\cdot, \cdot)$, i.e., $\Theta_W(X, Y)$ with the understanding that it means the subspace spanned by the columns of the matrix argument.

The following two lemmas are important to our later analysis, and for proofs and more details, the reader is referred to [12,16].

Lemma 1. ([12] *Lemma 3.2*). *Let $\mathcal{X}$ and $\mathcal{Y}$ be two subspaces in $\mathbb{R}^N$ with equal dimensional $dim(\mathcal{X}) = dim(\mathcal{Y}) = k$. Suppose $\theta_W^{(1)}(\mathcal{X}, \mathcal{Y}) < \pi/2$. Then, for any set $y_1, y_2, \cdots, y_{k_1}$ of the basis vectors in $\mathcal{Y}$ where $1 \leq k_1 \leq k$, there is a set $x_1, x_2, \cdots, x_{k_1}$ of linearly independent vectors in $\mathcal{X}$ such that $\Pi_W x_j = y_j$ for $1 \leq j \leq k_1$, where Π_W is the W-orthogonal projector onto $\mathcal{Y}$.*

Lemma 2. ([16] *Proposition 3.1*). *The matrix MK has N position eigenvalues $\lambda_1^2 \geq \lambda_2^2 \geq \cdots \geq \lambda_N^2$ with $\lambda_j > 0$. The corresponding right eigenvectors $\xi_1, \cdots, \xi_N$ can be chosen K-orthonormal, and the corresponding left eigenvectors $\eta_1, \cdots, \eta_N$ can be chosen M-orthonormal. In particular, for given $\{\xi_j\}$, one can choose $\eta_j = \lambda_j^{-1} K\xi_j$, and for given $\{\eta_j\}$, $\xi_j = \lambda_j^{-1} M\eta_j$, for $j = 1, \cdots, N$.*

3. Weighted Block Golub-Kahan-Lanczos Algorithm

3.1. Weighted Block Golub-Kahan-Lanczos Algorithm

In this section, we plan to introduce the weighted block Golub-Kahan-Lanczos algorithm (wbGKL) for LREP, which is a block version of the weighted Golub-Kahan-Lanczos algorithm [16]. Algorithm 1 gives the process of recursively generating the M-orthonormal matrix $\mathcal{X}_n$, the K-orthonormal matrix $\mathcal{Y}_n$, and the block bidiagonal matrix $\mathcal{B}_n$. Giving $Y_1 \in \mathbb{R}^{n \times n_b}$ with $Y_1^T K Y_1 = I_{n_b}$, denoting $E_n^T = [0_{n_b \times (n-1)n_b}, I_{n_b}]$, and

$$\mathcal{X}_n = [X_1, \cdots, X_n], \quad \mathcal{Y}_n = [Y_1, \cdots, Y_n], \quad \mathcal{B}_n = \begin{bmatrix} A_1 & B_1 & & \\ & A_2 & \ddots & \\ & & \ddots & B_{n-1} \\ & & & A_n \end{bmatrix},$$

then we have the relation from Algorithm 1:

$$K\mathcal{Y}_n = \mathcal{X}_n \mathcal{B}_n, \quad M\mathcal{X}_n = \mathcal{Y}_n \mathcal{B}_n^T + Y_{n+1} B_n^T E_n^T, \tag{1}$$

and

$$\mathcal{X}_n^T M \mathcal{X}_n = I_{nn_b} = \mathcal{Y}_n^T K \mathcal{Y}_n.$$

Remark 1. *In Algorithm 1, we only consider the case that $rank(\widetilde{X}_j) = rank(\widetilde{Y}_{j+1}) = n_b$, no further treatment is provided for the cases $rank(\widetilde{X}_j) < n_b$ or $rank(\widetilde{Y}_{j+1}) < n_b$. Because K and M are both symmetric positive definite, thus the two W in **Step 2** are both reversible.*

Remark 2. *With j increasing in **Step 2**, the M-orthogonality of $\mathcal{X}_j$ and the K-orthogonality of $\mathcal{Y}_j$ will slowly lose. Thus, in practice, we can add a re-orthogonalization process in each iteration to eliminate the defect. The same strategy is executed in the following algorithms.*

Algorithm 1: wbGKL

---

1. Choose Y_1 satisfying $Y_1^T K Y_1 = I_{n_b}$, and set $W = I_{n_b}$, $B_0 = I_{n_b}$, $X_0 = 0_{n \times n_b}$. Compute $F = KY_1$.

2. For $j = 1, 2, \cdots, n$

$\quad \widetilde{X}_j = FW - X_{j-1}B_{j-1}$

$\quad F = M\widetilde{X}_j$

$\quad$ Do Cholesky decomposition $\widetilde{X}_j^T F = W^T W$

$\quad A_j = W, W = inv(W), X_j = \widetilde{X}_j W \qquad \%W = inv(W)$ means $W = W^{-1}$

$\quad \widetilde{Y}_{j+1} = FW - Y_j A_j^T$

$\quad F = K\widetilde{Y}_{j+1}$

$\quad$ Do Cholesky decomposition $\widetilde{Y}_{j+1}^T F = W^T W$

$\quad B_j = W^T, W = inv(W), Y_{j+1} = \widetilde{Y}_{j+1} W$

$\quad$ End

From (1), we have

$$\begin{bmatrix} 0 & M \\ K & 0 \end{bmatrix} \begin{bmatrix} \mathcal{Y}_n & 0 \\ 0 & \mathcal{X}_n \end{bmatrix} = \begin{bmatrix} \mathcal{Y}_n & 0 \\ 0 & \mathcal{X}_n \end{bmatrix} \begin{bmatrix} 0 & \mathcal{B}_n^T \\ \mathcal{B}_n & 0 \end{bmatrix} + \begin{bmatrix} Y_{n+1} \\ 0 \end{bmatrix} B_n^T E_{2n}^T$$

with $E_{2n}^T = [0_{n_b \times (2n-1)n_b}, I_{n_b}]$. Then, the approximate eigenpairs of $\mathbf{H}$ can be obtained by solving a small eigenvalue problem of $\begin{bmatrix} 0 & \mathcal{B}_n^T \\ \mathcal{B}_n & 0 \end{bmatrix}$. Suppose $\mathcal{B}_n$ has an singular value decomposition

$$\mathcal{B}_n = \Phi \Sigma_n \Psi^T, \tag{2}$$

where $\Phi = [\phi_1, \phi_2, \cdots, \phi_{nn_b}]$, $\Psi = [\psi_1, \psi_2, \cdots, \psi_{nn_b}]$, $\Sigma_n = [\sigma_1, \sigma_2, \cdots, \sigma_{nn_b}]$ with $\sigma_1 \geq \cdots \geq \sigma_{nn_b} > 0$. Thus, we can take $\pm\sigma_j (1 \leq j \leq nn_b)$ as the Ritz values of $\mathbf{H}$ and

$$\tilde{z}_j = \frac{1}{\sqrt{2}} \begin{bmatrix} \mathcal{Y}_n \psi_j \\ \pm \mathcal{X}_n \phi_j \end{bmatrix}, \quad 1 \leq j \leq nn_b,$$

as the corresponding $\mathbf{K}$-orthonormal Ritz vectors, where $\mathbf{K} = \begin{bmatrix} K & 0 \\ 0 & M \end{bmatrix}$.

3.2. Convergence Analysis

In this section, we first consider the convergence analysis when using the first few σ_j as approximations to the first few λ_j. Then, the similar theories are presented if using the last few σ_j as approximations to the last few λ_j. Since a block Lanczos method with a suitable block size which is not smaller than the size of an eigenvalue cluster can compute all eigenvalues in the cluster. Now, we are considering the i-th to $(i + n_b - 1)$-st eigenpairs of LREP, in which the k-th to ℓ-th eigenvalues form a cluster as in the following figure with $1 \leq i \leq k \leq \ell \leq i + n_b - 1 \leq nn_b$ and $k \leq n$.

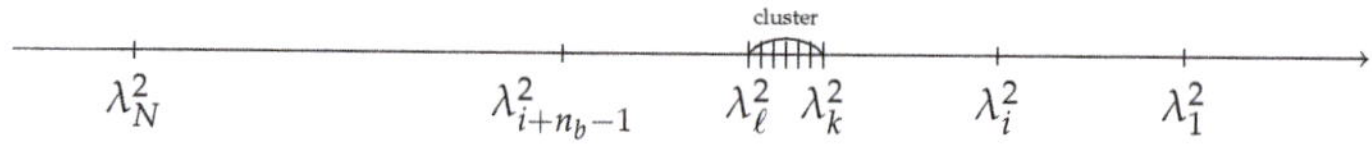

Here, the squares of the eigenvalues for LREP are listed. Hence, motivated by [12,17], we analyze the convergence of the cluster eigenvalues and their corresponding eigenspace, and give the error bounds of the approximate eigenpairs belonging to eigenvalue cluster together, instead of separately for each individual eigenpair.

We first give some notations and equations, which are critical in our main theorem. Note that from (1), we get

$$MK\mathcal{Y}_n = \mathcal{Y}_n \mathcal{B}_n^T \mathcal{B}_n + Y_{n+1} B_n^T A_n E_n^T. \tag{3}$$

Since (2) is the singular value decomposition of $\mathcal{B}_n$, thus the eigenvalues of $\mathcal{B}_n^T \mathcal{B}_n$ are σ_j^2 with the associated eigenvectors ψ_j for $1 \leq j \leq nn_b$.

From Lemma 2, if we let $\Xi = [\xi_1, \cdots, \xi_N]$, and $\Gamma = [\eta_1, \cdots, \eta_N]$, then $\Gamma = K\Xi\Lambda^{-1}$, and

$$MK\Xi = \Xi\Lambda^2. \tag{4}$$

Write Ξ and Λ^2 as

$$
\Xi = \begin{bmatrix} \overset{i-1}{\Xi_1} & \overset{n_b}{\Xi_2} & \overset{N-n_b-i+1}{\Xi_3} \end{bmatrix}, \quad
\Lambda^2 = \begin{matrix} \\ i-1 \\ n_b \\ N-n_b-i+1 \end{matrix}
\begin{bmatrix} \overset{i-1}{\Lambda_1^2} & \overset{n_b}{} & \overset{N-n_b-i+1}{} \\ & \Lambda_2^2 & \\ & & \Lambda_3^2 \end{bmatrix}.
$$

Let $\check{\Xi}_2 = \Xi_{(:,k:\ell)}$ and $\check{\Lambda}_2^2 = diag(\lambda_k^2, \cdots, \lambda_\ell^2)$. Denote C_j the first kind Chebyshev polynomial with j-th degree, and $0 \leq j \leq n$.

In the following, we assume

$$\theta_K^{(1)}(Y_1, \Xi_2) < \pi/2, \tag{5}$$

i.e., $rank(Y_1^T K\Xi_2) = n_b$, then from Lemma 1, we have $\exists\, Z \in \mathbb{R}^{N \times (\ell-k+1)}$ with $\mathcal{R}(Z) \subseteq \mathcal{R}(Y_1)$, s.t.,

$$\Xi_2 \Xi_2^T KZ = \check{\Xi}_2. \tag{6}$$

Theorem 1. *Suppose* $\theta_K^{(1)}(Y_1, \Xi_2) < \pi/2$, *and* Z *satisfy* (6), *then we have*

$$\|diag(\lambda_k^2 - \sigma_k^2, \cdots, \lambda_\ell^2 - \sigma_\ell^2)\|_F \leq (\lambda_k^2 - \lambda_N^2) \frac{\pi_{i,k,\ell}^2}{C_{n-k}^2(1+2\gamma_{i,\ell})} \|\tan^2 \Theta_K(\check{\Xi}_2, Z)\|_F \tag{7}$$

with

$$\gamma_{i,\ell} = \frac{\lambda_\ell^2 - \lambda_{i+n_b}^2}{\lambda_{i+n_b}^2 - \lambda_N^2}, \quad \pi_{i,k,\ell} = \frac{\displaystyle\max_{i+n_b \leq j \leq N} \prod_{m=1}^{k-1} |\sigma_m^2 - \lambda_j^2|}{\displaystyle\min_{k \leq t \leq \ell} \prod_{m=1}^{k-1} |\sigma_m^2 - \lambda_t^2|},$$

and

$$\|\sin \Theta_K(\check{\Xi}_2, \mathcal{Y}_n \Psi_{(:,k:\ell)})\|_F \leq \frac{\pi_{i,k}\sqrt{1 + c^2\|A_n^T B_n\|_2^2/\delta^2}}{C_{n-i}(1+2\gamma_{i,\ell})} \|\tan \Theta_K(\check{\Xi}_2, Z)\|_F \tag{8}$$

with constant c *lies between 1 and* $\pi/2$, *and* $c = 1$ *if* $k = \ell$, *and*

$$\delta = \min_{\substack{k \leq j \leq \ell \\ p < k \text{ or } p > \ell}} |\lambda_j^2 - \sigma_p^2|, \quad \pi_{i,k} = \prod_{j=1}^{i-1} \frac{\lambda_j^2 - \lambda_N^2}{\lambda_j^2 - \lambda_k^2}.$$

Particularly if $\sigma_{k-1}^2 \geq \lambda_k^2$, *then*

$$\pi_{i,k,\ell} = \prod_{m=1}^{k-1} \frac{|\sigma_m^2 - \lambda_N^2|}{|\sigma_m^2 - \lambda_k^2|}.$$

Proof. Multiplying L^T from left, (4) can be rewritten as $L^T ML(L^T \Xi) = (L^T \Xi)\Lambda^2$, so, $(\lambda_j^2, L^T \zeta_j)$ is the eigenpair of $L^T ML$, for $j = 1, \cdots, N$, and $L^T \zeta_1, \cdots, L^T \zeta_N$ are orthonormal. Do the same process to (3), we have

$$L^T ML\mathcal{V}_n = \mathcal{V}_n \mathcal{B}_n^T \mathcal{B}_n + \mathcal{V}_{n+1} B_n^T A_n E_n^T, \tag{9}$$

where $\mathcal{V}_n = L^T \mathcal{Y}_n$, $\mathcal{V}_{n+1} = L^T Y_{n+1}$, and $\mathcal{V}_n^T \mathcal{V}_n = I_{nn_b}$, which can be seen as the relation generalize by using standard Lanczos process to $L^T ML$. Thus, $\sigma_1^2, \cdots, \sigma_{nn_b}^2$ are the Ritz values of $L^T ML$, with the corresponding orthonormal Ritz vectors $\mathcal{V}_n \psi_1, \cdots, \mathcal{V}_n \psi_{nn_b}$.

Premultiplying L^T to Equation (6), we have $L^T \Xi_2 \Xi_2^T L(L^T Z) = L^T \breve{\Xi}_2$. Consequently, the conditions of the block Lanczos convergence Theorem 4.1 and Theorem 5.1 in [17] are satisfied. Thus, using the results Theorem 5.1 in [17], one has

$$\|diag(\lambda_k^2 - \sigma_k^2, \cdots, \lambda_\ell^2 - \sigma_\ell^2)\|_F \leq (\lambda_k^2 - \lambda_N^2)\frac{\pi_{i,k,\ell}^2}{C_{n-k}^2(1 + 2\gamma_{i,\ell})}\|\tan^2 \Theta(L^T \breve{\Xi}_2, L^T Z)\|_F.$$

Then the bound (7) can be easily got by using ([21] Theorem 4.2)

$$\Theta(L^T \breve{\Xi}_2, L^T Z) = \Theta_K(\breve{\Xi}_2, Z). \tag{10}$$

Let $\Pi_n = \mathcal{V}_n \mathcal{V}_n^T$, then Π_n is the orthogonal projection onto $\mathcal{K}_n(L^T ML, L^T Z)$, thus from (9), we have

$$\begin{aligned}
\|\Pi_n L^T ML(I - \Pi_n)\|_2 &= \|\mathcal{V}_n \mathcal{V}_n^T L^T ML(I - \mathcal{V}_n \mathcal{V}_n^T)\|_2 \\
&= \|\mathcal{V}_n(\mathcal{B}_n^T \mathcal{B}_n + E_n A_n^T B_n \mathcal{V}_{n+1}^T) - \mathcal{V}_n \mathcal{B}_n^T \mathcal{B}_n \mathcal{V}_n^T\|_2 \\
&= \|\mathcal{V}_n A_n^T B_n \mathcal{V}_{n+1}^T\|_2 \\
&= \|A_n^T B_n\|_2.
\end{aligned}$$

Consequently, applying the results of Theorem 4.1 in [17], we get

$$\begin{aligned}
\|\sin \Theta(L^T \breve{\Xi}_2, \mathcal{V}_n \Psi_{(:,k:\ell)})\|_F &\leq \frac{\pi_{i,k}\sqrt{1 + \|\Pi_n L^T ML(I - \Pi_n)\|_2^2/\delta^2}}{C_{n-i}(1 + 2\gamma_{i,\ell})}\|\tan \Theta(L^T \breve{\Xi}_2, L^T Z)\|_F \\
&= \frac{\pi_{i,k}\sqrt{1 + \|A_n^T B_n\|_2^2/\delta^2}}{C_{n-i}(1 + 2\gamma_{i,\ell})}\|\tan \Theta(L^T \breve{\Xi}_2, L^T Z)\|_F.
\end{aligned}$$

Then the bound (8) can be derived by using $\Theta(L^T \breve{\Xi}_2, \mathcal{V}_n \Psi_{(:,k:\ell)}) = \Theta_K(\breve{\Xi}_2, \mathcal{Y}_n \Psi_{(:,k:\ell)})$ and (10). $\quad\square$

Theorem 1 is used to bound the errors of the approximate eigenvalues to an eigenvalue cluster including the multiple eigenvalues. It can be also applied to the single eigenvalue case, the following corollary is derived by setting $k = \ell = i$, except the left equality of (10), which needs to be proved.

Corollary 1. *Suppose $\theta_K^{(1)}(Y_1, \Xi_2) < \pi/2$, then for $1 \leq i \leq nn_b$, there exits a vector $y \in \mathcal{R}(Y_1)$, s.t., $\Xi_2 \Xi_2^T y = \zeta_i$, and*

$$\lambda_i^2 - \sigma_i^2 \leq (\lambda_i^2 - \lambda_N^2)\frac{\pi_{i,j}^2}{C_{n-i}^2(1 + 2\gamma_i)}\tan^2 \theta_K(\zeta_i, y)$$

with

$$\gamma_i = \frac{\lambda_i^2 - \lambda_{i+n_b}^2}{\lambda_{i+n_b}^2 - \lambda_N^2}, \quad \pi_{i,j} = \max_{i+n_b \leq j \leq N} \prod_{m=1}^{i-1} \frac{|\sigma_m^2 - \lambda_j^2|}{|\sigma_m^2 - \lambda_i^2|},$$

and

$$\left((1 - \frac{\sigma_i^2}{\lambda_i^2}) + \frac{\sigma_i^2}{\lambda_i^2} \sin^2 \theta_M(\eta_i, \mathcal{X}_n \phi_i) \right)^{1/2} \tag{11}$$

$$= \sin \theta_K(\xi_i, \mathcal{Y}_n \psi_i) \leq \frac{\pi_i \sqrt{1 + \|A_n^T B_n\|_2^2 / \delta^2}}{C_{n-i}(1 + 2\gamma_i)} \tan \theta_K(\xi_i, y)$$

with

$$\delta = \min_{i \neq j} |\lambda_j^2 - \sigma_i^2|, \quad \pi_i = \prod_{j=1}^{i-1} \frac{\lambda_j^2 - \lambda_N^2}{\lambda_j^2 - \lambda_i^2}.$$

Proof. We only proof the left equality of (11). From (4) and Lemma 2, we have $\Xi = MK\Xi\Lambda^{-2} = M\Gamma\Lambda^{-1}$. If we let $Z_1 = (\mathcal{Y}_n \psi_i)^T K \xi_i$, and $Z_2 = (\mathcal{X}_n \phi_i)^T M \eta_i$, then we can get $Z_1 = \frac{\sigma_i}{\lambda_i} Z_2$ by using $K\mathcal{Y}_n \Psi = \mathcal{X}_n B_n \Psi = \mathcal{X}_n \Phi \Sigma_n$. Thus

$$\begin{aligned}
\sin^2 \theta_K(\xi_i, \mathcal{Y}_n \psi_i) &= 1 - \cos^2 \theta_K(\xi_i, \mathcal{Y}_n \psi_i) \\
&= 1 - Z_1^T Z_1 \\
&= 1 - \frac{\sigma_i^2}{\lambda_i^2} Z_2^T Z_2 \\
&= 1 - \frac{\sigma_i^2}{\lambda_i^2} \cos^2 \theta_M(\eta_i, \mathcal{X}_n \phi_i) \\
&= 1 - \frac{\sigma_i^2}{\lambda_i^2} + \frac{\sigma_i^2}{\lambda_i^2} \sin^2 \theta_M(\eta_i, \mathcal{X}_n \phi_i).
\end{aligned}$$

Then,

$$\sin \theta_K(\xi_i, \mathcal{Y}_n \psi_i) = \left(1 - \frac{\sigma_i^2}{\lambda_i^2} + \frac{\sigma_i^2}{\lambda_i^2} \sin^2 \theta_M(\eta_i, \mathcal{X}_n \phi_i) \right)^{1/2}.$$

$\square$

Next, we are going to consider the last few σ_j to approximate as the last few λ_{N-nn_b+j}, $j = k, \cdots, \ell$, and λ_{N-nn_b+k} to $\lambda_{N-nn_b+\ell}$ form a cluster in $\lambda_{\hat{i}}$ to $\lambda_{\hat{i}+n_b-1}$, which is described in the following figure, where $N + 1 - nn_b \leq \hat{i} \leq \hat{k} \leq \hat{\ell} \leq \hat{i} + n_b - 1 \leq N$, $nn_b - \ell + 1 \leq n$, $\hat{k} \triangleq N - nn_b + k$, and $\hat{\ell} \triangleq N - nn_b + \ell$.

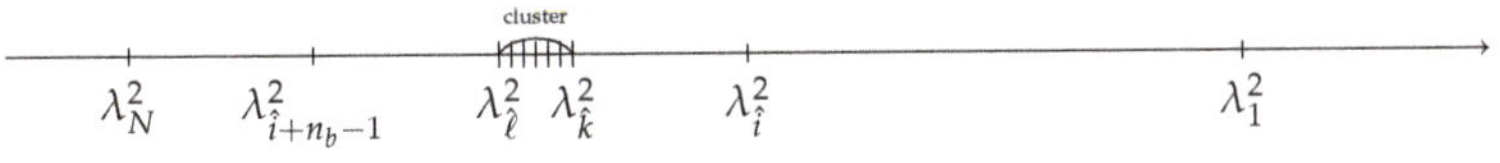

Similar to the above discussion for the first few eigenvalues, we can also obtain the error bounds of the approximate last few eigenpairs belongs to eigenvalue cluster together. We use the same notion, except $\hat{\Lambda}_2^2 = diag(\lambda_{\hat{k}}^2, \cdots, \lambda_{\hat{\ell}}^2)$ and $\hat{\Xi}_2 = \Xi_{(:,\hat{k}:\hat{\ell})}$. Assuming $\theta_K^{(1)}(Y_1, \Xi_2) < \pi/2$, then from Lemma 1, there $\exists \hat{Z} \in \mathbb{R}^{N \times (\ell-k+1)}$ with $\mathcal{R}(\hat{Z}) \subseteq \mathcal{R}(Y_1)$, s.t.,

$$\Xi_2 \Xi_2^T K \hat{Z} = \hat{\Xi}_2. \tag{12}$$

Theorem 2. *Suppose* $\theta_K^{(1)}(Y_1, \Xi_2) < \pi/2$ *and* $\hat{Z}$ *satisfy* (12)*, then we have*

$$
\begin{aligned}
&\|diag(\sigma_k^2 - \lambda_{\hat{k}}^2, \cdots, \sigma_\ell^2 - \lambda_{\hat{\ell}}^2)\|_F \\
&\leq (\lambda_1^2 - \lambda_{\hat{\ell}}^2) \frac{\hat{\pi}_{\hat{i},\hat{k},\ell}^2}{C_{n-N+\ell-1}^2 (1 + 2\hat{\gamma}_{\hat{i},\hat{k}})} \| \tan^2 \Theta_K(\hat{\Xi}_2, \hat{Z})\|_F
\end{aligned}
\tag{13}
$$

with

$$
\hat{\gamma}_{\hat{i},\ell} = \frac{\lambda_{\hat{i}-1}^2 - \lambda_{\hat{k}}^2}{\lambda_1^2 - \lambda_{\hat{i}-1}^2}, \quad
\hat{\pi}_{\hat{i},\hat{k},\ell} = \frac{\displaystyle \max_{1 \leq j \leq \hat{i}-1} \prod_{m=\ell+1}^{nn_b} |\sigma_m^2 - \lambda_j^2|}{\displaystyle \min_{\hat{k} \leq t \leq \hat{\ell}} \prod_{m=\ell+1}^{nn_b} |\sigma_m^2 - \lambda_t^2|},
$$

and

$$
\| \sin \Theta_K(\hat{\Xi}_2, \mathcal{Y}_n \Psi_{(:,k:\ell)})\|_F \leq \frac{\hat{\pi}_{\hat{i},\ell} \sqrt{1 + \hat{c}^2 \|A_n^T B_n\|_2^2 / \hat{\delta}^2}}{C_{n+\hat{i}+n_b-N-2}^2 (1 + 2\hat{\gamma}_{\hat{i},\hat{k}})} \| \tan \Theta_K(\hat{\Xi}_2, \hat{Z})\|_F
\tag{14}
$$

with constant $\hat{c}$ *lies between* 1 *and* $\pi/2$*, and* $\hat{c} = 1$ *if* $k = \ell$*, and*

$$
\hat{\delta} = \min_{\substack{k \leq j \leq \ell \\ p < k \; or \; p > \ell}} |\lambda_j^2 - \sigma_p^2|, \quad
\hat{\pi}_{\hat{i},\ell} = \prod_{j=\hat{i}+n_b}^{N} \frac{\lambda_1^2 - \lambda_j^2}{\lambda_{\hat{\ell}}^2 - \lambda_j^2}.
$$

Remark 3. *Similar to Corollary* 1*, Theorem* 2 *can also be applied to the single eigenvalue case, here we omit the detail.*

Remark 4. *In Theorem* 1 *and* 2*, we use the Frobenius norm to estimate the accuracy of eigenpairs approximations, in fact, any unitary invariant norm can be used to measure.*

Remark 5. *Compared with the single-vector type of the weighted Golub-Kahan-Lanczos method in* [16]*, our convergence results show the superiority of the block version. For instance, in Corollary* 1*, the convergence rate of the approximate eigenvalues* σ_j *is proportional to* $C_{n-i}^{-2}(1 + 2\gamma_i)$ *with* $\gamma_i = \frac{\lambda_i^2 - \lambda_{i+n_b}^2}{\lambda_{i+n_b}^2 - \lambda_N^2}$*, which is obviously better than* $C_{n-i}^{-2}(1 + 2\tilde{\gamma}_i)$ *with* $\tilde{\gamma}_i = \frac{\lambda_i^2 - \lambda_{i+1}^2}{\lambda_{i+1}^2 - \lambda_N^2}$ *in (*[16] *Theorem 3.4). While the additional cost caused from the block version can be paid by the improvements generated by* γ_i*, especially when the desired eigenvalues lie in a well-separated cluster* [12]*.*

4. Thick Restart

As the number of iterations increases, Algorithm 1 may encounter the dilemma that the amount of calculation and storage increases sharply and the numerical stability gradually weakens. In this section, we will apply the thick restart strategy [20] to improve the algorithm. After running n iterations, Algorithm 1 derives the following relations for LREP:

$$
\begin{cases}
K\mathcal{Y}_n = \mathcal{X}_n \mathcal{B}_n, \\
M\mathcal{X}_n = \mathcal{Y}_n \mathcal{B}_n^T + Y_{n+1} B_n^T E_n^T,
\end{cases}
\tag{15}
$$

with $\mathcal{X}_n^T M \mathcal{X}_n = I_{nn_b} = \mathcal{Y}_n^T K \mathcal{Y}_n$.

Recall the SVD (2), let Φ_k and Ψ_k be the first kn_b columns of Φ and Ψ, respectively, i.e.,

$$
\Phi_k = [\phi_1, \phi_2, \cdots, \phi_{kn_b}], \quad \Psi_k = [\psi_1, \psi_2, \cdots, \psi_{kn_b}].
$$

Thus it follows that

$$
\mathcal{B}_n \Psi_k = \Phi_k \Sigma_k \quad \text{and} \quad \mathcal{B}_n^T \Phi_k = \Psi_k \Sigma_k,
\tag{16}
$$

where $\Sigma_k = diag(\sigma_1, \cdots, \sigma_{kn_b})$.

By using the approximate eigenvectors of $\mathbf{H}$ for thick restart, we post-multiply Ψ_k and Φ_k to the Equation (15), respectively, and get

$$\begin{cases} K\mathcal{Y}_n\Psi_k = \mathcal{X}_n\mathcal{B}_n\Psi_k, \\ M\mathcal{X}_n\Phi_k = \mathcal{Y}_n\mathcal{B}_n^T\Phi_k + Y_{n+1}B_n^T E_n^T\Phi_k, \end{cases} \tag{17}$$

From (16), and let $\widehat{\mathcal{Y}}_k = \mathcal{Y}_n\Psi_k$, $\widehat{\mathcal{X}}_k = \mathcal{X}_n\Phi_k$, $\widehat{\mathcal{B}}_k = \Sigma_k$, $\widehat{Y}_{k+1} = Y_{n+1}$, $U^T = E_n^T\Phi_k$, $\widehat{B}_k = B_n$, then (17) can be rewritten as

$$\begin{cases} K\widehat{\mathcal{Y}}_k = \widehat{\mathcal{X}}_k\widehat{\mathcal{B}}_k, \\ M\widehat{\mathcal{X}}_k = \widehat{\mathcal{Y}}_k\widehat{\mathcal{B}}_k^T + \widehat{Y}_{k+1}\widehat{B}_k^T U^T, \end{cases} \tag{18}$$

and $\widehat{\mathcal{X}}_k^T M\widehat{\mathcal{X}}_k = I_{kn_b} = \widehat{\mathcal{Y}}_k^T K\widehat{\mathcal{Y}}_k$.

Next, $\widehat{X}_{k+1}$ and $\widehat{Y}_{k+2}$ will be generalized. Firstly, we compute

$$\begin{aligned} \widetilde{X}_{k+1} &= K\widehat{Y}_{k+1} - \widehat{\mathcal{X}}_k\widehat{\mathcal{X}}_k^T MK\widehat{Y}_{k+1} \\ &= K\widehat{Y}_{k+1} - \widehat{\mathcal{X}}_k U\widehat{B}_k. \end{aligned}$$

From the second equation in (18), we know $\widetilde{X}_{k+1}^T M\widehat{\mathcal{X}}_k = 0$. Do Cholesky decomposition $\widetilde{X}_{k+1}^T M\widetilde{X}_{k+1} = W^T W$, and set $\widehat{A}_{k+1} = W$, $W = inv(W)$. Compute $\widehat{X}_{k+1} = \widetilde{X}_{k+1}W$, and let

$$\widehat{\mathcal{X}}_{k+1} = [\widehat{\mathcal{X}}_k, \widehat{X}_{k+1}], \quad \widehat{\mathcal{B}}_{k+1} = \begin{bmatrix} \widehat{\mathcal{B}}_k & U\widehat{B}_k \\ 0 & \widehat{A}_{k+1} \end{bmatrix},$$

we have

$$K\widehat{\mathcal{Y}}_{k+1} = \widehat{\mathcal{X}}_{k+1}\widehat{\mathcal{B}}_{k+1} \quad \text{with} \quad \widehat{\mathcal{X}}_{k+1}^T M\widehat{\mathcal{X}}_{k+1} = I_{(k+1)n_b}. \tag{19}$$

Secondly, from the above equation, we can compute

$$\begin{aligned} \widetilde{Y}_{k+2} &= M\widehat{X}_{k+1} - \widehat{\mathcal{Y}}_k\widehat{\mathcal{Y}}_k^T KM\widehat{X}_{k+1} - \widehat{Y}_{k+1}\widehat{Y}_{k+1}^T KM\widehat{X}_{k+1} \\ &= M\widehat{X}_{k+1} - \widehat{Y}_{k+1}\widehat{A}_{k+1}^T. \end{aligned}$$

Again using (19), it is easily got that $\widetilde{Y}_{k+2}^T K\widehat{\mathcal{Y}}_{k+1} = 0$. Similarly, do Cholesky decomposition $\widetilde{Y}_{k+2}^T K\widetilde{Y}_{k+2} = W^T W$, and let $\widehat{B}_{k+1} = W^T$, $W = inv(W)$. Compute $\widehat{Y}_{k+2} = \widetilde{Y}_{k+2}W$, and let $\widehat{\mathcal{Y}}_{k+1} = [\widehat{\mathcal{Y}}_k, \widehat{Y}_{k+1}]$, we get

$$M\widehat{\mathcal{X}}_{k+1} = \widehat{\mathcal{Y}}_{k+1}\widehat{\mathcal{B}}_{k+1}^T + \widehat{Y}_{k+2}\widehat{B}_{k+1}^T E_{k+1}^T \quad \text{with} \quad \widehat{\mathcal{Y}}_{k+1}^T M\widehat{\mathcal{Y}}_{k+1} = I_{(k+1)n_b}.$$

Continue the same procedure for $\widehat{X}_{k+2}, \cdots, \widehat{X}_n$ and $\widehat{Y}_{k+3}, \cdots, \widehat{Y}_{n+1}$, we can obtain the new M-orthonormal matrix $\widehat{\mathcal{X}}_n \in \mathbb{R}^{N\times nn_b}$, the new K-orthonormal matrix $\widehat{\mathcal{Y}}_n \in \mathbb{R}^{N\times nn_b}$, and the new matrix $\widehat{\mathcal{B}}_n \in \mathbb{R}^{nn_b\times nn_b}$, and relations

$$\begin{cases} K\widehat{\mathcal{Y}}_n = \widehat{\mathcal{X}}_n\widehat{\mathcal{B}}_n, \\ M\widehat{\mathcal{X}}_n = \widehat{\mathcal{Y}}_n\widehat{\mathcal{B}}_n^T + \widehat{Y}_{n+1}\widehat{B}_n^T E_n^T, \end{cases} \tag{20}$$

with $\widehat{\mathcal{X}}_n^T M\widehat{\mathcal{X}}_n = I_{nn_b} = \widehat{\mathcal{Y}}_n^T K\widehat{\mathcal{Y}}_n$, and

$$\widehat{\mathcal{B}}_n = \begin{bmatrix} \widehat{\mathcal{B}}_k & U\widehat{B}_k & & \\ & \widehat{A}_{k+1} & \widehat{B}_{k+1} & \\ & & \ddots & \widehat{B}_{n-1} \\ & & & \widehat{A}_n \end{bmatrix}.$$

Note that $\widehat{\mathcal{B}}_n$ is no longer a block bidiagonal matrix. Algorithm 2 is our thick-restart weighted block Golub-Kahan-Lanczos algorithm for LREP.

Remark 6. *Actually, from the construction of $\widehat{\mathcal{B}}_n$, we can know the procedure for getting $\widehat{X}_{k+2}, \cdots, \widehat{X}_n$ and $\widehat{Y}_{k+3}, \cdots, \widehat{Y}_{n+1}$ is the same as applying Algorithm 1 to $\widehat{Y}_{k+2}$ for $n - k - 1$ iterations, thus we use Algorithm 1 directly in restarting* **Step 2** *of the following Algorithm 2.*

Algorithm 2: wbGKL-TR

---

1. Given an initial guess Y_1 satisfying $Y_1^T K Y_1 = I_{n_b}$, a tolerance *tol*, an integer k that the k blocks approximate eigenvectors we want to add to the solving subspace, an integer n the block dimension of solving subspace, as well as w_ℓ the desired number of eigenpairs;
2. Apply Algorithm 1 from the current point to generate the rest of $\mathcal{X}_n$, $\mathcal{Y}_{n+1}$, and $\mathcal{B}_n$. If it is the first cycle, the current point is Y_1, else Y_{k+2};
3. Compute an SVD of $\mathcal{B}_n$ as in (2), select $w_\ell (w_\ell \leq nn_b)$ wanted singular values σ_j, and their associated left singular vectors ϕ_j and right singular vectors ψ_j. Form the approximate eigenpairs for **H**, if the stopping criterion is satisfied, then stop, else continue;
4. Generate new $\widehat{\mathcal{X}}_{k+1}$, $\widehat{\mathcal{Y}}_{k+2}$ and $\widehat{\mathcal{B}}_{k+1}$:
Compute $\widehat{\mathcal{Y}}_k = \mathcal{Y}_n \Psi_k$, $\widehat{\mathcal{X}}_k = \mathcal{X}_n \Phi_k$, $\widehat{B}_k = \Sigma_k$, $\widehat{Y}_{k+1} = Y_{n+1}$, $U^T = E_n^T \Phi_k$, $\widehat{B}_k = B_n$;
Compute $\widetilde{X}_{k+1} = K \widehat{Y}_{k+1} - \widehat{\mathcal{X}}_k U \widehat{B}_k$, do Cholesky decomposition $\widetilde{X}_{k+1}^T M \widetilde{X}_{k+1} = W^T W$, set
$\quad \widehat{A}_{k+1} = W$, $W = inv(W)$, $\widehat{X}_{k+1} = \widetilde{X}_{k+1} W$;
Compute $\widetilde{Y}_{k+2} = M \widehat{X}_{k+1} - \widehat{Y}_{k+1} \widehat{A}_{k+1}^T$, do Cholesky decomposition $\widetilde{Y}_{k+2}^T K \widetilde{Y}_{k+2} = W^T W$, set
$\quad \widehat{B}_{k+1} = W^T$, $W = inv(W)$, $\widehat{Y}_{k+2} = \widetilde{Y}_{k+2} W$;
Let $\mathcal{X}_{k+1} = \widehat{\mathcal{X}}_{k+1} = [\widehat{\mathcal{X}}_k, \widehat{X}_{k+1}]$, $\mathcal{B}_{k+1} = \widehat{\mathcal{B}}_{k+1} = \begin{bmatrix} \widehat{B}_k & U \widehat{B}_k \\ 0 & \widehat{A}_{k+1} \end{bmatrix}$, $\mathcal{Y}_{k+2} = \widehat{\mathcal{Y}}_{k+2} = [\widehat{\mathcal{Y}}_k, \widehat{Y}_{k+1}, \widehat{Y}_{k+2}]$,
and go to **Step 2**.

Remark 7. *In* **Step 3**, *we compute the harmonic Ritz pairs after n iterations. In practice, we do the computation for each iterations $j = 1, \cdots, n$. When restarting, the information chosen to add to the solving subspaces are the wanted w_ℓ singular values of $\mathcal{B}_n$ with their corresponding left and right singular vectors. Actually, we use MATLAB command "sort" to choose the w_ℓ smallest ones or the w_ℓ largest ones, and which singular values to choose depends on the desired eigenvalues of* **H**.

In the end of this section, we list the computational costs in a generic cycle of four algorithms, which are weighted block Golub-Kahan-Lanczos algorithm, thick-restart weighted block Golub-Kahan-Lanczos algorithm, block Lanczos algorithm [12], and thick-restart block Lanczos algorithm [12], and denoted by **wbGKL**, **wbGKL-TR**, **BLan**, and **BLan-TR**, respectively. The detail pseudocodes of **BLan** and **BLan-TR** are be found in [12].

The comparisons are presented in Tables 1 and 2. Here, we denote "block vector" a $N \times n_b$ rectangular matrix, denote "mvb" the product number of a $N \times N$ matrix and a block vector. "dpb" denotes the dot product number of two block vectors X and Y, i.e., $X^T Y$. "saxpyb" denotes the number of adding two block vectors or multiplying a block vector to a $n_b \times n_b$ small matrix. "Ep($2n \times 2n$)(with sorting)" means the number of $2n \times 2n$ size eigenvalue problem with sorting eigenvalues and their corresponding eigenvectors in one cycle. Similarly, "Sp($n \times n$)" denotes the number of $n \times n$ size singular value decomposition in one cycle. Because **wbGKL** and **BLan** are non-restart algorithms, we just count the first n Lanczos iterations.

Table 1. Main computational costs per cycle **wbGKL** and **wbGKL-TR**.

	wbGKL	wbGKL-TR (1-st Cycle)	wbGKL-TR (Other Cycle)
mvb	$2n+1$	$2n+1$	$2(n-k)$
dpb	$2n+1$	$2n+1$	$2(n-k)$
saxpyb	$8n$	$8n$	$8(n-k)+2k(2n+1)$
block vector updates	$2n+2$	$2n+2$	$2n+2$
Ep($2n \times 2n$)(with sorting)	0	0	0
Sp($n \times n$)	1	1	1

Table 2. Main computational costs per cycle **BLan** and **BLan-TR**.

	BLan	BLan-TR (1-st Cycle)	BLan-TR (Other Cycle)
mvb	$2n+1$	$2n+1$	$2(n-k)$
dpb	$2n+1$	$2n+1$	$2(n-k)$
saxpyb	$6n$	$6n$	$6(n-k)+2k(2n+1)$
block vector updates	$2n+2$	$2n+2$	$2n+2$
Ep($2n \times 2n$)(with sorting)	1	1	1
Sp($n \times n$)	0	0	0

5. Numerical Examples

In this section, two numerical experiments are carried out by using MATLAB 8.4 (R2014b) on a laptop with an Intel Core i5-6200U CPU 2.3 GHz memory 8 GB under the Windows 10 operating system.

Example 1. *In this example, we check the bounds established in Theorem 1 and 2. For simplicity, we take $N = 100$, the number of weighted block Golub-Kahan-Lanczos steps $n = 20$, $K = M$ as diagonal matrix $diag(\lambda_1, \lambda_2, \cdots, \lambda_N)$, where*

$$\lambda_1 = 11 + \rho, \quad \lambda_2 = 11, \quad \lambda_3 = 11 - \rho,$$

$$\lambda_{N-2} = 1 + \rho, \quad \lambda_{N-1} = 1, \quad \lambda_N = 1 - \rho,$$

$$\lambda_j = 5 + \frac{5(N-j+1)}{N-3}, \quad j = 4, \cdots, N-3,$$

and $i = k = 1$, $\ell = 3$, $\hat{i} = \hat{k} = N-2$, $\hat{\ell} = N$, $n_b = 3$. There are three positive eigenvalue clusters: $\{\lambda_1, \lambda_2, \lambda_3\}$, $\{\lambda_4, \cdots, \lambda_{N-3}\}$, or $\{\lambda_{N-2}, \lambda_{N-1}, \lambda_N\}$. Obviously, $\Xi = \Gamma = K^{-\frac{1}{2}}$.

We seek two groups of the approximate eigenpairs, the first is related to the first cluster, the second is related to the last cluster, i.e., $\{\sigma_1, \sigma_2, \sigma_3\}$ approximate $\{\lambda_1, \lambda_2, \lambda_3\}$, and $\{\sigma_{nn_b-2}, \sigma_{nn_b-1}, \sigma_{nn_b}\}$ approximate $\{\lambda_{N-2}, \lambda_{N-1}, \lambda_N\}$. In order to see the affect that generated from ρ to the upper bounds of the approximate eigenpairs errors in weighted block Golub-Kahan-Lanczos method for LREP, we change the parameter $\rho > 0$ to overmaster the tightness among eigenvalues within $\{\lambda_1, \lambda_2, \lambda_3\}$ and $\{\lambda_{N-2}, \lambda_{N-1}, \lambda_N\}$. First, we choose the same matrix Y_0 as in [12,17], i.e.,

$$Y_0 = \begin{bmatrix} 1 & 0 & 0 \\ 0 & 1 & 0 \\ 0 & 0 & 1 \\ \frac{1}{N} & sin1 & cos1 \\ \vdots & \vdots & \vdots \\ \frac{N-n_b}{N} & sin(N-n_b) & cos(N-n_b) \end{bmatrix}.$$

Obviously, $rank(Y_0) = n_b$ and $rank(Y_0^T K \Xi_{(:,1:3)}) = n_b$. Since K symmetric positive definite, thus do Cholesky decomposition $Y_0^T K Y_0 = W^T W$, let $Y_1 = Y_0 W^{-1}$, hence, Y_1 satisfies (5), i.e., $Y_1^T K \Xi_{(:,1:3)}$ is singular. We take $Z = Y_1 (\Xi_{(:,1:3)}^T K Y_1)^{-1}$, then Z satisfies (6). We execute the weighted block Golub-Kahan-Lanczos method with full re-orthogonalization for LREP in MATLAB, and check the bounds in (7), (8), (13), and (14). Since the approximate eigenvalues are $\{\sigma_1, \sigma_2, \sigma_3\}$ and $\{\sigma_{nn_b-2}, \sigma_{nn_b-1}, \sigma_{nn_b}\}$, thus $\pi_{i,k,\ell} = \pi_{i,k} = \hat{\pi}_{i,k,\ell} = \hat{\pi}_{i,\ell} = 1$, $c = \hat{c} = 1$, and we measure the following two groups of errors:

$$\varepsilon_{11} = \|diag(\lambda_1^2 - \sigma_1^2, \lambda_2^2 - \sigma_2^2, \lambda_3^2 - \sigma_3^2)\|_F,$$

$$\varepsilon_{21} = \frac{\lambda_1^2 - \lambda_N^2}{C_{n-1}^2 (1 + 2\gamma_{1,3})} \| \tan^2 \Theta_K(\Xi_{(:,1:3)}, Z)\|_F,$$

$$\varepsilon_{31} = \| \sin \Theta_K(\Xi_{(:,1:3)}, \mathcal{Y}_n \Psi_{(:,1:3)})\|_F,$$

$$\varepsilon_{41} = \frac{\sqrt{1 + \|A_n^T B_n\|_2^2 / \delta^2}}{C_{n-1}(1 + 2\gamma_{1,3})} \| \tan \Theta_K(\Xi_{(:,1:3)}, Z)\|_F,$$

and

$$\varepsilon_{12} = \|diag(\sigma_{N-2}^2 - \lambda_{N-2}^2, \sigma_{N-1}^2 - \lambda_{N-1}^2, \sigma_N^2 - \lambda_N^2)\|_F,$$

$$\varepsilon_{22} = \frac{\lambda_1^2 - \lambda_N^2}{C_{n-1}^2 (1 + 2\hat{\gamma}_{N-2,N})} \| \tan^2 \Theta_K(\Xi_{(:,N-2:N)}, Z)\|_F,$$

$$\varepsilon_{32} = \| \sin \Theta_K(\Xi_{(:,N-2:N)}, \mathcal{Y}_n \Psi_{(:,nn_b-2:nn_b)})\|_F,$$

$$\varepsilon_{42} = \frac{\sqrt{1 + \|A_n^T B_n\|_2^2 / \hat{\delta}^2}}{C_{n-i}(1 + 2\hat{\gamma}_{N-2,N})} \| \tan \Theta_K(\Xi_{(:,N-2:N)}, Z)\|_F.$$

Actually, ε_{21} and ε_{41} are upper bounds of ε_{11} and ε_{31}, and ε_{22} and ε_{42} are upper bounds of ε_{12} and ε_{32}. Tables 3 and 4 report the results of ε_{ij}, $i = 1, \cdots, 4$, $j = 1, 2$ with the parameter ρ goes to 0. From the two tables, we can see that the bounds for the eigenvalues lie in a cluster and their corresponding eigenvectors are sharp, and they are not sensitive to ρ when ρ goes to 0.

Table 3. ε_{11}, ε_{31} together with their upper bounds ε_{21}, ε_{41} of Example 1.

ρ	ε_{11}	ε_{21}	ε_{31}	ε_{41}
10^{-1}	4.0295×10^{-13}	2.6773×10^{-10}	1.2491×10^{-10}	2.6260×10^{-6}
10^{-2}	5.1238×10^{-14}	5.4555×10^{-11}	6.1184×10^{-11}	1.1407×10^{-6}
10^{-3}	7.1054×10^{-14}	4.6711×10^{-11}	5.7698×10^{-11}	1.0520×10^{-6}
10^{-4}	2.4449×10^{-13}	4.5993×10^{-11}	5.7370×10^{-11}	1.0436×10^{-6}
10^{-5}	2.1552×10^{-13}	4.5922×10^{-11}	5.7338×10^{-11}	1.0427×10^{-6}

Table 4. ε_{12}, ε_{32} together with their upper bounds ε_{22}, ε_{42} of Example 1.

ρ	ε_{11}	ε_{21}	ε_{31}	ε_{41}
10^{-1}	7.1089×10^{-16}	6.0352×10^{-11}	1.9393×10^{-10}	8.8823×10^{-7}
10^{-2}	1.3688×10^{-15}	3.5913×10^{-11}	1.9562×10^{-10}	6.8797×10^{-7}
10^{-3}	3.9968×10^{-15}	3.4113×10^{-11}	1.9580×10^{-10}	6.7081×10^{-7}
10^{-4}	4.8495×10^{-15}	3.3938×10^{-11}	1.9582×10^{-10}	6.6912×10^{-7}
10^{-5}	8.1221×10^{-15}	3.3920×10^{-11}	1.9582×10^{-10}	6.6895×10^{-7}

Example 2. *In this example, we are going to test the effectiveness of our weighted block Golub-Kahan-Lanczos algorithms. Four algorithms are tested, i.e.,* **wbGKL**, **wbGKL-TR**, **BLan**, *and* **BLan-TR**. *We choose 3 test problems used in [12,13], which are listed in Table 5. All the matrices K and M in the problems are symmetric positive definite. Specifically, Test 1 and Test 2, which are derived by the turboTDDFT command in QUANTUM*

ESPRESSO [22], are from the linear response research for Na2 and silane (SiH4) compound, respectively. The matrices K and M in Test 3 are from the University of Florida Sparse Matrix Collection [23], where the order of K is N = 9604, and M is the leading N × N principal submatrix of finan512.

Table 5. The matrices K and M in Test 1–3.

Problems	N	K	M
Test 1	1862	Na2	Na2
Test 2	5660	SiH4	SiH4
Test 3	9604	fv1	finan512

We aim to compute the smallest 5 positive eigenvalues and the largest 5 eigenvalues, i.e., λ_i for $i = 1, \cdots, 5, N - 4, \cdots, N$, together with their associated eigenvectors. The initial guess is chosen as $V_0 = eye(N, n_b)$ with block size $n_b = 3$, where eye is the MATLAB command. The same as in Example 1, since K is symmetric positive definite, thus do Cholesky decomposition $Y_0^T K Y_0 = W^T W$, let $Y_1 = Y_0 W^{-1}$, hence, Y_1 satisfies $Y_1^T K Y_1 = I_{n_b}$. In **wbGKL-TR** and **BLan-TR**, we select $n = 30$, $k = 20$, i.e., the restart will occur once the dimension of the solving subspace is larger than 90, and the information of 60 Ritz vectors are kept. For **wbGKL** and **BLan**, because there is no restart, then we compute the approximate eigenpairs when the Lanczos iterations equals to $30 + 10 \times (j - 1)$, $j = 1, 2, \cdots$, hence, the Lanczos iterations are as the same amount as in **wbGKL-TR** and **BLan-TR**. The following relative eigenvalue error and relative residual 1-norm for each 10 approximate eigenpairs are calculated:

$$
e(\sigma_j) := \begin{cases} \dfrac{|\lambda_j - \sigma_j|}{\lambda_j}, & j = 1, \cdots, 5, \\[2mm] \dfrac{|\lambda_{n+j-k} - \sigma_j|}{\lambda_{n+j-k}}, & j = nn_b - 4, \cdots, nn_b, \end{cases}
$$

$$
r(\sigma_j) := \frac{\|\mathbf{H}\tilde{z}_j - \sigma_j \tilde{z}_j\|_1}{(\|\mathbf{H}\|_1 + \sigma_j)\|\tilde{z}_j\|_1}, \quad j = 1, \cdots, 5, nn_b - 4, \cdots, nn_b,
$$

where the "exact" eigenvalues λ_j are calculated by the MATLAB code *eig*. The calculated approximate eigenpair $(\sigma_j, \tilde{z}_j)$ is regarded as converged if $r(\sigma_j) \leq tol = 10^{-8}$.

Tables 6 and 7 give the number of the Lanczos iterations (denote by *iter*) and the CPU time in seconds (denote by *CPU*) for the four algorithms, and Table 6 is for the smallest 5 positive eigenvalues, Table 7 is for the largest 5 eigenvalues. From Table 6, one can see that, no matter the smallest or the largest eigenvalues, the iteration number of the four algorithms are competitive, but **wbGKL** and **wbGKL-TR** cost significant less time than **BLan** and **BLan-TR**, especially, **wbGKL-TR** consumes the least amount of time. Because **BLan** and **BLan-TR** need to compute the eigenvalues of $\begin{bmatrix} 0 & T_n \\ D_n & 0 \end{bmatrix}$, which is a nonsymmetric matrix, thus the two algorithms slower than **wbGKL** and **wbGKL-TR**. Due to the saving during the orthogonalization procedure and solving a much smaller $\mathcal{B}_n$, **wbGKL-TR** is the faster algorithm.

Table 6. Compute 5 smallest positive eigenvalues for Test 1–3.

Algorithms	Test 1		Test 2		Test 3	
	CPU	*iter*	*CPU*	*iter*	*CPU*	*iter*
wbGKL	1.5070	149	25.7848	319	15.9308	379
wbGKL-TR	1.0746	179	20.3593	359	5.1302	589
BLan	4.6739	149	87.1670	349	43.9506	379
BLan-TR	2.1243	163	39.1306	393	19.9677	592

Table 7. Compute 5 largest eigenvalues for Test 1–3.

Algorithms	Test 1		Test 2		Test 3	
	CPU	*iter*	*CPU*	*iter*	*CPU*	*iter*
wbGKL	0.6387	79	12.4658	179	1.0639	109
wbGKL-TR	0.5284	79	9.9093	179	0.8774	109
BLan	1.4634	79	27.4028	179	6.7574	109
BLan-TR	1.0151	82	18.3415	186	4.1298	113

The accuracy of the last two approximate eigenpairs in Test 1 are shown in Figure 1. From the figure, we can see that, for the last two eigenpairs, **wbGKL** and **BLan** require almost the same iterations to obtain the same accuracy, and the case of **wbGKL-TR** and **BLan-TR** also need almost the same iterations, which are one or two more restarts than **wbGKL** and **BLan**. On one hand, without solving a nonsymmetric eigenproblem, **wbGKL** and **wbGKL-TR** can save much more time than **BLan** and **BLan-TR**. On the other hand, since the dimension of the solving subspace for **wbGKL-TR** is bounded by nn_b, the savings in the process of orthogonalization and a much smaller singular value decomposition problem is sufficient to cover the additional restart steps.

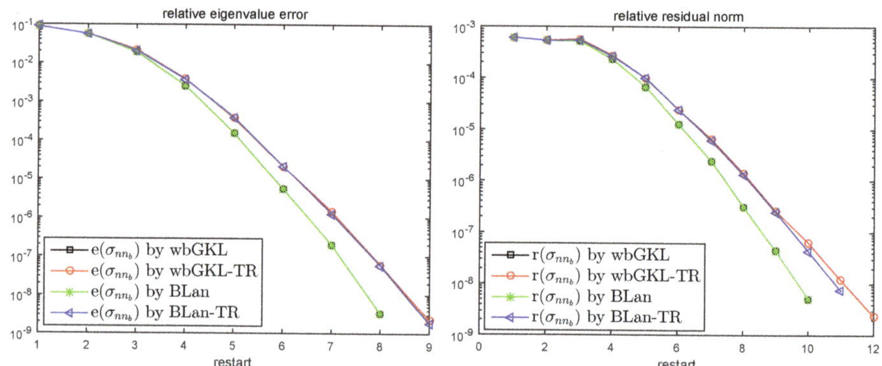

Figure 1. Errors and residuals of the 2 smallest positive eigenvalues for Test 1 in Example 2.

6. Conclusions

In this paper, we present a weighted block Golub-Kahan-Lanczos algorithm to solve the desired small portion of smallest or largest positive eigenvalues which are in a cluster. Convergence analysis is established in Theorems 1 and 2, and bound the errors of the eigenvalue and eigenvector approximations belonging to an eigenvalue cluster. These results also show the advantages of the block algorithm over the single-vector version. To make the new algorithm more practical, we introduced a thick-restart strategy to eliminate the numerical difficulties caused by the block method. Numerical examples are executed to demonstrate the efficiency of our new restart algorithm.

Author Contributions: Conceptualization, G.C.; Data curation, H.Z. and Z.T.; Formal analysis, H.Z. and Z.T.; Methodology, H.Z.; Project administration, H.Z. and G.C.; Resources, H.Z.; Visualization, H.Z. and Z.T.; Writing—original draft, H.Z.; Writing—review and editing, Z.T. and G.C.

Funding: This work was financial supported by the National Nature Science Foundation of China (No. 11701225, 11601081, 11471122), Fundamental Research Funds for the Central Universities (No. JUSRP11719), Natural Science Foundation of Jiangsu Province (No. BK20170173), and the research fund for distinguished young scholars of Fujian Agriculture and Forestry University (No. xjq201727).

Conflicts of Interest: The authors declare no conflict of interest.

References

1. Casida, M.E. Time-Dependent Density Functional Response Theory for Molecules. In *Recent Advances in Density Functional Methods*; Chong, D.P., Ed.; World Scientific: Singapore, 1995.
2. Onida, G.; Reining, L.; Rubio, A. Electronic excitations density functional versus many-body Green's function. *Rev. Mod. Phys.* **2002**, *74*, 601–659. [CrossRef]
3. Rocca, D. Time-Dependent Density Functional Perturbation Theory: New algorithms with Applications to Molecular Spectra. Ph.D. Thesis, The International School for Advanced Studies, Trieste, Italy, 2007.
4. Shao, M.; da Jornada, F.H.; Yang, C.; Deslippe, J.; Louie, S.G. Structure preserving parallel algorithms for solving the Bethe-Salpeter eigenvalue problem. *Linear Algebra Appl.* **2016**, *488*, 148–167. [CrossRef]
5. Ring, P.; Ma, Z.; Giai, V.N.; Vretenar, D.; Wandelt, A.; Gao, L. The time-dependent relativistic mean-field theory and the random phase approximation. *Nucl. Phys. A* **2001**, *694*, 249–268. [CrossRef]
6. Bai, Z.; Li, R.-C. Minimization principles for the linear response eigenvalue problem I: Theory. *SIAM J. Matrix Anal. Appl.* **2012**, *33*, 1075–1100. [CrossRef]
7. Bai, Z.; Li, R.-C. Minimization principles for the linear response eigenvalue problem II: Computation. *SIAM J. Matrix Anal. Appl.* **2013**, *34*, 392–416. [CrossRef]
8. Bai, Z.; Li, R.-C. Minimization principles and computation for the generalized linear response eigenvalue problem. *BIT Numer. Math.* **2014**, *54*, 31–54. [CrossRef]
9. Li, T.; Li, R.-C.; Lin, W.-W. A symmetric structure-preserving ΓQR algorithm for linear response eigenvalue problems. *Linear Algebra Appl.* **2017**, *520*, 191–214. [CrossRef]
10. Teng, Z.; Li, R.-C. Convergence analysis of Lanczos-type methods for the linear response eigenvalue problem. *J. Comput. Appl. Math.* **2013**, *247*, 17–33. [CrossRef]
11. Teng, Z.; Lu, L.; Li, R.-C. Perturbation of partitioned linear response eigenvalue problems. *Electron. Trans. Numer. Anal.* **2015**, *44*, 624–638.
12. Teng, Z.; Zhang, L.-H. A block Lanczos method for then linear response eigenvalue problem. *Electron. Trans. Numer. Anal.* **2017**, *46*, 505–523.
13. Teng, Z.; Zhou, Y.; Li, R.-C. A block Chebyshev-Davidson method for linear response eigenvalue problems. *Adv. Comput. Math.* **2016**, *42*, 1103–1128. [CrossRef]
14. Zhang, L.-H.; Lin, W.-W.; Li, R.-C. Backward perturbation analysis and residual-based error bounds for the linear response eigenvalue problem. *BIT Numer. Math.* **2014**, *55*, 869–896. [CrossRef]
15. Zhang, L.-H.; Xue, J.; Li, R.-C. Rayleigh-Ritz approximation for the linear response eigenvalue problem. *SIAM J. Matrix Anal. Appl.* **2014**, *35*, 765–782. [CrossRef]
16. Zhong, H.; Xu, H. Weighted Golub-Kahan-Lanczos bidiagonalizaiton algorithms. *Electron. Trans. Numer. Anal.* **2017**, *47*, 153–178.
17. Li, R.-C.; Zhang, L.-H. Convergence of the block Lanczos method for eigenvalue clusters. *Numer. Math.* **2015**, *131*, 83–113. [CrossRef]
18. Chapman, A.; Saad, Y. Deflated and augmented Krylov subspace techniques. *Numer. Linear Algebra Appl.* **1996**, *4*, 43–66. [CrossRef]
19. Lehoucq, R.B.; Sorensen, D.C. Deflation techniques for an implicitly restarted Arnoldi iteration. *SIAM J. Matrix Anal. Appl.* **1996**, *17*, 789–821. [CrossRef]
20. Wu, K.; Simon, H. Thick-restart Lanczos method for large symmetric eigenvalue problems. *SIAM J. Matrix Anal. Appl.* **2000**, *22*, 602–616. [CrossRef]
21. Knyazev, A.V.; Argentati, M.E. Principal angles between subspaces in an A-based scalar product: Algorithms and perturbation estimates. *SIAM J. Sci. Comput.* **2002**, *23*, 2008–2040. [CrossRef]
22. Giannozzi, P.; Baroni, S.; Bonini, N.; Calandra, M.; Car, R.; Cavazzoni, C.; Ceresoli, D.; Chiarotti, G.L.; Cococcioni, M.; Dabo, I.; et al. QUANTUM ESPRESSO: A modular and open-source software project for quantum simulations of materials. *J. Phys. Condens. Matter* **2009**, *21*, 395502. [CrossRef]
23. Davis, T.; Hu, Y. The University of Florida Sparse Matrix Collection. *ACM Trans. Math. Softw.* **2011**, *38*, 1:1–1:25. [CrossRef]

Article

Green's Classifications and Evolutions of Fixed-Order Networks

Allen D. Parks

Electromagnetic and Sensor Systems Department, Naval Surface Warfare Center Dahlgren Division, 18444 Frontage Road Suite 327, Dahlgren, VA 22448-5161, USA; allen.parks@navy.mil

Received: 31 August 2018; Accepted: 20 September 2018; Published: 25 September 2018

Abstract: It is shown that the set of all networks of fixed order n form a semigroup that is isomorphic to the semigroup B_X of binary relations on a set X of cardinality n. Consequently, B_X provides for Green's $\mathcal{L}$, $\mathcal{R}$, $\mathcal{H}$, and $\mathcal{D}$ equivalence classifications of all networks of fixed order n. These classifications reveal that a fixed-order network which evolves within a Green's equivalence class maintains certain structural invariants during its evolution. The "Green's symmetry problem" is introduced and is defined as the determination of all symmetries (i.e., transformations) that produce an evolution between an initial and final network within an $\mathcal{L}$ or an $\mathcal{R}$ class such that each symmetry preserves the required structural invariants. Such symmetries are shown to be solutions to special Boolean equations specific to each class. The satisfiability and computational complexity of the "Green's symmetry problem" are discussed and it is demonstrated that such symmetries encode information about which node neighborhoods in the initial network can be joined to form node neighborhoods in the final network such that the structural invariants required by the evolution are preserved, i.e., the internal dynamics of the evolution. The notion of "propensity" is also introduced. It is a measure of the tendency of node neighborhoods to join to form new neighborhoods during a network evolution and is used to define "energy", which quantifies the complexity of the internal dynamics of a network evolution.

Keywords: network classification; network evolution; network symmetries; Green's symmetry problem; network invariants; network internal dynamics; symmetry ensembles; propensities; energy

1. Introduction

Symmetry is a principle which has served as a guide for the spectacular advances that have been made in modern science, especially physics. For example, the continuous translational symmetry of ordinary space and time guarantees the invariance of the laws of physics under such translations. Thus, any mathematical expression describing a physical system, whether subatomic or macroscopic, must be invariant under space and time translations.

Group theory is the mathematical language used to describe symmetry and its associated invariant properties (recall that an abstract group is a set S of elements together with a law of composition " $\circ$ " such that for $x, y, z \in S$ (i) $x \circ y \in S$; (ii) $x \circ (y \circ z) = (x \circ y) \circ z$; (iii) there is an identity element $e \in S$ such that $x \circ e = e \circ x = x$; and (iv) for $x \in S$ there is an inverse $x^{-1} \in S$ such that $x \circ x^{-1} = x^{-1} \circ x = e$). As a simple example, the set S of $0°$, $90°$, $180°$, and $270°$ rotations in the plane of a square about its fixed center under "composition of rotations" form a symmetry group for the square ($0°$ is the identity element and the inverse of an $X° \in S$ rotation is a $360° - X°$ rotation). Each of these rotations is a symmetry which brings the square into coincidence with itself, i.e., they preserve the invariant shape of the square. A much more complicated example are the so called gauge symmetries of the standard model of physics which classify and describe three fundamental forces of nature (i.e., the electromagnetic, weak, and strong forces) in terms of groups (specifically, the unitary group $U(1)$ of degree 1 and the special unitary groups $SU(2)$ and $SU(3)$ of degree 2 and 3, respectively).

In recent years, the notion of generalized symmetry has been introduced to further describe graph symmetry [1,2]. The generalized symmetries of a graph are a generalization of the notion of the automorphism group of a graph and are derived from the application of Green's equivalence relations to the endomorphism monoid of the graph (the automorphism group is a subgroup of the graph's endomorphism monoid). Since these symmetries and invariant properties are strictly associated with a single graph, they do not address properties that remain fixed when the connection topology of the graph changes.

An important problem in network theory is identifying those properties of networks that remain fixed (invariant) as the network's connection topology changes with time. It was shown in [3] that the set of all networks (i.e., all connection topologies) on a fixed number of nodes also forms a semigroup. There it was also shown that the application of Green's equivalence relations to this semigroup partitions the associated set of networks into equivalence classes, each of which contains many fixed node number networks with various connection topologies, such that all networks within each class share some identifiable invariant connectivity property. If the connection topology of a network changes such that its initial and final configurations are in the same equivalence class, then the initial and final configurations share a common invariant property. It follows that, in this context, Green's equivalence classifications can be useful for identifying invariant properties of networks which evolve within an equivalence class. Such connectivity invariants can be used, for example, to identify important actors in evolving social networks and to select communication network reconfigurations that will retain a desired connectivity between specific node sets.

Transformations between networks within an equivalence class which preserve the associated invariant connectivity properties are called "Green's symmetries". Here, in addition to reviewing the Green's classification of networks [3], the "Green's symmetry problem" is introduced and defined. This problem is to determine (by calculation) the ensemble set of all the Green's symmetries which evolve an initial network configuration into a final configuration within a fixed Green's $\mathcal{R}$ equivalence class or within a fixed Green's $\mathcal{L}$ equivalence class. As discussed below, each such symmetry encodes information about the internal dynamics of the evolution, i.e., how node neighborhoods in the initial network configuration are joined to form node neighborhoods in the final configuration such that the invariant properties are preserved.

Since the cardinality of such ensembles can be large, the statistical notion of propensity is introduced. This quantity provides measures of the overall tendency of node neighborhoods in an initial network configuration to associate and form node neighborhoods in the final network configuration. Propensities are used to define "propensity energies", which quantify the overall complexity of the internal dynamics of a network evolution, and "energies of evolution", which quantify the complexity of internal dynamical activity for an evolution produced by a specific ensemble symmetry.

The objective of this paper is to motivate the application of Green's symmetry principles to network science by demonstrating how Green's equivalence relations can be applied to: classify networks; identify associated structural invariants; determine symmetries that preserve these invariants; and define associated measures that quantify aspects of the internal dynamics of network evolutions. The remainder of this paper is organized as follows: To make this paper reasonably self-contained, the relevant definitions and terminology from semigroup theory are summarized in the next section (for additional depth and clarification the reader is invited to consult such standard references as [4,5]). The semigroup B_X of all binary relations on a finite set X and the semigroup B_n of $n \times n$ Boolean matrices are defined and shown to be isomorphic to one another in Section 3. The semigroup of networks N_V on a fixed set V of nodes is introduced and is shown to be isomorphic to B_V in Section 4. This isomorphism provides for the Green's equivalence classifications of N_V given in Section 5. Green's evolutions of networks and their associated invariant properties are discussed in Section 6. The "Green's symmetry problem" is defined in Section 7 and its satisfiability and computational complexity are discussed in Section 8. The information encoded in symmetries as internal dynamics is detailed in Section 9. Symmetry "ensembles" and their "propensities" and "energies" are introduced in Section 10. A simple example illustrating aspects of the theory is presented in Section 11. Concluding remarks comprise the final section of this paper.

2. Semigroups

A *semigroup* $S \equiv (S, \circ)$ is a set S and an associative binary operation " $\circ$ " called multiplication defined upon the set (contrast this with the above definition of a group and note that a group is a semigroup endowed with the additional special properties given by items (*iii*) and (*iv*)). The one-sided right (one-sided left) multiplication of $x \in S$ by $y \in S$ is the product $x \circ y \in S$ ($y \circ x \in S$). An element $e \in S$ is an *identity* if $x \circ e = e \circ x = x$ for $x \in S$. An identity can be adjoined to S by setting $S^1 = S \cup \{e\}$ and defining $x \circ e = e \circ x = x$ for $x \in S^1$. Semigroup $S \equiv (S, \circ)$ and the semigroup $T \equiv (T, *)$ on set T with associative binary operation " $*$ " are *isomorphic* (denoted $S \approx T$) when there is a bijective map (i.e., an isomorphism) $\theta : S \to T$ such that $\theta(x \circ y) = \theta(x) * \theta(y)$ for all $x, y \in S$.

The well-known $\mathcal{L}$, $\mathcal{R}$, $\mathcal{H}$, and $\mathcal{D}$ Green's equivalence relations on a semigroup S partition S into a highly organized "egg box" structure using their relatively simple algebraic properties. In particular, the equivalence relation $\mathcal{L}(\mathcal{R})$ on S is defined by the rule that $x\mathcal{L}y$ ($x\mathcal{R}y$) if and only if $S^1 x = S^1 y$ ($x S^1 = y S^1$) for $x, y \in S$ and the equivalence relation $\mathcal{H} = \mathcal{L} \cap \mathcal{R}$ is similarly defined so that $x\mathcal{H}y$ if and only if $x\mathcal{L}y$ and $x\mathcal{R}y$. The relations $\mathcal{L}$ and $\mathcal{R}$ commute under the composition " $\bullet$ " of binary relations and $\mathcal{D} \equiv \mathcal{L} \bullet \mathcal{R} = \mathcal{R} \bullet \mathcal{L}$ is the smallest equivalence relation containing $\mathcal{L}$ and $\mathcal{R}$.

For $x \in S$ and $X \in \{\mathcal{L}, \mathcal{R}, \mathcal{H}, \mathcal{D}\}$ denote the X class containing x by $X(x)$ where $X = L, R, H,$ or D when $X = \mathcal{L}, \mathcal{R}, \mathcal{H},$ or $\mathcal{D}$, respectively. Thus, xXy if and only if $X(x) = X(y)$. If $x, y \in S$ and $R(x) = R(y)(L(x) = L(y))$, then there exist elements s (t) in S^1 such that $xs = y$ ($tx = y$) (hereafter the juxtaposition xy will *also* be used for the multiplication $x \circ y$).

3. The Semigroups B_n and B_X

The semigroup B_n of Boolean matrices is the set of all $n \times n$ matrices over $\{0, 1\}$ with Boolean composition $\gamma = \alpha \circ \beta$ defined by

$$\gamma_{ij} = \vee_{k \in J} \left(\alpha_{ik} \wedge \beta_{kj} \right), \tag{1}$$

as the semigroup multiplication operation. Here $J = \{1, 2, \cdots, n\}$, where $n \geq 1$ is a counting number, $\wedge$ denotes Boolean multiplication (i.e., $0 \wedge 0 = 0 \wedge 1 = 1 \wedge 0 = 0$, $1 \wedge 1 = 1$), and $\vee$ denotes Boolean addition (i.e., $0 \vee 0 = 0$, $0 \vee 1 = 1 \vee 0 = 1 \vee 1 = 1$).

The rows (columns) of any $\alpha \in B_n$ are Boolean row (column) n, vectors, i.e., row (column) n, tuples over $\{0, 1\}$, and come from the set $V_n(W_n)$ of all Boolean row (column) n-vectors. These vectors can be added coordinate-wise using Boolean addition. If $u, v \in V_n(W_n)$, then $u \sqsubseteq v$ when the ith coordinate $u_i = 1$ implies the ith coordinate $v_i = 1$, $1 \leq i \leq n$ ($\sqsubseteq$ is a partial order).

Let $\mathbf{0}(\mathbf{1})$ be either the zero (unit) row or zero (unit) column vector (the context in which $\mathbf{0}(\mathbf{1})$ is used defines whether it is a row or column vector). The matrix with $\mathbf{0}$ in every row, i.e., the zero matrix, is denoted by " $\mathcal{Z}$ " and the matrix with $\mathbf{1}$ in every row is denoted by " ω ". For $\alpha \in B_n$, the row space $\Gamma(\alpha)$ of α is the subset of V_n consisting of $\mathbf{0}$ and all possible Boolean sums of (one or more) nonzero rows of α. $\Gamma(\alpha)$ is a lattice $(\Gamma(\alpha), \sqsubseteq)$ under the partial order $\sqsubseteq$. The row (column) *basis* $r(\alpha)$ ($c(\alpha)$) of α is the set of all row (column) vectors in α that are not Boolean sums of other row (column) vectors in α. Please note that each vector in $r(\alpha)$ ($c(\alpha)$) must be a row (column) vector of α. The vector $\mathbf{0}$ is never a basis vector and the empty set $\varnothing$ is the basis for the $\mathcal{Z}$ matrix [6,7].

The semigroup B_X of binary relations on a set X of cardinality n (denoted $|X| = n$) is the power set of $X \times X$ with multiplication $a = bc$ being the "composition of binary relations" defined by

$$a = \{(x, y) \in X \times X : (x, z) \in b, \ (z, y) \in c, \ \text{when } z \in X\}. \tag{2}$$

It is easy to see that a bijective index map $f : X \to J$ induces an isomorphism $\lambda : B_X \to B_n$ defined by $\lambda(a) = \alpha$, where $\alpha_{ij} = 1$ if $\left(f^{-1}(i), f^{-1}(j)\right) \in a$ and is 0 if $\left(f^{-1}(i), f^{-1}(j)\right) \notin a$. B_n is therefore the Boolean matrix representation of B_X [8].

4. The Semigroup N_V

A network E of order n is the pair $E = (V, C)$, where V is a nonempty set of nodes with $|V| = n$, and the binary relation $C \subseteq V \times V$ is the set of directed links connecting the nodes of the network. Thus, E is both a digraph and a binary relation. If $(x, y) \in C$, then node $x(y)$ is an in(out)-neighbor of node $y(x)$. The *in-neighborhood of* $x \in V$ is the set $I(E; x)$ of all in-neighbors of x and the *out-neighborhood of* $x \in V$ is the set $O(E; x)$ of all out-neighbors of x.

Let N_V be the set of networks on V and define "multiplication of networks" by $EF = G \equiv (V, C^\#)$, where $E = (V, C)$, $F = (V, C')$, and

$$C^\# = \left\{ (x, y) \in V \times V : (x, z) \in C, (z, y) \in C', \text{ when } z \in V \right\}. \tag{3}$$

Lemma 1. N_V *is a semigroup that is isomorphic to* B_V.

Proof. The operation "multiplication of networks" is the same as the operation "composition of binary relations". Since it is clearly an associative binary operation on N_V, then N_V is a semigroup under the operation "multiplication of networks". Also, the bijective map $\varphi : N_V \to B_V$ defined by $\varphi(E) = C$ preserves multiplication. Thus, φ is a semigroup isomorphism and $N_V \approx B_V$. $\square$

Lemma 2. *If* $|V| = n$, *then* $N_V \approx B_n$.

Proof. This follows from the facts that $N_V \approx B_V$ (Lemma 1) and $B_V \approx B_n$ [8]. $\square$

Thus, B_n is also a Boolean matrix representation of N_V.

5. Green's Equivalence Classifications of N_V

Let $\theta : N_V \to B_n$ be the isomorphism of Lemma 2 and $f : V \to J$ be an associated index bijection. If α_{i*} is the ith Boolean row vector and α_{*j} is the jth Boolean column vector in the matrix $\alpha = \theta(E)$ corresponding to network E, then α_{i*} encodes the out-neighbors of node $f^{-1}(i)$ in E as the set

$$O\left(E; f^{-1}(i)\right) = \left\{ f^{-1}(k) : \alpha_{ik} = 1, k \in J \right\} \tag{4}$$

and α_{*j} encodes the in-neighbors of node $f^{-1}(j)$ in E as the set

$$I\left(E; f^{-1}(j)\right) = \left\{ f^{-1}(j) : \alpha_{kj} = 1, k \in J \right\}. \tag{5}$$

When $\alpha_{i*} \in r(\alpha)$ and $\alpha_{*j} \in c(\alpha)$, then $O_r\left(E; f^{-1}(i)\right) \equiv O\left(E; f^{-1}(i)\right)$ is a basis out-neighborhood and $I_c\left(E; f^{-1}(j)\right) \equiv I\left(E; f^{-1}(j)\right)$ is a basis in-neighborhood for network E. Thus, a basis neighborhood in E is a nonempty neighborhood in E which is not the set union of other neighborhoods in E.

Let $O_r(E)$ be the set of basis out-neighborhoods and $I_c(E)$ be the set of basis in-neighborhoods in network E. Also, define $P(E)$ as the set whose elements are $\varnothing$ and the sets generated by the closure under set union of the out-neighborhoods in E and let $(P(E), \subseteq)$ be the poset ordered by the set inclusion relation " $\subseteq$ ". Thus, when $\theta(E) = \alpha$, it may be formally stated that:

Lemma 3. $(P(E), \subseteq)$ *is a lattice that is isomorphic to* $(\Gamma(\alpha), \sqsubseteq)$.

Proof. The proof for this Lemma is the same as that given as the proof of Lemma 3.3 in [3]. $\square$

In what follows, $(P(E), \subseteq)$ will be referred to as the Π *lattice* for E.

The following major theorem provides complete $\mathcal{L}, \mathcal{R}, \mathcal{H}$, and $\mathcal{D}$ equivalence classifications of all fixed-order networks:

Theorem 1. *Let* $E, F \in N_V$. *Then*

i. $L(E) = L(F)$ *if and only if* $O_r(E) = O_r(F)$;
ii. $R(E) = R(F)$ *if and only if* $I_c(E) = I_c(F)$;
iii. $H(E) = H(F)$ *if and only if* $O_r(E) = O_r(F)$ *and* $I_c(E) = I_c(F)$;
iv. $D(E) = D(F)$ *if and only if* $(P(E), \subseteq)$ *and* $(P(F), \subseteq)$ *are lattice isomorphic.*

Proof. The proof of this result is the same as the proof of Theorem 3.4 in [3]. $\square$

Thus, the Green's $\mathcal{L}, \mathcal{R}$, and $\mathcal{H}$ equivalence classifications of the networks in N_V depend entirely upon their having (generally distinct) nodes with identical out-neighborhoods, identical in-neighborhoods, and both identical out-neighborhoods and in-neighborhoods, respectively, whereas the $\mathcal{D}$ equivalence classification of networks in N_V depends entirely upon their having isomorphic Π lattices which are generated by their out-neighborhoods. As an illustration of this theorem the reader is invited to consult the simple example given in [3] which corresponds to the complete Green's equivalence classification of (and the associated "egg box" structure for) all order two networks.

6. Green's Evolutions of Fixed-Order Networks

For $E, F \in N_V$, let $E \to F$ denote the evolution of a network during a time interval $[t_1, t_2]$, where E is the initial network at t_1 and F is the final network at $t_2 > t_1$. If $L(E) = L(F)(R(E) = R(F))[H(E) = H(F)] \{D(E) = D(F)\}$, then the evolution $E \to F$ is a *Green's* $\mathcal{L(R)}[\mathcal{H}]\{\mathcal{D}\}$ *evolution*. It is important to note that since $\mathcal{D} = \mathcal{L} \bullet \mathcal{R} = \mathcal{R} \bullet \mathcal{L}$ and $\mathcal{H} = \mathcal{L} \cap \mathcal{R}$, then $\mathcal{L}$ and $\mathcal{R}$ evolutions are also $\mathcal{D}$ evolutions, whereas $\mathcal{H}$ evolutions are both $\mathcal{L}$ and $\mathcal{R}$ evolutions, as well as $\mathcal{D}$ evolutions.

Theorem 2. *The following statements are true for network evolutions in* N_V:

i. $\mathcal{L}$ *evolutions preserve basis out-neighborhood sets and* Π *lattice isomorphism;*
ii. $\mathcal{R}$ *evolutions preserve basis in-neighborhood sets and* Π *lattice isomorphism;*
iii. $\mathcal{H}$ *evolutions preserve basis out-neighborhood and in-neighborhood sets and* Π *lattice isomorphism;*
iv. $\mathcal{D}$ *evolutions preserve* Π *lattice isomorphism.*

Proof. This is a direct and obvious consequence of the definitions of Green's evolutions and Theorem 1. $\square$

To illustrate this theorem, consider the order two networks $\psi \equiv (V, C_\psi)$ and $\mu \equiv (V, C_\mu)$ in the example in [3], where $V = \{a, b\}$, $C_\psi = \{(a, a)\}$, and $C_\mu = \{(a, a), (b, a)\}$. As can be seen from the associated Green's equivalence classification performed there, since $L(\psi) = L(\mu)$ and $D(\psi) = D(\mu)$, the evolution $\psi \to \mu$ is both a Green's $\mathcal{L}$ evolution and a Green's $\mathcal{D}$ evolution. Theorem 2 (*i*) is satisfied, since, from Table 1 and the discussion in [3], it is also seen that $O_r(\psi) = \{\{a\}\} = O_r(\mu)$ and that the Π lattices are isomorphic undirected paths of length 1.

7. The Green's Symmetry Problem

In general, a symmetry associated with a "situation" is defined as an "immunity to change" for some aspect of the "situation". For a "situation" to have a symmetry: (a) the aspect of the "situation" remains unchanged when a change is performed; and (b) it must be possible to perform the change, although the change does not actually have to be performed [9].

Recall from Section 2 that for an $\mathcal{R}(\mathcal{L})$ evolution $E \to F$ in N_V, there exists at least one $A \in N_V$ ($T \in N_V$) such that $EA = F$ ($TE = F$). Although $A(T)$ does not have to be applied to E, it can produce the desired evolution when applied as a right (left) multiplication of E. In so doing, this multiplication not only preserves $I_c(E)$ ($O_r(E)$), but also E's Π lattice structure. Thus, (a) and (b) above are satisfied and both $I_c(E)(O_r(E))$ and the associated Π lattice structure can be considered as the invariant properties associated with the *symmetries* A (T) which produce the evolution. Symmetries such as A (T) are *Green's* $\mathcal{R}(\mathcal{L})$ *symmetries*.

The *"Green's symmetry problem"* is defined here as the determination of all symmetries that produce an evolution from an initial to a final network within an $\mathcal{R}$ or an $\mathcal{L}$ class such that each symmetry preserves the structural invariants required by Theorem 2. As will be discussed below, such symmetries encode information about which node neighborhoods in the initial network can be joined to form neighborhoods in the final network such that the structural invariants required by the evolution are preserved.

8. Satisfiability and Computational Complexity of the Green's Symmetry Problem

The Green's symmetry problem for an evolution is $m-$ *satisfiable* if there are m symmetries which can produce the evolution.

Theorem 3. *The Green's symmetry problem is at least* $1-$ *satisfiable for both Green's* $\mathcal{R}$ *and* $\mathcal{L}$ *evolutions.*

Proof. Semigroup theory guarantees the existence of at least one Green's symmetry in N_V that can produce a Green's $\mathcal{R}$ evolution and at least one Green's symmetry in N_V that can produce a Green's $\mathcal{L}$ evolution. $\square$

8.1. Green's $\mathcal{R}$ Evolutions

The isomorphism established in Lemma 2 provides for computational solutions to the Green's symmetry problem. In particular, if $E \to F$ is a Green's $\mathcal{R}$ evolution, then, since E and F are known, the equation $EA = F$ can be solved for A for each $i, j \in J$ using the disjunctive normal form logical expression

$$\vee_{k \in J} \left(E_{ik} \wedge A_{kj} \right) = F_{ij}, \tag{6}$$

where use is now made of the Boolean matrix representations of E, F, and A. This expression for fixed j and all $i \in J$ defines a system of $|J|$ equations for node j.

This system of equations is *column-j satisfied* if there exists a column vector $A_{*j} \in W_n$ for which (1) is a true statement for each $i \in J$. For each $j \in J$, let G_{*j} be the set of all A_{*j} for which the associated system of equations is satisfied and define $\gamma \equiv \prod_{j \in J} |G_{*j}|$. Clearly, if $\gamma > 0$, then $EA = F$ is column-j satisfied for each $j \in J$ and *the evolution* $E \to F$ *is* γ-*satisfiable*. Each instantiation of A is represented by a Boolean matrix in B_n which has an $x \in G_{*j}$ as its jth column.

Let $M_i = \{k \in J : E_{ik} = 1\}$ index the unit valued entries in the row vector $E_{i*} \in V_n$.

Lemma 4. *Let* $F_{ij} = 0$ *for some* $i, j \in J$ *and* $M_i \neq \varnothing$. *If* $A_{*j} \in W_n$ *column- j satisfies* $EA = F$, *then* A_{*j} *has* $A_{kj} = 0$ *when* $k \in M_i$.

Proof. Assume for some $j \in J$ that $A_{*j} \in W_n$ column-j satisfies $EA = F$. If $F_{ij} = 0$ and $M_i \neq \varnothing$ for some $i \in J$, then (1) is true and zero valued for A_{*j} and that i value, and the following implication chain is valid:

$$\vee_{k \in J} \left(E_{ik} \wedge A_{kj} \right) = 0 \Rightarrow \vee_{l \in J - M_i} \left(0 \wedge A_{lj} \right) \vee_{k \in M_i} \left(1 \wedge A_{kj} \right) = 0 \Rightarrow \vee_{k \in M_i} \left(1 \wedge A_{kj} \right) = 0 \Rightarrow A_{kj} = 0,$$

$k \in M_i$. However, since $A_{*j} \in W_n$ column-j satisfies $EA = F$, it must also satisfy (1) for all $k \in J \Rightarrow A_{*j}$ has $A_{kj} = 0$ when $k \in M_i$. $\square$

Corollary 1. *If* $E = \omega$, *then* $A_{*j} = \mathbf{0}$.

Proof. $E = \omega \Rightarrow M_i = J \Rightarrow \vee_{k \in J} \left(1 \wedge A_{kj} \right) = 0 \Rightarrow A_{kj} = 0, k \in J \Rightarrow A_{*j} = \mathbf{0}$. $\square$

The computational complexity $\mathcal{C}_{\mathcal{R}}$ of the Green's symmetry problem for Green's $\mathcal{R}$ evolutions is the number of remaining combinations of $A_{kj} \in \{0,1\}$ values which must be checked for $EA = F$ satisfiability after the $A_{kj} = 0$ assignments specified by Lemma 4 have been made. Assume that $E \neq \omega, z$ and for each $j \in J$ let $Q(j) = \{i \in J : F_{ij} = 0\}$ index the zero valued Boolean equations of form (1).

Theorem 4. $\mathcal{C}_{\mathcal{R}} = \sum_{j \in J} \left[2^{n - |\cup_{i \in Q(j)} M_i|} \right]$.

Proof. For each $j \in J$, the set $\cup_{i \in Q(j)} M_i$ (which can possibly be empty) indexes all row locations $k \in J$ in A_{*j} for which $A_{kj} = 0$ in every A_{*j} that column-j satisfies $EA = F$. The set $J - \cup_{i \in Q(j)} M_i$ indexes all $k \in J$ for which A_{kj} must be evaluated to determine the column-j satisfiability of an associated A_{*j}. Since there are $Z_j = 2^{n - |\cup_{i \in Q(j)} M_i|}$ such evaluations for each $j \in J$, then for all $j \in J$ there are a total of $\mathcal{C}_{\mathcal{R}} = \sum_{j \in J} Z_j$ evaluations required to determine all $A_{*j} \in W_n$ which column-j satisfy $EA = F$. $\square$

8.2. Green's $\mathcal{L}$ Evolutions

If $E \to F$ is a Green's $\mathcal{L}$ evolution, then, since $TE = F$, it can be solved for T for each $i, j \in J$ using the disjunctive normal form logical expression

$$\vee_{k \in J} \left(T_{ik} \wedge E_{kj} \right) = F_{ij}, \tag{7}$$

which, for fixed i and all $j \in J$, defines a system of $|J|$ equations for node i. This system is *row-i satisfied* if there exists a row vector $T_{i*} \in V_n$ for which (2) is a true statement for each $j \in J$. For each $i \in J$, let H_{i*} be the set of all T_{i*} for which the associated system of equations is row-i satisfied and define $\delta \equiv \prod_{i \in J} |H_{i*}|$. If $\delta > 0$, then $TE = F$ is row-i satisfied for each $i \in J$ and *the evolution $E \to F$ is $\delta-$ satisfiable*. Each instantiation of T is represented by a Boolean matrix in B_n which has a $y \in H_{i*}$ as its ith row.

Let $K_j = \{k \in J : E_{kj} = 1\}$ index the unit valued entrees in the column vector $E_{*j} \in W_n$.

Lemma 5. Let $F_{ij} = 0$ for some $i, j \in J$ and $K_j \neq \varnothing$. If $T_{i*} \in V_n$ row-i satisfies $TE = F$, then T_{i*} has $T_{ik} = 0$ when $k \in K_j$.

Proof. Assume for some $i \in J$ that $T_{i*} \in V_n$ row, i satisfies $TE = F$. If $F_{ij} = 0$ for some $j \in J$ and $K_j \neq \varnothing$, then (2) is true and zero valued for T_{i*} and that j value, and the following implication chain is valid:
$$\vee_{k \in J} \left(T_{ik} \wedge E_{kj} \right) = 0 \Rightarrow \vee_{l \in J - K_j} (T_{il} \wedge 0) \vee_{k \in K_j} (T_{ik} \wedge 1) = 0 \Rightarrow \vee_{k \in K_j} (T_{ik} \wedge 1) = 0 \Rightarrow T_{ik} = 0, k \in K_j.$$
However, since T_{i*} row-i satisfies $TE = F$, it must also satisfy (2) for all $j \in J \Rightarrow T_{i*}$ has $T_{ik} = 0$ when $k \in K_j$. $\square$

Corollary 2. If $E = \omega$, then $T_{i*} = 0$.

Proof. $E = \omega \Rightarrow K_j = J \Rightarrow \vee_{k \in J} (T_{ik} \wedge 1) = 0 \Rightarrow T_{ik} = 0, k \in J \Rightarrow T_{i*} = 0.$ $\square$

The computational complexity $\mathcal{C}_{\mathcal{L}}$ of the Green's symmetry problem for Green's $\mathcal{L}$ evolutions is the number of remaining combinations of $T_{ik} \in \{0,1\}$ values which must be checked for $TE = F$ satisfiability after the $T_{ik} = 0$ assignments specified by Lemma 5 have been made. Assume that $E \neq \omega, z$ and for each $i \in J$ let $Y(i) = \{j \in J : F_{ij} = 0\}$ index the zero valued Boolean equations of form (2).

Theorem 5. $\mathcal{C}_{\mathcal{L}} = \sum_{i \in J} \left[2^{n - |\cup_{j \in Y(i)} K_j|} \right]$.

Proof. For each $i \in J$, the set $\cup_{j \in Y(i)} K_j$ (which can possibly be empty) indexes all column locations $k \in J$ for which $T_{ik} = 0$ in every T_{i*} that row-i satisfies $TE = F$. The set $J - \cup_{j \in Y(i)} K_j$ indexes all $k \in J$ for which T_{ik} must be evaluated to determine the row-i satisfiability of an associated T_{i*}. Since there are $Z_i = 2^{n - |\cup_{j \in Y(i)} K_j|}$ such evaluations for each $i \in J$, then for all $i \in J$ there are a total of $\mathcal{C}_{\mathcal{L}} = \sum_{i \in J} Z_i$ evaluations required in order to determine all $T_{i*} \in V_n$ which row-i satisfy $TE = F$. $\quad\square$

9. Symmetries: Instantiations of Internal Dynamics

Since Green's symmetries are themselves effectively elements of B_n, they correspond to special binary relations between network nodes that encode aspects of the internal dynamics of a Green's evolution $E \to F$. In particular, they generally identify many-to-one correspondences between neighborhood sets in E that are joined by set union to produce a neighborhood in F. Each of these correspondences occurs in such a way as to preserve the structural invariants required by Theorem 2. These correspondences are the *internal dynamics* of the evolution.

Consider a Green's $\mathcal{R}$ evolution $E \to F$ where each symmetry A satisfies $EA = F$ and is one instantiation of a possible set of symmetries which produce the evolution and preserve the required invariants. If $j \in J$ is a column in A with a 1 in each of the rows in the set $\Psi_j = \{i_1, i_2, \cdots, i_k\}$ and zeros in every other row location (i.e., there are $|\Psi_j| = k$ 1's and $n - k$ 0's), then this column encodes an internal dynamic of the evolution where the in-neighborhoods of nodes $i_1, i_2, \cdots, i_k$ in E are joined together as $\cup_{i \in \Psi_j} I(E_{*i})$ and associated with the in-neighborhood $I(F_{*j})$ in F according to

$$\cup_{i \in \Psi_j} I(E_{*i}) \subseteq I(F_{*j}). \tag{8}$$

This expression is called a Ψ_j internal $\mathcal{R}$ dynamic of the evolution and the set Ψ_j is *the associated motion of the dynamic*. Clearly, for the special case where $\Psi_j = \{i\}$,

$$I(E_{*i}) = I(F_{*i}).$$

If $E \to F$ is a Green's $\mathcal{L}$ evolution, a symmetry T which produces the invariant preserving evolution satisfies $TE = F$. If i is a row in T with a 1 in each of the column locations in $\Phi_i = \{j_1, j_2, \cdots, j_l\}$, then this row encodes an internal dynamic of the evolution where the out-neighborhoods of nodes $j_1, j_2, \cdots, j_l$ in network E are joined by set union and associated with the out-neighborhood $O(F_{i*})$ in network F according to

$$\cup_{j \in \Phi_i} O(E_{j*}) \subseteq O(F_{i*}). \tag{9}$$

This expression is a Φ_i internal $\mathcal{L}$ dynamic of the evolution and the set Φ_i is the associated motion of the dynamic. When $\Phi_i = \{j\}$, then

$$O(E_{j*}) = O(F_{i*}).$$

These notions will be clarified below using a simple example.

10. Symmetry Ensembles, Propensities, and Energies

Since the symmetry which produces a Green's evolution is not necessarily unique, it can be unclear as to how to assign a specific symmetry to an evolution. However, the collection of symmetries obtained from Green's symmetry problem, i.e., the *symmetry ensembles*, can be used to construct *propensities*. Propensities can be viewed as weighted symmetries which, in some sense, represent their respective ensembles.

Let $I_{\mathcal{R}}$ $(I_{\mathcal{L}}) \neq \varnothing$ index the symmetries which are solutions to the Green's symmetry problem for some Green's $\mathcal{R}(\mathcal{L})$ evolution $E \to F$. The sets

$$\mathcal{E}_{\mathcal{R}} = \left\{ A^{(i)} : i \in I_{\mathcal{R}}, \ EA^{(i)} = F \right\}$$

and

$$\mathcal{E}_{\mathcal{L}} = \left\{ T^{(i)} : i \in I_{\mathcal{L}}, T^{(i)}E = F \right\}$$

are the associated symmetry ensembles. The propensities associated with each ensemble are defined as

$$\overline{A} \equiv |I_{\mathcal{R}}|^{-1} \sum_{i \in I_{\mathcal{R}}} A^{(i)}$$

and

$$\overline{T} = |I_{\mathcal{L}}|^{-1} \sum_{i \in I_{\mathcal{L}}} T^{(i)}.$$

Thus, $\overline{A}_{*j}$ is a measure of the tendency of the nodes in column j in network E to form motions Ψ_j that associate in-neighborhoods in E with in-neighborhoods in network F according to the internal dynamic (3). Similarly, $\overline{T}_{i*}$ is a measure of the tendency of nodes in row i in E to form motions Φ_i that associate out-neighborhoods in E with out-neighborhoods in F according to the internal dynamic (4).

Propensities can be used to associate energies with both ensembles and specific symmetries. These energies quantify in a directly proportional manner the complexity level of the internal dynamical activity that is associated with an evolution. The *propensity energies* provide a representative measure of the "overall" complexity of internal dynamical activity for an evolution based upon ensemble propensity. The propensity energies for ensembles $\mathcal{E}_{\mathcal{R}}$ and $\mathcal{E}_{\mathcal{L}}$ are defined as

$$\mathfrak{E}_{\mathcal{R}} \equiv \sum_{i,j \in J} \overline{A}_{ij}$$

and

$$\mathfrak{E}_{\mathcal{L}} \equiv \sum_{i,j \in J} \overline{T}_{ij},$$

respectively.

The *energies of evolution* for the specific symmetries in an ensemble measure the complexity of internal dynamical activity for an evolution produced by a specific symmetry in an ensemble. In particular, if $A^{(k)} \in \mathcal{E}_{\mathcal{R}}$ and $B^{(k)} \in \mathcal{E}_{\mathcal{L}}$, then the associated energies of evolution are defined as

$$\mathfrak{E}_{\mathcal{R}}\left[A^{(k)}\right] \equiv \sum_{i,j \in J} A_{ij}^{(k)} \overline{A}_{ij}$$

and

$$\mathfrak{E}_{\mathcal{L}}\left[T^{(k)}\right] \equiv \sum_{i,j \in J} T_{ij}^{(k)} \overline{T}_{ij}.$$

The following Lemma guarantees that the energy of evolution for a symmetry never exceeds the propensity energy for the associated ensemble.

Lemma 6. *For any Green's $\mathcal{R}$ or $\mathcal{L}$ evolution, $\mathfrak{E}_x \geq \mathfrak{E}_x[y]$, where $y = A^{(k)}$ or $T^{(k)}$ when $x = \mathcal{R}$ or $\mathcal{L}$.*

Proof. $A_{ij}^{(k)}, T_{ij}^{(k)} \in \{0,1\} \Rightarrow \overline{A}_{ij} \geq A_{ij}^{(k)} \overline{A}_{ij}, \overline{T}_{ij} \geq T_{ij}^{(k)} \overline{T}_{ij} \Rightarrow \sum_{i,j \in J} \overline{A}_{ij} \geq \sum_{i,j \in J} A_{ij}^{(k)} \overline{A}_{ij}, \sum_{i,j \in J} \overline{T}_{ij} \geq \sum_{i,j \in J} T_{ij}^{(k)} \overline{T}_{ij} \Rightarrow \mathfrak{E}_{\mathcal{R}} \geq \mathfrak{E}_{\mathcal{R}}[A^{(k)}], \mathfrak{E}_{\mathcal{L}} \geq \mathfrak{E}_{\mathcal{L}}[T^{(k)}].$ $\square$

Recall that internal $\mathcal{R}$ and $\mathcal{L}$ dynamics are strictly defined by their motions. These motions also have energies that provide a measure of the level of internal dynamical activity induced by the motion. Since the symmetries A and T encode $\mathcal{R}$ and $\mathcal{L}$ internal dynamics with motions Ψ_j and Φ_i, respectively, then the associated *energies of motion* are the quantities

$$\mathfrak{E}_{\mathcal{R}}\left[A; \Psi_j\right] \equiv \sum_{i \in \Psi_j} A_{ij} \overline{A}_{ij}$$

and

$$\mathfrak{E}_{\mathcal{L}}[T; \Phi_i] \equiv \sum_{j \in \Phi_i} T_{ij} \overline{T}_{ij}.$$

The energies of motion are related to their energies of evolution by the following theorem:

Theorem 6 (Conservation of Energy of Evolution). *The energy of evolution of a Green's symmetry is conserved by the energies of motion of its internal dynamics.*

Proof. Let $A \in \mathcal{E}_\mathcal{R}$ and set M index all the Ψ_j internal $\mathcal{R}$ dynamics encoded by A. Then $\sum_{j \in M} \mathfrak{E}_\mathcal{R}[A; \Psi_j] = \sum_{j \in M} \sum_{i \in \Psi_j} A_{ij} \overline{A}_{ij} = \sum_{i,j \in J} A_{ij} \overline{A}_{ij} = \mathfrak{E}_\mathcal{R}[A]$, where use has been made of the fact that $\sum_{j \in M} \sum_{i \in \Psi_j}$ is equivalent to $\sum_{i,j \in J}$ because $A_{ij} = 0$ when $i \in J - \Psi_j$ and $j \in J - M$. It is similar for the $\mathcal{L}$ dynamics. $\square$

11. Example

Let $E \to F$ be a Green's $\mathcal{R}$ evolution in N_V, $V = \{1,2\}$ (or equivalently in B_2), where (in B_2)

$$E = \begin{bmatrix} 0 & 0 \\ 1 & 0 \end{bmatrix}, \; F = \begin{bmatrix} 0 & 0 \\ 1 & 1 \end{bmatrix},$$

with $I_c(E) = \{2\} = I_c(F)$ (note that this evolution corresponds to the $\tau \to \lambda$ Green's $\mathcal{R}$ evolution in [3]). Theorem 3 guarantees the existence of at least one A such that

$$EA = \begin{bmatrix} 0 & 0 \\ 1 & 0 \end{bmatrix} \circ \begin{bmatrix} a_{11} & a_{12} \\ a_{21} & a_{22} \end{bmatrix} = \begin{bmatrix} 0 & 0 \\ 1 & 1 \end{bmatrix} = F.$$

The disjunctive normal form logical expression (1) for this equation yields the following system of equations

$$(0 \wedge a_{11}) \vee (0 \wedge a_{21}) = 0 \quad (0 \wedge a_{12}) \vee (0 \wedge a_{22}) = 0$$

$$(1 \wedge a_{11}) \vee (0 \wedge a_{21}) = 1 \quad (1 \wedge a_{12}) \vee (0 \wedge a_{22}) = 1$$

which can be used to solve the associated Green's symmetry problem.

For the two equations in the second row of this system to be satisfied requires the assignment $a_{11} = 1 = a_{12}$. By inspection it is seen that the complete system is satisfied when, in addition to these assignments, a_{21} and a_{22} each assume both values from the set $\{0,1\}$. Thus,

$$G_{*1} = G_{*2} = \left\{ \begin{bmatrix} 1 \\ 0 \end{bmatrix}, \begin{bmatrix} 1 \\ 1 \end{bmatrix} \right\}$$

so that $\gamma = |G_{*1}||G_{*2}| = 2 \cdot 2 = 4 = |I_\mathcal{R}|$ and the evolution $E \to F$ is 4-satisfiable. The associated symmetry ensemble is the set

$$\mathcal{E}_\mathcal{R} = \left\{ \begin{bmatrix} 1 & 1 \\ 0 & 0 \end{bmatrix}, \begin{bmatrix} 1 & 1 \\ 0 & 1 \end{bmatrix}, \begin{bmatrix} 1 & 1 \\ 1 & 0 \end{bmatrix}, \begin{bmatrix} 1 & 1 \\ 1 & 1 \end{bmatrix} \right\} \equiv \left\{ A^{(1)}, A^{(2)}, A^{(3)}, A^{(4)} \right\}.$$

To calculate the computational complexity of this Green's symmetry problem, refer to Section 8.1 and observe that $M_1 = \varnothing$, $M_2 = \{1\}$, and $Q(1) = \{1\} = Q(2)$. Application of Theorem 4 yields $\mathcal{C}_\mathcal{R} = 2^{2-|M_1|} + 2^{2-|M_1|} = 2^2 + 2^2 = 8$, i.e., four combinations of value assignments must be checked for each j since, according to Lemma 4, a_{ij} values cannot be assigned when $F_{ij} = 0$ because $M_i = \varnothing$.

The propensity and propensity energy for the ensemble are

$$\overline{A} = \begin{bmatrix} 1 & 1 \\ \tfrac{1}{2} & \tfrac{1}{2} \end{bmatrix}$$

and $\mathfrak{E}_\mathcal{R} = 3$, respectively, and the energies of evolution are $\mathfrak{E}_\mathcal{R}[A^{(1)}] = 2$, $\mathfrak{E}_\mathcal{R}[A^{(2)}] = 2 = \mathfrak{E}_\mathcal{R}[A^{(3)}]$, and $\mathfrak{E}_\mathcal{R}[A^{(4)}] = 3$. Please note that this validates Lemma 6. These energies also indicate that $A^{(1)}$ produces the least energy of evolution in the sense that the evolution involves simpler internal dynamical activity than evolutions produced by the other symmetries in the ensemble.

To illustrate this further, first observe that $I(E_{*1}) = \{2\}$, $I(E_{*2}) = \varnothing$, and $I(F_{*1}) = \{2\} = I(F_{*2})$ (here the jth column vector is set directly equal to the nodes in the in-neighborhood of node j). It is also easily determined that the motions of the dynamics for: $A^{(1)}$ are $\Psi_1 = \{1\} = \Psi_2$; $A^{(2)}$ are $\Psi_1 = \{1\}$ and $\Psi_2 = \{1,2\}$; $A^{(3)}$ are $\Psi_1 = \{1,2\}$ and $\Psi_2 = \{1\}$; and $A^{(4)}$ are $\Psi_1 = \{1,2\} = \Psi_2$. By inspection it is found that each of these motions satisfies (3). Using $A^{(4)}$ as an example, it is seen that (3) yields the correct set theoretic relationship $I(E_{*1}) \cup I(E_{*2}) \subseteq I(F_{*1}) \cup I(F_{*2})$ or $\{2\} \cup \varnothing \subseteq \{2\} \cup \{2\}$ or $\{2\} \subseteq \{2\}$ for both Ψ_1 and Ψ_2. Also note that the internal dynamics for $A^{(1)}$ are simpler than those for the other symmetries in the ensemble, in the sense that both of the $A^{(1)}$ motions are singleton sets, whereas at least one of the motions for the other symmetries is a doubleton set. This is consistent with the fact mentioned above that $A^{(1)}$ produces the least energy of evolution.

Now consider the energies of motion for each ensemble symmetry. They are easily calculated from the theory and are found to be:

$$\mathcal{E}_{\mathcal{R}}\left[A^{(1)}; \Psi_1\right] = 1 = \mathcal{E}_{\mathcal{R}}\left[A^{(1)}; \Psi_2\right];$$

$$\mathcal{E}_{\mathcal{R}}\left[A^{(2)}; \Psi_1\right] = 1, \ \mathcal{E}_{\mathcal{R}}\left[A^{(2)}; \Psi_2\right] = 1;$$

$$\mathcal{E}_{\mathcal{R}}\left[A^{(3)}; \Psi_1\right] = 1, \ \mathcal{E}_{\mathcal{R}}\left[A^{(3)}; \Psi_2\right] = 1;$$

and

$$\mathcal{E}_{\mathcal{R}}\left[A^{(4)}; \Psi_1\right] = 1 = \mathcal{E}_{\mathcal{R}}\left[A^{(4)}; \Psi_2\right].$$

Thus, the motions associated with an $A^{(1)}$ evolution are the least energetic since

$$\mathcal{E}_{\mathcal{R}}\left[A^{(1)}; \Psi_j\right] \leq \mathcal{E}_{\mathcal{R}}\left[A^{(k)}; \Psi_j\right], \ k = 2,3,4; j = 1,2.$$

This is also consistent with the fact that an $A^{(1)}$ induced evolution is the least energetic and involves the least complex internal dynamics.

Finally, observe that these results validate Theorem 6. In particular,

$$\mathcal{E}_{\mathcal{R}}\left[A^{(1)}; \Psi_1\right] + \mathcal{E}_{\mathcal{R}}\left[A^{(1)}; \Psi_2\right] = 2 = \mathcal{E}_{\mathcal{R}}\left[A^{(1)}\right];$$

$$\mathcal{E}_{\mathcal{R}}\left[A^{(2)}; \Psi_1\right] + \mathcal{E}_{\mathcal{R}}\left[A^{(2)}; \Psi_2\right] = 2 = \mathcal{E}_{\mathcal{R}}\left[A^{(2)}\right];$$

$$\mathcal{E}_{\mathcal{R}}\left[A^{(3)}; \Psi_1\right] + \mathcal{E}_{\mathcal{R}}\left[A^{(3)}; \Psi_2\right] = 2 = \mathcal{E}_{\mathcal{R}}\left[A^{(3)}\right];$$

and

$$\mathcal{E}_{\mathcal{R}}\left[A^{(4)}; \Psi_1\right] + \mathcal{E}_{\mathcal{R}}\left[A^{(4)}; \Psi_2\right] = 3 = \mathcal{E}_{\mathcal{R}}\left[A^{(4)}\right].$$

12. Concluding Remarks

The research documented in [3] was inspired by earlier research performed by Konieczny [6] and Plemmons et al. [7]. This paper has reviewed the results developed in [3], i.e., that the set of all networks on a fixed number of nodes can be classified using the Green's equivalence relations of semigroup theory and that all networks within a Green's equivalence class have a common structural invariant (neighborhoods or poset relationships between node sets generated by neighborhoods). By extension, it was deduced in this paper from these results that if a network evolves from an initial network configuration to a final network configuration such that both the initial and final networks are in the same Green's equivalence class, then the structural invariants for the class are preserved by the evolution. In addition, the Green's symmetry problem was also defined in this paper. This problem is to determine by computation all symmetries which produce a network evolution within a Green's $\mathcal{R}$ or a Green's $\mathcal{L}$ equivalence class (i.e., a symmetry ensemble). These symmetries were shown to be solutions to special Boolean equations whose form is dictated by semigroup theory. Each such symmetry encodes information about the internal dynamics of the associated evolution and an ensemble associated with an evolution was used to define propensities and energies which quantify aspects of the internal dynamics of the evolution. However, it should be noted that a practical limitation exists for solving the Green's symmetry problem. This occurs because the cardinality of symmetry ensembles associated with large real networks can be quite large, thereby requiring the use of considerable computational resources to solve such problems (see future research suggestions below).

In conclusion, it is believed that the results of this paper are new and not in general use (perhaps having the closest resemblance to these results are the applications of Green's relations to social networks [10] and automata theory, e.g., [11]). However, the results of this paper are important and should be of general interest to network science researchers and those working in areas of applied network theory. In addition to applications similar to those mentioned in Section 1 (actor identification in social networks and communication network reconfiguration), contemporary areas of frontier research, such as identifying emerging scientific disciplines, e.g., [12], analyzing brain connectivity, e.g., [13–15], and finding symmetries in engineering processes [16], could also benefit from the results of this paper.

Before closing it is worthwhile to mention several directions for related future research. First, because of the computational resources required to solve the Green's symmetry problem, it would be useful to investigate how sampling and statistics can be used to obtain symmetry sub-ensembles that effectively yield the same information about propensities and energies as the associated full ensemble. A second research area involves understanding symmetries and their computation for network evolutions occurring within Green's $\mathcal{H}$ and $\mathcal{D}$ equivalence classes. A third and potentially very interesting research area concerns determining the relationships (if any) between the theory developed in this paper and the relatively new theory of persistence that is used to analyze large data sets, e.g., [17].

Funding: This research was supported by a Grant from the Naval Surface Warfare Center Dahlgren Division's In-house Laboratory Independent Research program.

Conflicts of Interest: The author declares no conflict of interest. The funder had no role in the design of the study; in the collection, analyses, or interpretation of data; in the writing of the manuscript, and in the decision to publish the results.

References

1. Fan, S. Generalized symmetry of graphs. *Electron. Notes Discrete Math.* **2005**, *23*, 51–60.
2. Fan, S. Generalized symmetry of graphs—A survey. *Discret. Math.* **2009**, *309*, 5411–5419. [CrossRef]
3. Parks, A. Green's Symmetries in Finite Digraphs. *Symmetry* **2011**, *3*, 564–573. [CrossRef]
4. Clifford, A.; Preston, G. *The Algebraic Theory of Semigroups*; American Mathematical Society: Providence, RI, USA, 1961; Volume 1.
5. Howie, J. *An Introduction to Semigroup Theory*; Academic Press, Inc.: New York, NY, USA, 1976; ISBN 75-46333.
6. Konieczny, J. Green's Equivalences in Finite Semigroups of Binary Relations. *Semigroup Forum* **1994**, *48*, 235–252. [CrossRef]
7. Plemmons, R.; West, T. On the Semigroup of Binary Relations. *Pac. J. Math.* **1970**, *35*, 43–753. [CrossRef]
8. Lallement, G. *Semigroups and Combinatorial Applications*; John Wiley and Sons: New York, NY, USA, 1979; ISBN 0-471-04379-6.
9. Rosen, J. *Symmetry Rules: How Science and Nature are Founded on Symmetry*; Springer-Verlag: Berlin, Germany, 2008.
10. Boyd, J. *Social Semigroups: A Unified Theory of Scaling and Blockmodelling as Applied to Social Networks*; George Mason University Press: Fairfax, VA, USA, 1991; ISBN 0-913969-34-6.
11. Colcombet, T. Green's Relations and Their Use in Automata Theory. In *Language and Automata Theory and Applications. LATA 2011. Lecture Notes in Computer Science*; Dediu, A., Inenaga, S., Martin-Vide, C., Eds.; Springer: Berlin, Germany, 2011; Volume 6638. [CrossRef]
12. Sun, X.; Kaur, J.; Milojević, S.; Flammini, A.; Menczer, F. Social Dynamics of Science. *Sci. Rep.* **2013**, *3*, 1069. [CrossRef] [PubMed]
13. Vaiana, M.; Muldoon, S. Multilayer Brain Networks. *J. Nonlinear Sci.* **2018**, 1–23. [CrossRef]

14. Schmälzle, R.; O'Donnell, M.; Garcia, J.; Cascio, C.; Bayer, J.; Bassett, D.; Vettel, J.; Falk, M. Brain connectivity dynamics during social interaction reflect social network structure. *Proc. Natl. Acad. Sci. USA* **2017**, *114*, 5153–5158. [CrossRef] [PubMed]

15. Petri, G.; Expert, P.; Turkheimer, F.; Carhart-Harris, R.; Nutt, D.; Hellyer, P.; Vaccarino, F. Homological scallolds of brain functional networks. *J. R. Soc. Interface* **2014**, *11*, 20140873. [CrossRef] [PubMed]

16. Parks, A. *Process Symmetries: System Laws for Operational Processes*; FY18 ILIR/IAR Midyear Review; University of Mary Washington Dahlgren Campus: Dahlgren, VA, USA, 2018.

17. Perea, J. A Brief History of Persistence. *arXiv*. 2018. Available online: https://arxiv.org/abs/1809.03624 (accessed on 21 September 2018).

MDPI

St. Alban-Anlage 66

4052 Basel

Switzerland

Tel. +41 61 683 77 34

Fax +41 61 302 89 18

www.mdpi.com

Mathematics Editorial Office

E-mail: mathematics@mdpi.com

www.mdpi.com/journal/mathematics